EFFECTS OF
ACID PRECIPITATION
ON TERRESTRIAL
ECOSYSTEMS

NATO CONFERENCE SERIES

I Ecology
II Systems Science
III Human Factors
IV Marine Sciences
V Air—Sea Interactions
VI Materials Science

I ECOLOGY

EFFECTS OF ACID PRECIPITATION ON TERRESTRIAL ECOSYSTEMS

Edited by
T.C. Hutchinson
and M. Havas
University of Toronto
Toronto, Ontario, Canada

Published in coordination with NATO Scientific Affairs Division

PLENUM PRESS · NEW YORK AND LONDON

Library of Congress Cataloging in Publication Data

Nato Conference on Effects of Acid Precipitation on Vegetation and Soils, Toronto, 1978.
  Effects of acid precipitation on terrestrial ecosystems.

  (NATO conference series: I, Ecology; v.4)
  Proceedings of a conference held in Toronto May 21–27, 1978.
  Includes index.
  1. Acid precipitation (Meteorology)—Environmental aspects—Congresses. 2. Plants, Effect of acid precipitation on—Congresses. I. Hutchinson, Thomas C., 1939-
II. Havas, M. III. North Atlantic Treaty Organization. Scientific Affairs Division.
IV. Title. V. Series.
QH545.A17N37 1978   574.5'222   79-21816
ISBN 0-306-40309-9

Proceedings of the NATO Conference on Effects of
Acid Precipitation on Vegetation and Soils, held in
Toronto, Ontario, Canada, May 21–27, 1978.

© 1980 Plenum Press, New York
A Division of Plenum Publishing Corporation
227 West 17th Street, New York, N.Y. 10011

Preface

This volume contains papers presented at a NATO Advanced
Research Institute, sponsored by their Eco-Sciences Panel, on
"The effects of acid precipitation on vegetation and soils," held at
Toronto, Canada from May 22-26, 1978. The organizing expenses and
greater part of the expenses of the speakers and chairmen were
provided by N.A.T.O. The scientific programme was planned by
T. C. Hutchinson together with an international planning committee
of G. Abrahamsen (Norway), G. Likens (U.S.A.), F.E. Last (U.K.),
C.O. Tamm (Sweden) and B. Ulrich (W. Germany).

Many of the dimensions of the 'acid rain' problem are common
to countries of northern Europe and North America. The developing
awareness over the past ten years of the international nature of
the acid rain phenomenon has lead to studies documenting damaging
effects on susceptible freshwater bodies. Large areas of the
Canadian Pre-Cambrian Shield, with its extension into the United
States, and the granitic areas of southern Norway and Sweden contain
lakes which are in the process of acidification. The biological
resources of these affected areas are of considerable national
concern. However, while clearly damaging effects of acidification
on freshwater systems have been well documented, the impact of acid
precipitation on terrestrial systems has not been so well understood.
The Advanced Research Institute was held specifically to bring
together international experts to present, examine and debate the
scientific research relevant to such terrestrial ecosystems. The
soil and its biota are a wonderfully complex system. The broad
inter-disciplinary nature of the symposium included presentation of
position papers by experts whose major interest has not necessarily
been in acid precipitation. These provided an invaluable framework
in which discussion could take place of acid precipitation effects
and from which predictions could be developed. The identification
of sensitive sites and soils was one of the topics dealt with in
Toronto, as well as the nature of atmospheric inputs to the soil
and from the soil to freshwater bodies.

The symposium was of value to those who took part in it through
the exchange of ideas and through clarification of issues. Many

new contacts were made and have increased the flow of knowledge. A
post-symposium excursion to the Sudbury area of Ontario allowed
delegates the opportunity to examine forest ecosystems which have
been subjected to extremes of sulphur dioxide and acidification
stress in this century.

As organizers and editors we should like to record our gratitude
to all participants who made the meeting so stimulating, and the
book possible. We thank also the N.A.T.O. Eco-Sciences Panel who
proposed such a meeting and then so helpfully supported it to
completion. We thank also the staff of the Guild Inn Toronto who
provided a superb atmosphere for the meeting and Mrs. Doreen Jones
who did much of the secretarial work. The Department of Botany of
the University of Toronto has been largely responsible for running
the symposium and for the editorial production of the book for
Plenum. Dr. Bill Freedman provided some able editorial assistance.
Professor F.K. Hare has been most supportive both through his role
as a member of the N.A.T.O. Eco-Sciences Panel and as Director of
the Institute for Environmental Studies at Toronto. Invaluable
logistic support at the conference was provided by Bill Gizyn.
Cynthia Jordan has done much of the typing of manuscripts with
great skill. N.R. Sprankling and F. Blair have helped in managing
finances.

                                        Thomas C. Hutchinson

                                        Magda Havas

Contents

INTRODUCTION

T. C. Hutchinson

As standards of living rise in the industrialised nations and populations grow everywhere, the second half of the 20th century seems likely to be remembered as the time of awakening in the value we place on an unspoiled environment. It is also a time in which large scale regional problems of pollution became evident and the demands for resources were at last seen in the finiteness of the earth's supply. The concern amongst scientists, members of the public and politicians over the spread, on a regional basis, of acid rain episodes has been accelerating throughout the 1970's, following the initial evaluation of network data in Scandinavia. The elimination of fish populations in extensive lake systems in Norway and Sweden, as well as in eastern North America has struck a particular chord of concern. At a time when more people than ever have access to recreational lake and cottage areas than ever before, and sports fishing is at an all time high, the people find that scientists warn of increasing devastation via acidification to freshwater systems, even in very remote areas from major industrial sources. The international nature of the problem has become evident. Nations receiving more 'acidity' than they export have become particularly concerned, since they are also frequently just those countries of northern latitudes with poorly buffered lake systems which are especially vulnerable to additions of acidity. The long distance transport of pollutants is now a clearly recognized fact. The sulphur content of the polluted air masses has been especially identified as causal, though greater emphasis is now being placed on the substantial and increasing contribution of emitted nitrogen oxides to the problem.

Inevitably it seems, the need for energy generation in all forms lies at the root of the acid precipitation problem. In the absence of a strongly committed and effective energy conservation policy, the increasing energy demand seems likely to cause much more coal to be consumed in the future. While a further spread of acidified lakes seems presently very probable, a good deal of concern has been expressed about the health of our terrestrial ecosystems and the susceptibility of forests and agricultural crops to acid precipitation. On this subject, at least, a good deal of

1

confusion has reigned, due largely to the diversity and complexity
of the ecosystems concerned, and to the fact that over much of the
regions receiving acid rains and acid snowmelt water, the soils are
and have been naturally acidic. The natural farming practices in
higher rainfall areas are to compensate for nutrient leaching and
development of acidity by liming and fertilising. In forests, the
majority of plants are long lived perennials so that, unless effects
of acid precipitation are dramatic and severe, changes will be slow
and perhaps masked by the natural annual fluctuations in climate.
These latter differences from year to year and decade to decade can
be very large.

This symposium was organized to examine our scientific
knowledge of the normal functioning of terrestrial ecosystems and
the presently documented effects of acid precipitation on this
functioning. Climatologists, meteorologists, plant physiologists,
forest ecologists, soil scientists, microbiologists, plant
pathologists and limnologists all had a role to play in the meetings
since their knowledge is an essential to our understanding of how
airborne acid additions may influence forest function and stability.
We also examined the interactions between terrestrial and aquatic
systems.

The presentations generally follow the course of acid rain or
snow from the atmosphere to the foliage of the plants, thence to
the soil, through the soil either to influence plant nutrition
through root uptake and microbial interactions, or via drainage
from the soil to the watershed streams and rivers to the lakes.
The order of papers presented in Parts 1 to VI of this volume is
largely that of the sessions of the conference. Summary reports
of the papers and discussions are then given in Part VII. Finally,
an overall summary of the conference conclusions and recommendations
is given. The aim has always been to provide a forum for a proper
assessment of the present and anticipated impact of acid precipitation on terrestrial ecosystems, and for identification of those
systems and soils which may be especially sensitive to this type
of stress.

ADDRESS TO CONFERENCE DELEGATES*

Dr. Frank Maine, M.P.

Parliamentary Secretary

Minister of State for Science and Technology
Government of Canada

In extending the good wishes of the Federal government to all
the delegates I am aware that you, as scientists, have been
concentrating your attention on the scientific aspects of acid
precipitation. You want to identify and characterize the various
sources of atmospheric pollution; you wish to know what components
give rise to acid precipitation, how they change with time, and
under what conditions they are leached out of the air; you are
interested in how they react with plants and farmland, and what the
ultimate consequences are for living things. Even more important,
you want to know what are the gaps in our knowledge and how we can
fill them. However, as a politician, I am more concerned with what
can be done about the problem, about what practical arrangements
are most effective, and how scientific considerations can best be
integrated into decision-making.

As you are all well aware, industrial atmospheric pollution and
its harmful effects on terrestrial ecosystems are not new problems.
On your field excursion to Sudbury you will see a sobering example
of the kinds of effects such pollution can have on the natural
environment. Anyone who has visited the industrial Midlands of
England, or the Ruhr in Germany, knows that when the factories and
smelters were constructed to establish our present industrial base,
virtually no thought was given to eliminating or even reducing the
noxious emissions from the ubiquitous smoke-stacks. Fumes of all

*This paper is an edited version of a luncheon address given by
Dr. Frank Maine, Parliamentary Secretary to the Honourable
Judd Buchanan, Canadian Minister of State for Science and Technology,
on the final day of the conference.

colours were allowed to billow across the countryside in whatever
direction the wind took them. As often as not, it took them into
someone else's backyard. The problem was particularly severe in
certain localized regions where, because of the prevailing winds
and the proximity of large industrial plants to an international
frontier, one country could export its unwanted atmospheric
pollution on a fairly reliable basis. Naturally enough, this was
not always appreciated by its neighbour.

I should like to describe for you a specific example of such a
situation, in which the search for a solution, which took several
years, resulted in a landmark decision in international law, and
which also shows how scientific advice was an essential component
of a process which involved legal, economic and political
negotiations. It involved the operation of a smelter by the
Consolidated Mining and Smelting Company of Canada at Trail,
British Columbia, on the Columbia River about eleven miles (17.6
kilometres) from the international boundary with the United States.
The sulphur-bearing ores roasted at Trail gave rise to sulphur
dioxide which was vented to the air, drifted down the river valley,
and crossed the frontier in sufficient amount to cause damage to
the adjacent agricultural areas. There was no question about the
fact that the effects were harmful; there was less agreement about
the nature and extent of the damage, however, and about its
evaluation in monetary terms.

When similar problems had arisen elsewhere in Canada or the
United States, the customary procedure had been to negotiate
individual damage claims and for the company to acquire smoke
easements by purchase or otherwise. In the Trail case, however,
there were legal barriers which prevented the company from acquiring
such easements in the State of Washington; and besides, the company
felt that the claims from south of the border were excessive.

To find out how this dispute was resolved, it is necessary to go
back in time, to the signing in 1909 of the Boundary Waters Treaty
between the United States and Canada. This remarkable treaty was
established primarily to prevent disputes regarding the management
and control of the inland waters occurring along the frontier between
the two countries. The treaty is remarkable because of the unique
agency it established to deal with all matters under its jurisdic-
tion. This agency is known as the International Joint Commission,
or IJC for short, and has established a reputation for impartiality
and an enviable record of action in the common interest.

It consists of six commissioners, three of them appointed by
the U.S. President and three by the Governor in Council in Canada.
However, the six act as a single body, rather than as national
delegates under instructions from their respective governments, and
decisions are by a simple majority. One of their principal roles
involves the launching of studies to obtain precise information on

the factors affecting the existing or proposed use of boundary
waters; they therefore have a substantial requirement for expert
advice. For this purpose, the Commission is authorized by the two
governments to call upon the best-qualified experts in the public
services of the two countries and to assemble them into
"international boards" which again act as one body under joint
chairmen. On the whole, this system has proved highly-effective as
a means of mobilizing the variety of talent and experience required.

This mechanism, then, was in existence at the time the Trail
smelter case arose in the mid 1920's, and the two governments
agreed in 1928 to refer the questions at issue to the Commission.

The investigation was exhaustive and was carried out with the
help of very eminent university scientists as well as that of
experts from the public services of the two countries. Public
hearings were held and witnesses were invited to present their case.
The Commission submitted a unanimous report to the two governments
on February 28, 1931. I emphasize the word "unanimous". It shows
the judicial integrity of the six members of the Commission,
particularly the United States members, who were under the most
intense pressure from interests in their country to recognize the
very substantial claims that had been made on the U.S. side - claims
which Canada considered excessive.

The Canadian Government accepted the Commission's recommendations; however, they were rejected by the U.S. partly because the
amount of compensation was considered inadequate. After lengthy
negotiations, the two governments signed a Convention in 1935 under
which the dispute was referred to a three-member Tribunal, which
comprised a neutral chairman from Belgium, one U.S. member and one
Canadian, and was charged with finding a just and permanent settle-
ment. Each government also designated a scientist to assist the
Tribunal.

I shall not go into all the details of the Tribunal's
activities. It inspected at first hand both the affected area and
the smelter itself, and held several public hearings. Pending a
final report, it imposed temporary restrictions on the Company and
directed a detailed study of the emissions and their effects over
three complete growing seasons. The Tribunal described this part
of the investigation as "probably the most thorough study ever made
of any area subject to pollution by industrial smoke".

The Tribunal reported its final decision to the two governments on March 11, 1941, fourteen years after the original U.S.
complaint. It is this report of the arbitration Tribunal that has
made the "Trail Smelter" famous in international law. I should like
to quote briefly from this historic ruling. The Tribunal concluded
that: "....under the principles of international law...no state has

the right to use or permit the use of its territory in such a manner
as to cause injury by fumes in or to the territory of another or the
properties or persons therein, when the case is of serious
consequence and the injury is established by clear and convincing
evidence...."

As well as fixing the sum to be paid as compensation, The
Tribunal imposed in perpetuity, a regime of control over the
emission of sulphur dioxide fumes from the smelter. The capital
cost to the Company was approximately $20 million - a very consi-
derable sum. In order to comply with the regime, it was compelled
to remove from its stack effluent more sulphur dioxide than was
taken from the stacks of all other smelters on the North American
continent combined! Fortunately, the Company found a market for
the major product of its smoke abatement program, namely, sulphuric
acid, and based a substantial fertilizer business on it.

The Tribunal also decided that an indemnity would have to be
paid in the unlikely event of future damage occurring. It is worth
noting however, that it did not insist on the absolute cessation
of damage by the Trail Smelter. This would have been a dangerous
precedent which could well have shut down the Trail operation
altogether, but which, in addition, would also have had embarrassing
implications for U.S. companies in other border areas like Detroit
and Buffalo. As I stated earlier, this was a landmark decision,
based on studies and investigations which were pursued with great
thoroughness, sincerity and dedication over many years and which
led to a satisfactory ending for both sides to the dispute. It
confirmed the credibility of the International Joint Commission whose
rulings have been accepted in subsequent similar cases, most notably
in the Detroit - Windsor area.

What the example of the Trail Smelter Case brings out clearly
is the way in which the scientific understanding of what was
happening had to be interwoven into a complex legal, economic and
ultimately political process. In this case the international context
was at its least complicated, since only two states were involved;
in other parts of the world, for example in Europe, the number of
dimensions can expand significantly.

The Trail smelter case is an excellent example of how, with
goodwill on both sides, joint measures can be devised to resolve
disputes occasioned by the fact that environmental phenomena don't
respect political frontiers. However, to be effective, this kind
of action on the political level presupposes agreement on the
scientific facts behind the phenomena in question. This is where the
role of such organizations as the NATO Science Committee and its
subsidiary bodies become so important.

The NATO Science Committee itself fosters the goodwill necessary for joint action. Its basic aim is to encourage the establishment of lasting international contacts among the members of the scientific community in the member countries of the Alliance and thus to maintain and improve their collective strength. It tries to ensure that scientific advances in one country can be used for the benefit of all. In order to do this effectively, a foresighted decision was taken, right at the beginning, that the program should be run by the scientists themselves, not by bureaucrats. Your discussions here this week are an excellent example of this principle. The subject matter is at the frontier of science and is of major concern to all industrialized countries. You are all practising scientists and experts in your field. You are drawing on knowledge and experience in many areas of science and from many countries. The results of your deliberations and the insights you have gained will be shared by all. Hopefully, they will lead to coordinated research programs. The contacts you have made will be maintained and may result in cooperative research and joint studies. All this is a model of the scientific ideal, and effectively demonstrates the success of the NATO Science Program.

I am happy to think that a Canadian, the late Lester B. Pearson, was one of the "Three Wise Men" whose recommendations to the North Atlantic Council in 1956 led to the establishment of this most successful program. I am also proud of the fact that Canadian scientists have always played a prominent role in the various specialized bodies responsible for the successful operation of the program over the last twenty years. Two years ago, the Ministry of State for Science and Technology carried out a searching review of the NATO Science Program and of our own participation in it. The conclusions of that study were entirely favourable. The program itself, in spite of funding problems due to inflation, is largely meeting the aims set out for it in 1958. The scientific calibre has been maintained at a very high level. The concept behind the program is unique since it makes possible forms of international cooperation which would otherwise not exist. Our review included a number of recommendations, starting with a strong plea for continued Canadian support of the program. I can assure you that such support is not in doubt and that Canada will continue its active and enthusiastic participation.

WET AND DRY DEPOSITION OF SULPHUR AND NITROGEN COMPOUNDS FROM

THE ATMOSPHERE

David Fowler

Institute of Terrestrial Ecology
Bush Estate, Penicuik
Midlothian, EH26 0QB, U.K.

INTRODUCTION

The distance scales of atmospheric transport of primary air pollutants and their products are regional or continental, and on these scales oxidized compounds of sulphur and nitrogen are the principle agents generating acidity in rain. The following brief review of removal mechanisms is confined to these compounds.

Though large in number, individual removal mechanisms may be placed in two broad categories: wet deposition, which includes all pollutant material reaching the earth's surface in precipitation; and dry deposition, comprising the processes of adsorption of particulate and gaseous material by land or water surfaces. The dominant gaseous sulphur compound in the atmosphere is sulphur dioxide. Though hydrogen sulphide and dimethyl sulphide may be important in the global sulphur cycle as natural sources, their removal from the atmosphere probably follows oxidation to sulphur dioxide or sulphate. Atmospheric oxides of nitrogen in the gaseous state include N_2O, NO, NO_2, N_2O_5 and HNO_3. The major sink for N_2O is diffusion into the stratosphere where it plays an important role in photochemical processes (Junge[1]). Reaction products eventually re-enter the troposphere as particulate nitrate and NO_x. Nitric oxide is oxidized to nitrogen dioxide by atmospheric ozone, and away from sources there is much more NO_2 than NO. It is also likely that removal from the atmosphere follows oxidation to NO_2. For gaseous HNO_3 and N_2O_5, so few measurements are available that it is difficult to predict their average contributions to total gaseous oxides of nitrogen, though several authors have suggested that on average HNO_3 is the major gaseous component of tropospheric

NO_x (Crutzen[2]; McConnell and McElroy[3]). In the sections to
follow on dry deposition, the lack of published measurements of
tropospheric concentrations of oxides of nitrogen introduce
considerable uncertainity to estimates of the total acidic input.

For particulate sulphur and nitrogen compounds, sulphate and
nitrate are the dominant species, but because of the hygroscopic
nature of $SO_4{}^{2-}$ and $NO_3{}^-$ aerosols they are generally present in
droplet form (Garland[4]).

DRY DEPOSITION

Gases

In vegetation, pollutant gases may be absorbed on cuticles of
plants, in intercellular fluids following uptake through stomata,
or beneath the vegetation on a particular component of the soil.
The chemical affinity of the gas for each of the components of
these surfaces, acting in parallel, will determine the affinity of
the surface as a whole for the gas. Overall rates of deposition
onto the surface also depend on rates of gas transport to the surface, as effected by turbulent diffusion (molecular diffusion is an
important transfer mechanism only within the boundary layer adjacent to the surface distances of the order of 1 mm). What we
require is the total flux to the surface. This requires a
sufficient understanding of the mechanisms of deposition and the
factors affecting them to enable accurate prediction of deposition
rates for given atmospheric and surface conditions.

Of a wide spectrum of available methods for measuring dry
deposition rates (Droppo and Hales[5]) the micrometeorological
methods have particular value in studying individual components of
the deposition process. These have been widely applied in field
and wind tunnel measurements of SO_2 fluxes to various surfaces.

The total flux (Fp) of pollutant gas to suitable surfaces
(extensive uniform areas of vegetation/soil/water) is proportional
to the gradient of concentration of the gas within the turbulent
boundary layer that develops over these surfaces $(\partial \chi / \partial z)$. The
constant of proportionality (Kp) is the eddy diffusivity, which is
assumed identical for heat and mass transfer, and is obtained from
vertical gradients of horizontal wind speed and temperature above
the same surface.

$$Fp = Kp \frac{\partial \chi}{\partial z} \qquad \ldots (1)$$

Integrating (1) with respect to z between z_1 and z_2,

$$F_p = \frac{\chi_p (z_2) - \chi_p (z_1)}{\displaystyle\int_{z_1}^{z_2} \frac{dz}{K_p}} \qquad \ldots (2)$$

This is a form analogous to Ohm's law: flux = potential difference/ resistance (Monteith[6]). Now if we assume that the surface concentration is zero, and that the lower limit of the integral is the surface, this may be written:

$$r_{t(z)} = \frac{\chi_p(z)}{F_p} \qquad \ldots (3)$$

where $r_{t(z)}$ is the total resistance to transfer between a height (z) in the atmosphere and the absorbing surface. The reciprocal of the total resistance has dimensions of velocity and is identical to the velocity of deposition (Vg),:

i.e. $\dfrac{1}{r_{t(z)}} = V_g(z) = \dfrac{F_p}{\chi_p(z)} \qquad \ldots (4)$

$r_{t(z)}$ consists of surface (r_c) and atmospheric (r_a) terms for surface affinity, and transfer processes within the boundary layer respectively. The atmospheric terms may be estimated from the wind and temperature profile, assuming that transfer of gases through the turbulent boundary layer is analogous to the transfer of momentum. The aerodynamic resistance to momentum transfer between z_1 and z_2 is defined by:

$$\tau = \rho (U z_2 - U z_1) \qquad \ldots (5)$$

where τ = momentum flux, $N \cdot m^{-2}$
  u = horizontal windspeed, $m \cdot s^{-1}$
  ρ = air density, $g \cdot m^{-3}$

Taking z_1 as the height where u = 0,

$$r_{a(z_2)} = \frac{U z_2}{U_*^2} \qquad \ldots (6)$$

U_* = friction velocity, $m \cdot s^{-1}$, estimated from wind profile.

Because bluff-body forces acting on components of the surface assist
momentum transfer whereas the transfer of SO_2 molecules is restricted
to molecular diffusion, the aerodynamic resistance to momentum
transfer is smaller than the aerodynamic resistance to SO_2 transfer
for rough surfaces (eg. vegetation). In practice a correction for
this is made by adding a further boundary layer resistance r_b in
series with r_a in the manner described by Chamberlain[7]

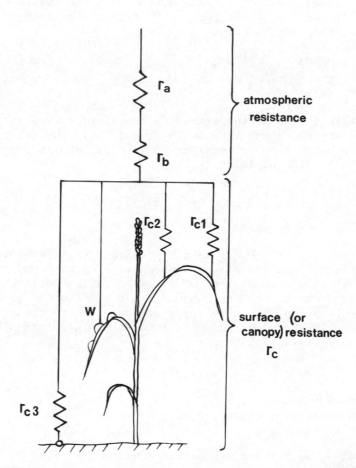

Figure 1. Resistance to dry deposition of pollutant gases in
a cereal crop. Surface resistance comprising r_{c1} – the stomatal
component, r_{c2} – the cuticular component and r_{c3} – the soil component
W refers to the situation with pure water on foliage when normal
paths of uptake are short-circuited.

where $r_b = (B \, U_*)^{-1}$ ...(7)

Experimental evidence suggests that the factor B^{-1} varies little with windspeed or plant geometry for most crop canopies and appropriate values for the most important sulphur and nitrogen pollutant gases are SO_2:7, HNO_3:7, NO_2:6.

If measurements show r(a+b) equal to r_t then the surface is behaving as a perfect sink. However, for most surfaces (natural bodies of water excepted) a residual r_c (known as surface or canopy resistance) is present. Fig. 1 illustrates these resistances and shows diagrammatically the major sinks for pollutant gases in vegetation. The relationships between surface roughness, windspeed, atmospheric stability and aerodynamic resistance are well documented (Thom[8]), so that for known surface resistances, rates of deposition may be predicted from vegetation height and windspeed (Fig. 2).

Results of published work may now be used to summarize our knowledge of dry deposition rates on different natural surfaces.

Uptake by natural waters. For the three gases SO_2, HNO_3 and NO_2 the surface resistance is controlled by the solubility in water, the subsequent reactions of dissolved gas with water and the removal of both dissolved gas and reaction products from the surface layers of the water.

Liss and Slater[9] suggested that for uptake of these gases by oceans and inland water surfaces, resistances in the liquid phase (surface resistance) were negligible by comparison with atmospheric resistance. These predictions have been confirmed for SO_2 in field measurements by Garland[10] and Whelpdale and Shaw[11] (though periodic formation of a surface film of organic material may prevent perfect sink behaviour of some inland waters) and in laboratory experiments by Brimblecombe and Spedding[12]. No field or laboratory experiments have been reported for HNO_3 or NO_2 gas, though it is probable that HNO_3 will behave in much the same way as SO_2. In the case of NO_2, which enters solution less readily, surface processes may also be important.

Soil. For such a heterogeneous material, general predictions of surface resistance are not possible. For certain soil, such as calcareous ones, the expectation of negligible surface resistance has been confirmed for SO_2 in field measurements by Garland[13] and by Payrissat and Beilk[14]. They showed that surface resistance to SO_2 deposition increased as soil pH decreased from negligible resistance values at pH 7.0 to about 3.0 $s \cdot cm^{-1}$ for soils at pH 3.5. For other than calcareous soils, those which are wet provide a better sink than those which are dry.

For the nitrogen oxides, the lack of measurements is a major problem since soil is not only a sink for oxides of nitrogen, but also a source. In addition to the well known release of N_2O from soils, NO_2 is also reported to be liberated. Makarov[15] reported values of 0.3 to 7 x 10^{-9} g NO_x – N·m^{-2}·s^{-1} in pot experiments. The upper limit of this range is of the same order as dry depos-

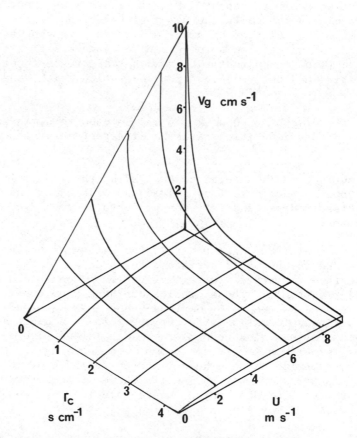

<u>Figure 2</u>. Variation of deposition velocity with windspeed and canopy resistance for a crop height of 1.0 m (neutral stability assumed). Reference height 2.0 m.

ition rates of NO_2 with an ambient concentration of 10 μg m^{-3} and
Vg 0.5 cm·s^{-1}. Since HNO_3 is a very reactive gas, it is probable
that deposition rates will be at least as large as those for SO_2
(i.e. about 1 cm·s^{-1}), though the net flux of nitrogen oxides for
soil is uncertain.

Vegetation. Rates of dry deposition of SO_2 on a wide range of
vegetation have been reported (Garland et al.[16];Holland et al.[16a];
Fowler and Unsworth[17]). From these measurements,surface resistance is
shown to control rates of deposition in most conditions. From
a compilation of published values, Garland[10] showed surface resistance to increase with vegetation height. As bulk aerodynamic
resistance decreases with vegetation height and bulk canopy resistance to water vapour transfer increases with vegetation height
(an adaptation towards stricter control of water loss in tall
vegetation) this implies two things. First, that surface resistance increases as a proportion of total resistance with increasing
vegetation height (Fig. 3) and second, that stomata are also
important sites for the uptake of SO_2. The latter receives support
from the laboratory measurements of Spedding[18] and the field
measurements of Fowler[19] and Garland and Branson[20]. From measurements over a cereal crop Fowler[19,21] evaluated the various components of surface resistance (Fig. 1). For a growing crop with
stomata open, 30% of the total SO_2 flux was to the cuticles of the
plants and 70% was to the sub-stomatal cavity. The flux to soil
beneath the canopy was negligible in these conditions. In the
longer term, however, because stomata are only open for a limited
period each day and cuticular uptake is continuous, the latter
process accounts for more than half of the SO_2 deposited on the
crop during the growing season (April-August). Measurements of
SO_2 uptake by pine needles using a cuvette method (Garland and
Branson[20]) indicated cuticular uptake, but the subsequent fate of
this gas was uncertain and the importance of cuticular uptake for
forests is still uncertain. The relative importance of these two
sinks for other vegetation will probably be different, as stomatal
behaviour and cuticle composition are very variable between species
(Meidner and Mansfield[22]).

Water on vegetation has a pronounced effect on rates of
deposition. In the case of dew, surface resistance decreases to
zero and enhanced deposition rates (which may exceed V_g = 1 cm·s^{-1})
continue until the pH of the layer of water decreases to about 3.5
(Fowler and Unsworth[17]). For gaseous HNO_3 and NO_2 much of the
following comment requires examination by experiment. As HNO_3 gas
is very reactive and soluble in water, it would be expected to be
absorbed in the intercellular fluids of leaves after diffusion
through stomata. Its molecular diffusivity is almost identical
to SO_2. Thus this component of canopy resistance (r_{c1}) would
probably be the same as that for SO_2. HNO_3 gas would also be
absorbed by cuticles of plants, with r_{c2} being equal to or less

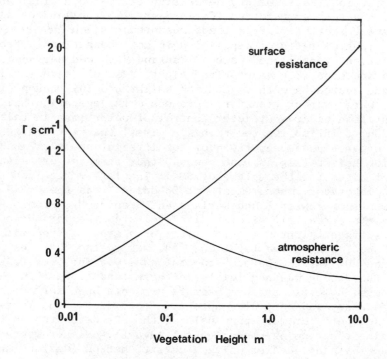

<u>Figure 3</u>.   Changes in mean surface and atmospheric
resistance to SO_2 deposition with vegetation height.

than that for SO_2. A velocity of deposition of HNO_3 gas of about
1.0 cm·s^{-1} for short vegetation would be expected on this basis.
For NO_2, measurements by Hill and Chamberlain[23] indicate a deposition velocity of about 0.8 cm·s^{-1}. This is similar to that for SO_2,
and implies a small cuticular resistance, since NO_2 is not
expected to be as soluble in intercellular fluids as SO_2. Table
1 summarizes mean deposition rates for the gaseous sulphur and
nitrogen oxides discussed in this section.

Particles

   Although the same processes of turbulent diffusion are
responsible for transporting particles (considering particles of
< 50 µm diameter) through the atmosphere to the viscous boundary
layer over elements of the surface, the processes of deposition
of gaseous and particulate material at the surface differ
considerably. Particles of > 1 µm are transported through the sub-
layer by impaction and gravitational forces, while smaller
particles are transported by Brownian diffusion. Also, when
particles have been transported through the viscous sub-layer they
may be returned to the free atmosphere by bounce-off or blow-off.
These processes of immediate and delayed removal from surfaces are
considered in detail by Chamberlain[24].

   Chamberlain[24] showed the relative efficiency of the different
mechanisms for transporting particles to a short grass surface
(Fig. 4). This shows the importance of increasing particle size
in the range 1 to 10 µm for impaction processes, and the gradual
increase in the importance of Brownian diffusion processes for
the deposition of sub-micron particles. For smoother surfaces the
deposition rate for 0.1 to 1.0 µm particles is typically an order
of magnitude smaller.

   Most of the particulate sulphate and nitrate aerosol lies in
the region 0.1 to 1.0 µm where dry deposition processes are least
efficient and the deposition velocity for this material is generally
less than 0.1 cm·s^{-1}.

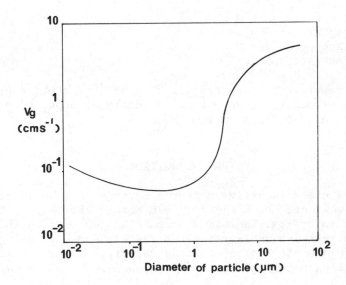

Figure 4. Deposition velocity of particles to short grass
from Chamberlain[24], ($U_* = 1.4$ m·s^{-1}).

Table 1. Dry deposition for atmospheric and gaseous S and N compounds.

	SURFACE	LIMITED BY	MEAN v_g cm·s^{-1}	RANGE
SO$_2$ (g)	Water	A	0.4	0.1-0.8
	Acid Soil	S	0.3	0.1-0.5
	Alkaline Soil	A	0.6	0.3-1.0
	Short Vegetation 0.1 m	S/A	0.5	0.1-0.7
	Medium Vegetation 1 m	S	0.8	0.1-1.5
	Tall Vegetation 10 m	S	0.5	0.1-2.0
HNO$_3$ (g)	Water	A	0.4	0.1-0.8
	Most Soils	A	0.8	–
	Most Vegetation	A	1.0	–
*NO (g)	Water	A	0.4	–
	Most Soils	S	0.5	–
	Most Vegetation	S	0.4	–
SO$_4^{2-}$ NO$_3$ (p)	All Surfaces	A	0.1	–

g   = Gas
p   = Particle
A   = Atmospheric processes
S   = Surface processes
S/A = Surface/Atmosphere

* Recent measurements reported verbally at this meeting by Dr. G. Gravenhurst suggest a very small velocity of deposition for NO$_2$ on water and soil (i.e. 0.1 cm·s^{-1}).

WET DEPOSITION

Of the elements arriving at the earth's surface in rain, some have been collected by cloud droplets before they began their descent as raindrops. Individual mechanisms comprising this process are collectively known as rainout. The processes removing gases and particulate material by falling raindrops are known as washout. Figure 5 illustrates the various processes involved in wet deposition.

Rainout

Gases. Molecules of SO_2 gas, following uptake by cloud droplets ionized in solution according to:

$$SO_2 + H_2O \quad {}^{K_1}\rightleftharpoons HSO_3^- + H^+ \qquad K_1 = 0.013 \text{ mol} \cdot 1^{-1}$$

$$HSO_3^- \quad {}^{K_2}\rightleftharpoons SO_3^{2-} + H^+ \qquad K_2 = 10^{-7} \text{ mol} \cdot 1^{-1}$$

and assuming irreversible change, most of the sulphur in solution is present as the bisulphite ion[25]. In this very simple situation the equilibrium, given by Postma[25] is:

$$H_e \simeq \frac{H K_1 M}{C (\infty)} \qquad \ldots (8)$$

Where H_e is the effective solubility, C is the concentration of dissolved SO_2 after infinite time, M is the molecular weight of SO_2 and H is the ratio of dissolved undissociated SO_2 to the concentration in air. Garland[10] has calculated the times necessary for different droplet sizes to attain equilibrium in different ambient SO_2 concentrations. The time is generally of the order of seconds, so that cloud droplets are generally close to equilibrium concentrations. The proportion of average cloud water sulphur due to dissolved SO_2 is, however, quite small because the long residence times of cloud droplets allow oxidation of dissolved SO_2 to SO_4^{2-} by a variety of mechanisms (van den Heuval and Mason[26]; Kellog et al.[27]). Some of the more recently identified mechanisms involving hydrogen peroxide and ozone seem particularly efficient (Penkett et al.[28]).

The removal of NO_2 by rainout probably follows gas phase conversion to HNO_3 gas, which according to Robinson and Robbins[29] is rapidly removed by reaction with atmospheric ammonia and absorbed into hygroscopic particles. At relative humidities exceeding 98%, condensation of dilute nitric acid droplets will occur (10°C and HNO_3 gas concentration 0.1 ppb),indicating an efficient rainout mechanism for HNO_3 gas in cloud.

Particles. Mechanisms leading to the capture of SO_4^{2-} and NO_3^- containing particulate material by cloud droplets include impaction, interception, Brownian diffusion, thermo and diffusiophoresis. In addition, particulate material, especially sulphate and nitrate containing aerosols act as cloud condensation nuclei. Impaction and interception processes seem much less important within clouds than below cloud base (the latter contributing to washout). From work by Mason[30] the attachment of particles by

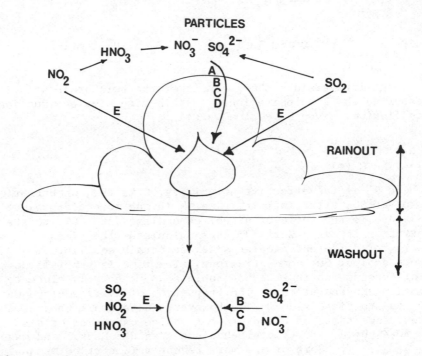

A   Cloud condensation nuclei
B   Impaction and interception
C   Thermo and diffusiophoresis
D   Brownian difffusion
E   Solution and oxidation of gas

    Figure 5. Major processes leading to the removal of sulphur
and nitrogen compounds by precipitation.

Brownian diffusion for a simple model cloud has been estimated
by Garland[10]. Attachment times of 70 and 10^5 hours were estimated for 0.1 μm and 1.0 μm particles respectively, indicating
that this process is not an important rainout mechanism. Phoretic
effects were found to be even less important than Brownian
diffusion and the only in-cloud particle removal process yielding
a significant proportion of sulphur in precipitation is the

rainout of cloud condensation nuclei. Condensation of water vapour
in clouds on particulate material and subsequent coagulation of
cloud water droplets has been treated in detail by cloud physicists
(Mason[30]). Sulphur and nitrogen-containing aerosols form a major
component of cloud condensation nuclei globally and are even more
important in the atmospheres of industrial countries. Though a
number of separate mechanisms result in the final sulphate content
of rain at the earth's surface, an idea of the efficiency of the
rainout of cloud condensation nuclei may be obtained from measurements of the concentration in rain of fossil fuel-derived material
that only exists in the particulate form in the atmosphere and which
has a similar size distribution to that of SO_4^{2-} and NO_3^- aerosols.
Measurements of appropriate elements have been reported by Cawse[31]
in the form of washout ratios where:

$$\text{washout ratio} = \frac{\text{Concentration per gramme of cloud water}}{\text{Concentration per gramme of air}}$$

Washout ratios, for most of the fossil-fuel derived elements that
only exist in particulate form, are about 300. Garland[10] concluded
that appropriate values for SO_4^{2-} and NO_3^- would probably be larger
because soluble particles are more efficient cloud condensation
nuclei. Rainout of cloud condensation nuclei contributes 90% or
more of the rainout sulphur; Brownian diffusion being the only other
significant mechanism. Since so little information is available
for estimation of rainout of NO_3^- aerosol, it is assumed for the
present discussion that similar rates of removal by each of the
rainout processes apply to SO_4^{2-} and NO_3^- particles.

Washout

The absorption of pollutant gases by falling rain has been
examined by Chamberlain[32] assuming irreversible uptake, and more
recently by Hales[33] who considered reversible uptake. In general,
the concentration of SO_2 in rain decreases with increasing drop
size. This partly compensates for increases in rainfall rate, as
droplet size and rainfall rate are positively correlated. Expressed
as a washout coefficient (mass fraction of a compound removed each
second) Chamberlain[32] gave values of 10^{-4} and 3×10^{-4} for rainfall rates of 1 and 10 $mm \cdot h^{-1}$. The washout of HNO_3 gas is likely
to be of a similar order to that of SO_2 and as NO_2 is less readily
taken up by raindrops, its removal by washout is likely to be
quite small.

Particles. Washout of particulate material is achieved through
processes of impaction and interception by falling raindrops and
by Brownian diffusion and phoretic effects, though these last two
processes are of little practical importance to the washout of SO_4^{2-}

Table 2. Principle wet deposition processes for sulphur and nitrogen compounds in the atmosphere assuming (i) 10 $\mu g \cdot m^{-3}$ SO_4^{2-} aerosol and 2 $\mu g \cdot m^{-3}$ of NO_3^- aerosol composed of (a) 80% submicron particles, typical diameter 0.02 μm (b) 20% larger particles typically of 2 μm diameter (ii) 10 $\mu g \cdot m^{-3}$ SO_2, rainfall rate 1 $mm \cdot h^{-1}$. (From Garland[10]

			RAINOUT				WASHOUT			
		Process	Lifetime hours		Concentration in rain mg l^{-1} SO_4^{2-} or NO_3^-		Process	Lifetime hours		Concentration in rain mg l^{-1} SO_4^{2-} or NO_3^-
			a	b	a	b		a	b	a	b
particle	SO_4^{2-}	Cloud condensation nuclei (CCN)	1	10	3	10	Impaction & interception	10^3	8	10^{-2}	0.2
		Diffusiophoresis	-	-	10^{-2}	10^{-3}	Brownian diffusion	10^4	10^5	10^{-3}	10^{-5}
		Brownian Diffusion cloud droplets	70	10^5	10^{-3}	10^{-5}					
	NO_3^-	CCN	1	10	0.6	2.0	Impaction & interception	10^3	8	2×10^{-3}	4×10^{-2}
gas	SO_2	Solution & oxidation in droplets			3		uptake by falling rain	10		1	
	HNO_3	Solution by cloud droplets			0.2		"	10		0.2	
	NO_2	Solution & oxidation by cloud droplets			0.1		"				

and NO_3^- aerosol. The total contribution to wet deposition by particle washout is less than 10% (Garland[10]) for SO_4^{2-}. A similar figure is assumed to be appropriate for NO_3^- particles.

The contribution by the various wet removal processes discussed in this section to the sulphur content of rain is given in table 2. From information currently available equivalent detail is not possible for the nitrogen compounds.

SUMMARY

Principle routes for removal of the sulphur and nitrogen compounds by wet and dry deposition have now been described. For some of these routes, rates of deposition are well established and the mechanisms by which the process occurs are understood (especially in the case of sulphur compounds). However, uncertainties have also been revealed in a number of important aspects.

Summarizing 1. Dry deposition of the gases; SO_2, NO_2 and HNO_3 is effected by turbulent transfer to the surface. For SO_2, deposition velocities for most terrestrial surfaces are in the range of 0.3 to 1.0 $cm \cdot s^{-1}$. The average for the countryside is 0.8 $cm \cdot s^{-1}$, the deposition process being limited by surface rather than atmospheric processes. For NO_2, the process is also limited by the affinity of the surface, (Vg is in the range 0.2 to 1.0 $cm \cdot s^{-1}$). HNO_3 gas is deposited readily on most surfaces. It is limited by atmospheric transfer. No mean Vg is available, but it is probably close to 1.0 $cm \cdot s^{-1}$.

2. Particles SO_4^{2-} and NO_3^- - containing particulate material have a deposition velocity of 0.1 $cm \cdot s^{-1}$ or less. Dry deposition is, therefore, an inefficient removal mechanism for this material.

3. Wet deposition. The major wet removal process for both sulphur and nitrogen compounds considered in this text is the rainout of cloud condensation nuclei, though solution and oxidation of SO_2 is also important. Of the other mechanisms, only the uptake of SO_2 and interception of particulate SO_4^{2-} and NO_3^- by falling rain make a significant contribution to total wet deposition.

Important weaknesses in the discussion require:

a. Measurements of rates of dry deposition of NO, NO_2 and HNO_3 on a range of natural surfaces.

b. Extensions of dry deposition measurements of SO_2 to vegetation other than cereals and grass.

Table 3.   Relative contributions of Wet and Dry Deposition.

	Deposition Velocity V_g	Sub-Urban Area 3-30 km from source		Agricultural Area 30-300 km from source		Forest Area 300-3000 km from source	
	$cm\ s^{-1}$	$conc^n$ $\mu g\ m^{-3}$	deposition $kg\ ha^{-1}\ y^{-1}$	$conc^n$ $\mu g\ m^{-3}$	deposition $kg\ ha^{-1}\ y^{-1}$	$conc^n$ $\mu g\ m^{-3}$	deposition $kg\ ha^{-1}\ y^{-1}$
DRY GAS SO_2	0.8	100	126.2	20	25.2	5	6.3
NO_2	0.5	30	14.4	5	2.4	2	1.0
ptle SO_4^{2-}	0.1	20	2.1	10	1.1	10	1.1
ptle NO_3^{-}	0.1	5.0	0.4	2	0.1	2	0.1
	rainfall $mm\ y^{-1}$	$conc^n$ $\mu g\ ml^{-1}$	deposition $kg\ ha^{-1}\ y^{-1}$	$conc^n$ $\mu g\ ml^{-1}$	deposition $kg\ ha^{-1}\ y^{-1}$	$conc^n$ $\mu g\ m^{-1}$	deposition $kg\ ha^{-1}\ y^{-1}$
WET RAIN SO_4^{-2}	900	5	15.0	4	12.0	3	9.0
NO_3^{-}	900	1.5	3.0	0.7	1.4	0.5	1.0
DRY/WET	S		8.5		2.2		0.8
*	N		4.9		1.8		1.1

    c.   Studies of wet removal processes for NO_2 and HNO_3 gas NO_3^- particles.

    d.   Further measurements of ambient concentrations of SO_2, NO_2, NO and especially HNO_3, and SO_4^{2-} and NO_3^- aerosols in rural areas.

Relative contributions of wet and dry deposition
for different regions:-

Though individual processes contributing to nitrate in rain
have received little attention by comparison with those for
sulphur compounds, there are extensive measurements of the nitrate
concentration in precipitation, especially for N. America and W.
Europe. These measurements may now be used in conjunction with
mean concentrations of other sulphur and nitrogen compounds in the
atmosphere, and with dry deposition rates to estimate the relative
contributions of wet and dry deposition processes to total
sulphur and nitrogen deposition in a range of different situations,
(Table 3). The table shows the dominance of dry deposition close
to source areas and the increasing contribution made by wet deposition as one moves away from the source. It is also clear that
for all areas considered both processes yield a significant input
to the ground and should, therefore, be included in field studies
of the effects of deposited sulphur and nitrogen compounds.

ACKNOWLEDGEMENTS

Financial support for this work by the U.K. Department of
the Environment is gratefully acknowledged.

REFERENCES

1.   C.E. Junge, Quart, J. Roy. Met. Soc. 98:711 (1972).

2.   P.J. Crutzen, Tellus, 26:47 (1974).

3.   J.C. McConnell and M.B. McElroy, J. Atmos. Sci. 30:1465 (1973).

4.   J.A. Garland, Atmos. Environ. 3:347 (1969).

5.   J.G. Droppo and J.M. Hales, Proceedings of Symposium, Atmosphere - Surface Exchange of Particulate and Gaseous Pollutants, Richland, Washington (1974).

6.   J.L. Monteith, "Principles of Environmental Physics"
        Edward Arnold, London (1973).

7.   A.C. Chamberlain, Quart. J. Roy. Met. Soc. 94:318 (1968).

8.   A.S. Thom, in: "Vegetation and Atmosphere" J.L. Monteith,
        ed., Acad. Press (1975).

9.   P.S. Liss and P.G. Slater, Proceedings of Symposium
        Atmosphere – Surface Exchange of Particulate and
        Gaseous Pollutants, Richland, Washington (1974).

10.   J.A. Garland, "Dry and Wet Removal of Sulphur from the Atmos-
        phere," in Atmos. Environ. 12: 1-3: 349-362.

11.   D.M. Whelpdale and P.W. Shaw, Tellus, 26:196 (1974).

12.   P. Brimblecombe and D.J. Spedding, Nature, 236:225 (1972).

13.   J.A. Garland, Proceedings of Symposium, Atmosphere – Surface
        Exchange of Particulate and Gaseous Pollutants,
        Richland, Washington (1974).

14.   M. Payrissat and S. Beilke, Atmos. Environment 9:211 (1975).

15.   B.N. Makarov, Pochvodenie 1:49 (1969).

16.   J.A. Garland, W.S. Clough, and D. Fowler, "Deposition of Sulphur
        Dioxide on Grass," in Nature. 242: 256-257 (1973).

16a. P.K. Holland, J. Sugden, and K. Thornton, "The Direct Deposition
        of SO_2 from the Atmosphere Pt 2 Measurements at Ringinglow
        Bog, Report No. NW/SSD/RN/P1/1/74 (1974).

17.   D. Fowler and M.H. Unsworth, "Dry Deposition of SO_2 on Wheat,"
        in Nature. 249: 389-390 (;974).

18.   D.J. Spedding, Nature 224:1229 (1969).

19.   D. Fowler, "Dry Deposition of SO_2 on Agricultural Crops," in
        Atmos. Environ. 12: 1-3: 369-373 (1978).

20.   J.A. Garland and J.R. Branson, Tellus 29:445 (1977).

21.   D. Fowler, Uptake of Sulphur Dioxide by Crops and Soil. PhD
        thesis, University of Nottingham.

22.   H. Meidner and T.A. Mansfield, "Physiology of Stomata"
        McGraw Hill, London (1969).

23.   A.C. Hill and E.M. Chamberlain, Proceedings of Symposium,
      Atmosphere – Surface Exchange of Particulate and
      Gaseous Pollutants, Richland, Washington, (1974).

24.   A.C. Chamberlain, "The Movement of Particles in Plant Communi-
      ties," in Vegetation and the Atmosphere, ed. Monteith.
      Vol. 1, Academic Press: 155-201 (1975).

25.   A.K. Postma, AEC Symposium Series 22 CONF-700601 – Effect of
      Solubility of Gases on Their Scavenging by Raindrops,
      Precipitation Scavenging, NTIS Springfield, Virginia:
      247-259 (1970).

26.   A.P. van den Heuval and B.J. Mason, Quart. J. Roy. Met.
      Soc. 89:271 (1963).

27.   W.W. Kellog, R.D. Cadle, E.R. Allen, A.L. Lazarus and
      E.A. Martell, Science 175:587 (1972).

28.   S.A. Penkett, B.M.R. Jones and K.A. Brice, Atmos. Environ.
      12: (1978).

29.   E. Robinson and R.C. Robbins, J. APCA. 20(5):303 (1970).

30.   B.J. Mason, "Cloud Physics" Clarendon Press, Oxford (1971).

31.   P.A. Cawse, AERE-R 7669. H.M.S.O. London (1974).

32.   A.C. Chamberlain, Int. J. Air. Pollut. 3:63 (1960).

33.   J.M. Hales, Atmos. Environ. 12: (in press) (1978).

AN ESTIMATION OF THE ATMOSPHERIC INPUT OF ACIDIFYING

SUBSTANCES TO A FOREST ECOSYSTEM

Peringe Grennfelt, Curt Bengtson, Lena Skarby

Swedish Water and Air Pollution Research Laboratory

Gothenburg, Sweden

INTRODUCTION

The term 'acid precipitation' gives the impression that the acid
in rain and snow is the only input of acidifying substances to an
ecosystem. This is wrong. Acidifying substances, i.e. compounds
that might increase the hydrogen ion activity in the ecosystem,
will be deposited also in other forms. Moreover, the substances
are not necessarily acidifying immediately on entering the
ecosystem. This means that the acidifying properties may not be
developed until after deposition. Sulphur dioxide will e.g. be
less acidifying than its oxidation product, sulphuric acid. To
avoid misunderstandings in the following we will use the expression
'deposition of acidifying substances'.

     To evaluate the effects of the deposition of acidifying sub-
stances to the ecosystem the atmospheric input must be know
qualitatively and quantitatively, together with the pathways by
which the deposition takes place. The deposition mechanism for
acidifying substances are extremely complex and involve different
compounds, different forms for the deposition (gas, particles,
rain), and different pathways for further transport, incorporation
and turnover in the ecosystem. Moreover, the factors affecting
these different flows vary, and a change in one factor might
change the deposition flows radically. Consequently, it is
extremely difficult to make a detailed deposition budget which
both describes the deposition of the different acid compounds to
a terrestrial ecosystem and the pathways for deposition. So far,
a broad approach to this problem has not been made. Instead, most
of the efforts have been concentrated on two processes only: sulphur input by precipitation and sulphur dioxide deposition to

vegetation during dry conditions in daylight.

The aim of this contribution is:

(1) to describe briefly the essential pathways for deposition
    of nitrogen and sulphur compounds, and

(2) to give a rough estimate of the total input of these com-
    pounds to a coniferous forest ecosystem in southern Sweden.

The description will cover only man-produced nitrogen and
sulphur pollutants. The input of nitrogen through nitrogen fixation or the different turnover reactions in the soil leading to
e.g. reemission will therefore not be considered here.

In order to reduce the number of factors to be handled, we
have chosen the coniferous ecosystem since the conifers are winter-
green. Furthermore, effects of acidification may be expected to
occur earlier in a coniferous forest than in a decidous forest.

THE OCCURRENCE OF ACIDIFYING SUBSTANCES IN THE ATMOSPHERE

What compounds should be considered in a budget for deposition of acidifying substances? It is obvious that the anthropogenic nitrogen and sulphur compounds and their reaction products
are the most important.[1] In addition to these compounds, hydrochloric acid seems to be the only substance that might give an
acid input in the form of strong acid. The possible influence of
hydrochloric acid will be discussed below.

The acidifying substances occur in the atmosphere either as
gases or as suspended material (solids, liquids, or aqueous solutions). The suspended material covers a very wide size range:
from the smallest particles (e.g. condensation nuclei), via larger
particles, fog and mist to rain and snow. The sulphur and nitrogen
compounds of interest occur in all these forms.

The only gaseous sulphur compounds occurring in significant
concentrations in the atmosphere are sulphur dioxide and hydrogen
sulphide. Other compounds, such as carbonyl sulphide, carbon di-
sulphide and dimethyl sulphide may occur but in concentrations at
least one order of magnitude lower.[2]

Hydrogen sulphide and sulphur dioxide are either deposited
directly as gases or oxidized to sulphates in the atmosphere
and then deposited with precipitation or particles. The sulphur
dioxide oxidation will occur mainly as a photochemical or cataly-

tic reaction. The photochemical reaction appears in gas phase
but the reaction product (sulphur trioxide or sulphuric acid)
will rapidly take up water vapour and form a droplet. The cata-
lytic reaction takes place in water phase only, which means
that the gas has to be absorbed to a particle or a raindrop
before oxidation.

The mean atmospheric concentrations of sulphur dioxide,
hydrogen sulphide and particle-borne sulphate in a rural area
of southern Sweden are presented in Table 1.

Table 1

Typical concentrations of acidifying substances in gas phase
and on particles in the atmosphere in southern Sweden.

Compound		Concentration nmole/m^3	References
Gases	SO_2	200	3, 4
	H_2S	5	5
	NO	50	6
	NO_2	150	6
	HNO_2	3	7
	HNO_3	40	8
	NH_3	10	9
Particles	SO_4^{2-}	60	4
	NO_3^-	20	4
	NH_4^+	100	4

The picture is more complex for nitrogen compounds. Anthropogenic nitrogen oxides are mainly emitted as nitrogen monoxide.
In the atmosphere this compound is oxidized to nitrogen dioxide
and further to other products of which nitric acid is predominant.
Nitrogen monoxide and nitrogen dioxide exist as gases while nitric acid (nitrates) occurs both in gas phase and in suspended
material (particles and precipitation). The presence of nitric
acid in gas phase has been almost neglected so far.[8] However,
recent data indicate that it is important and that the concentration in rural areas in Sweden is approx. 40 n mole/$m^{3x)}$.[8] This
means that there is more nitrate in gas phase than on particles.
Fig. 1 shows the concentration of gaseous and particulate nitrates at a rural station in Sweden.

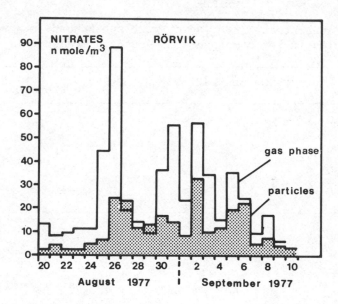

Figure 1. Concentrations of particulate and gaseous nitrates
at a rural station (Rörvik) on the Swedish west coast during
Aug. 20 - Sept. 10 1977.

x) In this paper concentration of atmospheric pollutants is denoted
in n mole/m^3. For conversion to ppb at STP, multiply by 0.0224.

Nitrite will probably occur in gas phase only. Measurements in England indicate that the nitrite concentration is much lower than the concentration of gaseous nitrate.[7]

In addition to the oxidized nitrogen compounds, ammonia is an important constituent of the atmosphere. By forming neutral ammonium sulphate and ammonium nitrate aerosols it will neutralize acid particles in the atmosphere. Consequently, the concentration of free ammonia in the atmosphere depends on the acidity of the particles, i.e. in acid atmospheres the concentration is very low (less than 1 ppb) while concentrations up to 20 ppb have been observed in alkaline atmospheres.[9, 10]

In Table 1 typical concentrations of nitrogen compounds in gas phase and on particles in Sweden are presented.

DEPOSITION PROCESSES

The vegetation is the primary recipient for most pollutants on their way from the atmosphere into the terrestrial ecosystem. However, the vegetation is not a passive participant in the transfer of pollutants into the ecosystem but will take an active part in the deposition process. Both the amount and the chemical composition of deposited pollutants are strongly dependent on the properties of the vegetation surfaces.

Deposition of any pollutant to vegetation is determined mainly by the following factors:

- atmospheric concentration and form of the compound
- atmospheric turbulence and wind speed
- canopy characteristics (composition, height, density, roughness, leaf area index)
- surface wetness (snow, wet, dry)
- stomates (size and number, opened or closed).

In a deposition budget it is necessary to exclude the least important factors, e.g. stomatal openings will not be important for deposition of particles.

Gases

Although parameters such as atmospheric turbulence and canopy density are very important for the deposition velocity of an acidifying gas, we assume that the main parameters affecting the deposition velocities are the surface wetness and the stomatal openings.

Since the deposition process for the acidifying components involves dissolution of the components in an aqueous solution, the existence of water on vegetation surfaces will strongly influence the deposition velocities.[11],[12] However, information is scarce on the frequencies of surface wetness during a year, the amount of water on the leaves in different climatic situations, and the chemical composition of these aqueous solutions.

The importance of snow has not been considered in earlier deposition budgets. For example, in northern regions such as Canada and Scandinavia, the long duration of snow cover is probably of importance. In forest districts of southern Sweden, a typical figure of snow cover duration is 20% of the year. In northern Sweden, the corresponding figure might be 50%. These figures derive from meteorological observations and of course do not represent the time during which e.g. a forest canopy is covered with snow.

Several investigators have already concluded that the stomatal openings are limiting for the deposition velocity of sulphur dioxide during dry conditions (reviewed by 13). The influence of stomates on the deposition of other gaseous pollutants is less well known but the stomatal activity is probably of major importance also for these compounds.[14],[15],[16],[17]

Estimated figures of the duration of different vegetation surface situations are given in Table 2.

Table 2

Duration in percent of the year of different vegetation surface situations of importance for the deposition of gaseous pollutants.

Vegetation surface situations	Percent of the year
Snow cover	20
No snow:	
Stomata closed	45
rain	9
wet without rain	10
dry	26
Stomata open	35
rain	3
wet without rain	2
dry	30

Particles

The smallest particles in the atmosphere behave like gases
and Brownian diffusion is an important process for the deposition
of these particles. However, Brownian diffusion becomes less important for larger particles and, instead, inertia and gravitational forces predominate. The size of the particles is, consequently, crucial for the deposition velocity. To accurately estimate the deposition of a certain compound it is therefore necessary to know both the size distribution of the particles and the
deposition velocities for the different size fractions. However,
the concentration of acidifying substances on particles is normally lower than in gas phase and besides, the deposition velocities are normally lower for particles than for gases.[18]
Consequently, in a rough deposition budget we believe it is
sufficient to treat the particles as belonging to one group, i.e.
it is not necessary to estimate the contribution of different
size fractions.

Mist and Fog

The importance of mist and fog in deposition processes is
largely unknown. Few investigations exist that deal with the
amount of water deposited to a forest canopy during mist and fog
situations as well as with the concentration of acidifying substances in the water. Rough estimates indicate that this deposition might be important at places where fog and mist are common.[19]
In Sweden, a typical figure of the fog frequency is a total of 3%
of the year.

Precipitation

Deposition by precipitation (rain, snow) is the best known
part of the total input of acidifying substances. Data obtained
from precipitation networks are used in the deposition budget.
These data normally include only sulphate and data concerning
nitrogen compounds are therefore insufficient. Furthermore, the
precipitation collectors are not suitable for measuring snowfall
and, as a result, the data on deposition with snow are uncertain.

DEPOSITION BUDGET

Based on the inventory of acidifying substances and the deposition processes discussed above, a deposition budget for a
coniferous forest ecosystem was made including all known sulphur
and nitrogen compounds of expected significance. The yearly deposition in molar terms $(m\ mole/m^2\ yr)$ as well as the total amounts
of nitrogen and sulphur $(kg/ha\ yr)$ are summarized in Table 3.

Table 3

Typical quantities of atmospheric sulphur and nitrogen compounds deposited to a coniferous forest ecosystem in rural areas of southern Sweden.

		Deposition mmole/m^2 yr	S-deposition kg/ha yr	N-deposition kg/ha yr
Gases	SO_2	32	10.2	
	H_2S	<0.5	<0.2	
	NO	<3		<0.4
	NO_2	24		3.4
	HNO_2	<0.5		<0.1
	HNO_3	8		1.1
	NH_3	<2		<0.3
Particles	SO_4^{2-}	9.5	3.0	
	NH_4^+	16		2.2
	NO_3^-	3.2		0.4
Mist and fog	SO_4^{2-}	2.5	0.8	
	NH_4^+	1.7		0.2
	NO_3^-	1.7		0.2
Precipitation	SO_4^{2-}	30	9.6	
	NH_4^+	20		2.8
	NO_3^-	20		2.8
			23.6-23.8	13.1-13.9

To obtain values of the yearly deposition of the different gases, we started with sulphur dioxide for which literature data on deposition velocities during different situations are available.[13] From the different vegetation surface situations in Table 2 we have obtained five "deposition classes". For each class given in Table 4, a deposition velocity is assumed, based on literature data. The deposition velocities are presented in Table 4. These data are "qualified guesses" of probable deposition velocities.

When treating the other gaseous compounds, in cases where no literature data exist, we have assumed that their deposition velocities depend on their solubilities in water.[14] Moreover, we have assumed that there is a fixed relation between their deposition velocities and the deposition velocity of sulphur dioxide independent of deposition class. The ratios chosen are 0.5 for hydrogen sulphide, 0.06 for nitrogen monoxide,[17] 0,67 for nitrogen dioxide,[17] 0,67 for nitrous acid, 1.0 for nitric acid, and 0.5 for ammonia.

When calculating the yearly deposition of gaseous compounds we determined the deposition for each vegetation surface situation (see Table 2) using vertical deposition velocities according to Table 4 and mean concentrations for the different situations when such data were available. The different contributions to the total deposition are presented in Table 3.

Table 4

Estimated vertical deposition velocities for sulphur dioxide to a forest canopy.

"Deposition class"	Deposition velocity, m/s	References
Snow cover	0.005	20
Stomata closed, dry canopy	0.001	
Stomata closed, wet canopy	0.10	
Stomata open, dry canopy	0.008	11, 21
Stomata open, wet canopy	0.15	11

Concerning particles, we have not gone into details as very little is presently known about how the deposition velocities depend on different parameters. We have used a deposition velocity of 0.005 m/s for all particles and typical concentration data for the southern parts of Sweden (see Tables 1 and 3).

For _fog and mist_, we have used the calculations of Unsworth[19] but translated them to the situation in Sweden. We used the fog frequency 3% and a deposition velocity of 0.20 m/s.[22] The deposition of sulphate, nitrate, and ammonium is presented in Table 4.

For _precipitation_, we used representative rainwater data for southern Sweden.[4]

DISCUSSION

From this presentation it is clear that the different atmospheric transformations and the different pathways through which the deposition takes place give a very complex picture. The chemistry of deposition to a leaf surface is presented in Figure 2. As will be seen from this figure, we are not dealing with flows only from the atmosphere to the vegetation but, under certain conditions, some compounds will be reemitted to the atmosphere. Some of the flows are speculative, _e.g._ the flow of nitrogen dioxide to the surface and further through the cuticule has not been confirmed. Furthermore, the flows of ammonium and ammonia have so far not been thoroughly investigated.

As shown in Figure 2 and in the budget (Table 4) it is obviously not possible to simplify the deposition of acidifying substances to include only sulphur dioxide or the sulphate content in the precipitation. To adequately estimate the total deposition it is necessary to include several other parts of the input. The nitrogen compounds will probably give an input that, on a molar basis, is somewhat higher than that of sulphur (100 m mole $/m^2$ yr for nitrogen and 75 m mole/m^2 yr for sulphur).

An important question is if there are any compounds in the atmosphere, apart from the nitrogen and sulphur compounds, that can give a significant input of acid to a coniferous forest ecosystem. If we consider strong acids only, hydrochloric acid is probably the only compound of interest. However, the anthropogenic emission of hydrochloric acid is small compared to emissions of sulphur and nitrogen, and since most of the chloride existing in the atmosphere comes from neutral sea salt, it does not give any input of acids.

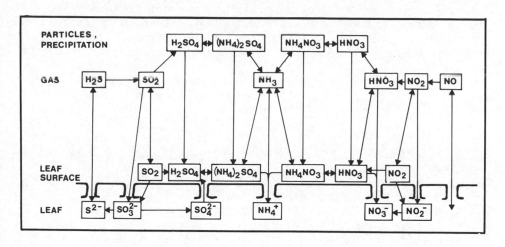

Figure 2. The flows of different nitrogen and sulphur compounds between the atmosphere and a leaf surface.

The budget includes only one part of the deposition problem: the deposition routes for different compounds and the quantities deposited to the vegetation surfaces. To improve our knowledge on the effects of acid precipitation it is probably necessary to include also the different ways of incorporation and the fate in the ecosystem of these compounds. This means that factors such as erosion on leaf surface, reemission, uptake and exudation of different compounds have to be considered in a complete model. The ability of the vegetation surfaces to absorb different compounds will change drastically depending on the amounts of water deposited on the leaves; this mechanism will probably also make it possible for volatile compounds such as ammonia and nitric acid to be reemitted.

For deposition models, the influence of a water phase on the canopy is in fact one of the most important issues to be solved. The problem is not only how to define the situation and how to calculate the duration of water covering the leaves, but parameters such as pH of the water phase and hygroscopy of the salts on the surfaces will also affect the deposition velocity.

There are several other factors in addition to the factors
in our budget that must be considered in a complete deposition
model.   Among these we can mention atmospheric turbulence, canopy
characteristics and surface chemistry.   However, it is always necessary to make "best estimations" of the deposition in order to give
priority and feed-back to deposition research and to produce input
data for research on ecosystem effects and modelling.

REFERENCES

1. R. Söderlund, and B.H. Svensson, Ecol. Bull. (Stockholm) 22: 23
    (1976).
2. F.J. Sandalls, and J.A. Penkett, Atm. Environm. 11: 197 (1977).
3. C. Brosset, private communication.
4. OECD Programme on Long Range Transport of Air Pollutants (1977).
5. P. Grennfelt, unpublished data.
6. P. Grennfelt, Swedish Water and Air Pollution Research Institute
    Publication B 418 (1978).
7. R.A. Cox, R.G. Derwent, and F.J. Sandalls, AERE, Report R 8324,
    Harwell, England, 1 (1976).
8. P. Grennfelt, Swedish Water and Air Pollution Research Institute
    Publication B 412 (1978).
9. M. Ferm, private communication.
10. O.T. Denmead, J.R. Simpson, and I.R. Freney, Science 185: 609
    (1974).
11. J.A. Garland, and J.R. Branson, Tellus 29: 445 (1977).
12. D. Fowler, Int. Symposium on Sulfur in the Atmosphere Dubrovnik
    7 - 14 Sept. (1977).
13. C. Bengtson, P. Grennfelt, and L. Skärby (in Swedish, English
    abstract) SNV PM 924, (1977).
14. A.C. Hill, J. Air Pollut. Contr. Assoc. 21: 341 (1971).
15. R.A. O'Dell, M. Taheri, and R.L. Kabel, J. Air Pollut. Contr.
    Assoc. 27: 1104 (1977).
16. H.H. Rogers, H.E. Jeffries, E.P. Stahel, W.W. Heck, L.A.
    Ripperton, and A.M. Witherspoon, J. Air Pollut. Contr. Assoc.
    27: 1192 (1977).
17. J.H. Bennett, and A.C. Hill, J. Air Pollut. Contr. Assoc. 23:
    203 (1973).
18. L. Granat, and H. Rhode, Ecol. Bull. (Stockholm) 22: 89 (1976).
19. M.H. Unsworth, Workshop on Methods in Acid Precipitation Studies,
    Edinburgh, Sept. 19-23 (1977).
20. H. Dovland, and A. Eliassen, Atm. Environm. 10: 783 (1976).
21. A. Martin, and F. R. Barber, SSD.MID.R. 44/75.   CEGB, Midlands
    Region, Rutcliffe on Soar, Nottingham, England. (1975).
22. A.C. Chamberlain, in: "Vegetation and the Atmosphere", Vol. 1,
    J.L. Monteith, ed., Academic Press, London, (1975).

SULFUR DIOXIDE ABSORBED IN RAIN WATER

Gode Gravenhorst[a], Sigfried Beilke[b], Martin Betz[c], and
Hans-Walter Georgii[c]

Inst. f. Atmosphärische Chemie[a], Kernforschungsanlage
Jülich, D-5170 Jülich; Umweltbundesamt[b], Pilot-station,
D-6000 Frankfurt; Inst. f. Meteorologie und Geophysik[c],
D-6000 Frankfurt

INTRODUCTION

The chemical composition of a rain drop is the integral result
of the incorporation of aerosol particles as well as the absorption
of trace gases. Both processes are effective from the beginning of
heterogeneous nucleation of water molecules which forms a cloud
element and continue until the rain drop reaches the ground.

Since there are always sufficient atmospheric particles to act
as condensation nuclei at supersaturation of less than 1% with
respect to water, all cloud elements originating around aerosol
particles and their chemical properties are from thereon
characterized by the composition of these cloud-active nuclei.
When cloud elements grow by condensation and coalescence into cloud-
and rain droplets, aerosol particles can still be incorporated by
Brownian diffusion or inertial impaction. At the same time, the
absorption of gaseous trace elements takes place both within the
clouds as well as below the clouds. One result of this interaction
between trace substances and droplets in the air is an acidification
of rain water. If only carbon dioxide (concentration ca. 330 ppmv)
were present as a trace substance besides water molecules, a pH
value of about 5.6 for rain water would be the result. In large
areas, however, lower values are found. This additional acidity is
due to the incorporation of acidic substances such as sulfuric or
nitric acid by the droplets. In this paper the contribution of
sulfur dioxide gas to precipitation acidity will be discussed.

S (IV) SPECIES IN THE DROPLET PHASE

If a water droplet is formed in an atmosphere with gaseous sulfur dioxide present as trace constituent, some of the SO_2 will be physically dissolved in the water according to Henry's law:

$$SO_2 \underset{\leftarrow}{\overset{H}{\rightarrow}} [SO_2 \cdot H_2O] \qquad (1)$$

Here SO_2 is the partial pressure of SO_2 in the air, $[SO_2 \cdot H_2O]$ is the concentration of physically dissolved SO_2 in the droplet phase and H the Henry constant. Since cloud- and rain droplets can be treated as dilute solutions, this equilibrium depends only on temperature. The $[SO_2 \cdot H_2O]$ dissociates into protons and bisulfite ions according to

$$[SO_2 \cdot H_2O] \underset{\leftarrow}{\overset{K_1}{\rightarrow}} [HSO_3^-] + [H^+] \qquad (2)$$

The bisulfite forms a sulfite ion and a proton:

$$[HSO_3^-] \underset{\leftarrow}{\overset{K_2}{\rightarrow}} [SO_3^=] + [H^+] \qquad (3)$$

$H = 1.76$ mol/l atm, $K_1 = 2.19 \times 10^{-2}$ mol/l, $K_2 = 7.90 \times 10^{-8}$ mol/l at $T = 288$ K (see Beilke and Gravenhorst[1]).

The equilibrium between sulfur dioxide in the gasphase and the sulfur (IV) species $[SO_2 \cdot H_2O]$, HSO_3^- and $SO_3^=$ in the droplet phase can be reached within a few seconds in droplets with radii smaller than 50 μm (Beilke and Gravenhorst[1]). In Fig. 1 the relative abundance of the S (IV) species in the droplet phase is plotted as a function of the droplet's pH. As seen in this figure, the equilibrium between the S (IV) species is shifted from sulfite ($SO_3^=$) to bisulfite (HSO_3^-) and physically dissolved SO_2 ($[SO_2 \cdot H_2O]$) as droplet's pH decreases; that is, as the droplet's acidity increases.

In the pH region 3-6 most relevant for atmospheric conditions the prevailant S (IV) species is HSO_3^-. In Fig. 2 the absolute concentration of the S (IV) species is shown in equilibrium with a SO_2 concentration of 10 ppbv as a function of pH value.

The amount of physically dissolved SO_2 is constant for all pH values; the concentrations of HSO_3^- and $SO_3^=$ are inversely proportional to $[H^+]$ and $[H^+]^2$, respectively.

As is obvious from relations (1), (2) and (3), the concentration
of S (IV) components is proportional to the atmospheric sulfur
dioxide concentration. This correlation was confirmed by measurements of the S (IV) species in actual rain samples (Fig. 3). The
S (IV) concentration was determined by sampling the rain in polyethylene bottles containing a tetrachlormercurate (TCM) solution.
The mixed rain water and TCM-solution was analyzed according to
West and Gaeke[2]. The amount of SO_2 absorbed depends, however, not
only on SO_2-gasphase concentration, but also on the rain water pH
and temperature. Therefore, the equilibrium pressure of SO_2 around
rain drops was calculated according to the measured S (IV)
concentration, pH value and temperature of the rain water and
compared with the actual SO_2-gasphase concentration at the sampling
site (Fig. 4). In general, the calculated values were lower than
the measured ones by a factor of approximately 2.7. Thus the
collected rain had not reached equilibrium with the ground level
SO_2-concentration. These measurements taken in Frankfurt indicate
that strong vertical SO_2 gradients are common. An approximated
profile is shown in Fig. 5 suggesting that the measured SO_2
concentration at the mount Kl. Feldberg (about 20 km from the city
of Frankfurt) is representative for the atmosphere at 700 m above
ground. The transfer of SO_2 to rain drops falling through strong
vertical SO_2 gradients does not seem to be fast enough to establish
equilibrium. This effect has been demonstrated by model calculations
of Barrie[3]. In regions where the sulfur dioxide is well mixed
within the atmospheric boundary layer sulfur dioxide should be close
to equilibrium (Hales[4]).

SULFATE FORMED THROUGH OXIDATION OF ABSORBED SULPHUR DIOXIDE

The sulfur dioxide which is absorbed from the atmosphere into
a droplet can also be irreversibly oxidized to sulfate. The
absorbed SO_2 will therefore not totally exist in the form of S (IV)
species, but also in the form of sulfate. This heterogeneous
oxidation of sulfur dioxide was discussed in detail by Beilke and
Gravenhorst[1] and will be treated here only so far as it is relevant
for the decrease of S (IV) species in rain water. In order to
estimate the amount of absorbed SO_2 oxidized to sulfate in rain
water, the disappearance of S (IV) species in rain water was
measured in natural rain samples. The decrease of S (IV) concentration in a closed liquid system of collected rain water can be
characterized by

$$- \frac{d\ S\ (IV)}{dt} = k_4\ S\ (IV) \tag{4}$$

The k_4 values derived from the relative decrease of S (IV)
concentration with time are plotted in Fig. 6 as a function of
pH (dots 295 K, circles 278 K). K_4 increases with increasing pH and

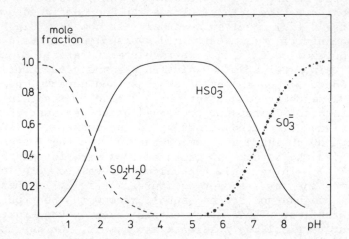

Fig. 1. Mole fraction of sulfur (IV) species in equilibrium at T = 298 K as a function of aqueous solution pH.

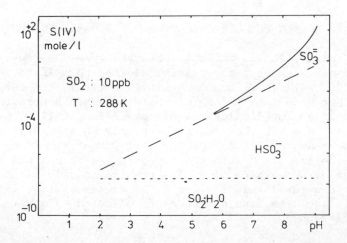

Fig. 2. Cumulative equilibrium concentration of S (IV) species in water as a function of pH.

temperature. The upper limit for these conditions is a k_4 value of about 10^{-4} sec^{-1}. It represents a transformation rate of about 35%* (in liquid phase) from S (IV) species into sulfate. For representative conditions (pH 4.0 and T = 283 K) oxidation rates of the order of 7% h^{-1}* (in liquid phase) were measured.

If oxidation of absorbed SO_2 in falling rain proceeds at the rate measured here the amount of sulfur dioxide oxidized to sulfate during rain development and precipitation would be negligible. The total amount of SO_2 absorbed from the gasphase can therefore be determined within reasonable certainty by measuring the sum of the S (IV) species in rain water at the time of rainfall.

Another yet uncertain parameter to describe the fate of absorbed sulfur dioxide and formation of sulfate could be deduced from the S (IV) concentration with time. If the formation of sulfate in solution could be described by

$$\frac{d \ [SO_4^=]}{dt} = k_5 \ [SO_3^=] \eqno(5)$$

The value of k_5 for natural rain water can be estimated from the S (IV) oxidation measurements. The $SO_3^=$ concentration was therefore deduced from the measured S (IV) concentration with the help of relations (1) – (3). The sulfate formation rate in this closed liquid system of rain water was set equal to the decrease of S (IV) concentration with time. The thus derived k_5 value could be approximated in the pH range 3–5 as a function of temperature:

$$k_5 = 1.95 \times 10^{16} \ \exp \ (-11676/T) \eqno(6)$$

This value for natural rain water is faster by a factor 50 to 500 than for uncatalyzed oxidation. In Table 1 the k_5 value calculated with equation (6) for T = 298 K and representing conditions for natural rain water in a closed system is compared with values used by other authors for sulfate formation rates. The oxidation of sulfite ions in natural rain water proceeds faster than assumed in models to predict sulfate formation rates in liquid systems.

The faster oxidation in natural rain water could be due to dissolved metals acting as catalysts for this reaction. The average concentrations of two dissolved metals in these rain water samples were: Fe: 3.8 $\pm$ 1.3 x 10^{-6} mole/l and Mn: 2.6 $\pm$ 0.5 x 10^{-7} mole/l. The concentrations of dissolved Fe and Mn are lower than the cata-

---

*   This rate of SO_2 oxidation in only the liquid phase has to be distinguished from the gasphase removal rate by liquid phase oxidation.

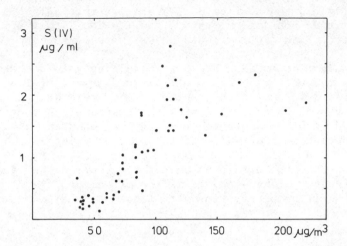

Fig. 3.   S (IV) concentration in rain water as a function of SO₂⁻
           gasphase concentration at the rain sampling site (Frankfurt,
           Germany, S (IV) expressed as SO₂).

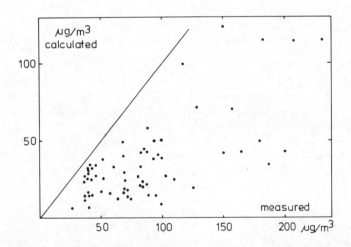

Fig. 4.   Equilibrium SO₂ concentration above rain water calculated
           from measured S (IV), pH and temperature as a function of
           measured atmospheric SO₂ concentration at the rain sampling
           site in Frankfurt.

lytic effective concentrations for each component found by Barrie and Georgii[5]. Synergistic interactions could be responsible for this faster SO_3 oxidation in natural rain water. The possibility of an even faster oxidation in the atmosphere caused by absorbed H_2O_2 molecules was discussed by Penkett et al.[6]. In a closed system such as that used here, however, a continued addition of oxidizing agents like H_2O_2 is not possible in contrast to the situation in the atmosphere. The above measurements only indicate therefore the ability of catalysts to oxidize dissolved SO_2 in rain water.

In summary these S (IV) oxidation measurements in natural rain water suggest, though the SO_2 absorbed is oxidized faster than in "pure" systems, that the main part of the sulfur will remain in the S (IV) state for times comparable with average lifetimes of precipitation droplets. The amount of sulfur dioxide absorbed in rain water can therefore be measured and compared with the total amount of incorporated sulfur.

The capability of rain water to act as a sink for atmospheric sulfur dioxide was discussed e.g. by Georgii[7], Beilke[8], Barrie and Georgii[5], Davis[9], Gravenhorst et al.[10], Beilke and Gravenhorst[1]. The knowledge of absorbed S (IV) concentrations in precipitation allows one to determine:

(i)   the fraction of the total sulfur in rain water that originates
       from absorbed SO_2

(ii)   the fraction of atmospheric sulfur dioxide which was taken
       out of the atmosphere by precipitation.

To estimate the importance of sulfur dioxide absorption in rain water as an atmospheric sink for SO_2, the total amount of SO_2 in the reservoir from which SO_2 is taken out by precipitation has to be known. Since S (IV) concentrations in rain water were measured at Frankfurt, the amount of SO_2 in a vertical column over Frankfurt was deduced from Fig. 5. As a yearly average about 5.2×10^{-4} SO_2 moles are present in a column (base 1 m^2). A mean S (IV) concentration of 6.9×10^{-6} moles per liter were measured for rain samples collected in Frankfurt at different seasons of the year. This concentration is taken, therefore, to be close to a representative annual average. Taking this value and a precipitation amount of 4 $1/m^2$ for one single rainfall then 2.8×10^{-5} moles m^{-2} of absorbed SO_2 are deposited in rain water during one precipitation event. This means that about 5% of the SO_2 present in such a column is removed from the atmosphere by such a rainfall. This small decrease can hardly be detected by atmospheric SO_2 concentration measurements. A measured 10% decrease of the SO_2 concentration from the onset of rainfall within the first hour for several occasions in Frankfurt can be due to airmass changes accompanying the beginning of rain.

Table 1.   Reaction rate constant, k_5, for sulfate formation due to
            catalyzed S (IV) oxidation by O_2 at T = 298 K for
            natural rain water in a closed system compared with rates
            for the uncatalyzed SO_2-oxidation.

$$+ \frac{d\ [SO_4^=]}{dt} = k_5\ [SO_3^=]$$

Author	$k_5\ sec^{-1}$
Scott and Hobbs[11] from the measurements of van den Heuvel and Mason[12]	0.0017
Miller and de Pena[13] (pH 2-4)	0.003
Brimblecomb and Spedding[14] (pH 4-6)	0.006 - 0.0037 pH 4 pH 6
Penkett et al.[15] (pH 6, T = 293 K)	0.0005
Beilke et al.[16] (pH 3-6)	0.00036 - 0.0011 pH 3 pH 6
this investigation (pH 3-5)	0.19

Table 2.   Relative contribution of S (IV) sulfur to total sulfur
            in rain water at the time of collection in industrialized
            areas.

Author	location	ratio (%) S (IV) sulfur to total sulfur
Jost[17]	Frankfurt	15 ± 8
Fricke et al.[18]	South Germany	17 ± 12
this investigation	Frankfurt	7 ± 4

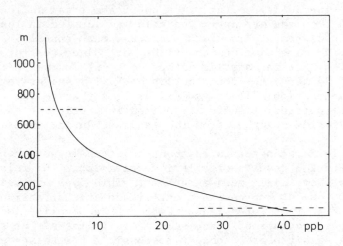

Fig. 5.   Approximated vertical SO$_2$ profile over Frankfurt (annual
          average) deduced from ground measurements in Frankfurt
          and a mountain station approximately 20 km from Frankfurt.

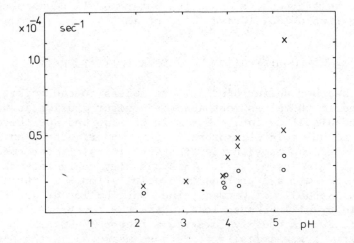

Fig. 6.   k$_4$ values measured in rain water as a function of pH and
          temperature (crosses 298 K, circles 278 K).

In order to compare the concentration of total S content in rain water with the amount of sulfur absorbed from the SO_2 gasphase the total S was determined with an isotopic dilution method (Klockow et al.[19]). The S (IV) sulfur fraction of the total sulfur represents 7 $\pm$ 4 % on the average. The rain was sampled in the industrialized area of Frankfurt where the mean annual SO_2 concentration at the same time was about 40 ppb. Since the total sulfur concentration in rain water is rather constant in the Federal Republic of Germany, the contribution of S (IV) sulfur to the total sulfur concentration will in general be less than 7-17%. Therefore, only a small portion of sulfur deposited in precipitation belongs to sulfur dioxide absorbed from the gasphase during rainfall.

In Table 2 the percentages found in other measurements are compared with the results of this investigation. The relative proportions show fairly good agreement. This does not mean that the absorbed SO_2-sulfur represents a certain constant fraction of the total sulfur in rain water independent of the sampling location. Similar concentrations of trace substances in the air at the rain water sampling sites are probably responsible for these agreements.

To calculate a SO_2 removal rate coefficient, the number of SO_2 moles present in a vertical column is divided by the number of SO_2 moles absorbed in precipitation per year. The thus determined removal coefficient for this process is on the order of 0.1% h^{-1} or 2.9 x 10^{-7} sec^{-1}. It is close to the values calculated for similar conditions (rain water pH: 4.0) in an one-dimensional chemical model for the troposphere comparing wet removal and homogeneous gasphase oxidation of SO_2 (Gravenhorst et al.[10]).

CONTRIBUTION TO FREE ACIDITY IN RAIN

The absorbed sulfur dioxide contributes to the acidity of rain. At atmospheric pH values most absorbed sulfur dioxide is present in the form of bisulfite ions. One mole of absorbed SO_2 therefore produces ca. one mole of protons in rain water. The annual average of S (IV) species deposited in Frankfurt in rain water amounts to ca. 0.5 x 10^{-2} moles m^{-2} a^{-1}. This quantity represents 6% of the 9 x 10^{-2} moles m^{-2} a^{-1} protons in precipitation. In rural areas of the Federal Republic of Germany, the contribution of absorbed SO_2 to the acidity will be even less. Here the concentration of atmospheric SO_2 is roughly lower by a factor 10 but the pH values do not differ systematically from those measured in the industrial area of Frankfurt. If, however, the pH values increase in remote areas and compensate the decrease in sulfur dioxide concentration by an increased absorption, the amount of SO_2 brought down by rain can be as high as in industrial regions. Most rain water analyses show pH values between 3 and 6. In this region the amount of absorbed sulfur dioxide is nearly inversely proportional to the

proton concentration. The influence on dissolved SO_2 in rain by a
factor of 10 can therefore be counterbalanced by a one unit increase
in pH.

In order to estimate this counterbalance atmospheric sulfur
dioxide concentrations and proton concentrations in rain water are
compared for different geographical regions. In Fig. 7 an attempt
is made to give a synoptic picture for pH values of rain water on
the Northern Hemisphere. It is based on available, but by far not
sufficient, pH measurements at different sampling stations taken
from a survey of precipitation chemistry data by Böttger[20]. A
rough pattern can only be given especially because remote sampling
sites are very sparse, average pH values instead of average proton
concentrations were frequently reported and the analysed samples
most often represent bulk precipitation which includes dry deposition of gases and particles and not only rain water.

Background atmospheric SO_2 concentrations are hardly reported
because analytical methods are not yet sophisticated enough. Thus
SO_2 concentration values for remote areas should be treated as
preliminary ones. They suggest SO_2 concentrations on the order of
1–10 ppbv in remote continental areas and 0.1 ppbv over oceans.

With the available information on the pH and SO_2-gasphase
concentration, the amount of absorbed SO_2 per liter of rain water
for different geographical regions was approximated by calculating
the sum of S (IV) species present in rain water at equilibrium
conditions. The thus derived S (IV) concentrations are plotted as
a function of pH value for different atmospheric SO_2 concentrations
at T = 288 K (Fig. 7). For T = 278 K SO_2 would be absorbed by a
factor of 2.5 more and at T = 298 K by a factor of 2.3 less than
indicated. The shaded areas are suggested to represent characteristic conditions for oceanic regions, remote continental landscapes
and industrialized areas. The horizontal line AB at constant S (IV)
concentration of 1×10^{-5} mole/l (Fig. 8) indicates that the same
amount of SO_2 can be absorbed in all three situations. For this
special case, the relative contribution of adsorbed SO_2 to the
total amount of sulfur in rain will be smallest in the industrialized
zone because there the total sulfur concentration will be higher
than in the two other areas. In general the total amount of SO_2
deposited in rain at one location depends on the pH values of rainwater, SO_2-gasphase concentration, temperature and precipitation
amount. Such set of data is very limited and no attempt is made
here to specify the deposition rates of absorbed SO_2. However,
upper and lower boundary values for rainwater S (IV) concentrations
in three regions differing with respect to SO_2-gasphase concentration and rain water pH value are suggested.

In large areas of the world, as in Frankfurt and other parts
of the Federal Republic of Germany, a great fraction of the acidity

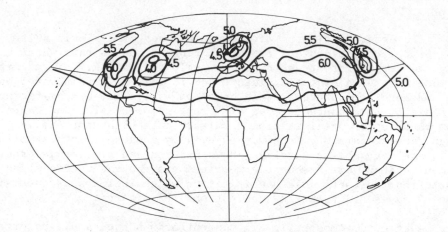

Fig. 7.   Estimated distribution of the pH value for rain water in
          the Northern Hemisphere.

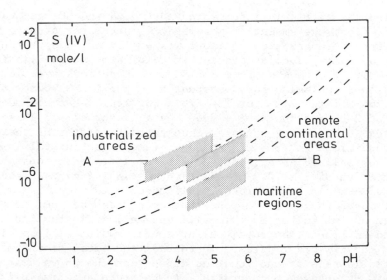

Fig. 8.   S (IV) concentration in water in equilibrium with atmos-
          pheric SO_2 as a function of pH at T = 288 K. The hatched
          fields indicate probable conditions to be expected at
          different geographical locations (the dashed curves
          represent SO_2 concentrations of 10 ppb, 1 ppb and 0.1 ppb).

in rain water is due to incorporation of acidic aerosol particles
as condensation nuclei into cloud elements and not to SO_2-gasphase
absorption by rain drops. In addition, more acidity can reach the
ground via dry deposition of acidic gaseous compounds such as SO_2
or HNO_3 than by precipitation. In the Federal Republic of Germany,
for example, ca. 0.7×10^{10} moles of free protons are deposited in
rain during one year assuming a yearly amount of rainfall of 750
$1 \text{ m}^{-2} \text{ a}^{-1}$ and a pH of 4.0, whereas ca. 1.5×10^{10} moles of SO_2 are
estimated to be absorbed on the ground (Persecke[21]).

In order to understand the acidity of rain water and possible
acidifications of soils and fresh water systems all atmospheric
processes - not only SO_2 absorption by rainwater - leading to dry
and wet deposition have to be included in the investigation.

ACKNOWLEDGEMENT

We thank Dr. L. Barrie[2], Toronto for helpful discussions and
A. Böttger, Jülich, for providing the pH values.

The investigation was partly supported by Deutsche
Forschungsgemeinschaft, through Sonderforschungsbereich 73
"Atmosphärische Spurenstoffe".

REFERENCES

1. S. Beilke and G. Gravenhorst, Heterogeneous SO_2-oxidation in
the droplet phase, Atm. Environm. 12:231 (1978).

2. P. W. West and G. C. Gaeke, Fixation of sulfur dioxide as
disulfitomercurate and subsequent colorimetric
estimation, Anal. Chem. 28(12):1816 (1956).

3. L. A. Barrie, An improved model of reversible SO_2-washout
by rain, paper presented at the Int. Symp. on Sulfur
in the Atmosphere, Dubrovnik, Jugoslawien, Sept. 7-14
(1977).

4. J. M. Hales, West removal of sulfur compounds from the
atmosphere, paper presented at the Int. Symp. on Sulfur
in the Atmosphere, Dubrovnik, Jugoslavia, Sept. 7-14
(1977).

5. L. A. Barrie and H. W. Georgii, An experimental investigation
of the absorption of sulfur dioxide by water drops
contacting heavy metal ions, Atm. Environm. 10:743 (1976).

6. S. A. Penkett, K. A. Brice and A. E. J. Eggleton, A study of
    the rate of oxidation of sodium sulfite solution by
    hydrogen peroxide and its importance to the formation of
    sulfate in cloud-and rain water, paper presented at
    the European Sulfur Symposium, Ispra (1976).

7. H. W. Georgii, Untersuchungen uber Ausregnen und Auswaschen
    atmospharischer Spurenstoffe durch Wolken und Niederschlag,
    Berichte d. Dt. Wetterdienstes 100:5 (1965).

8. S. Beilke, Untersuchungen uber das Auswaschen atmospharischer
    Spurenstoffe durch Niederschlage, Berichte d. Inst. f.
    Meterorologie und Geophysik, Univ. Frankfurt, Nr. 19
    (1970).

9. T. D. Davis, Precipitation scavenging of sulphur dioxide in an
    industrial area, Atm. Environm. 10:879 (1976).

10. G. Gravenhorst, Th. Jan en-Schmidt, D. H. Ehhalt and E. P. Roth,
    The influence of clouds and rain on the vertical
    distribution of sulfur dioxide in a one-dimensional steady-
    state model, Atm. Environm. 12:691 (1978).

11. W. D. Scott, and P. V. Hobbs, The formation of sulfate in water
    droplets, J. Atm. Sci. 24:54 (1967).

12. A. P. van den Heuvel, and B. J. Mason, The formation of ammonium
    sulfate in water droplets exposed to gaseous sulfur dioxide
    and ammonia, Q. J. Roy. Met. Soc. 89:217 (1963).

13. J. M. Miller, and R. G. de Pena, Contribution of scavenged sulfur
    dioxide to the sulfate content of rain water, J. Geophys.
    Res. 77:5905 (1972).

14. P. Brimblecombe, and D. J. Spedding, The catalyzed oxidation of
    micromolar aqueous sulfur dioxide-I, Atmospheric Environm.
    8:937 (1974).

15. S. A. Penkett, B. M. R. Jones, and A. E. J. Eggleton, Rate of
    oxidation of sodium sulfite solution by oxygen and by
    ozone, paper presented at the European Sulfur
    Symposium, Ispra (1975).

16. S. Beilke, D. Lamb, and J. Miller, On the uncatalyzed oxidation
    of atmospheric SO_2 by oxygen in aqueous systems,
    Atm. Environm. 9:1083 (1975).

17. D. Jost, Aerological studies on the atmospheric sulfur budget,
    Tellus XXVI (1-2):206 (1974).

18. W. Fricke, H. W. Georgii, and G. Gravenhorst, Application of a new sampling device for cloud-water analysis. Some problems of cloud physics, collected papers, Gidrometeoizdat, Leningrad, p.200 (1978).

19. D. Klockow, H. Denzinger and G. Ronicke, Anwendung der substochiometrischen Isotopenverdunnungsanalyse auf die Bestimmung von atmospharischem Sulfat und Chlorid in "Background" Luft. Chemie-Ing. Techn. 46(19):831 (1974).

20. A. Böttger, Atmospharische Bilanz von Stickoxiden und Ammoniak, unpublished Diplomarbeit, Institut fur atmospharische Chemie, Kernforschungsanlage D-5170 Jülich (1978).

21. C. Persecke, Die Gesamtschwefel-Deposition in der Bundesrepublik Deutschland auf der Grundlage von Me daten des Jahres 1974, Deiplomarbeit, Institut fur Meteorologie u. Geophysik, Frankfurt (1978).

DIFFICULTIES IN MEASURING WET AND DRY DEPOSITION ON FOREST CANOPIES

AND SOIL SURFACES

James N. Galloway and Geoffrey G. Parker

Department of Environmental Sciences

University of Virginia, Charlottesville, Va. 22903

Our ability to determine the effects of wet and dry deposition of acidic substances on soils and forests is directly dependent on:

1. The identification of the important wet- and dry-deposition processes.

2. The determination of the nature of the effect, i.e. cumulative versus episodic, relative to the depositional process.

3. The quantitative measurement of the deposition rates of each process.

The removal mechanisms of material from the atmosphere are complex. They are generally placed in two categories: wet and dry deposition. Each of these categories has subcategories that are usually identified by mechanistic differences, such as, gravitational versus diffusion removal of particles.

The generally accepted subcategories for wet and dry deposition are:

1. Wet deposition

    a. *Incident wet deposition* - water that is gravitationally transferred from the atmosphere to earth surfaces.

    b. *Throughfall* - water that has passed through a leaf canopy.

c. *Net throughfall* – composition of the water that has passed through a leaf canopy with the contribution from incident precipitation removed.

2. Dry deposition

a. *Dry fallout* – material that gravitationally settles out of the atmosphere. This material is >10–30 μm in diameter and is primarily composed of soil and sea-salt particles.

b. *Aerosol impaction* – material that is too small to gravitationally settle but rather is impacted or deposited onto surfaces. This material is commonly <10 μm and is usually formed by atmospheric gas–phase reactions that create submicron aerosols (i.e., $[NH_4]_2SO_4$, NH_4NO_3, etc.).

c. *Gaseous absorption* – gases that are absorbed by foliage or soils (i.e., SO_2, NO_x, CO_2, etc.).

After identifying the mechanisms that deposit acids, it is necessary to determine the relationship of the depositional processes with the effects. In a general sense, two effects can be identified. Episodic effects are short term and involve the rapid deposition of high concentrations of material (in this case, acids) to the receiving surface. Cumulative effects involve the long-term deposition of acidic material to surfaces. An example of an episodic versus a cumulative effect would be the effect on leaf surfaces of a single wet deposition event having a pH of 2.5 versus the input of pH 4.0 wet deposition over a 12-month period. The point is this, different types of effects require different types of measurements of deposition and hence different strategies for surmounting sampling difficulties.

The difficulties in measuring wet and dry deposition are quite different. The reason is that wet deposition is more easily collected. Dry deposition, however, is a combination of several complex processes (gravitational settling, aerosol impaction, and gaseous exchange); and thus it is difficult for the collection technique to duplicate the natural collection mechanism. Because of this relative ease of collection of wet deposition, research on the methodologies is more advanced. These advances are documented in reports from several investigations on collector design, choice of sampling period. sample stability, etc. (Granat[1], WMO[2]; Berry et al.[3]; Galloway and Likens[4]).

The general consensus of these reports is:

1. A wet-deposition collector should be of simple and rugged
    construction, able to withstand wide variations of
    temperature and wind conditions.

2. Due to the low ionic strength of wet deposition,
    collectors should be scrupulously cleaned.

3. Wet deposition should be collected separately from dry
    deposition due to chemical and physical interaction
    between the two components.

4. If wet-deposition samples are not taken on an event
    basis, then air-mass trajectory analysis and the episodic
    effects of wet-deposition composition on aquatic and
    terrestrial ecosystems are not possible to determine. In
    addition, chemical and biological changes may occur in the
    collector if the sample is not collected after the event.
    This is especially true in regions having wet deposition
    with pH >4.5.

5. Chemical preservation of wet-deposition samples is
    difficult due to the dilute nature of the samples. There-
    fore, physical preservation of the samples is recommended;
    i.e., storage at low temperatures.

6. The variability of the composition of wet deposition on a
    spatial scale can be large. Therefore, care must be taken
    in locating the collector relative to the local sources.
    Similarly, the objectives of the research program must
    allow for this inherent variability.

Research on the measurement of wet deposition is continuing.
Difficulties that require further definition are:

1. Severe effects of an episodic nature (as opposed to
    cumulative) on aquatic and terrestrial ecosystems have
    been documented in laboratory and field studies. However,
    these effects are quite dependent on the spatial and
    temporal variability of the wet-deposition composition
    for single storms. Therefore further work is necessary
    to understand the relationship between episodic effects
    and the variability of compositional fields of
    deposition.

2. The effects of wet deposition on leaf surfaces have been
    documented in controlled environments. However, before
    these effects can be documented in the field, the physical
    and chemical interactions of the individual drops with
    the leaf surface have to be investigated. For example,
    after a surface is moistened in the first period of the

storm, do the following drops run off without interaction
with the leaf? Similarly, what is the effect of
evaporation of droplets and, hence, concentration of ionic
species on the leaf surface.

3.   The composition of wet deposition alters as it passes
     through the forest canopy. Therefore, to assess effects
     on leaves, it is necessary to realize that incident wet
     deposition will have a different effect than throughfall
     on leaf surfaces. To do this, it is necessary to assess
     how much of the canopy is exposed to incident wet
     deposition as opposed to throughfall.

4.   The techniques for measurement of throughfall composition
     are not as advanced as are those for the collection of
     incident wet deposition. There are a variety of
     difficulties that increase the sampling problem.

     a.   The spatial variation of throughfall composition is
          greater than for wet deposition. Therefore, more
          intensive sampling networks are required. Currently
          there are two techniques: First is the placement of
          individual collectors under the canopy; second is
          the placement of integrating collectors under the
          canopy. This type of sampler is typically a piece of
          plastic pipe cut in half lengthwise and placed
          horizontally under the canopy. Due to the integrating
          nature of this collector (Legton et al. [5]) some
          of the spatial variability in the composition of
          throughfall is eliminated.

     b.   Contamination of the sample by litterfall is
          prevalent and requires constant attention to the
          sampling system.

     It is clear that there are still substantial difficulties
remaining in the measurement of wet deposition. These difficulties,
however, are slight when compared to those encountered in the
measurement of dry deposition. For not only do the same problems
exist for dry as well as wet deposition, but in addition the basic
mechanisms of dry deposition are unknown or if known, difficult to
measure.

     Dry deposition is defined as the process that removes material
from the atmosphere when it is not precipitating. Note that this
is a negative definition, defining what it is not rather than what
it is. This reflects our ignorance of the process. As previously
mentioned, the three subcategories of dry deposition are dry fallout, aerosol impaction, and gaseous absorption. All three processes
require different measurement systems, each with its own inherent

difficulties. There is, however, a commonality in the difficulties.
All of the depositional processes involve the dry transfer of small
particles from the atmosphere to surfaces (foliage, soils). Therefore, to absolutely measure this transfer rate (i.e., dry
deposition), it is necessary to duplicate the surface in the chosen
collector. This is virtually an impossible task. The surface
structure of foliage and soils is quite complex and is essentially
nonduplicatable. Therefore, any measure of dry-deposition rates
will be estimates at best and guesses at worst. This inability to
measure is critical and will limit our understanding of the effects
of dry deposition on aquatic and terrestrial ecosystems.

There are several techniques that have been used to estimate
rates of partial or total dry deposition.

Dry fallout has been measured by placing open containers in
the field. Rates measured in this manner are either over- or
underestimates of dry fallout, depending on what is taken to be
the absolute (Feely et al. [6]; Volchok[7]). These rates
can be significant when compared to wet deposition (Figure 1).

Gaseous absorption rates have been measured by at least two
processes: field uptake experiments by vegetation and atmospheric
losses as calculated by budgetary considerations.

Absolute aerosol-impaction rates are difficult to measure
directly. What is typically done is to measure the atmospheric
concentration and to use a depositional velocity to calculate a
dry-deposition rate. The selection of the correct depositional
velocity is difficult due to surface effects of the receiving area
and the dependency of the depositional velocity on a variety of
environmental conditions.

The need for research in dry deposition is clear. Some
specific questions are:

1. What is the relative importance of dry fallout, gaseous
    absorption, and aerosol impaction in acidic dry
    deposition?

2. What is the most suitable instrumentation to measure
    the components of dry deposition?

3. What factors (related to depositional mechanisms)
    influence the spatial variability of the depositional
    and composition fields?

4. What are the errors associated with the use of depositional
    velocity to calculate dry-deposition rates?

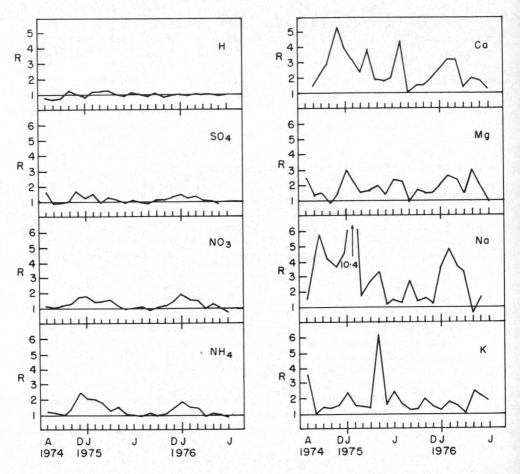

Fig. 1   The relative importance of dry fallout versus wet deposition.
The values represent the parameter, R, plotted as a function
of time.

$$R = \frac{\text{Monthly weighted concentration in bulk deposition}}{\text{Monthly weighted concentration in wet deposition}}$$

If R > 1, dry fallout has occurred;
   R = 1, no dry fallout has occurred;
   R < 1, an interaction between wet deposition and dry
        fallout has occurred.

5.   How sensitive is our understanding of the effects of
     acidic dry deposition to errors in dry-deposition rates?

So far, this discussion has been on the difficulties in the
individual measurement of individual processes. However, in the
field, atmospheric deposition of acidic substances is not a
simple-process phenomena but is rather a complex assortment of
processes. To illustrate the difficulties that are encountered in
the real-world measurement of atmospheric deposition, a case study
has been chosen. The example is the determination of the
components of atmospheric deposition that contribute to the
composition of throughfall.

The composition of throughfall is determined by three
primary processes:

1.   Incident wet deposition;

2.   dry deposition washed off the leaf surface by
     the incident wet deposition;

3.   material leached out of the leaf due to the chemical
     and physical interactions with the leaf surface by the
     previous two processes.

In this case study, we discuss current techniques used to
determine the importance of the individual processes and suggest
avenues of future work.

The determination of the incident wet deposition is relatively
easy; it requires the collection of precipitation in a suitable
collector, either above or outside the forest.

Direct methods for separating the leaching and dry-deposition
components usually involve the collection of bulk precipitation
(wet deposition and dry fallout) under inert surfaces, broadly
intended to simulate canopy areas; artificial fir trees
(Schlesinger and Reiners[8]), stacks of plastic sheeting
(Nihlgard[9]),vertical plexiglass baffles (Etherington[10]),
or polyethylene screening (Hart and Parent[11]). In these cases,
actual rainwater impinges on the "canopy". Occasionally inert
surfaces, such as, Teflon plates, petri dishes or filter paper
(White and Turner[12]), are exposed to dry deposition, retrieved
and washed in the laboratory. In any case, the materials mobilized
from such surfaces estimate the "washoff" contribution to net
throughfall. The remaining portion is taken as the amount due to
canopy leaching.

Such methods poorly simulate the action of actual plant surfaces.
The exact leaf geometry and chemistry, both on large (leaf shape,

arrangement, and wetness) and small (roughness and chemical binding
sites) scales are critical in determining deposition velocities,
especially for particles too small for rapid sedimentation. Such
an approach also assumes that washoff, as measured using inert
surfaces, and leaching, from actual canopies, will be strictly
additive to net throughfall. This neglects interactions, such as,
foliar uptake of material and chemical conversions at the real
leaf surface.

Another approach involves the controlled washing of excised
plant parts. Where samples are periodically taken, increases in
wash-water concentrations are taken to be due to dry fallout,
assuming constant leaching contributions over short intervals
(Lindberg et al. [13]). Leaf washing is common in tracer
experiments estimating the portion of known deposition which cannot
be mobilized by water (e.g. Wedding et al.[4]). White and Turner[12],
washed leaves and branches of several tree species exposed
to known aerosol concentrations to calibrate the trapping
efficiency of filter paper used to assess monthly "capturability"
of real surfaces. Since leaching only requires that surfaces be
wettened (Tukey [15]), such washings contain materials robbed from
plant tissue. Experiments utilizing washing of natural surfaces
out to be used in conjunction with strict controls on leaching
levels.

Indirect methods of partitioning net throughfall are equally
crude. Mayer and Ulrich [16], suggested that the washoff fraction
might be equal to the net throughfall in leafless deciduous forests
since leaching will be at a minimum. In computing a yearly income
by "canopy filtering", they assume a similar trapping efficiency for
the fully expanded summer canopy. This is not reasonable since
1) nonleafy plant parts can be leached (Tukey [15]), and 2) trapping
efficiencies will increase with canopy elaboration.

Using linear regressions of weekly throughfall deposition on
incident wet deposition, Miller et al. [17] estimated the leaching
fraction as equal to the positive y-intercept of the relation. This
treatment implies that very small storms will consist largely of
leachate; washing studies with tracers have shown that soluble
deposits are quickly released on wetting, if at all. Furthermore,
the relation of net-throughfall deposition on gross-incident
deposition is probably curvilinear and convex upwards, with an
intercept very close to the origin, as found by Bernhard-Renversat[18],
and the authors. Other investigators have suggested that the
relative leachabilities of several ionic species might indicate
their importance in net throughfall (Henderson et al.[19]),
especially when the foliar levels of these species are taken into
account. Such an approach could yield a minimum importance of
leaching but will be of little use for elements well represented
both in leachates and aerosols, such as, Ca, Na, Si and S.

Other indirect approaches involve consideration of the changes in the ion ratios in wet deposition passing through canopies (Attiwill[20]) or estimations of reasonable deposition velocities (Raybould et al.[21]). Knowledge of the factors controlling these processes is still too tentative for use in finely partitioning net-throughfall chemistry.

Clearly, all approaches presently used to separate net-throughfall components have drawbacks which compromise their usefulness as general methods. Any two methods might yield different results. Research is needed to accurately determine:

1. The magnitude of the particle load trapped and retained by plant surfaces;

2. the portions of the deposit which can be mobilized by wet deposition onto the surface;

3. how much of the wet- and dry-fallout elements are retained in plant tissue. Total nitrogen in wet deposition usually exceeds that in net throughfall (e.g., Carlisle et al.[22]), suggesting foliar uptake. It is quite possible that elements are similarly retained; both dry and wet deposition might be underrepresented in throughfall.

4. The magnitude and role of plant surface exudates and precipitates in throughfall composition. Such substances derive from the organism yet will most likely behave like dry deposits.

Considering the potential for confounding sources in throughfall, it would be useful to adopt some uniform methods for its collection; for example,

1. As throughfall shows great heterogeneity, researchers ought to take pains to quantify that spatial variability; it is an important feature of the phenomenon.

2. The stability of throughfall solutions must be evaluated. Although wet-fall stability has been previously discussed, little is known about the chemical constancy of these richer solutions.

3. A definition of throughfall ought to be adopted; one which addresses the problem of litterfall particles and their exclusions by filters, both before and after collection.

In summary, our understanding and elucidation of the effects

of acid deposition are absolutely dependent on the quantitative
measurement of the rates of wet and dry deposition. Of special
concern are deposition effects on forests where the determination
of imputs are further complicated by the biotic interactions of this
leafy, inefficient bulk collector. Without knowledge of inputs of
acids, determination of effects is impossible. Of all the processes
involved in atmospheric deposition, those of dry deposition have
the greatest difficulties and, therefore, require the most research.

REFERENCES

1. L. Granat, On the variability of rainwater composition and
        errors in estimates of areal wet deposition, Procs. Symp.
        Precip. scavenging, Champaign, Ill (1974).

2. WMO, WMO Operation Manual for Sampling and Analysis Techniques
        for Chemical Constituents in Air and Precipitation, WMO
        Publ. No. 299, Geneva, Switz., (1974).

3. R. L. Berry, D. M. Whelpdale, and H. A. Wiebe, An evaluation
        of collectors for precipitation chemistry sampling.
        Rept. WMO Expert Meeting Wet and Dry Deposition. AES,
        Downsview, Canada. 17-21 Nov., (1975).

4. J. N. Galloway, and G. E. Likens, Calibration of collection
        procedures for the determination of precipitation chemistry,
        Water, Air, Soil Pollut. 6:241 (1976).

5. L. Legton, E. R. L. Reynolds, and F. B. Thompson, Interception
        of rainfall by trees and moorland vegetation, in: "The
        Measurement of Environmental Factors in Terrestrial Ecology"
        R. M. Wadsworth, ed., Blackwell Scientific Publ. (1968).

6. H. W. Feely, H. L. Volchok, and L. Toonkel, Trace metals in
        atmospheric deposition, U.S. Dept. of Energy Rept. HASL-308
        (1976).

7. H. L. Volchok, Dry particulate deposition inferred from global
        and local fallout studies, Rept. WMO Expert Meeting Dry
        Deposition, Gothenburg, Sweden, 18-22 April, (1977).

8. W. H. Schlesinger, and W. A. Reiners, Deposition of water and
        cations on artificial foliar collectors in fir Krummnolz
        of New England mountains, Ecology 55:378 (1974).

9. B. Nihlgard, Precipitation, its chemical composition and effect
        on soil water in a beech and spruce forest in south
        Sweden, Oikos 21:208 (1970).

10. J. H. Etherington, Studies of nutrient cycling and
     productivity in oligotrophic ecosystems. I. Soil potassium
     and windblown sea spray in south Wales dune grassland,
     J. Ecol. 55:743 (1967).

11. G. E. Hart, and D. R. Parent, Chemistry of throughfall under
     Douglas fir and Rocky Mountain juniper, Am. Midland Nat.
     92:191 (1974).

12. E. J. White, and F. Turner, A method of estimating income of
     nutrients in catch of airborne particles by a woodland
     canopy, J. Appl. Ecol. 7:441 (1970).

13. S. E. Lindberg, D. S. Shriner, R. R. Turner, and L. K. Mann,
     Environmental rate of emission from coal combustion
     plants. The role of vegetation aerosol scavenging, in:
     "Envi. Sci. Div. Annual Rept."[S. I. Auerbach, dir.]
     ORNL-TM-5257 (1976).

14. J. B. Wedding, R. W. Carlson, J. J. Stuckel, and F. A. Bazzaz,
     Aerosol deposition on plant leaves, in: "Procs. 1st
     Intern. Symp. Acid Precip. and Forest Ecosystems", USDA
     Forest Service General Tech. Report NE-23 (1976).

15. H. B. Tukey, The leaching of substances from plants,
     Ann. Rev. Plant Physiology 21:305 (1970).

16. R. Mayer, and B. Ulrich, Conclusions on the filtering action
     of forests from ecosystem analysis, Oecol. Plant 9:157
     (1972).

17. H. G. Miller, J. M. Cooper, and J. D. Miller, Effect of
     nitrogen supply on nutrients in litterfall and crown
     leaching in a stand of Corsican pine, J. Appl. Ecol.
     13:233 (1976).

18. F. Bernhard-Reversat, Nutrients in throughfall and their
     quantitative importance in rain forest mineral cycles,
     in: "Tropical Ecological Systems-Trends in Terrestrial
     and Aquatic Research", F. B. Golley and E. Medina, eds.,
     Springer-Verlag, N.Y. (1975).

19. G. S. Henderson, W. F. Harris, D. E. Todd, and T. Grizard,
     Quality and chemistry of throughfall as influenced by
     forest-type and season, J. Ecol. 65:365 (1977).

20. P. M. Attiwill, The chemical composition of rainwater in
     relation to cycling of nutrients in mature eucalyptus
     forest, Plant and Soil 24:390 (1966).

21.   C. C. Raybould, M. H. Unsworth, and P. J. Gregory, Sources
      of sulphur collected below a wheat canopy (1977).

22.   A. Carlisle, A. H. F. Brown, and E. J. White, The organic
      matter and nutrient elements in the precipitation
      beneath a Sessile oak (Quercus petraea), J. Ecol. 54:87
      (1966).

WET AND DRY DEPOSITION OF SULFUR AT HUBBARD BROOK

John S. Eaton,[1] Gene E. Likens,[1] F. Herbert Bormann[2]

[1]Section of Ecology and Systematics
Division of Biological Sciences
Cornell University, Ithaca, New York 14853

[2]School of Forestry and Environmental Studies
Yale University, New Haven, Connecticut 06511

INTRODUCTION

In recent years ecologists and others have become increasingly interested in chemical inputs to natural ecosystems, not only in wet deposition (rain and snow), but also in the form of particulate and gaseous deposition. This interest is illustrated by several recent conferences and symposia relative to the subject of "acid precipitation" and its effects (e.g. Dochinger and Seliga,[1]; AMBIO,[2]. Most of the available data pertain to wet deposition, however, the importance of dry deposition in some ecosystems is increasingly recognized (e.g. Whitehead and Ferth,[3]; White et al.,[4]; Elwood and Henderson,[5]; Likens et al.,[6]).

Studies of chemical input in precipitation to forested ecosystems have been carried out at the Hubbard Brook Experimental Forest, a northern hardwood forest in north-central New Hampshire, since 1963. In addition, losses from the ecosystem in drainage water have been continuously measured and standing stocks have been determined in the various ecosystem compartments or "pools" (e.g. living and dead biomass, forest floor, soil, etc.; Likens et al.,[6]). Further studies have determined the rates of transfer of chemicals between the various compartments within the ecosystem. In this paper we shall briefly summarize our current understanding of some aspects of the biogeochemistry of sulfur as it relates to the dry deposition of gases and aerosols in the Hubbard Brook Experimental Forest. Our current understanding of the biogeo-

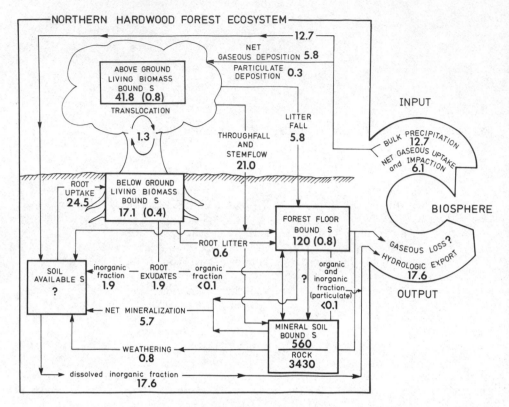

Figure 1.   Annual sulfur budget for an aggrading forested ecosystem
             at Hubbard Brook. Standing crop values are in kg/ha and
             sulfur fluxes are in kg/ha-yr. Values in parentheses
             represent annual accretion rates in kg/ha (from Eaton
             et al.,[7] (1978).

chemistry of sulfur for this northern hardwood forest is depicted
in Figure 1.

    Using this extensive background of information (Figure 1) and
an ecosystem approach, we are able to sum the net fluxes between
compartments and across the ecosystem boundary to estimate (by
difference) that 6.1 kg/ha of sulfur is added to the ecosystem each
year from dry deposition. We also have made measurements of sulfur
dioxide and particulate sulfur concentrations in the atmosphere at
Hubbard Brook. Based on these data we estimate that 0.3 kg S/ha-yr

should originate from the dry deposition of particulate sulfur and
5.8 kg S/ha-yr should originate from the dry deposition of sulfur
dioxide (Eaton et al.,[7]). Corresponding deposition velocities
of 0.1 cm/sec for particulates and 0.9 cm/sec for SO_2 would be
necessary for these inputs.

Collections of bulk precipitation (wet and dry deposition) have
been made at Hubbard Brook since 1963. Some measurements of wet
precipitation were done between 1965 and 1975. However, since
1975 a more intensive effort has been made to partition the
contribution of wet and dry components in bulk precipitation for the
Hubbard Brook ecosystem.

There are three basis types of precipitation collectors
currently in use for studies of precipitation chemistry (Galloway
and Likens,[8]). They are:

(1)   Bulk Precipitation Collector: This type is a continuously
      open container (often a bucket or funnel with a collection
      bottle) that collects both wet (e.g. rain, snow, sleet) and
      dry components (e.g. dust, aerosols, gases).

(2)   Wet Precipitation Collector: This collector differs from the
      bulk collector in that the collecting surface is covered at
      all times except when wet precipitation is occurring.

(3)   Wet/Dry Precipitation Collector: This collector consists of
      two separate containers, one for wet components and one for
      dry components. These containers have a cover or covers so
      arranged that the wet collector is open and the dry collector
      is closed during wet precipitation events, and at all other
      times the dry collector is open and the wet collector is
      covered.

All three of these types of precipitation collectors are used
at Hubbard Brook. Estimates of wet deposition inputs have been
determined from samples obtained in a HASL-type wet/dry precipitation
collector (Photograph 4, Likens et al.,[6]). Dry deposition can
be estimated using either material from the dry sample of the
HASL collector or it can be obtained by difference between the bulk
collector and the wet collector. We have had some difficulty with
spiders, insects, etc., taking up residence in our dry collector so
we have used the latter method.

Bulk precipitation samples are collected on a weekly basis.
Analyses for sulfate were performed by liquid ion chromatography
(Small et al.,[9]). Long-term studies at Hubbard Brook show that
the concentration of sulfate in bulk precipitation varies widely
from storm to storm. However, the mean annual concentration of
SO_4-S in bulk precipitation is 0.96 ± 0.04 (Likens et al.,[6]).

Sulfate is the predominant anion in bulk precipitation both on a
mass basis and on an ionic basis. On an ionic basis, sulfate
accounts for 62% of the anionic strength (Likens et al.,[6]).
Summer rain on the average has a higher concentration of sulfate-S
(1.1 mg/liter) than winter snow (0.63 mg/liter).

We have measured SO_2 at Hubbard Brook between 1973 and 1975
using lead candles and the West-Gaåke method. These data show a
strong seasonal variation with a winter maximum of about 10 $\mu g/m^3$
and a summer minimum of 0.5 $\mu g/m^3$. The weighted annual mean SO
concentration for the two years is 2.5 $\mu g/m^3$ (Eaton et al.,[7]).

Atmospheric aerosols have been measured for two years with a
modified two-stage Lundgren impactor using an after-filter. With
this instrument, we separated particles into three size classes:
0.1-0.65 μm, 0.65-3.6 μm and 3.6-20 μm. Concentrations of sulfur
aerosols at Hubbard Brook ranged from 200 ng/m^3 to 2000 ng/m^3 with
the highest concentrations generally occurring during midsummer.
The two-year mean concentration of sulfur aerosols was 980 ng/m^3.
However, 73% of the sulfur occurred in the smaller-sized aerosols
(0.1-0.65 μm) and only 2% of the sulfur occurred in the larger
aerosols (3.6-20 μm) (Eaton et al.,[7]).

The mean annual input of SO_4-S in bulk precipitation for the
three years, 1975 to 1978, was 11.0 $\pm$ 0.67 kg/ha (mean and standard
error). Of this 10.4 $\pm$ 0.53 occurred as wet deposition and
0.65 $\pm$ 0.24 kg/ha was as dry deposition (Table 1). Thus 95% of the
sulfur input in bulk precipitation occurred as wet components and
5% as dry components.

Dry deposition in the bulk precipitation collectors at Hubbard
Brook showed a strongly seasonal component. During the winter
months the proportion of dry components of sulfur is relatively
much larger than it is during the summer (Table 1). In fact, 60%
of the annual dry deposition as measured by the bulk precipitation
collector, takes place during the four winter months (December-
March). Based on atmospheric concentrations and deposition
velocities, SO_2 should account for 96% of the gravitational dry
deposition of sulfur at Hubbard Brook (Eaton et al.,[7]). Eaton
et al.,[7] also found that SO_2 concentrations reached a maximum
in the atmosphere during the winter months. Thus, if we assume a
uniform deposition velocity throughout the year, calculations based
on the ambient concentrations of SO_2 in the atmosphere would suggest
that dry deposition of SO_2 during the winter months (December-March)
should be about 60% of the annual dry deposition of sulfur.

Bulk precipitation collectors are poor collectors of dry
deposition, particularly aerosols less than 0.65 μm in diameter and
gases which dominate at Hubbard Brook. Vegetation, on the other
hand, with its much more elaborate architecture and greater surface

TABLE 1. Seasonal input of sulfur in wet and dry components of
          bulk precipitation for the Hubbard Brook ecosystem.
          Input data are seasonal means with standard errors for
          three years of precipitation (1975–1978).

	Input (kg/ha)	% of Seasonal Input
RAIN		
Bulk Precipitation	9.04 ± 1.02	100
Wet Component	8.77 ± 0.79	97
Dry Component	0.26 ± 0.34	3
SNOW		
Bulk Precipitation	2.02 ± 0.35	100
Wet Component	1.63 ± 0.24	81
Dry Component	0.39 ± 0.11	19

area is a much better collector. Vegetation also can obtain a
percentage of its sulfur requirement by absorbing and incorporating
sulfur dioxide directly from the atmosphere (e.g. Olsen,[10];
Hoeft et al.,[11]).

     There is an added problem with our bulk precipitation
collector. During the summer months these collectors are funnels
(Likens et al.,[12]) which are dependent on rainfall to wash the
dry deposition into the collection vessel. (We do not wash down
the sides of the collector assembly at the end of the the period
as a part of the collection procedure.) However, the amount of
rainfall at Hubbard Brook is nearly constant on a monthly basis
throughout the year and on the average, rainfall (> trace amounts)
occurs about once every three days (Likens et al.,[6]). Therefore,
we estimate that the dry deposition during the rainfall period
(April–November) would be increased by about 0.1 kg/ha if the
collector assembly were washed down at the end of each collection
period. During the winter straight-sided barrels are used and the
majority of the dry deposition is collected as a part of the bulk
sample.

The total dry deposition for the Hubbard Brook ecosystem would be 6.8 kg/ha-yr (6.1 kg/ha-yr of net dry sulfur deposition which is needed to balance the budget, Figure 1, plus 0.65 kg/ha-yr from the dry components of bulk precipitation). Thus, measurements of dry deposition of sulfur with bulk precipitation collectors at Hubbard Brook (about 0.75 kg/ha-yr) would represent only about 11% of the total annual dry deposition to the ecosystem.

ACKNOWLEDGEMENTS

This is a contribution to the Hubbard Brook Ecosystem Study. Financial support was provided by the National Science Foundation. The Hubbard Brook Experimental Forest is operated by the U.S.D.A. Forest Service, Upper Darby, Pennsylvania.

REFERENCES

1. L.S. Dochinger and T.A. Seliga, (eds.). "Proceedings of the First International Symposium on Acid Precipitation and the Forest Ecosystem". USDA Forest Service General Technical Report NE-23. Northeastern Forest Experiment Station, Upper Darby, Pennsylvania. 1,074 pp. (1976).

2. AMBIO, Report from the International Conference on the Effects of Acid Precipitation in Telemark, Norway, June 14-19, Ambio, 5(5-6):199 (1976).

3. H.C. Whitehead and J.H. Feth. Chemical composition of rain, dry fallout and bulk precipitation at Menlo Park, California, 1957-1959, J. Geophys. Res., 69:3319 (1964).

4. E. White, R.S. Starkey and M.J. Saunders. An assessment of the relative importance of several chemical sources to the waters of a small upland catchment, J. Appl. Ecol., 8:743 (1971).

5. J.W. Elwood and G.S. Henderson. Hydrologic and chemical budgets at Oak Ridge, Tennessee, pp. 31-51, in: "Proc. INTECOL Symp. on Coupling of Land/Water Systems", A.D. Hasler, ed., 1971, Leningrad. Springer-Verlag, New York Inc. (1975).

6. G.E. Likens, F.H. Bormann, R.S. Pierce, J.S. Eaton and N.M. Johnson. Biogeochemistry of a Forested Ecosystem, Springer-Verlag, New York Inc., 146 pp. (1977).

7. J.S. Eaton, G.E. Likens and F.H. Bormann. The input
    gaseous and particulate sulfur to a forested ecosystem,
    Tellus (in press) (1978).

8. J.N. Galloway and G.E. Likens. The collection of precipitation
    for chemical analysis, _Tellus_, 30:71 (1978).

9. H. Small, T.S. Stevens and W.C. Bauman. Novel ion exchange
    chromatographic method using conductimetric detection,
    Anal. Chem., 47:1801 (1975).

10. R.A. Olsen. Absorption of sulfur dioxide from the atmosphere
    by cotton plants, _Soil Sci._, 84:107 (1956).

11. R.G. Hoeft, D.R. Keeny and L.M. Walsh. Nitrogen and
    sulfur in precipitation and sulfur dioxide in the
    atmosphere in Wisconsin, _J. Environ. Qual._, 1:203 (1972).

12. G.E. Likens, F.H. Bormann, N.M. Johnson and R.S. Pierce.
    The calcium, magnesium, potassium and sodium budgets
    for a small forested ecosystem, _Ecology_, 48:772 (1967).

ACID SNOW – SNOWPACK CHEMISTRY AND SNOWMELT

Hans M. Seip

Central Institute for Industrial Research

Oslo 3, Norway

INTRODUCTION

Areas in which acid precipitation falls to a considerable degree as snow, exhibit special acidification problems during snowmelt, as illustrated by the particularly low pH-values found in many rivers and lakes during this period in Norway.

This paper deals with aspects of snowmelting which may be important for acidication of soil or natural waters. These effects are, of course, not determined only by the composition and properties of the snow itself. The weather conditions during the melting period and the properties of the ground are also of great importance. In particular the contact between meltwater run-off and the soil and vegetation must be considered. An estimate of the amount of surface run-off compared to sub-surface run-off is valuable. However, even the former may be considerably influenced by contact with the ground. A detailed picture of the water flow both within the snowpack and upon or through the soil seems necessary to predict the effects on leaching and water quality. Unfortunately our knowledge of these processes is at present far from satisfactory.

Since the number of aspects connected to snowmelt is large, some of them can only be mentioned briefly. Many of the results presented are from the Norwegian SNSF project "Acid Precipitation – Effects on Forest and Fish", and no systematic search in the literature has been made.

IMPURITIES IN SNOW

The first problem to be discussed is how impurities are stored in a snowpack prior to the melting period.

Ice (Ih) forms solid solutions with other inorganic compounds only to a limited degree; the most soluble compounds are NH_4F, HF and NH_3[1,2]. The segregation of impurities from ice formed by freezing of dilute solutions is well known[3-5]; by rapid freezing an increased amount of impurities may be included in the ice. Solubility limits have been given as 2.10^{-3} mol/l for HF[6] and $1-2.10^{-4}$ mol/l for HCl[3].

The impurities in a snowpack may either be associated with the snow before the snowflakes reach the ground, or originate from dry deposition. The deposition velocity of say SO_2 on snow surfaces is low[7-9]. At least in most parts of Norway the contribution from dry deposition during winter is assumed to be small. Since the larger part of the impurities in the snowpack is associated with the formation of snow crystals, this process will be described very briefly (cf. Hobbs[2], chap. 10).

If the temperatute of a cloud, which consists of small water droplets, decreases below about -12°C, ice nuclei begin to form (about one per liter at -15° to -20°C). Since the cloud is considerably supersaturated with respect to ice, the crystals grow by deposition from the vapour phase. This process will slow down as the crystal size increases, and increase in mass by aggregation with other ice particles or by accretion of supercooled cloud droplets will dominate. Normally the final snowflake has been formed to a large extent by freezing of supercooled water. Since this process is rapid, impurities may be included in the crystal, probably to a greater extent than corresponding to equilibrium conditions.

Freshly deposited snow has a density of about 0.01 to 0.25 kg/dm^3. However, with time the snowflakes become rounded in shape and break up into smaller grains while the strength and density of the snow increase due to transfer of water molecules through the vapour phase and volume diffusion through the ice. The rearrangement is particularly fast if the temperature is close to zero. Even at the beginning of the snowmelting period it seems likely that the dominant part of the impurities is found in a brine at the crystal surfaces, primarily at grain boundaries. This leads to a freezing point depression and provides the snowpack (or a glacier) with a network of veins so that soluble impurities may drain out[10,11]. It should be mentioned that this idea is not entirely new; Buchanan published in 1887 a thorough study ("On Ice and Brines" with special relevance to ice from seawater[12].

Evidence for a liquid-like layer on ice has been obtained by
nuclear magnetic resonance spectroscopy[13-15]. It is perhaps somewhat surprising that the liquid has been observed down to about
-10°C also for "pure" ice. A layer of this kind was, in fact,
suggested by Faraday, and Fletcher[1] has discussed theoretical
reasons for its existence.

FRACTIONATION OF CHEMICAL IMPURITIES DURING SNOWMELT

Several experiments have been carried out to study the
concentration of various ions in successive fractions of meltwater
from a snowsample[16-19]. It is found that the first fractions of
the meltwater contain higher amounts of ions than the bulk snow.

Fig. 1 shows some results from laboratory experiments by
Johannessen and Henriksen[18]. Lysimeters consisting of 56 cm long
plastic tubes with 13 cm diameter, were filled with roughly homogenized snow. A 1.1 atm. pressure of N_2 was applied to the top of
the lysimeter, and a cooling mantle kept at 2.3 - 3.0°C. The
meltwater was collected in 25 ml aliquots. The concentration
factor plotted in Fig. 1, is defined as the concentration of a given
ion in fraction i divided by the concentration in the bulk snow. In
Fig. 1 this factor is roughly 4 - 5 in the first fractions.

The melting conditions during this experiment are somewhat
different from those found in nature. To reproduce field conditions
somewhat closer the experiments have been repeated with cylinders
with well isolated walls and supply of radiation energy to the snow
surface by a lamp. The concentration factors varied considerably
with the rate of the melting; values above 8 were obtained in a very
slow melting process[19].

Johannessen and Henriksen[18] have also carried out field
experiments. Lysimeters consisting of a polyethylene cylinder (30
cm high and 53.5 cm in diameter) were buried in the ground in the
fall, such that the top extended 3 cm above the ground surface, and
were covered by a perforated polyethylene shutter. During the
winter the snow accumulated on the top of the shutter, and during
periods of snowmelt the meltwater ran through the cylinder and was
collected in a polyethylene bottle. The concentration factors
obtained in an experiment of this type, are given in Fig. 2.

The laboratory and field experiments indicate that 50 - 80 %
of the pollutants are released when the first 30 % of the snow melts.
The very first meltwater may have concentrations about 5 times the
values found in bulk snow; in extreme cases the factors may perhaps
reach ten.

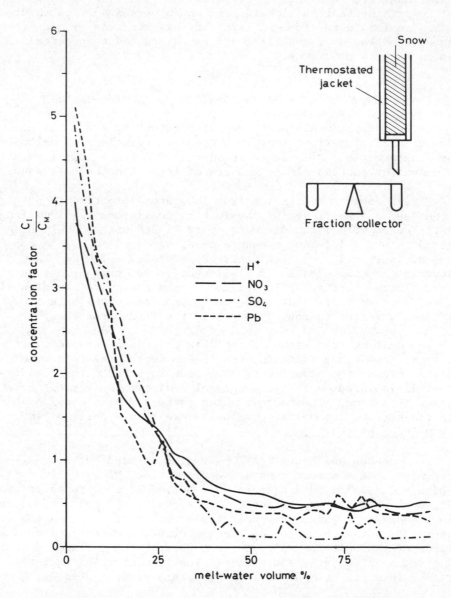

Fig. 1 Concentration factors obtained in a
melting experiment carried out in the
laboratory. pH of the snow was 4.91

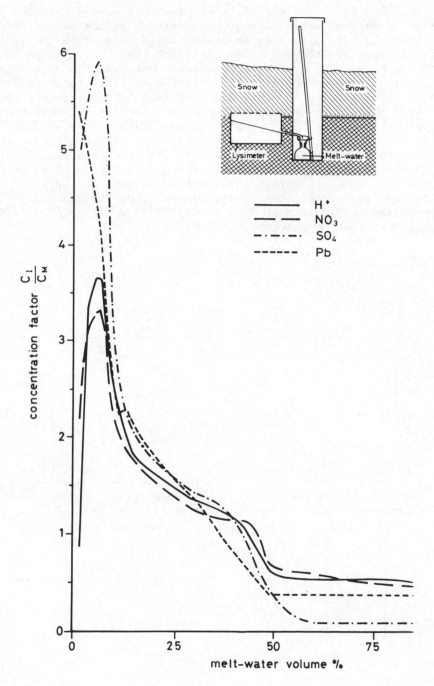

Fig. 2 Concentration factors in meltwater obtained in a field lysimeter.

The observed fractions of impurities in a snowpack during
melting seems reasonable, since, as mentioned in the second section,
the impurities are likely to be concentrated in a brine on crystal
surfaces. The environmental importance of the fractionation is
difficult to assess. Very low pH-values have been found just below
the ice in lakes during the melting period as illustrated in Fig. 3
from an investigation by Hagen and Langeland[20]. Johannessen and
Henriksen studied water formed in the snow cover just above the ice
on some lakes during mild periods[21]. pH-values as low as 3 and
sulphate concentrations up to nearly 100 mg/l were observed. These
results are probably partly a result of the mentioned fractionation,
though the water was in some areas rather coloured showing contact
between meltwater and soil. The possible effects on the water
quality in rivers and lakes is discussed further in the next section.
If the first meltwater gets into contact with soil and vegetation,
effects must also be expected.

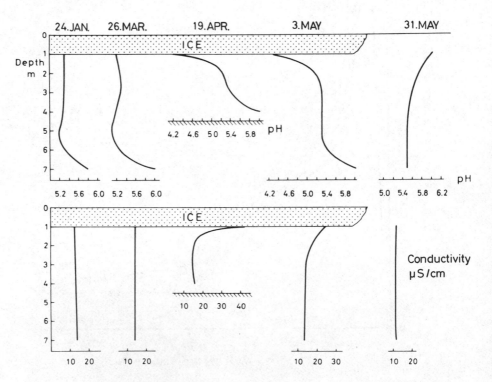

Fig. 3   pH and conductivity obtained in a small lake during
         winter and spring 1970.

CONCENTRATIONS OF IMPURITIES IN SNOW SAMPLES

No complete survey will be attempted; rather just a few examples from Norway will be given.

Elgmork et al.[22] studied snow samples from southern Norway during the winters 1968-71. They found a distinct stratification of the snow; gray bands were considerably more acid and had larger amounts of sulphur and heavy metals than white bands. pH values down to 3.3 and sulphate and lead concentrations up to 25.4 mg/l and 98 µg/l respectively, were reported. The chemical stratification in the snow was well preserved during periods with no snow-melting.

More recently, fairly extensive regional surveys have been carried out in Norway each year from 1975 to 1978, usually in March[23,24]. Fig. 4 shows results for pH in snow samples in 1976[24]. Low pH-values and high values for excess sulphate are found in the south-eastern part of Norway and, perhaps somewhat surprising, in Finnmark in the northernmost part of the country. The regional distribution in southern Norway is in fairly good agreement with measurements made directly of the precipitation during the same winter. However, the average concentrations in the precipitation are higher than in the snow samples, approximately by a factor of 2. In many areas in southern Norway there had been one or more thawing periods during the winter, and some of the impurities had probably leached out. Results from last winter (snowsamples taken in March 1978) show in general lower pH-values than given in Fig. 4 and are closer to the values usually found in precipitation.

Four years of such surveys have shown that this is a simple method to obtain a qualitatively satisfactory picture of the regional distribution of pollutants. As a quantitative estimate of the total amount of pollutants deposited during the winter, the results must be treated with some caution, particularly if there have been thawing periods prior to the sampling. The results may also be useful for predicting possible adverse effects during snowmelt.

DEGREE AND EFFECTS OF CONTACT BETWEEN MELTWATER AND THE GROUND

Evaporation from a snow surface is usually small[25,26] and is neglected in the following discussion. To estimate the effect of meltwater on terrestrial systems, and on rivers and lakes, some knowledge of the contact between run-off and ground is necessary. In extreme cases one might imagine no contact such that the melt-water reaches rivers and lakes unchanged, or, on the other hand, sufficient contact for the composition of the run-off to be completely dominated by the terrestrial system.

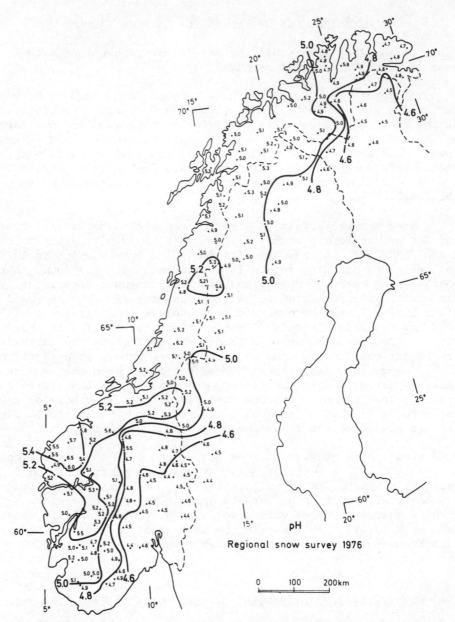

Fig. 4   pH values in snow samples obtained in Norway in 1976.

The flow of meltwater within the snowpack is considered first.
A snowpack is often stratified with distinct ice layers, caused by
previous thaw periods. In particular there is often an ice layer
just above the soil surface. As indicated in Fig. 5, the result
is a lateral movement of water along ice layers, while there is a

vertical flow through channels. Such conditions may affect the
run-off and reduce the soil contact[27]. A few simple experiments
with coloured water showed that the run-off does indeed follow such
layers at least for some meters. However, it seems likely that
dips or cracks in the ice layers will limit the horizontal flow to
such distances, though there seems to be a lack of quantitation
information on this point.

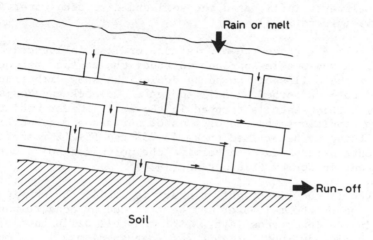

Fig. 5 Water movement in a stratified snowpack.

        The amount of surface run-off and the contact between melt-
water and soil, are determined by a number of factors. Among the
most important ones is the degree of soil freezing, which in turn
depends on several conditions, e.g. air temperature, thickness of
snow cover, texture of soil and type of vegetation. If the snow
starts to accumulate before the soil is frozen, there may be
practically no ground frost at all. In areas with frozen soil
during winter an increase in the soil temperature is often found
before the snowmelt[28,29]. There are further large variations within
an area. Komarov and Makarova state that with an average freezing
depth less than 60 cm, patches of thawed soil are normally found
prior to snowmelt[28].

Some investigations of the degree of surface run-off during snowmelt have been reported in the literature. Dunne and Black studied the melting of snowpacks on frozen ground in sloping pastures[30]. Almost one half of the meltwater left the plots as overland flow. Discharge rates, total volumes, and timing of this portion of the run-off were strongly controlled by incoming short-wave radiation. Other investigators working in very permeable ground or unfrozen soil during snowmelt, concluded that the amount of surface run-off is small[31].

Dincer et al.[32] measured the tritium and oxygen-18 content of precipitation, snowpack and run-off in a small mountain basin in Northern Czechoslovakia. The results showed that about two thirds of the meltwater infiltrated the soil and displaced water in this reservoir (cf. ref. 33).

Even the surface run-off is more or less influenced by soil contact. A reasonable approach to study the contact between meltwater and soil seems to be by doping the snow with tracers expected to be adsorbed by soil contact. Holecek and Noujaim used ^{125}I which they state is trapped in the soil. Under undisturbed, pristine conditions the proportions of surface and subsurface flow were in this way determined to be 45-75% and 55-25% respectively[34]. Experiments using [45]Ca to determine the meltwater-soil contact are carried out in Norway this spring (1978)[35].

Even if the direct soil contact is small, the chemical composition of the meltwater may be affected by dead remnants of plants in the lower snow layers. If the cell walls have been broken, soluble compounds (both organic and inorganic) may leach out. Basic cations may be removed in this way, and the pH of the run-off will be affected.

The number of studies designed to investigate the change in meltwater chemistry caused by soil contact, seems rather limited. Zeman and Slaymaker[36] studied ionic throughputs in snowmelt, glacier meltwater and baseflow, and ionic outputs in discharge waters of an alpine basin in the British Colombia Coast Mountains. They found large increases in the ionic concentrations in snow run-off measured only 1 m or less from the edge of the snowpack compared to meltwater percolating through the snowpack (see Table 1).

Rueslåtten and Jørgensen[37] compared the concentrations of ions in the snow with those in the run-off (see Fig. 6) during the last stage of the snowmelting period. The immediate increase in the H^+ concentration (compare the results for Ø1 and Ø4 and for N 1 and N 2 in Fig. 6) was explained by leaching of organic acids and cation exchange processes in the vegetation and in the humic layer. Unfortunately the samples were not analysed for Cl^- and SO_4^{2-} or on total organic carbon, which makes a discussion of the mechanism

Table 1

Weighted mean ionic concentrations in snowmelt and mean ionic
concentrations in snow run-off waters.

Dissolved	Snowmelt		Snow run-off	
constituents	mg 1^{-1}	%	mg 1^{-1}	%
Na	0.02	5.9	0.78	13.3
K	0.015	4.4	0.21	3.7
Ca	0.02	5.9	1.08	18.6
Mg	0.02	5.9	0.24	4.1
Cl	0.20	58.7	0.24	4.1
NO_3	0.03	8.8	0.03	0.5
SiO_2	0.026	7.6	3.23	55.6
NH_4	0.005	1.5	0.005	0.1
PO_4	0.005	1.5	0.005	0.1
Total	0.34		5.81	
pH	5.27		6.57	

somewhat speculative. The later decrease in the H^+ concentration
and corresponding increase in the concentrations of Ca^{2+} or Na^+
were supposed to originate from the bedrock surface through
weathering reactions.

Njøs analysed the snow run-off on various soils/vegetations
in laboratory experiments[38]. The pH of the snow was 4.6. The pH
even of surface run-off, varied drastically with the soil/
vegetation from about 4 for raw humus with heather (Caluna vulgaris)
to above 6 for loam without vegetation. It is difficult to
reproduce realistic conditions to snowmelting in a laboratory
experiment; in particular edge effects may impair the usefulness of
the results. One should therefore not draw too definite conclusions,
though in agreement with the investigations mentioned above, the
study indicates that the interaction with the soil/vegetation is
important for the composition of the run-off during snowmelt. Uhlen
found, on the other hand, that surface run-off from cultivated
frozen soil without vegetation had a pH-value close to that of the

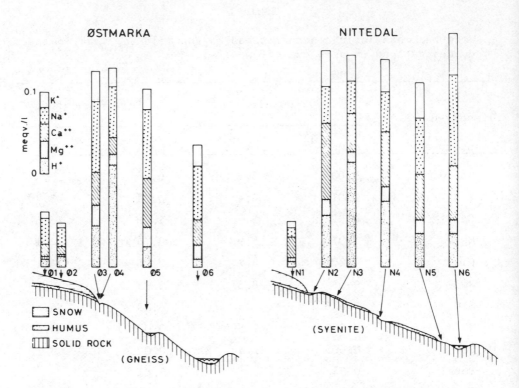

Fig. 6   Content of H, Mg, Ca, Na and K in mequ/l in meltwater and
         a schematic drawing of the sampling sites.

precipitation, probably because of an ice layer which reduced
drastically the contact between water and soil. However, small
amounts of plant remnants influenced the run-off considerably[39].

Abrahamsen et al.[35] studied the snowmelt from a series of
natural plots with areas ranging from 30 m^2 to nearly 300 m^2. The
plots are partly covered by shallow soil (usually less than 40 cm
deep) with vegetation mainly of heather or grass. Samples of snow
and run-off were taken frequently throughout the melting period.
The average pH of the snow in the area was about 4.3 prior to the
snowmelt. There is a considerable increase in pH in the run-off
during the melting period for all the plots. Fig. 7 shows the
water flow and the pH in the run-off for one of these plots. The
increase in pH from less than 4.0 to about 4.5 is probably mainly

due to the fractionation of ions in the snowpack during melting.
Preliminary results indicate that the sulphate content decreased
as expected; about 180 μeq/1 was found Apr. 10th and about 60 μeq/1
a fortnight later.

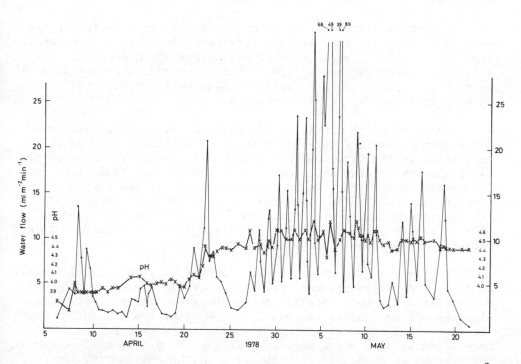

Fig. 7   Variations in water flow and pH from a natural plot (79 m^2)
during snowmelt. pH was measured in the field; subsequent
laboratory measurements gave slightly different (usually
higher) values.

VARIATION IN COMPOSITION OF RIVER WATER DURING SNOWMELT

The pH-values of rivers and lakes often show a drop when the
snowmelt starts. We will here discuss just one example from a small
brook in Telemark county in southern Norway; an area with acid
precipitation (average pH is about 4.3), and the overburden is on

average extremely shallow. The variations in water flow, conductivity, pH, and concentrations of some components are given in Fig. 8[40]. When the snowmelt starts near the end of April the very first run-off has rather high Ca-concentrations, perhaps indicating that a part of this water has been in the soil for some time. After a narrow maximum the pH drop is in this case fairly pronounced. The sulphate concentration shows a small maximum, but decreases after a few days; the calcium content decreases even more drastically in the same period.

No detailed study of the composition of the snow in the area was made. However, snow samples were taken in March some kilometers from the brook[23]. The snow cover was nearly 90 cm deep and pH = 4.6. The concentrations of some other components in the meltwater were: SO_4:1.2 mg/l, Cl:0.6 mg/l, Na:0.46 mg/l, Ca:0.01 mg/l, Mg:0.05 mg/l and NO_3 (as N): 190 μg/l.

The drop in pH during the first stage of the snowmelt is probably more or less related to the high concentrations of impurities in the first meltwater (cf. Section III). However, the water quality is also affected by soil contact as illustrated by the Ca-concentrations. In fact, it has been claimed that considerable run-off in many areas necessarily must cause acid rivers since a major part of the water under such circumstances is only in contact with the acid topsoil and does not penetrate through deeper parts of these soil profiles[41].

During the last stage of the snowmelt only very small amounts of ionic impurities are expected to be left in the snow. The Ca-concentration in the brook is low, perhaps because of fast water flow through the basin. However, the pH value remains rather low. One reason for this may be that the sulphate concentration in the run-off is about 2 mg/l; most of this must be due to soil contact. There seems to be a loss of sulphate from the soil during the snowmelt period. This sulphate has been stored in the soil, but originates from atmospheric deposition. Since the base saturation of the soil in the area is very low (usually below 10%), the sulphate anions are accompanied mainly by H^+ cations.

CONCLUSION

Studies of snowpack chemistry, melting and run-off are important for several reasons. Analyses of the amount of impurities in snow samples give at least semi-quantitative estimates of the deposits during the accumulation period. The environmental problems of acid snow may be amplified by the fractionation of impurities during snowmelt resulting in high concentrations in the first melt-water. The lack of exact knowledge seems most pronounced with respect to details of run-off of meltwater both within the snowpack

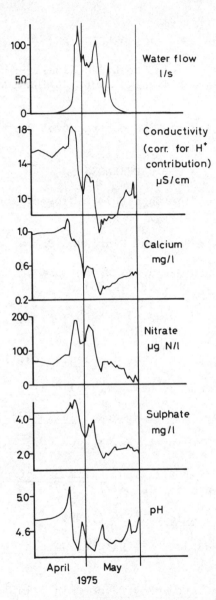

Fig. 8 Variation in water flow and chemical composition of the water in a small brook during snowmelt, 1975.

and on or through the ground. Improved information in these areas
is necessary in order to understand the effects on the ecosystems
of melting of acid snow.

ACKNOWLEDGEMENT

The author is grateful to Bjørn Alsaker-Nøstdahl, Eddy Hansen,
and Hans Gran for valuable suggestions during the preparation of
this paper.

This publication is SNSF contribution FA 30/78.

REFERENCES

1. N. H. Fletcher, "The Chemical Physics of Ice."
    Cambridge University Press (1970).

2. P. V. Hobbs, "Ice Physics," Clarendon Press, Oxford (1974).

3. G. W. Gross, Chen-Ho Wu, I. Bryant and C. McKee,
    J. Chem. Phys. 62:3085 (1975).

4. C. W. Gross, P. M. Wong, and K. Humes, J. Chem. Phys. 67:
    5264 (1977).

5. Y. Mizuno, and D. Juroiva, J. Glasiol. 9:117 (1970).

6. H. Haltenorth, and J. Klinger, Solid State Commun. 21:533 (1977).

7. D. M. Whelpdale, and R. W. Shaw, Tellus 26:196 (1974).

8. H. Dovland, and A. Eliassen, Atmos. Environm. 10:783 (1976).

9. H. Dovland, and A. Eliassen, Estimates of dry deposition on
    snow, SNSF-project IR 34/77 (1977).

10. A. Renaud, J. Glaciol. 1:320 (1949).

11. J. G. Paren, and J. C. F. Walker, Nature Phys. Science 230:77
    (1971).

12. J. Y. Buchanan, Proc. Roy. Soc. Edin., 14:129 (1887).

13. J. Clifford, Chem. Commun. 17:880 (1967).

14. J. D. Bell, R. W. Myatt, and R. E. Richards, Nature Phys.
    Science 230:91 (1971).

15. V. I. Kvlividze, V. F. Kiselev, A. B. Kurzaev, and
    L. A. Ushakova, Surface Science 44:60 (1974).

16. S. Oden, and J. Bergholm (unpublished results) (1973).

17. E. T. Gjessing, A. Henriksen, M. Johannessen and R. Wright,
    in:"Impact of Acid Precipitation of Forest and Freshwater
    Ecosystems in Norway", F. H. Braekke, ed,, SNSF-project,
    FR 6/76 (1976).

18. M. Johannessen, and A. Henriksen, Water Res. (in press).

19. N. Berg, and H. M. Seip, (unpublished results).

20. A. Hagen and A. Langeland, Environm. Pollut. 5:45 (1973).

21. M. Johannessen, and A. Henriksen, Studies of Snow and Surface
    Water on Ice-Covered Lakes in Fyresdal/Nissedal and
    Langtjern during the Winters 1974–75 (In Norwegian),
    SNSF-project, TN 24/76 (1976).

22. K. Elgmork, A. Hagen, and A. Langeland, Environ. Pollut.
    4:41 (1973).

23. E. Gjessing, T. Dale, M. Johannessen, C. Lysholm and
    R. F. Wright, Regional Snow Survey, Winter 1974–75
    (In Norwegian), SNSF-project TN 22/76 (1976).

24. A. Henriksen, M. Johannessen, E. Joranger, R. F. Wright and
    T. Dale, Regional Snow Survey, Winter 1975–76, SNSF-project,
    TN 28/76 (1976).

25. R. Lemmelä, and E. Kuusisto, Nordic Hydrology 5:64 (1974).

26. S. Furmyr, and A. Tollan, Results and Experiences from Snow
    Investigations in the Filefjell Area 1967-1974 (In
    Norwegian), Den norske komité for Den internasjonale
    hydrologiske dekade, Oslo (1975).

27. S. C. Colbeck, J. Glaciol. 19:571 (1977).

28. V. D. Komarov, and T. T. Makarova, Soviet Hydrology: Selected
    Papers, 243 (1973).

29. R. S. Heiersted, Frost i jord (Frost Action in Soil) 16:5 (1975).

30. T. Dunne and R. D. Black, Water Resources Res. 7:1160 (1971).

31. G. R. Stephenson, and R. A. Freeze, Water Resources Res. 10:284
    (1974).

32.   T. Dincer, B. R. Payne, T. Florkowski, J. Martinec and
      T. Tongiorgi, Water Resources Res., 6:110 (1970).

33.   J. Martinec, Water Resources Res. 11:496 (1975).

34.   G. R. Holecek, and A. A. Noujaim, Western Snow Conference,
      40th Annual Meeting, Phoenix, Az., Proceedings, F. T.
      Collins Co., Colorado State Univ., p.43-48 (1972).

35.   G. Abrahamsen, J. B. Dahl, E. Gjessing, H. M. Seip, and
      A. Stuanes (in prep.).

36.   L. J. Zeman, and H. O. Slaymaker, Arctic and Alpine Research
      7:341 (1975).

37.   H. G. Rueslåtten, and P. Jørgensen, Nordic Hydrology,
      (in press).

38.   A. Njøs (unpublished results).

39.   G. Uhlen, VII Int. Fertilizer Congress, Moscow, Vol. II:151
      (1976).

40.   M. Johannessen, and E. Joranger, Chemical Investigations of
      Water and Precipitation in Fyresdal/Nissedal 1/4-1973 -
      30/6-1975 (In Norwegian), SNSF-project TN 30/76 (1976).

41.   I. Th. Rosenqvist, "The Science of the Total Environ."
      (in press).

ION RELATIONSHIPS IN ACID PRECIPITATION AND STREAM CHEMISTRY

G. M. Glover, A. S. Kallend, A. R. W. Marsh and
A. H. Webb

Central Electricity Research Laboratories

Leatherhead, Surrey, England

INTRODUCTION

A complex chain of processes link the production of an anthropogenic emission and its action on the environment. The details vary
with the type of emission and circumstances, but our particular
field of interest, that of electric power plant emissions, gives
examples of the sort of processes which must be considered. Power
plant emissions are modified first by relatively rapid physio-
chemical processes in the near-field plume and then, as they become
diluted, by different slower reactions in the open atmosphere. The
latter include homogeneous gas phase reactions, liquid phase reactions in water droplets and heterogeneous processes on the surface
of atmospheric aerosols. The rates of the different processes are
greatly modified by changing humidity, the intensity of irradiation
by sunlight, or the level of catalysts for heterogeneous reaction.
Depending on the conditions, different reaction routes may dominate.
The products reach the ground by dry deposition of gases and solids
and by "washout" and "rainout" processes involving precipitation.
Again the relative importance of these processes is widely different
in different conditions. On contact with the ground - soil, vegetation or bare rock - the deposited material is rapidly modified
further by chemical reaction with the ground.

In view of the complexity of this chain of processes, a simple
relationship between emissions and their consequences is unlikely.
However, if we are to take effective action to protect our environment, it is essential that we understand at least the main features
of the processes involved. We shall consider here two areas of
uncertainty: 1) the factors controlling the composition of

precipitation, and 2) modification of precipitation by ground interaction. It will be shown that, even in these limited fields of
study, the relation of cause and effect is far from self-evident.

COMPOSITION OF PRECIPITATION

Factors Involved

The processes by which precipitation scavenges species from the
atmosphere are usually classified into two types, those associated
with cloud phenomena are called rainout, while those concerned with
removal by falling precipitation are termed washout. It is at
present possible to describe washout processes theoretically, but
no adequate theory of rainout exists. The washout of a soluble gas
involves gas and liquid phase mass transfer rates and the solution
equilibria of the gas. For SO_2, these solution equilibria involving
sulphurous acid are pH dependent. Low pH suppresses the solubility
of SO_2 and the ionization of the acid. Hales[1] described the fundamental theory of gas scavenging by rain and the washout of SO_2 by
rain falling through a uniform concentration of SO_2 as a function
of: (i) the size spectrum of droplets and hence the rate of raining,
(ii) the initial pH of the rain, (iii) the height of the SO_2 concentration and (iv) the absolute magnitude of the SO_2 concentration.
In general small rain drops of < 1 mm diameter are saturated with
SO_2, while larger droplets do not achieve saturation.

Fig. 1 shows the theoretical changes in the pH of rain falling
200 metres through various uniform SO_2 concentrations. If the rain
has an initial pH of 4.0 at cloud level, this will not be changed by
the washout of concentrations of SO_2 less than 100 $\mu g/m^3$. In remote
areas, where SO_2 concentrations are say $\sim$ 10 $\mu g/m^3$, rain initially
of pH 5.0 would be decreased to $\sim$ 4.9. The relatively few measurements of the pH of cloud water show a dependence on cloud type but
a value of $\sim$ 5.0 in remote areas has been observed[2].

The presence of NH_3 enhances the rate of removal of SO_2[3]. The
chemistry of the absorption of NO_x in aqueous solution is complex[4,5]
and estimates of final concentrations due to washout have not been
calculated but any resultant nitrite or nitrate from the washout of
NO_x will increase ammonia removal.

The calculations of washout of SO_2 and NH_3 gases and sulphate
aerosols reported by Marsh[3], suggest that not more than half of
the $SO_4^=$ in precipitation comes from SO_2 washout and that this contribution is dependent on local gas phase NH_3 concentrations. These
calculations are in agreement with the observations of Petrenchok
and Selezneva[2], who concluded that in remote areas washout accounted for $\sim$ 50% of the final concentrations.

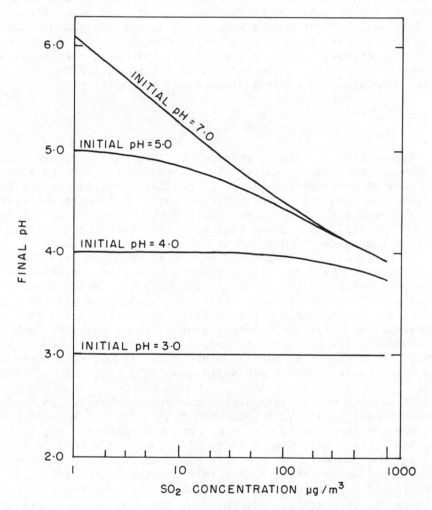

Fig. 1   The variation of the final pH of rain with concentration
of SO_2.

As emissions disperse, conversion of gases such as SO_2 and NO_x to aerosols occurs and at long distances from the source the aerosol phase should dominate. Aerosols should be more readily scavenged by rainout processes, such as condensation, than they are by the wash-out process of capture by falling droplets. The role of aerosols in rainout processes in remote areas should be more important than in the case of washout and the more remote the observation, the better should be the correlation between the aerosol and precipitation composition. In less remote areas with higher gas phase concentrations the composition of precipitation should be regarded as a mixture of washout and rainout processes involving both local and distant sources and the composition of precipitation need not correlate strongly with local gas and solid composition.

Observations

The acidity of rain is governed by the overall charge balance equation of the ions in solution. All ions contribute to determine the final pH but normally the pH is dominated by ions at greater or equal concetration to the hydrogen ions. Acid rain observations by the recent O.E.C.D. project, "Long Range Transport of Air Polluants, 1972-1975", have shown that acid rain is widespread in Europe and that the range of pH found at O.E.C.D. sites in remote areas, i.e. in S. Norway, is comparable with that found at relatively urban sites in the U.K. Fig. 2 shows a histogram comparison of daily pH readings in the two countries over the period October 1972 to March 1975. There is very little difference between the distribution. To account for the surprising acidity in the case of S. Norway, the total ionic composition of the precipitation has to be examined. The average composition of precipitation collected at O.E.C.D. sites in S. Norway is shown in Fig. 3. The dominant ions after correction for sea salts are $SO_4^=$, NO_3^- and NH_4^+.

Table 1 shows the very high correlations observed in S. Norway between the ions H^+, NH_4^+, $SO_4^=$ and NO_3^-. Most of the O.E.C.D. data are confined to H^+, $SO_4^=$ ion concentrations but, for 400 site days at 4 sites, NO1, NO8, NO9 and N10, analyses were made for NH_4^+ and NO_3^- as well. Table 1 is based on these 400 results. The regression coefficients were determined using micro-equivalent units of concentration. The relationships between $SO_4^=$, NO_3^- and NH_4^+ are not adequately represented by the average concentrations. Figure 4 shows the range of ratios of NH_4^+/NO_3^- excluding very low concentrations < 10 µeq/l. This particular ratio is chosen because the average concentrations are approximately equal to about 28 µeq/l and the mean ratio is $\sim$ 1.1. However, this apparent equality between NO_3^- and NH_4^+ only represents the mean of a very flat distribution of ratios with relatively few events actually having NH_4^+ equal to NO_3^-. This flat distribution is typical of all the ratio distributions and shows that no simple molecular compound is a major contribution to the composition of precipitation.

Table 1. Correlation between ions. Correlation Coefficent > 0.16 is > 99.9% significant.

Dep. Var.	Indep. Var.	Corr. Coefft.	b_{yx}	C
H^+	$SO_4^=$	0·80	0·62	5·8
H^+	NO_3^-	0·74	0·93	12·2
NH_4^+	$SO_4^=$	0·78	0·64	5·0
NH_4^+	NO_3^-	0·66	0·87	4·1
NO_3^-	$SO_4^=$	0·78	0·49	2·2

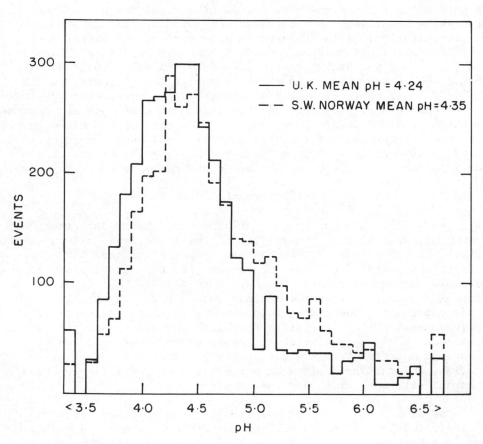

Fig. 2 Comparison of pH of rain in UK and S.W. Norway.

SURFACE WATERS

Ground Effects

As soon as precipitation reaches the ground (soil, vegetation, or bare rock), further chemical reaction occurs. The magnitude of the effect can be assessed from Fig. 5 which compares the weighted average composition of precipitation and river-water in the Tovdal region of S. Norway[7]. The discrepancy between these averages is an indication of the changes occurring within the catchment. These are of several kinds: first, water evaporation concentrates the precipitation; second, dry deposition of gases and solids introduces extra solutes; third, the precipitation interacts chemically and biochemically with the ground. The approximate magnitude of the first two effects is often assessed by considering the concentration changes of a relatively unreactive species such as chloride. Since there are no significant sources or sinks of chloride in such a catchment, chloride can be regarded as a conservative species against which the concentration of other species may be compared. River-water contains a concentration of chloride 52% higher than in precipitation. This is taken as a measure of the effect of water evaporation and dry deposition. Certain species such as Ca^{++}, Mg^{++} and K^+, are more than 52% higher in river-water and this suggests sources within the catchment. Other species such as NH_4^+, NO_3^- and most notably, H^+ are much less concentrated in river-water than in precipitation and this indicates that they are consumed by chemical or biochemical processes in the catchment. Although, in a natural system subject to large fluctuations, such comparisons must be used with caution, these effects are big enough to demonstrate the importance of ground influences beyond reasonable doubt.

Fieldwork

The significance of ground influences on the Tovdal river has been considered because the acidity of the Tovdal and certain other rivers and lakes in S. Norway has been reported to have increased over recent years[8,9] and this has given rise to concern over the possible ecological consequences[10], especially to fish life. In general, the acid content of surface waters has both strong and weak acid components. The strong acid anions, sulphate and nitrate, are derived mainly from precipitation, whilst the weak acids comprise humic and fulvic acids, which are produced by biodegradation of vegetable material[11], a small quantity of other acidic organic species, and the inorganic weak acids based on hydrated aluminium, iron and silica species, and the ammonium ion.

The strong acid content of Norwegian waters, particularly those of the Tovdal region has been determined before[12] but data have not been available on the weak acids content. A level of

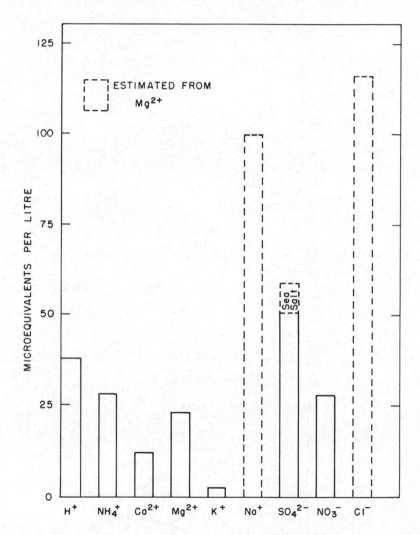

Fig. 3 Average composition of precipitation over S.W. Norway
        from September 1974 to March 1975.

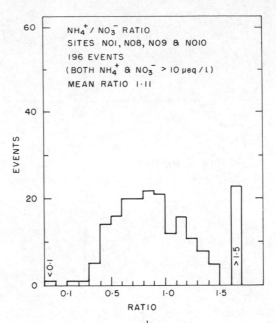

Fig. 4   The range of ratios of NH_4^+ to NO_3^- observed in precipitation over S. W. Norway.

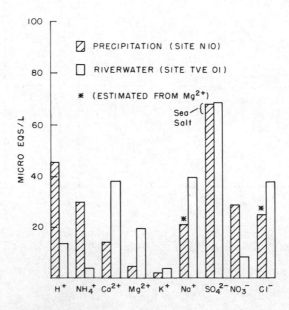

Fig. 5   Weighted average compositions of precipitation and river-water in the Tovdal region of S. W. Norway, April 1975 to October 1976.

$\sim$ 80 µeq/l can be inferred, however, from the reported organic carbon content ($\sim$4 mg/1)[12] using the values for the acidity of humic and fulvic acids given by Schnitzer and Khan[13]. Gamble[14] has reported that fulvic acids can be considered as a single dibasic weak acid with a first ionization constant in the range of 2.5 x 10^{-3} to 4.7 x 10^{-3}. From these values it can be calculated that the weak organic acids could dominate the hydrogen ion concentration down to a pH of about 4.2.

To shed light on this possibility we have examined the relative contributions of weak and strong acids to the pH of the surface waters of the middle Tovdal region during the spring snow-melt of 1977. The snow-melt period has been held to be critical[10] because the acid accumulated in snow during the winter may be released preferentially into the water courses in the early stages of melting.

The weak and strong acid concentrations were determined by a pH titration following the method due to Gran[15] and discussed by Johansson[16] for the determination of mixtures of weak and strong acids. The samples were thermostatted at 27 C and purged continuously with nitrogen to remove atmospheric carbon dioxide. In addition to the weak and strong acid determinations, the major chemical constituents of the samples were determined by standard procedures. The sampling period extended from March 20 to April 27 1977. This period was chosen to include the most probable date for the spring snow-melt. Samples were taken from the Tovdal river at Tveit Bridge near Ovre Ramse, from tributary streams and, to a lesser extent, from the snow-pack and from ground drainings.

Results

The main results of our measurements are shown in Fig. 6, which refers to the Tovdal River and Fig. 7, which relates to a tributary stream, the Ramse Brook. The figures show water depth, pH, weak and strong acid concentrations and also the contribution of the weak acid to the hydrogen ion concentration, obtained from the difference between the strong acid measurement and the total hydrogen ion concentration.

In examining these results, it is necessary to consider the weather conditions which prevailed. Immediately prior to the start of work on the 20 March, there had been a partial thaw. This was followed from the 20 March to the 21 April by a period of colder weather (Period I), during which the snow-melt proceeded more slowly and, as can be seen from Fig. 6, the river level fell. From the 22nd to the 27th of April, the weather became warmer, the rate of thaw increased and the river level rose again. We shall refer to this as period II.

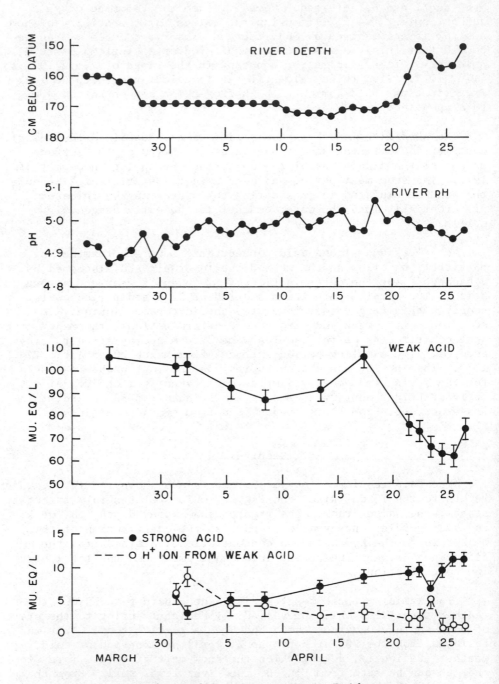

Fig. 6   Tovdal River at Tveit Bridge.

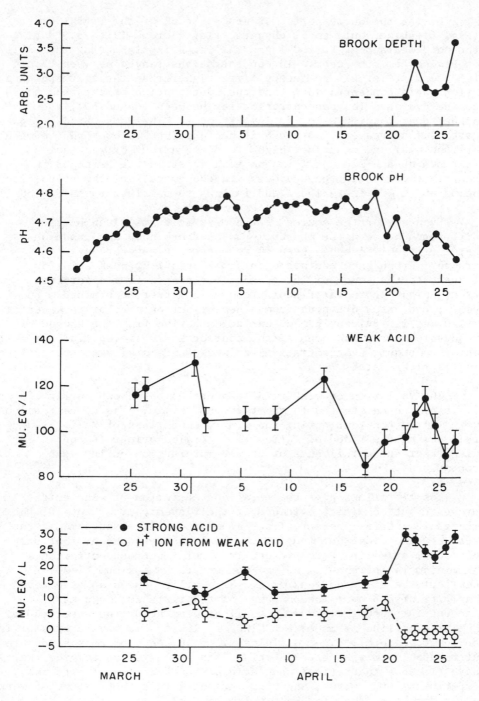

Fig. 7 Ramse Brook.

During the whole period of study the pH of the Tovdal River
(Fig. 6) showed only small changes, rising from 4.9 to 5.0 during
the cold weather of Period I and falling slightly during the melt
of Period II. The weak acid concentrations ranged between 62 and
106 µeq/l, (i.e. six to twenty times higher than the corresponding
strong acid concentrations) and the contribution of the weak acids
to the hydrogen ion concentration lay between 10 and 60%, the
higher contribution being during Period I. The Ramse Brook (Fig. 7)
gave similar results, although it had a slightly lower pH (between
4.5 and 4.8) and contained higher concentrations of weak acids
(between 86 and 130 µeq/l). The weak acids in the Ramse Brook
contributed less to the hydrogen ion concentration, the proportion
being about 40% in Period I and falling to nearly zero in Period II.

Evidence of the source of the weak acids is given by the
difference between the measurements on fresh snow and those on
water or snow which had been in contact with vegetation or the
ground. Fresh snow contained only $\sim$ 25 µeq/l of weak acid (due
largely to the ammonium ion content); old snow from a forested
area, which incorporated coniferous tree litter, contained $\sim$ 80
µeq/l; and water draining from a nearby barren area of rock contained
$\sim$ 110 µeq/l. Evidently the weak acids derive from the ground or
vegetable matter and even slight contact is sufficient for the meltwater to acquire concentrations of weak acids similar to those found
in the river-water.

The variation of weak acid concentrations shown in Figs. 6 and
7 is also consistent with their ground origin. The concentrations
were fairly constant during the wintry conditions of Period I and
fell during the flood of Period II. This behaviour is consistent
with the reduced influence of ground interaction during spate
conditions.

The titrations give the total concentration of weak acids
present, but no direct evidence of their chemical nature. Of the
inorganic species present, iron, aluminium, silica and, ammonia can
act as weak acids, but routine chemical analysis showed that only
aluminium or silica were present in significant concentration.
Aluminium is tetrabasic and may account for a maximum of 20 - 30
µeq/l of the titrated weak acid, while silica shows an effective
basicity of 0.8 over the pH range of the titration, and may
account for a further 20 - 30 µeq/l of weak acid (the exact contribution of silica is uncertain since at least some may be present in
an inactive form). The remaining 17 to 46 µeq/l of weak acids are
attributable to weak organic acids such as humic and fulvic. We
have fitted a theoretical weak acid/strong acid function to the
experimental pH curves, and this indicates that the mixture of weak
acids can be represented by a polybasic weak acid with a first
ionization constant in the range 10^{-6} to 5×10^{-7} and less well
defined weaker ionizations. It was not possible to determine

individual ionization constants and weak acid concentration more
precisely because of the long equilibration times in the middle
pH range, but no evidence for significant concentrations of weak
acids of the strength reported by Gamble[14] was detected. Using
the polybasic approximation, it can be calculated that the pH of
the river water was significantly buffered by the weak acids and
that in the absence of any strong acid, the Tovdal river would
have had a pH of about 5.2 to 5.3 during the period of study.

     Although the weak acid concentrations found in the present
work relate to a brief period of study, it seems likely that they
are reasonably typical, since other chemical species were near the
average for this period of the year. For example, the pH of routine
samples taken from the end of March to the end of April averaged
4.85 in 1977 compared with an average of 4.87 for the same period
in 1974 to 1976. Similarly, the sulphate concentration averaged
87 μeq/1 in 1977 compared with 72 μeq/1 over the three previous
years.

     The presence of such concentrations of weak acids is signifi-
cant. They contribute to the hydrogen ion concentration, at least
during some seasons of the year, and more importantly, they indicate
the magnitude of ground interaction effects. Even in a fairly barren
region like the Tovdal, where the soil cover is sparse, the chemical
composition of precipitation is rapidly modified by contact with the
ground. The mechanisms of the interaction are complex, but an
understanding is essential if we are to identify the reasons for
temporal changes in river chemistry.

                        CONCLUSIONS

     Between the production of an anthropogenic emission and its
action on the environment, the emitted products are modified by
a complex sequence of processes. We have examined two possible
stages of such a sequence. First, the factors controlling precipitation acidity were considered, then the processes which modify
the composition of precipitation after it has fallen in a catchment. Both stages reveal situations of considerable complexity.

     The acidity of precipitation is governed by the overall
charge balance equation of the ions in solution. The dominant
ions are usually SO_4^{2-}, NO_3^- and NH_4^+, after allowance is made for
sea salts.

     In remote areas with low concentrations of SO_2, theoretical
estimates suggest that the washout of the gas is unlikely to produce marked changes in the pH of rain, although the presence of
NH_3 enhances the final sulphate concentration. Observations in
such areas show a significant correlation between the ions SO_4^{2-},

NO_3^- and NH_4^+, but it is apparent that no single molecular species determines the acidity of precipitation.

In considering the modification of the composition of precipitation after it has fallen, observations on the Tovdal catchment in S. Norway are examined. The area is regarded as particularly significant because the acidity of the surface waters is reported to have increased over recent years.

Comparison of the compositions of precipitation and surface waters in this area shows that the surface waters gain significant concentrations of Ca^{++}, Mg^{++} and K^+ via processes occurring within the catchment, while NH_4^+, NO_3^- and notably H^+ are lost. Another constituent of surface waters is the weak acid derived from ground interaction. A short field study on the surface waters of the Tovdal region showed that weak acids were present at concentrations between 60 and 130 μeq/l, which was five to twenty times more than the strong acid content. The weak acids also have a significant buffering influence and were observed to contribute a maximum of 60% of the H^+ ion present in the surface waters.

ACKNOWLEDGEMENTS

We thank the Norwegian S.N.S.F. Project for hospitality and facilities during this work. We are also indebted to the Norsk Institutt for Vannforskning for routine chemical analysis of the the samples. This paper is published by permission of the Central Electricity Generating Board.

REFERENCES

1. J. M. Hales, Fundamentals of the theory of gas scavenging by rain, Atmos. Envir. 6:635 (1972).

2. O. P. Pentrenchok and E. S. Selezneva, Chemical composition of precipitation in regions of the Soviet Union, J. Geophys. Res. 75:3629 (1970).

3. A. R. W. Marsh, Sulphur and nitrogen contributions to the acidity of rain, Atmos. Envir. 12:401 (1978).

4. E. J. Koval and M. S. Peters, Reactions of aqueous nitrogen dioxide, Ind. Eng. Chem. 52:1011 (1960).

5. C. England and W. H. Corcoran, Kinetics and mechanisms of the gas phase reaction of water vapour and nitrogen dioxide, Ind. Eng. Chem. Fund. 13:373 (1974).

6. O.E.C.D. data from the Long Range Transport of Air Pollutants is issued by the Norwegian Institute for Air Research and given in a series of reports LRTAP 4/74, 4/75, 18/75, 20/75 and 2/76.

7. A. Henriksen, Norwegian S.N.S.F. Project (pers. comm.).

8. E. T. Gjessing, A. Henriksen, M. Johannessen and R. F. Wright, S.N.S.F. Research Report 6/76:65 (1976).

9. A. Henriksen, Vann. 1:69 (1972).

10. H. Leivestad and I. P. Muniz, Nature 259:391 (1976).

11. E. T. Gjessing, "Physical and Chemical Characteristics of Aquatic Humus" Ann Arbor Science Publ., Mich., (1976).

12. A. Henriksen, (pers. comm.).

13. M. Schnitzer and S. U. Khan, "Humic Substances in the Environment" Marcel Decker, p.37 (1972).

14. D. S. Gamble, Can. J. Chem. 48:2662 (1970).

15. G. Gran, The Analyst 77:661 (1952).

16. A. Johansson, The Analyst 15:535 (1970).

THE SULFUR BUDGET OF SWEDEN

Svante Odén[1] and Thorsten Ahl[2]

Department of Soil Sciences[1]
University of Agricultural Sciences
S-750 07 Uppsala, Sweden

Institute of Limnology[2]
Uppsala University
S-751 22 Uppsala, Sweden

ABSTRACT

Based on the discharge of sulfur by Swedish rivers (34 drainage basins) covering 78 % of Sweden, the total atmospheric fallout of sulfur has been computed for the year 1974. The figure amounts to about 590,000 tons, which is higher than the corresponding figure based on atmospheric chemical data. The increase of the discharge of sulfur caused by increased fallout amounts to 2-5 %, or an increase of about 20,000 tons for the whole of Sweden per year.

The fallout of sulfur from different parts of Sweden during this century has been reconstructed. The total figure amounts to about 60 million tons of sulfur.

INTRODUCTION

The regional acidification of Europe (Odén[1,2]) has now been widely accepted as being mainly caused by the anthropogenic emissions of sulfur into the atmosphere (OECD[3]). Consequently, the extent of this acidification is closely related to the fallout of sulfur, and much work has been devoted to determinations (wet fallout) and computations (dry fallout) of the total fallout of sulfur from meteorological and atmospheric chemical data. The results, however, vary considerably and this depends mainly on difficulties in the determinations and the necessary approximations

which have to be made. Without going further into details, the
following figures apply to Sweden, either as a whole or in part,
Table 1.

Table 1. Deposition of total excess sulfur over Sweden
according to different authors.

	Deposition of S	
	$kg \cdot ha^{-1} \cdot yr^{-1}$	$ton \cdot yr^{-1}$
1. Total Sweden, 1965[4]		206,000
South Lat. 61°N	6.5	119,500
North Lat. 61°N	3.3	86,500
2. Total Sweden, 1976[5]	11.2	500,000
3. Total Sweden, 1974[3]	10.0	447,000

As can be seen the figures increase with time, which basically
reflects improved knowledge and more careful computations, and not
a change in the actual situation. The latter amounts to an increase
of only 2 to 3 per cent per year.

A determination of the net, total fallout of sulfur can also
be made from the discharge of sulfur by the rivers. The word
'net' is used to denote that the amount of sulfur involved in local
recycling processes (the process of looping, Odén[2]) is not accounted
for. The total discharge of sulfur is a maximum figure with respect
to the atmospheric fallout and among others, corrections have to be
made for the industrial pollution of sulfur, sulfur due to
weathering and the agricultural use of fertilizers. The amount of
sulfur originating from the ocean water can also be accounted for.
By means of existing data the following computations have been
made:

1. The Swedish sulfur budget for 1974.

2. The transient parts of the Swedish sulfur budget
   from 1965 to 1974.

3. Reconstruction of the Swedish sulfur budget during this
   century.

TOTAL DISCHARGE OF SULFUR BY THE RIVER SYSTEMS

In 1961 a preliminary water chemical network was initiated in Sweden and Finland (17 river basins) with a limited analytical program. In 1965 this network was supervised by Ahl, who successively extended the network to 73 river basins (Ahl and Odén[6]). Water samples are taken every month and more than 20 different determinations are made on each sample. Part of the data from this network is used in this paper.

The drainage area covered by the network is given in Figure 1 along with the appropriate network stations. With the specific purpose to obtain figures of the total discharge of sulfur from Sweden, only those stations closest to the outflow of the rivers to the Baltic Sea or the North Sea have been utilized. To simplify the presentation, the country has been divided into four regions more or less homogeneous with respect to hydrological and atmospheric chemical conditions. Altogether data from 34 sampling stations are used and they cover 78 % of the area of Sweden. The variations in this respect for the different regions are given in Table 2.

Some river systems extend outside Sweden, altogether 24,700 km^2 or 5.2 % of the total drainage area. The discharge of the different elements have, consequently, been corrected for this external area by means of the discharge coefficient for each river system. On the other hand, 22 % of the land area is not incorporated in the network. The discharge contribution from this area is made by the mean discharge coefficient for each region.

Table 2. Area conditions in the computations for the sulfur budget of Sweden and data for measured discharges of sulfur in 1974 for the four regions illustrated in Figure 1.

Region	No. of rivers	Area, km^2 total[a]	studied[c]	%	Discharge of S[c] tons . yr^{-1}
I	17	287,664	236,706	82	219,200
II	5	66,509[b]	44,918	71	118,000
III	5	26,357	9,671	37	24,600
IV	7	67,470	55,293	82	127,700[d]
Total	34	448,000	348,588		489,500

[a] Areas within Sweden.
[b] Including Öland and Gotland (4,564 km^2).
[c] Corrected for areas outside Sweden. For region I 17,648 km^2, for region IV 7,052 km^2.
[d] Corrected by 14,700 tons caused by emissions from paper pulp plants.

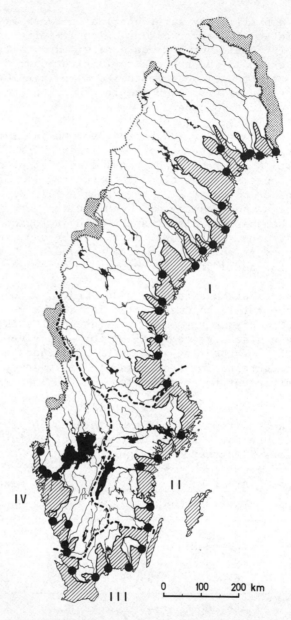

Figure 1. Map of the Swedish water quality network. The network stations used in this study is denoted by dots. Drainage areas outside Sweden is denoted by ⦂⦂⦂, areas not incorporated in the network by ///. Sweden has been divided in four regions, I–IV.

For 1974 the measured discharge of sulfur amounted to 489,500 tons. The figure for region IV has been corrected by the known amount of water pollution of sulfur from paper pulp plants (Ahl7)around Lake Vänern. For the other regions the industrial pollution has not been accounted for but is likely to be very small, since most industries are situated downstream of the net-work stations. Considering Sweden as a whole, the total discharge of sulfur amounts to 667,900 tons per year. This figure is slight-ly higher than the figure based on the trend for the period 1965-1974. There are several sources of this sulfur, such as atmosphe-ric fallout of industrial sulfur, sulfate-S from oceans, sulfur through weathering and sulfur from different fertilizers. Each part will be treated separately.

As shown previously (Odén[2]), the sulfate content in river waters is not constant with time. Excluding variations within the year, almost all station records show increasing values during the period 1965-1976. The linearized regression lines have been computed for the network stations operating from 1965 and the mean results for the four regions are given in Table 3. The computa-tions are made by an operation, where the variation of the dis-charge of water between different years is cancelled. The positive S-trends vary between 2.4 and 4.7 % per year and no river basin shows a negative trend.

The rate of change of the discharge coefficient (in $kg \cdot ha^{-1} \cdot yr^{-1}$) is also given in Table 3. The total increase of the dis-charge of sulfur amounts to 19,300 tons per year. When the ef-fects of fertilizer-S are also taken into account (cf. below), this figure increases to 20,800 tons per year. Since the effects of weathering and marine salt over so short a period as 10 years are negligible, this accelerating input of sulfur over Sweden must be due to increased anthropogenic emissions in Europe. The Swedish measures to reduce the domestic emissions have, conse-quently, not substantially reduced the fallout within Sweden. This was also to be expected, since the sulfur sources outside Sweden account for about 80 % of the fallout within Sweden (OECD3).

Table 3. Rate of change of the discharge of sulfur during the pe-riod 1965-1974 given in relative and absolute values.

Region	No. of rivers	S-trend $\% \cdot yr^{-1}$	$kg \cdot ha^{-1} \cdot yr^{-1}$
I	7	+2.4	+0.21
II	3	+3.6	+0.78
III	2	+4.7	+1.15
IV	3	+2.4	+0.57

SULFUR THROUGH WEATHERING

In ordinary bedrock and soil material the sulfur content is about 0.05 %. Some of this sulfur is released by weathering. This release was studied in soil samples (48) from 12 representative sites from the North to the South of Sweden (A_2, B and C horizons in mainly podsols). The acid weathering was made step-wise in a geometric time progression during one year using 0.1 M HCl. During one year of such an artificial weathering, 3 % of the original material dissolved at a maximum. Besides sulfur, 14 other elements were determined.

Some results are given in Figure 2. The release of sulfur from the Middle and North of Sweden amounts to about 2 g per kg weathered products and there is almost no change in the release of sulfur with increasing weathering intensity. Along the West coast of Sweden the release is higher and increases with increasing weathering intensity. The samples (6) from Skåne and Småland show a pattern which suggests that some S-rich, easily weatherable minerals occur in these parts of Sweden. At intensified weathering, the release of sulfur is reduced to the same range as for the other areas in Sweden.

Table 4 gives the release of sulfur due to weathering. The computations are based on the total discharge of silicon by the rivers and the S/Si ratio of the weathering products relevant for the different regions. The computed figures are very low and less than 0.5 % of the total discharge of sulfur. Even if some silicon is trapped in the soil (as amorphous SiO_2) or in the lake basins (as diatom SiO_2) the contribution to the sulfur budget due to weathering is almost negligible. Neither will increased weathering increase the discharge of sulfur to any measurable extent.

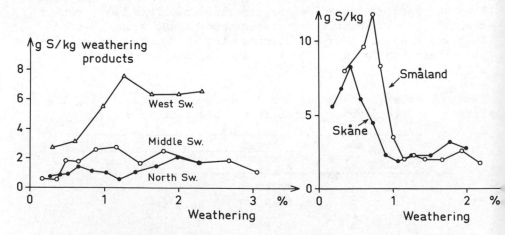

Figure 2. The release of sulfur by successive acid weathering of soils from different parts of Sweden.

Table 4. Computed figures for the release of sulfur due to weathering.

Region	Si_{river} tons $\cdot$ y^{-1}	Sulfur due to weathering	
		tons $\cdot$ yr^{-1}	kg $\cdot$ ha^{-1} $\cdot$ yr^{-1}
I	179,700	1,000	0.04
II	8,000	300	0.05
III	5,700	300	0.12
IV	17,000	900	0.13

MARINE SULFUR

Some of the sulfur in the river systems is due to cyclic salts originating from the oceans. It is possible to compute the amount of cyclic sulfur from the content of sulfur of ocean water in proportion to that of sodium, chloride or any other conservative element. We have used sodium since very little sodium is released by chemical weathering or added by any other large source of error. Chloride is somewhat hazardous because the frequent use of $CaCl_2$ as a road salt during the winter period.

The data for the different regions are given in Table 5. Marine sulfur amounts to a total of 33,100 tons, which is about 5% of the total discharge of sulfur by the rivers. The discharge per unit area varies with the different regions and reflects the distance from the North Sea. The figure for region II seems to be too high (by approximately 0.5 kg $\cdot$ ha^{-1} $\cdot$ yr^{-1}). This is probably

Table 5. Computed figures of the contribution of marine sulfur based on the discharge of Na within the different regions.

Region	Na tons $\cdot$ yr^{-1}	Marine sulfur*)	
		tons $\cdot$ yr^{-1}	kg $\cdot$ ha^{-1} $\cdot$ yr^{-1}
I	153,900	12,800	0.44
II	90,300	7,500	1.13
III	22,800	1,900	0.72
IV	131,000	10,900	1.62

*)
Calculated from $S_{marine} = 0.084$ (Na-0.02 Si) where $0.084 = \dfrac{S_{marine}}{Na_{marine}}$, and 0.02 Si equals the contribution of Na in the river systems due to weathering. The coefficient 0.02 is based on experimental studies on different Swedish tills.

due to the land rise, which leads to successive exposure and drainage of fossil salt water from the glacial period. Some sulfur may
also originate from acid sulfate soils.

    Since atmospheric fallout of sodium and chloride has been con-
stant during the last 20 years there is no reason to believe, that
the amount of marine sulfur is in a transient stage. Consequently,
the contribution of marine sulfur is assumed to be constant during
this century.

SULFUR IN FERTILIZERS

    Sulfur is added to cultivated areas by means of different fer-
tilizers such as ammonium sulfate (24 % S), potassium sulfate
(18 % S) and superphosphate (14 % S). From available statistics it
is possible to compute the input of fertilizer-S to the cultivated
areas in Sweden during the last 70 years. As can be seen from Figure 3, the figure is about 7 kg S/ha and year up to 1940 and increases to over 20 kg S/ha and year during the 1950's. This is
mainly due to the increased use of superphosphate and ammonium sulfate. Up to the present time there is a gradual decrease of the input of sulfur amounting to 0.5 kg per ha and year. Since the sulfate ion is not firmly bound to the soil colloids, this negative
trend will also influence the water chemistry especially from more
intensively cultivated drainage areas.

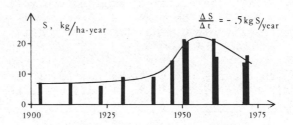

Figure 3. The consumption of fertilizer sulfur in Sweden during
this century. The figures are related to the agricultural area.
Corresponding figures for the different regions in 1970-72 are
given in Table 6.

Table 6. The input and the rate of change of the input of fertilizer sulfur of the four drainage regions of Sweden.

	Fertilizer-S	Rate of change
	$kg \cdot ha^{-1} \cdot yr^{-1}$	$kg \cdot ha^{-1} \cdot yr^{-1}$
I	0.3	-0.01
II	2.3	-0.08
III	4.0	-0.13
IV	1.3	-0.04

In 1970-72 the total use of fertilizer-S amounted to 44,300 tons/year. The distribution between the different regions is given in Table 6 along with computed figures for the rate of change. The figures are fairly low when plotted on bases of the drainage area of the different regions, and small variations from one year to another do not materially interfere with the budget figures for Sweden.

SULFUR IN THE ORGANIC MATTER

The analytical procedure behind the determinations of the total discharge of sulfur by the rivers does not include sulfur in organic matter (dissolved humus). The sulfur content in different humus fractions varies between 0.4 and 0.7 %. The mean value for the soluble fractions is about 0.5 %. The total discharge of organic matter has been reported to be 1,440,000 tons per year (Ahl and Odén[6]). This figure amounts to about 2,000,000 tons when the total drainage area of Sweden is taken into account. The discharge of S in the organic matter can thus be computed to about 10,000 tons per year. Most of it is released from region I. There are no transient conditions of this form of sulfur, and if any, it is likely to be negative, indicating that the amount of dissolved humus is reduced with time. This is due to the flocculating effects of both the increased acidity of the river waters and the increased concentration of mainly divalent cations.

THE TOTAL ATMOSPHERIC FALLOUT OF SULFUR

By means of the preceding data it is possible to compute the total atmospheric fallout of sulfur. The underlying figures are given in Table 7. The total discharge of sulfur by the rivers amounts to 667,900 tons acid is totally corrected by 79,900 tons of sulfur from known sources. Consequently, the net atmospheric deposition in

Table 7. The total discharge of sulfur from Sweden in 1974 with
distribution on different sources. The figures are given in tons
S per ha and year.

River discharge of SO₄-S	657,900
River discharge of organic sulfur	10,000
Total discharge of sulfur	667,900
Contribution from weathering, Table 4	- 2,500
Contribution from marine salts, Table 5	- 33,100
Contribution from fertilizers, Table 6	- 44,300
Atmospheric deposition in excess of marine salts	588,000

excess of marine salts amounts to 588,000 tons. This figure is much
higher than the fallout computed from atmospheric chemical data
(cf. Table 1). The actual difference, however, is still larger considering the effect of looping. Furthermore, some of the increased
addition of sulfate ions may be trapped in the B-horizon of the podsols and, consequently,may not appear in the river waters. We believe that this difference is real and may arise from any one of
the following causes:

1. The atmospheric chemical network does not account for elemental sulfur.

2. Sulfur in soil organic matter is in a transient stage due
   to increased mineralisation.

3. Sulfur in marine sediments and in acid sulfate soils are
   continuously being released.

Until further studies have been made, it is not possible to evaluate the magnitude of the different sources of the unaccounted
amount of sulfur.

     The net atmospheric deposition in excess of marine salts is
made up of two parts: one originating from anthropogenic emissions,
the other due to the non-anthropogenic background (sulfur from vol-
canoes, global dust, biogenic sulfur etc.). It is partly possible
to evaluate the two parts by means of old Swedish data, and furthermore, to describe the progression of the fallout of sulfur due to
anthropogenic emissions during this century. The situation is as
follows:

     The rate of change of the discharge of sulfur from the differ-
ent regions in Sweden can be reconstructed by use of the following
facts: the mean discharge of sulfur during the period 1965-1974,
the rate of change of the discharge of sulfur during the same period, and the discharge of sulfur for various years during the pe-

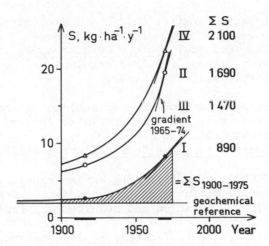

Figure 4. The probable development of the discharge of anthropogenic sulfur for region I, II and IV during this century. The gradients for the period 1965-1974 are denoted as well as the probable geochemical reference for region I. Σ S denotes the accumulated discharge of sulfur per ha.

riod 1909-1923. The latter data have been given by J.V. Eriksson[9], and have been recompiled by us with respect to the four regions.

    The results are given in Figure 4. The curved lines show the probable change of the discharge of sulfur from 1900 to the present time satisfying the three conditions mentioned above. From the curves it is clear, that the increase of the discharge of sulfur started before 1900 and that the discharge has increased 3 to 4 times in 75 years. A similar increase (2 to 3 times) has also been found in the Greenland ice cap during the corresponding period. The total, excess discharge of sulfur for the four regions amounts to the figures given in Figure 4. For Sweden as a whole the figure amounts to about 60 000 000 tons for the period 1900-1975.

    The non-anthropogenic background is only indicated in Figure 4. The figures vary from about 2 kg sulfur per ha and year (region I) to 5-8 kg for the other regions. This means that the fallout of anthropogenic sulfur amounts to 6 to 18 kg of S per ha per year in 1974. Going into the details within the different regions these figures vary still more. In the more exposed parts of Sweden the anthropogenic contribution is likely to be around 25 kg S per ha and year.

REFERENCES

1. S. Odén, The acidification of air and precipitation and its
       consequences on the natural environment. Swedish Nat.
       Sci. Res. Council, Ecology Committee, Bul. 1:68 (In
       Swedish.) Also: Tr-1172, Translation Consultants Ltd.,
       Arlington, Va., U.S. (1968).

2. S. Odén, The acidity problem – an outline of concepts.
       First Int. Symp. on Acid Precipitation and the Forest
       Ecosystem, Proc., Columbus, Ohio, U.S., p.1 (1975).
       Also: Water, Air, and Soil Pollution 6:137 (1976).

3. OECD, The OECD Programme on Long Range Transport of Air
       Pollutants: Measurements and Findings. Organization for
       Economic Co-operation and Development, Paris (1977).

4. B. Bolin, L. Granat, L. Ingstam, M. Johannesson, E. Mattsson,
       S. Odén, H. Rodhe, and C. O. Tamm, Air pollution across
       national boundaries. The impact on the environment of
       sulfur in air and precipitation. Sweden's case study for
       the United Nations conference on the human environment.
       P. A. Norstedt & Söner, Stockholm (1971).

5. L. Granat, (personal communication).

6. T. Ahl and S. Odén, River discharges of total nitrogen, total
       phosphorus and organic matter into the Baltic Sea from
       Sweden, Ambio Special Report 1:51 (1972).

7. T. Ahl, in: "Vänern en naturresurs", Liber förlag, Stockholm,
       (Swedish with English Summaries), p.95 (1978).

8. J. V. Eriksson, Den kemiska denudationen i Sverige. Medd. från
       statens meteorologisk-hydrografiska anstalt. Bd 5:3,
       pp.96 (Swedish with French Summary), (1929).

SOME SPECIAL FEATURES OF THE ECOPHYSIOLOGICAL EFFECTS OF AIR

POLLUTION ON CONIFEROUS FORESTS DURING THE WINTER

Paavo Havas and Satu Huttunen

Department of Botany, University of Oulu

SF-90101 Oulu 10, Finland

The ecophysiological effects of air-borne pollutants have mostly been investigated in the region of temperate deciduous forests. We now present some observations from the area of northern coniferous forests. It seems to us that this area is characterized by some special features of the effects of air pollution on coniferous trees. We deal with this question mainly at the eco-logical level (i.e. we have been observing areas influenced by several polluting agents simultaneously, not by only acidic rain, for example). This is the normal case, and it is exceedingly difficult - often possibly even futile - to endeavour to estimate the effect of each polluting component separately; it is more important to strive to estimate the combined effects of pollutants in different climatic conditions.

When considering the effects of air pollution the singular character of growing conditions in the north must be emphasized. Our research area is located in the northern part of the northern coniferous forest zone (about 65°N, in Finland), where the natural ecological stress, especially in wintertime, is great. The coniferous forest tree flora comprises only a small number of species of which Norway spruce (Picea abies) and Scots pine (Pinus silvestris) are the most common, both of which are known to be relatively sensitive to air pollution. In these northern forests the number of trees per unit area of land is small and their growth is extremely slow. Over wide areas of northern Finland the amount of stem wood is only c. 50 m^3/ha and early growth c. 1-1.5 m^3/ha. There seems little doubt that these trees and those of forest boundary zones, in general, are under a fairly strong natural ecological stress. The following features, for example, have often been emphasized: the critical influence of thermal relationships

123

(temperature, sun etc.) on tree growth and on seed production of
the coniferous trees, the great significance of winter conditions
to the tree's water balance and the effect of the long days on
photosynthesis. Attention has also been paid to problems of
nutrient availability (for example the slow cycling of nitrogen)
in these northern forests. Light, temperature, moisture and
nutrient conditions, for example, are in any case distinctive for
different areas and, in terms of their ecological effect, form
widely differing combinations. Relatively little attention has
been paid to the combined effect of these ecological variables and
to the measurement of natural ecological stress -- and indeed it is
difficult to delimit and measure them with any exactitude. Even so
it would be interesting to obtain an overall indication of the
degree of ecological stress to which, for example, the Finnish
coniferous forests are subjected as compared with the natural stress
in other coniferous forest areas with different climates. The
availability of reliable comparative meteorological data, and of
information about different tree species and the ecologically
limiting factors of their different proveniences now makes this
perfectly feasible. This type of widely-based information would be
of particular use in, for example, the mapping of threshold values
for the ecological effects of air pollution. There is probably no
simpler way of ascertaining the threshold values of polluting
effects, for the threshold values vary greatly, depending on the
plant species and the environmental conditions.

Speaking in general terms, I would claim that for the Finnish
coniferous forests the natural ecological stress is quite great but
that the effects of air pollution are, at the moment, relatively
small. Finnish air space receives c. 550,000 tonnes of sulphur per
year[1]. Finland receives more sulphur from outside its boundaries
than it disperses to other countries[2]. This long distance transported material has increased the acidity of rain water right up to
northern Finland although these changes are still very small. So
far it has not been possible in Finland to determine whether the
present degree of air pollution has any overall influence on soil
and/or tree growth, although in the immediate vicinity of certain
factories there are air pollution problems. In these areas it is
immediately obvious that even relatively low levels of pollution
can, under the prevailing climatic conditions, bring about forest
destruction. According to the latest estimates for Finland there
are, in the industrial areas, at least a couple of million cubic
metres of damaged trees[3]. It has also been shown that air pollution
has a detrimental effect on the lower vegetation of the forests,
particularly on certain dwarf shrubs[4,5]. Numerous classical
investigations show how in large urban areas the epiphytic lichens
normally to be expected on trees cannot exist.

In certain polluted areas a slowing down of tree growth has also
been demonstrated. In the industrial areas of the town of Oulu

(c. 65°N) a fairly definite slowing down in the radial increase of
tree trunks has been recorded[6], particularly in middle-aged trees
(less than 100 years). Frequently a thinning of the yearly growth
rings is found for up to 10 years before the tree dies. This type
of slowing down in growth in the vicinity of factories is due
partly to a gradual decrease in the number of needles (the needles
die because of the pollution) and partly to the diminished photosynthetic efficiency of the living needles. This retardation of
growth is not, however, readily demonstrable further away from the
industrial areas, neither is data of this effect from within
industrial areas very prolific.

One particularly significant feature has been noticed, at least
in certain northern industrial areas, namely that the visible
damage to the trees appears in the spring as the air warms up.   On
investigating the matter more closely we were able to establish
that the damage threshold has been exceeded during the winter[7,8].
If branches are brought into the warm during the winter then
obvious injuries appear on them in exactly the same way as happens
naturally later in the spring. Previously it was generally thought
that summer was the most likely time for damage from pollution to
take effect. This is generally the case in the more temperate areas,
because the metabolic activity of plants is greater in summer than
in winter, and the winter is not equally stressful for plants there
as it is in a colder climate.

How can critical conditions develop in winter and what causes
the winter injuries? We received the first indications regarding
factors we should pay particular attention to when we compared
needles of small trees which had been under the snow all winter with
those (in the same trees) which had had no snow cover to protect
them. It was quite clear that needle damage occurred only in needles
which had remained above the snow. The snow cover then acts as a
very efficient protection against the detrimental effects of air
pollution. (We do not wish to claim, however, that a snow cover
could not have a negative effect on certain plants: after all
throughout the winter the snow accumulates pollutants which, when
the snow melts in the spring, all suddenly come into contact with
the plants). But a snow cover has other important ecological
effects as well. It provides good protection against the cold and
against drought. On the other hand it is continuously dark beneath
the snow. In effect, conditions within the snow are 'oceanic' in a
region of continental climate. It is then that a snow cover is a
significant protection against pollution because within the snow
the needles do not come into contact with the pollutants or because
within the snow the natural ecological stress is lower -- or
perhaps both of these factors are operative.

In the industrial areas where the level of air pollution is
continually changing from year to year it is incredibly difficult

to determine the kind of winter most detrimental to the trees. It
does seem, however, that in Oulu the most widespread damage occurs
following those long and cold winters, in which there is hardly any
thawing periods. In this connection it should be pointed out,
however, that the most significant pollutants in this particular
factory area are fluorides, a certain amount of sulphur and nitrogen
oxides, the quantities of which in the air are fairly high. In the
same winter as that in which the trees in Oulu, were severely
injured similar types of injury were recorded in various other
countries from the vicinity of factories emitting fluoride to the
air. Experiments with pine trees (c. 2-4 m high, felled) moved
from the unpolluted area into the industrial area showed that within
as short a period as two months, especially if the winter was damp,
fluoride had accumulated in the needles in quantities in excess of
100 ppm dry weight[7]. Such a quantity of fluoride undoubtedly damages
the needles. Birch buds are also damaged by this level of fluoride
to the extent that leaves do not form in the following summer[9].
Sulphur may effect wintertime injuries on coniferous trees similar
to those caused by fluorides. The climatic conditions prevailing
during the winter and the quantity of pollutants accumulating in
the needles over the winter are the most crucial factors. Wintertime damage due to sulphur have also been reported by others; the
threshold of damage may be lower in winter than in summer [10, 11].

     Just over a year ago, a widescale research project was
initiated in Finland which aims to classify the ecophysiological
effects of air pollution on coniferous forests. This project will
pay particular attention to mechanisms causing damage during the
winter, with the major emphasis on the effect of sulphur. Various
other research projects have been started parallel to this. Their
aims are to provide comparable background information on natural
ecological stress during the winter with particular reference to
the tree's water balance. Since the research project is still in
its infancy and because much time has had to be invested in
developing suitable research methods the results are naturally at
this stage still to be interpreted with caution. We will therefore confine rather to using them as a basis for indicating which
factors appear to warrant closer scrutiny.

     Firstly the following features of the tree's winter water
balance should be noted. A long and cold winter can cause severe
water deficiency in coniferous needles, particularly in young
needles which have not formed sufficiently strong cuticle during
the previous summer, since the amount of water decreases as the winter
progresses. Trees growing in isolation with no protection from the
wind and which in spring are subjected to direct, strong sunshine
are at greatest danger[12]. Moreover the water reserves in the trunks
of small-sized trees are small in comparison with the amount of
water which is transpired from the needles. Throughout the winter
water is transferred from the trunk to the needles, given that

during those periods when the trunk is protected it manages to thaw.
Very small trees which have had a protective snow cover, and the
vegetation of the forest floor which remains under the snow, do not
usually experience any water balance problems in winter, nor even
in spring.

The situation outlined above is, broadly speaking, the normal
situation. It seems to us that in polluted areas the trees are even
more susceptible to water balance problems [7,8,13,14]. In these
areas the critical situation is usually reached in the spring, as
it is normal in northern areas. Our preliminary results show changes
in the water potential of pine needles during the spring in both
polluted and unpolluted areas near Oulu. In northern Finland the
most critical period usually is in March. The greater amount of
sulphur in the needles, the greater the water deficiency in the
needles in the spring seems to be. There is little sulphur in the
healthy trees and their water balance is relatively good while
needles containing large amounts of sulphur show a greater variation
in water balance and, frequently, quite a critical water situation.

Generally, the sulphur level in the Oulu area is not very high:
the amount of sulphur present in the air in the Oulu area (annual
mean value) is 40-60 $\mu g/m^3$ and the amount present in pine needles
is 0.11-0.26 % of dry weight (the normal background content of
sulphur in pine needles in northern Finland is 0.07--0.09 % of dry
weight). It seems, however, that fairly large amounts of sulphur
accumulate in the needles over the winter in polluted areas (e.g.
Oulu). According to the X-ray microanalyzer results [15, 16], sulphur
(similar to some other pollutants) first accumulates around the
transfusion tissues in winter. At the same time it also begins to
increase in the mesophyll of the needle. The amount accumulated
varies depending on the severity of the winter. If the amount
accumulated in a needle is great, an injury to the needle develops
in spring [17].

We have observed that the needles of coniferous trees are able
to absorb water from the surface of the needle during wintertime
thaws [7,18]. Such absorption is particularly common in late winter,
when the water deficiency in the needles is markedly great.
According to the preliminary results, certain pollutants are transported into the needles along with the water absorbed from the
surface [17]. This possibility is also indirectly suggested by the
results of the aforesaid microanalyses and the rapid accumulation
of fluoride in the needles of our test trees in winter.

Exact transpiration measurements may reveal whether needles in
polluted areas lose more water during the winter than healthy
needles. We are developing in Oulu, at the moment, a means of
measuring transpiration as accurately as possible and have made a
large number of measurements during this past winter. Unfortunately,

we have not been able to analyze the results in time for this
meeting, but we can say that the differences in transpiration, which
it is possible to demonstrate, are small. However, even small
differences would be sufficient to explain the differences which
have been demonstrated in water balance when it is the whole winter
which is in question.

We have also used electromicrographs of the surface of the
needles to investigate whether pollution possibly leads to corrosion
of the cuticle or of the protective mechanism of the stomata. This
work, too, is still in progress but it is evident that corrosion
does take place at least in the surroundings of the stomata[17].
However, it is not yet clear what the ultimate reason for these
changes in water balance observed in the needles of polluted areas
is. Nor do we know what effect these changes have in practice.
Are they just one of the many consequential features which appear
at the cell level in injured needles?

Our research project aims to investigate, on a broad comparative
basis, the mechanisms by which air pollution takes effect in cold
climates. Investigations are being carried out in a number of
localities from southern to northern Finland and with respect to a
number of plant groups, so that certain common mosses and dwarf
shrubs of the coniferous forest area are being considered in addition
to the coniferous trees themselves. The biochemical research
involves both biochemical and ecophysiological measurements of
enzymes and other metabolites under field conditions. For this,
measurements of the buffering capacity and the conductivity of the
injured cells have proved suitable for investigating changes during
the winter.

It is not possible here to describe in detail the biochemical
and electron-microscopic investigations now being carried out to
elucidate the polluting effects. Nevertheless, in order that certain
aspects of the project should be presented in this paper a little
more concretely than outlined above, we would like to give two
examples of these results at the ecological level specifically
supplied for this occasion.

The first example deals with the photosynthesis and respiration
of pine needles in polluted areas compared with an unpolluted area.
It seems apparent [17] that the ostensibly healthy needles of the
polluted area display certain special features of both photosynthesis
and respiration in early summer already. In the most badly polluted
area the needles are obviously under severe photosynthetic stress
for several weeks in the spring. In the less polluted areas the
rate of photosynthesis is somewhat variable when compared with that
of the control: at first the values are relatively high but later
in the spring these too show a certain amount of depression. The
relative respiration rate also shows a divergence from the norm as

spring proceeds. To begin with, the level of respiration is below
that of the control, but later in the early summer, it clearly
exceeds it. There are other examples of this type of shock effect
caused by pollution[7,19,20,21].

Our second example has been taken from the electron-microscopic
results of the project. Most of these analyses have been made on
spruce (Picea abies) needles. Briefly, the following differences
were noted between the needles of polluted and unpolluted areas[22, 23]:
1) Injury symptoms can be easily demonstrated in needles which have
been collected from the polluted areas but which outwardly appear to
be quite healthy. The symptoms are observable in the ultrastructure
of the mesophyll cells.
2) Subcellular injuries are already discernible in winter, that is
before the visible injuries appear in the spring.
3) Injuries observable in the fine structure vary to some extent in
quality depending upon which pollutant and which injury is being
considered. A number of different injuries have been photographed
of which some are visible in the outer membrane of the tylakoids of
chloroplasts and some in other parts of the cell.

All the features mentioned above, changes in the biochemistry of
the cell and in its fine structure, changes in the basic functions of
the plant, for example in photosynthesis, respiration and water
balance etc. are in one way or another interrelated. It is therefore
impossible at this stage of the project even to try a synthesis of
the effects of pollutants in the northern regions. We would expect,
however, that the research in progress at the moment which includes
such varied investigations at so many different levels should, before
too long, be able to give us a composite picture.

ACKNOWLEDGEMENTS

The following people participate in the scientific supervision
of the projects enumerated below: 1) The Academy of Finland finances
the project "The effects of air pollution on the terrestrial ecosystems especially northern coniferous forests". This project is
carried out during the years 1977-1980 under the supervision of
Dr. Satu Huttunen (Department of Botany, University of Oulu) and
Prof. Lauri Kärenlampi (Department of Environmental Hygiene,
University of Kuopio). 2) The project "The environmental effects
of air pollution under northern conditions" is going on for the years
1977-1978 and has been supported by a grant from Nessling's Foundation
to Dr. Satu Huttunen and her team: Dc. Satu Huttunen (air pollution
effects), Prof. Paavo Havas (water balance and wintering problems)
and Kari Laine, M.Sc. (development of methods).

REFERENCES

1. Ympäristötilasto – Environmental Statistics 1974. Centre Statistical Office of Finland.

2. The OECD Programme on long range transport of air pollutants. Measurements and Findings. Paris (1977).

3. S. Huttunen, Suomen Luonto 4–5: 264 (1977).

4. S. Huttunen, Acta Univ. Ouluensis, Ser. A, Scient. Rerum Naturalium 33, Biologica 2: (in: Part II) (1975).

5. K. Laaksovirta and J. Silvola, Ann. Bot. Fenn. 12: 81 (1975).

6. P. Havas and S. Huttunen, Biol. Conserv. 4(5): 361 (1972).

7. P. Havas, Acta Forest. Fenn. 121: 1 (1971).

8. P. Havas, Proc. of Nordic Symposium on Biological Parameters for Measuring Global Pollution, IBP i Norden 9: 69 (1972).

9. S. Huttunen, Aquilo, Ser. Bot. 13: 23 (1974).

10. J. Materna, IUFRO, IX Int. Tagung über die Luftverunreinigung und Forstwirtschaft, Tagungsbericht, Marianske Lazne (1974).

11. J. Materna and R. Kohout, Naturwiss. 50: 407 (1963).

12. W. Tranquillini, Ecol. Studies 19: 473 (1976).

13. S. Bortitz, and M. Vogl, Arch. F. Forstwesen 16: 663 (1967).

14. Th. Keller. Ber. Eidg. Anst. F. forstl. Versuchswesen Birmensdorf Nr67 (1971).

15. S. Huttunen, Aquilo, Ser. Bot. 12: 1 (1973).

16. S. Huttunen, Acta Univ. Ouluensis, Ser. A, Scient. Rerum Naturalium 33, Biologica 2: 1 (in:Part III) (1975).

17. S. Huttunen, personal communication (1978).

18. P. Havas, Rep. Kevo Subarctic Res. Stat. 8: 41 (1971).

19. S. Börtitz, Biol. Zbl. 87:489 (1968).

20. J.E. Hallgren and K. Gezelius, Doktorshandling i Umeå universitet, 33 pp. (in Part V) (1978).

21.   Th. Keller, Mitteilungen 53(4): 161 (1977).

22.   S. Soikkeli, Can. J. Bot. (in press) (1978).

23.   S. Soikkeli, and T. Tuovinen, Can. J. Bot. (in press) (1978).

REGIONAL AND LOCAL EFFECTS OF AIR POLLUTION, MAINLY SULPHUR

DIOXIDE, ON LICHENS AND BRYOPHYTES IN DENMARK

Ib Johnsen

Institute of Plant Ecology
University of Copenhagen
Øster Farimagsgade 2 D, DK-1353
Copenhagen K, Denmark

INTRODUCTION

Also in Denmark has the impact of SO_2 on the vegetation of lichens and bryophytes been of significance. Furthermore, the observation of the change in properties of these plants, especially epiphytic lichens, has caught increasing interest among the authorities as an environmental surveillance tool. In this paper a short description of regional and local effects observed in lichens and bryophytes of Denmark is given and the implications of these observations as well as suggestions for further research in this field.

Recent investigations (Søchting and Ramkaer,[1]) of road side trees throughout Denmark has revealed a distribution pattern of particularly sensitive epiphytic lichens reflecting a regional effect. The main species used were Physconia pulverulenta and Anaptychia ciliaris, and the degree of fertility was used as an additional parameter. The investigation comprised registration of epiphytic lichens on free-standing road side trees. In a phytosociological sense, the vegetation belonged to the association Physcietum ascendentis with a typical number of 20-25 species in background areas. Anaptychia ciliaris and Physconia pulverulenta is usually present and under optimum conditions richly fructifying. These two species, however, are extremely sensitive to air pollution and react already at low levels by reduction of fertility. The fertility is suppressed primarily in Anaptychia ciliaris, followed by Physconia pulverulenta. At higher levels, morphological changes may be observed, and, finally the species are eliminated. Typically, the two species are present at rural phorophytes

throughout Denmark, but their condition with respect to fertility
and morphological appearance is subject to strong variations.
Thus, fertile Anaptychia ciliaris is only rarely found in Denmark
today, even if apothecia were described as frequently occuring at
the beginning of this century in northern Germany. The
distribution of Anaptychia ciliaris in Denmark is shown in Figure
1. The different signatures indicate the optimum condition of the
plants registrated within each square. A similar pattern was
observed in Physconia pulverulenta (Figure 2), and it was
concluded, that the observations would indicate a regional effect
with respect to these particularly sensitive plants probably
reflecting a long range transport of acidifying atmospheric
substances, primarily sulphur dioxide from S-SE of Denmark.
Obviously, when approaching the Sound region, the regional pattern
is disturbed by the rather large sulphur dioxide emission in this
area.

More scanty information suggests a negative change in the
frequency of occurrence and healthiness of vegetations of the
oceanic lichen Lobaria pulmonaria (Søchting, pers. comm.) and
bryophyte Antitrichia curtipendula (Holmen and Mogensen, pers.
comm.) during the last decades. It is, however, difficult
to assign this trend exclusively to increasing air pollution levels.
A factor, that may induce substantial epiphytic vegetation changes,
is the forestry practice leading to elimination of older,
deciduous trees and severe changes in microclimatic conditions of
the phorophytes, e.g. by clear-cutting of neighbouring tree
stands. On the other hand, no long term changes in macroclimate
seem to offer a reasonable explanation of the above mentioned
observations.

Furthermore, the arguments presented by Barkman[2],
pointing at regional air pollution increase as a main factor in
explaining similar observations in the Netherlands are valid for
Denmark, too.

The vegetation of epiphytic lichens on tree twigs at the
canopy surfaces of Bornholm, SE Denmark, has been subject to
impoverishment during the last decades (Degelius, pers. comm.).
Species like Xanthoria lobulata is believed to have become rare
in Denmark on such twigs, and the very exposed habitat favoured
by these species infers, that increasing air pollution levels
may be a main factor adversely affecting the plants here.

The NE-SW air pollution gradient observed as well in sulphur
dioxide studies as in heavy metal studies over the Scandinavian
peninsula seems to change to a N/NW-S/SE direction over Denmark,
probably as a result of sulphur dioxide emission of largely
unknown magnitude due to oil and brown coal consumption in
Eastern Europe, particularly Poland and DDR. This N/NW-S/SE

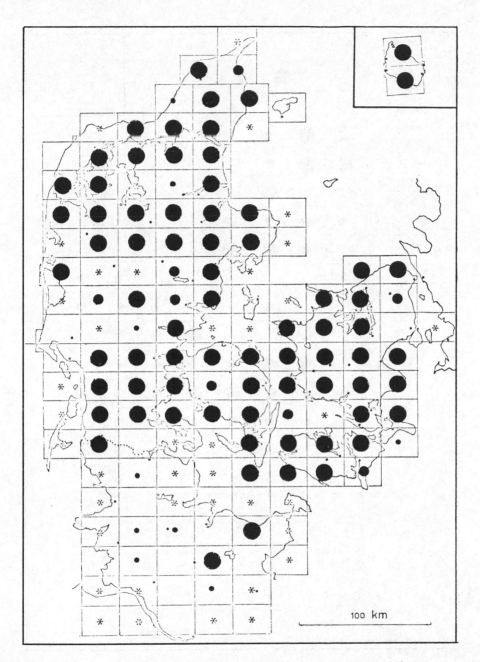

Figure 1. Map showing the optimum condition of Physconia pulverulenta registrations within each square: large circle: healthy, fertile; medium circle: healthy, sterile; small circle: unhealthy, sterile; no symbol: not observed; asterisk:square not covered.

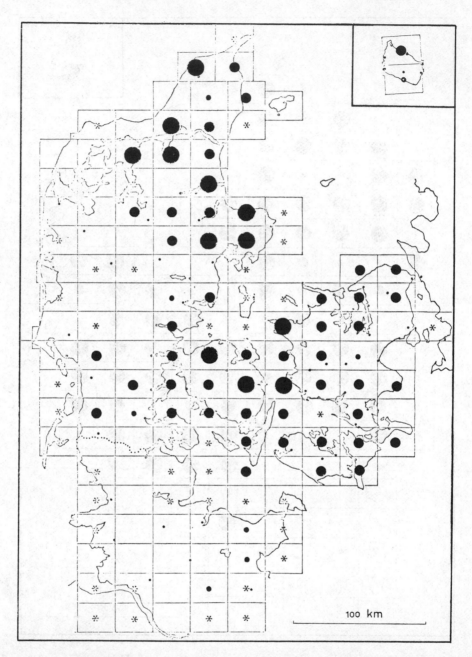

Figure 2. Map showing the optimum condition of Anaptychia ciliaris
registrations within each square: large circle: healthy, fertile;
medium circle: healthy, sterile; small circle: unhealthy, sterile; no
symbol: not observed: asterisk: square not covered.

direction of the pollution gradient has newly been supported by
heavy metal analyses of bulk precipitation collected in open
plastic funnels at 17 sites throughout Denmark in rural areas
(Hovmand,[3]) as well as by similar analyses of epiphytic
lichens and bryophytes (Lecanora conizaeoides, Hypogymnia physodes
and Hypnum cupressiforme) sampled at background sites throughout
Denmark.

    With respect to epigeic and epilithic lichens and bryophytes,
no adverse regional effects have been reported from Denmark.

    The general pattern of lichen and bryophyte elimination with
proximity to city centres have been described in a number of urban
areas in Denmark, the basic work being done in Copenhagen
(Johnsen and Søchting,[4]). The results have led to a selection
of indicator species useful in estimation of average sulphur
dioxide levels, analogous to the scale given by Hawksworth and
Rose,[5]. An important question to answer with respect to lichen
and bryophyte reaction to sulphur dioxide emissions is, whether the
well documented negative correlation may be explained by the toxic
effect of sulphur dioxide itself, the direct effect on the plants
themselves and/or the indirect effect via the substratum of the
simultaneously increasing hydrogen ion concentration. The only
point to be made here is, that a major distinction between regional
and local sulphur dioxide emission is due to the degree of
neutralisation of strong acid in areas with high levels of airborne
alkaline dust particles. An example of a strongly dust influenced
area is the city of Aalborg-Nørresundby with a rather large cement
production. Even under these circumstances the inner distribution
limits of the above mentioned epiphytic lichen indicator species
were in accordance with the limits obtained in Copenhagen with low
dust levels (Johnsen and Søchting,[6]). This implies for
these species a direct effect of sulphur dioxide air pollution
independent of the neutralisation of sulphur dioxide on alkaline
particles and the indirect effects on the bark such as relatively
high pH and buffer capacity.

    Transplant experiments according to the method introduced in
Germany by Schonbeck,[7] with Hypogymnia physodes have been
carried out in a number of sites in Denmark, most recently in the
city of Hirtshals, Northern Jutland (Ramkaer et al.,[8]) with
a heavy fishing industry, in order to attempt to calibrate an
emission-based computation of the sulphur dioxide
Adverse effects on the transplanted Hypogymnia physodes specimens
have been observed in Hirtshals as well as in Copenhagen; this is
noteworthy, as Hypogymnia physodes is considered a rather sulphur
dioxide tolerant species inhabiting acidic, oligotrophic bark.
It was concluded from the Copenhagen experiment (Søchting and
Johnsen,[9]), that the sulphur dioxide air pollution gradient
for the period September 1974 to August 1975 was reflected well

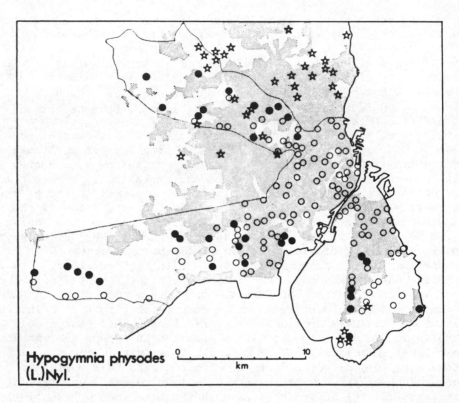

Figure 3. Distribution of Hypogymnia physodes in the Copenhagen area
Open circle: species absent 1971-2; full circle: species present
1971-2; asterisk: species present 1936-46. The line circumscribes th
1971-2 investigation area.

by this species, which was neither too sensitive nor too tolerant in relation to the sulphur dioxide levels of the investigation area.

In situ epiphytic Hypogymnia physodes was in an earlier study shown to have progressed towards the centre of Copenhagen from 1936-46 to 1971-72 (Søchting and Johnsen,[10]), (Figure 3) in contrast to the general pattern of retrogression. This observation was believed to be due to mainly the acidifying effect on the substrate of sulphur dioxide, making otherwise circumneutral bark more accesible for this species confined to more acid bark. The distribution increase in other species, e.g. Lecanora conizaeoides and Stereocaulon nanodes, has not been observed in Denmark. It is, however, most probable, that Lecanora conizaeoides has extended its distribution in Danish cities over the last half-century.

The main problems in interpretation of the regional patterns of epiphytic cryptogams are the following:

1. Possible effects of pesticides in agricultural land

2. Synergistic/antagonistic effects of sulphur dioxide in combination with other pollutants as nitrogen oxides, ozone, fluorides, heavy metals and particulate matter. No observed reaction in an indicator plant can be proven to be completely specific - which is a main weakness and strength in biological monitoring methods generally. Research to develop the highest degree of specificity is still strongly needed, and in relation to lichen and bryophyte vegetation of Denmark, more work enlightening the influence of pesticides, heavy metals and ozone on the regional distributions should be encouraged. As regards local effects, more knowledge is needed of the interference from nitrogen oxides, fluorides and heavy metals in the air on the lichen and bryophyte vegetation properties, mainly believed to be determined by the sulphur dioxide levels.

REFERENCES

1. U. Søchting and K. Ramkær, Report to the Danish EPA, Institute of Thallophytes, University of Copenhagen, (1978).

2. J.J. Barkman, in: "Proceedings of the first international congress on the influence of air pollution on plants and animals," Wageningen, (1966).

3. M.F. Hovmand, Ph.D. Thesis, Institute of Plant Ecology, University of Copenhagen, (1978).

4. I. Johnsen and U. Søchting, Oikos 24(3):344 (1973).

5.   D.L. Hawksworth and F. Rose, Nature, (London) 227:145 (1970).

6.   I. Johnsen and U. Søchting, Bryologist 79(1):86 (1976).

7.   H. Schonbeck, Staub 29:14 (1969).

8.   K. Ramkær, U. Søchting and I. Johnsen, Report to the Danish
      EPA, Institute of Thallophytes, University of Copenhagen,
      (1978).

9.   U. Søchting and I. Johnsen, Bull. Environ. Contam. Toxicol.
      19(1):1 (1978).

10.   U. Søchting and I. Johnsen, Bot. Tidsskr. 69:60 (1974).

SOME EFFECTS OF RAIN AND MIST ON PLANTS, WITH IMPLICATIONS FOR ACID PRECIPITATION

H. B. Tukey, Jr.

Dept. of Floriculture and Ornamental Horticulture

Cornell University, Ithaca, NY 14853

Rain and mist mean different things to different people. To the farmer, rain means water for growth of crops, affecting the movement of nutrients from the soil into the roots. To the plant pathologist, rain means a medium in which spores can germinate and grow, often with an increased incidence of disease.

But rain, mist and dew have other effects upon plants, with implications in the deposition of atmospheric contaminants and in acid precipitation. For as rain and mist percolate down through plant canopies, substances in plants can be washed out of the foliage, a phenomenon called leaching. Leaching has been reported in the literature for more than 250 years, but modern use of radioactive tracers has aided the investigation and evaluation of the process.

Substances leached from plants include a great diversity of materials (23). Inorganic nutrients leached include all the essential minerals and some other elements commonly found in plants, including both the macro- and micro-elements. In addition, large amounts of organic substances have been noted including free sugars, pectic substances, and sugar alcohols. All of the amino acids found in plants and many of the organic acids have been detected in leachates. Growth-regulating chemicals, such as the gibberellins (15), auxins (18) and abscisic acid (12) can be leached from plants as well as vitamins (38), alkaloids, and phenolic substances (19). The extensive list of substances, which can be leached from plants suggests that under proper conditions, many, if not all constituents, can be leached.

Of the inorganic nutrients leached from plants, K, Ca, Mg, and

Mn are usually leached in greatest quantities. Arens (1) calculated
that 62 kg of ash constituents, 39 kg of phosphoric acid
equivalents, and 5 kg of CaO were leached from each hectare of
sugar beets (Beta vulgaris L.) during 18-24 hr of rainfall. Losses
from apple foliage (Malus domestica Bork.) of 20-30 kg of K, 10.5
kg of Ca, and 9 kg of Na per hectare in one year were reported (4).
Fifty percent of the Ca and more than 80 percent of the K content
of apple leaves could be leached in 24 hrs. Tamm (29) reported
that 2-3 kg each per hectare of K, Na, and Ca were carried to the
ground by rain from spruce (Picea abies Karst.) and pine (Pinus
sylvestris L.) trees during 30 days in the autumn.

However, not all inorganic nutrients are leached with equal
facility. For example, in radioisotope studies (36), those
minerals which were readily leached from young leaves included
^{22}Na and ^{54}Mn, with losses of more than 25 percent of the nutrient
in the leaves during 24 hrs. Minerals which were moderately
leached (1 to 10 percent) included ^{45}Ca, ^{28}Mg, ^{42}K, and ^{90}Sr.
Minerals which were leached with difficulty (less than 1 percent)
were $^{55-59}$Fe, ^{65}Zn, ^{32}P, and ^{36}Cl.

Despite the large quantities of inorganic nutrients which can
be leached, organic substances, principally carbohydrates, account
for the major quantity of leached materials. Losses of carbohydrates
may be as great as 800 kg per hectare per year from apple trees
(Malus domestica Bork.) (4) and up to 6 percent of the dry weight
equivalent could be leached from young bean leaves (Phaseolus
vulgaris L.) during 24 hrs., mainly in the form of carbohydrates (32)

The list of plants from which substances have been leached
includes a wide variety of representative species of deciduous
and coniferous forest and shade trees, tree fruits, bush fruits
and grapes, tropical and subtropical fruits, plantation crops,
vegetable crops, grains, grasses, forage crops, greenhouse and
herbaceous ornamentals, tropicals and conservatory plants, and
woody ornamentals, now totaling 180 different plant species (35).
This does not include some 90 additional species studied in
different types of experiments (1). Thus, leaching is of widespread
occurrence in nature, and in fact, no plant has yet been studied
which cannot be leached to some degree at least. Of course, there
are differences in the ease and degree in which substances can be
leached from some plants as compared with others.

Although most of the experiments investigating leaching have
been concerned with foliage, other plant parts are also
susceptible to loss of substances. Thus, the stems and branches of
woody plants lose nutrients by leaching during both the growing and
the dormant season (4,21) and during propagation. Flowers and
flower parts can be leached; in the case of grains, leaching of
flowers and developing seeds caused severe economic losses (2).

There are many factors which influence the quantity and quality of substances leached from foliage, including factors associated directly with the plant as well as those associated with the environment (35). And not only are there differences among species in respect to leaching, but there are also differences among individual leaves of the same crop and even the same plant (26), depending on the physiological age of the leaf. For example, young, actively growing tissue is relatively immune to loss of mineral nutrients and carbohydrates, whereas more mature tissue approaching senescence is very susceptible to leaching. For example, losses of K from very young leaves during 24 hrs. may be less than 5 percent of the initial K content of the leaf (36), but as the leaves become mature, the loss of K may be 80 percent or greater (37).

Although young leaves appear delicate and fragile, they are less susceptible to leaching than are older leaves due in part to the wetting properties of the leaf (2). Leaves with a smooth, waxy surface, which are wetted with difficulty, are less susceptible to the leaching of aqueous solutions (32). On the other hand, leaves which are relatively large, flat-surfaced, pubescent, and wetted with ease, are more easily leached. Similarly, leaves of such tropicals as banana (Musa sp.), cacao (Theobroma cacao L.), and coffee (Coffea arabica L.), are easily wetted and easily leached despite the presence of a thick and continuous cuticle (31). By removing the waxy cuticle of cabbage (Brassica oleracea L.) artificially, loss by leaching has been increased (1). Use of surfactants (wetting agents) in the leaching solution also increases losses of metabolites by leaching (20).

Leaves from healthy and vigorous plants are much less susceptible to leaching than are leaves which are injured (10,34), whether injury be induced by microorganisms, insects and other pests, adverse climate, nutritional and physiological disorders, mechanical means or by pH (8,40).

The pH of the solution on leaves, including rain, can cause injury to plants if sufficiently low. For example, in studies of foliar uptake of phosphate compounds, chrysanthemum leaves were injured at pH values lower than 3, and penetration of the phosphate compounds was greatly enhanced when injury occurred (25). This correlates well with reports of injury induced by simulated acid rain (6,7,13,39). In addition, the nature of the solution other than pH applied to leaves may also have an effect. Young juniper foliage (Juniperus chinensis L.) was injured by dilute nutrient solutions with pH values well above 4 (24), probably due to soluble salts.

Susceptibility to injury is very specific, and depends upon many factors including the plant species, age of the foliage,

integrity of the cuticle, and environmental factors such as
temperature, light, and humidity, which may affect both
penetration through and also development and maintenance of the
cuticles. Thus, any study of the effects of acid precipitation
on plants would seem to require a study of effects upon the
cuticle, particularly the epicuticular waxes, as recent studies
have suggested (6,7).

Light has little effect on the loss of most minerals from
young bean leaves (Phaseolus vulgaris L.) with the exception of
^{32}P and ^{35}S, losses of which were three times as great in darkness
as in the light (32). Although loss of nutrients from mature
leaves is greater in the light than in darkness (1), leaching of
carbohydrates from young bean leaves directly parallels the light
intensity received by the plants, being greater during periods of
high light intensity and lower during periods of darkness (32).

Temperature also influences loss of metabolites by leaching,
especially organic compounds (32). The loss of mineral nutrients
is generally not greatly influenced by temperature (36) except in
some mature leaves (1,27) and when tissues are injured by
temperature extremes.

Rain water that falls on foliage contains varying amounts of
dissolved salts and thus may have an appreciable effect on the
quantity of nutrients lost by leaching. In general, K salts
increase the losses as compared to distilled water, whereas Na
salts increase losses only slightly, and Ca salts tend to inhibit
losses (32).

Leaves need only be wetted to be leached; increasing the
volume of the leaching solution had very little effect on the
leaching of cations. Dew is very important as a leaching agent
(1), especially in seasons and climates where rainfall is low, and
fog which accumulates on leaves and then drips to the soil beneath
may leach substances from plants in a very efficient manner (5).

Leaching losses of cations increase as a function of time.
After several days of continuous leaching, losses reached three
to eight times the amount of nutrients initially in the leaves
(16,26,36), indicating that leached nutrients are replaced by root
absorption and translocation from other plant parts. Thus, plants
with adequate nutrient supplies are capable of withstanding
losses of metabolites by brief infrequent rains. However, during
prolonged periods of rain, as during the wet season in the tropics,
the growth and yield of the plant may be severely limited by the
inability to root-absorb and replace nutrients in sufficient
quantity to overcome leaching losses.

Cations are leached from exchangeable cation pools involving

so-called "free space" areas within the plant. Little if any
cations are leached directly from within the cell or from residue
components such as cell walls (22). Leaching of cations involves
exchange reactions on the leaf surface in which cations on
exchange sites of the cuticle are exchanged by hydrogen from
leaching solutions, and/or cations move directly from the
translocation stream within the leaf into the leaching solution by
diffusion and mass flow through areas devoid of cuticle (22). The
leached cations combine with dissolved CO_2 in the leaching solution
to form alkaline carbonates which may accumulate to such a degree
as to produce whitish encrustations on leaves, as has been reported
on some greenhouse plants.

This explains the characteristic alkaline nature of plant
leachates (1), including the common report that bicarbonate is
the primary anion in leachates (9), and that anions are not leached
in amounts equivalent to cations. It explains why the volume of
the leaching solution need only be sufficient to wet the leaf
surface, and additional volume would increase an exchange only
slightly, perhaps not enough to be detected. This emphasizes the
effectiveness of dew, fog, and light rain of long duration in
leaching substances as compared with heavy rains of short
duration (16,26,27,32). The fact that changes in temperature, light,
and energy levels in the plant have little influence upon the
leaching of cations would be expected, as the physical processes
of cation exchange and diffusion would be affected by such
factors to such a small degree as to make detection difficult.

As has been pointed out, mature leaves and tissues are more
susceptible to leaching of organic and inorganic substances than
are young leaves. In explanation, young, vigorously growing
tissues accumulate cations from the exchangeable pool, thus reducing
the amount of exchangeable cations available for leaching. In
addition, cations accumulate within cells and cell walls from
where they are not readily leached. In mature tissues, there is a
reduced rate of cellular accumulation leaving more exchangeable
cations for foliar leaching. Similarly, organic metabolites and
inorganic anions are not leached from within cells but rather from
materials which are being exported to an area of utilization.
Substances which are accumulated in excess of requirements
(exchangeable nutrients) are leached quickly (1,26,32).

Evidently, stomata are not the primary pathway of nutrient
loss. However, materials may be deposited upon leaf surfaces
from within by active processes such as guttation from hydathodes
(3) and trichomes and secretion from nectaries (11), later to be
washed away by rain and dew.

Organic and inorganic substances leached from above-ground
parts can be reabsorbed directly by the roots or foliage of

either the same plant from which they are leached or adjacent
plants of the same or different species (1,4,33,36). Such cycling
and reutilization of substances is apparently a widespread natural
phenomenon, and is of particular importance in natural communities,
including provision of nutrients for plant growth, as is the case
of mosses beneath forest trees (28) and grass beneath shade trees.

The tropical rainforest with its population of understory
species and epiphytes is an ecosystem adapted to leaching.
Bromeliads in the rainforest are oriented to intercept leachates
from the overhead canopy by foliar absorption, whereas roots of
this group of plants are much less important in nutrient uptake
and seem concerned primarily with anchorage (31). In addition,
epiphytes which commonly grow on the foliage of many tropical
plants may play a role in nutrition of the host by absorbing or
fixing nutrients leached from overhead (31).

Inhibitors leached from one plant can completely or partially
suppress the germination and growth of other plants nearby. Known
as allelopathy, this phenomenon is particularly well documented
in desert communities but also is known in agricultural situations
(30).

Leaching of substances has been implicated in the yield,
quality, and nutritive value of economic food plants, including
grain crops (2), soft fruits (2,36) and tropical fruits. The
vitamin content of tropical plants is also greatly reduced by
leaching (38), and the nutritive value of forage crops is
influenced by rainfall prior to and during harvest(17).

Another example of the effect of rain and mist is in the
development of fall color in the foliage of some shade trees and
ornamental shrubs (19). The red color which is so prized in the
autumn is due to the accumulation of anthocyanin pigments which
are derived from sugars stored in the leaves. It was observed
that during a rainy autumn the intensity of red colors in maples
(Acer sp.) was not nearly so striking as during a year with less
rainfall. This suggested that something having to do with
synthesis of anthocyanins might be influenced by the rain,
perhaps sugars or some other precursor.

When Euonymus alatus plants were grown under mist,
anthocyanins did not develop, and the leaves remained green. In
control plants grown without mist, anthocyanins developed
normally. Analysis of the misted plants showed that levels of
sugars, precursors of anthocyanins, were lower in the misted
plants than in the non-misted. Further, the metabolism of the
misted plants was altered so that colorless leucoanthocyanins
were produced rather than the red-colored anthocyanins. Other
factors associated with anthocyanin synthesis, such as nutrition

and temperature, were not responsible for the observed effects (9).

Another interesting effect of mist is on the development of
dormancy in woody plants (19). Normally, temperate zone woody
plants begin preparation for winter in mid-summer. Growth slows,
sugars accumulate, leaves turn red and orange, and abscise in the
autumn, and the plant becomes dormant, requiring about 6 weeks
of cold temperatures to resume growth. However, when Euonymus
alatus plants were grown under intermittent mist during the late
summer and early fall, they did not become dormant and the leaves
did not become red and abscise. Instead, leaves remained green
and plants continued to grow through the winter season if warm
temperatures were maintained. In fact, the plants did not become
dormant for periods of up to 2 years if kept under the mist.
However, when the plants were removed from the mist, they immediately
responded as if it were the fall of the year; red foliage
developed, the leaves abscised, and the plants became dormant, even
if this occurred during the middle of the summer.

The growth regulating chemical abscisic acid (ABA) has been
correlated with dormancy in many woody plants. Levels of ABA in
both the misted and non-misted Euonymus were determined (12).
ABA in non-misted plants increased gradually during the late summer
and early fall and then increased rapidly at the time red color was
developing and the leaves were abscising. In contrast, in plants
under mist, ABA remained at a low level as long as the plants were
under the mist, but increased rapidly when the plants were removed
from the mist. Further, ABA was detected in the leachates from
the misted plants. This suggested that the delay in dormancy was
due to leaching of ABA by the mist.

Intermittent mist can also have an appreciable effect upon
rooting of cuttings during propagation. When Euonymus plants were
analyzed for substances associated with rooting, levels of auxins,
carbohydrates, enzymes involved in the formation of auxins, rooting
cofactors, and other flavanoid compounds were all much higher in
plants and cuttings grown under mist than in plants which did not
receive the mist (18). It was no wonder that cuttings under mist
rooted so much more quickly than did similar cuttings without the
mist. Therefore, mist affects rooting not only by cooling leaves
and reducing transpiration, but also by leaching substances from
the cuttings, and inducing cuttings to form natural compounds which
stimulate the rooting process (18).

Another effect of rain and mist is in flowering. It is
commonly observed in the warm humid tropics that plants often
flower in relation to rainfall. In the chrysanthemum which is
induced to flower by short days, misting the plants can delay
flowering if mist occurs before and immediately following the
application of short days. In the case of the Japanese morning

glory (<u>Pharbitis</u> <u>nil</u>) the results are even more striking. If
Pharbitis, which requires only one short day (one long night) to
flower, is grown under intermittent mist, flowering can be completely
inhibited regardless of the photoperiod treatment (14).

These changes in metabolism induced by rain and mist can be
observed in other crop plants growing throughout the world. For
example, in the San Joaquin Valley of California, commercial grape
growers use overhead irrigation to reduce the mid-day temperatures
during summer. The leaf temperatures are indeed reduced, but in
addition, the flowering patterns of the grapes are changed
somewhat, the sugar content of the fruit is increased and the
resulting wine product is of higher quality. This is important
because the use of overhead irrigation to ameliorate environments
of crop plants is increasing, including fruit and vegetable crops
as well as landscape plants grown commercially in containers.

In the Pacific Northwest and other fruit regions, fruit crops
are being grown in closely planted hedgerows with overhead
irrigation. Large increases in both production and quality have
been noted as compared with plants without overhead irrigation.
Tea of high quality is produced in areas of India where mist and
fog are prevalent; leaching by the mist is recognized as an
important factor. Gardeners and nurserymen comment on the
luxuriant growth of woody plants in areas with rain and mist, such
as the British Isles or the Pacific Northwest. In the warm humid
tropics, lush growth is often associated with the warm temperatures.
However, if those same tropical plants are grown at warm temperatures
but without the constant bathing of the rain, growth does not occur
to the same degree. Thus, rain and mist have many effects upon
plants, not only providing water for growth, but also by leaching
of substances, and by influencing the development of fall color,
root initiation and development, dormancy, and flowering. It is
important to consider these effects when determining the influence
of acid precipitation on plant growth.

REFERENCES

1. K. Arens, <u>Jahrb. Wiss. Bot.</u> 80:248 (1934).

2. N. Cholodny, <u>Ber. Deut. Bot. Ges.</u> 50:562 (1932).

3. L. D. Curtis, <u>Plant Physiol.</u> 19:1 (1944).

4. S. Dalbro, <u>Proc. Int. Hort. Congr. (Paris), 14th,</u> p. 770 (1956).

5. R. del. Moral and C. H. Muller, <u>Bull. Torrey Bot. Club,</u> 91:327
        (1964).

6. L. S. Evans, N. F. Gmur and F. da Costa, Amer. J. Bot. 64:903 (1977).

7. L. S. Evans, N. F. Gmur and J. J. Kelsch, Environ. Exp. Bot. 17:145 (1977).

8. J. A. W. Fairfax and N. W. Lepp, Nature 255:324 (1975).

9. D. J. Greendale and P. H. Nyc, Abstr. Int. Bot. Congr. 10th, Edinburgh, p. 248 (1964).

10. R. J. Helder, Handb. Pflanzenphysiol. 2:468 (1956).

11. R. J. Helder, Handb. Pflanzenphysiol. 6:978 (1958).

12. D. D. Hemphill, Jr. and H. B. Tukey, Jr., J. Amer. Soc. Hort. Sci. 98:416 (1973).

13. J. S. Jacobson and P. van Leuken, in: Proc. 4th Intern. Clean Air Congr. p. 124 (1977).

14. K. Kimura and H. B. Tukey, Jr., Ohara Instit. Landwirts. Biol. 15:61 (1971).

15. P. C. Kozel and H. B. Tukey, Jr., Am. J. Bot. 55:1184 (1968).

16. T. Lausberg, Jahrb. Wiss. Bot. 81:769 (1935).

17. J. A. LeClerc and J. F. Breazeale, Yearb. U.S. Dept. Agr. 1908:389.

18. C. I. Lee and H. B. Tukey, Jr., J. Amer. Soc. Hort. Sci. 96:731 (1971).

19. C. I. Lee and H. B. Tukey, Jr., J. Amer. Soc. Hort. Sci. 97:97 (1972).

20. H. F. Linskens, Planta 41:40 (1952).

21. H. A. I. Madgwick and J. D. Ovington, Forestry 32:14 (1959).

22. R. A. Mecklenburg, H. B. Tukey, Jr., and J. V. Morgan, Plant Physiol. 41:610 (1966).

23. J. V. Morgan and H. B. Tukey, Jr., Plant Physiol. 39:590 (1964).

24. E. T. Paparozzi, M. S. Thesis, Cornell Univ., Ithaca N.Y. (1978).

25. D. W. Reed and H. B. Tukey, Jr., J. Amer. Soc. Hort. Sci. 103:337 (1978).

26. K. Schoch, Ber. Schweiz. Bot. Ges. 65:205 (1955).

27. G. Stenlid, Encycl. Plant Physiol. 4:615 (1958).

28. C. O. Tamm, Oikos 2:60 (1950).

29. C. O. Tamm, Physiol. Plant 4:184 (1951).

30. H. B. Tukey, Jr., Bot. Rev. 35:1 (1969).

31. H. B. Tukey, Jr., in: "A Tropical Rainforest" H. T. Odum,
     ed., U.S. Atomic Energy Comm., Washington, D.C. (1969).

32. H. B. Tukey, Jr., Ann. Rev. Plant Physiol. 21:305 (1970).

33. H. B. Tukey, Jr., and R. A. Mecklenburg, Am. J. Bot. 51:737
     (1964).

34. H. B. Tukey, Jr., and J. V. Morgan, Physiol. Plant 16:557
     (1963).

35. H. B. Tukey, Jr., and J. V. Morgan, Proc. 16th Int. Hort.
     Congr. (Brussels) 4:146 (1964).

36. H. B. Tukey, Jr., H. B. Tukey and S. H. Wittwer, Proc. Am.
     Soc. Hort. Sci. 71:496 (1958).

37. T. Wallace, J. Pomol. Hort. Sci. 8:44 (1930).

38. R. Wasicky, Sci. Pharm. 26:100 (1958).

39. T. Wood and F. H. Bormann, Environ. Pollut. 7:259 (1974).

40. T. Wood and F. H. Bormann, Ambio 4:169 (1975).

EXPERIMENTAL STUDIES ON THE PHYTOTOXICITY OF ACIDIC PRECIPITATION:

THE UNITED STATES EXPERIENCE

Jay S. Jacobson

Division of Biomedical and Environmental Research

U.S. Department of Energy, Washington, D.C. 20545, U.S.A.

INTRODUCTION

The existence of acidic precipitation in the eastern U.S.A. presents a difficult issue for national energy and pollution control policies. We need to determine whether the effects of acidic precipitation are sufficiently serious to require changes in fuel usage and energy or emission control strategies. Because decisions made today have effects a decade or more in the future, we must predict the influence of current policies on this environmental problem. A thorough understanding of the phenomenon and its consequences is required to form a sensible basis for policy decisions.

The repeated occurrence of acidic precipitation in the U.S.A. is well established and increased emphasis on the sampling and analysis of precipitation (Table 1) and in related studies on the sources, distribution, transport, and transformations of sulfur and nitrogen oxides (Ballantine, Beadle, and Jacobson[1]) will provide a more complete understanding of the problem in the next several years. The subjects are complex and one illustration of this complexity is the variety of procedures required for the collection of precipitation depending on the specific aims of the study (Table 2).

Lakes, rivers, the land, and vegetation are direct recipients of rain and snow and alteration of lake ecosystems seems to be among the first demonstrable effects of acidic precipitation. Although the occurrence of changes in components of terrestrial ecosystems is suspected (Dochinger and Seliga[2]), these effects have not been proven. One aspect of this problem, the responses of vegetation to direct contact with acidic precipitation has been described pre-

151

Table I. Existing Precipitation Networks in the U.S.A.[a]

Network	Region	Sites	Last Report
World Meteorological Organization-National Oceanic and Atmospheric Administration-U.S. Environmental Protection Agency	Nationwide	10	1975
Multistate Atmospheric Power Production Pollution Study (U.S. Department of Energy)	Northcentral and northeastern USA	8	1978
Environmental Measurements Laboratory (U.S. Department of Energy)	Nationwide	7	Expected in 1978
Tennessee Valley Authority	Southeastern USA	12	Expected in 1978

[a]A network is defined as a group of five or more sites that sample
precipitation on a continuous basis using the same, standardized
methods and publish results on a regular basis. During 1978, three
new networks will begin operations. These include the Sulfate
Regional Experiment supported by the Electric Power Research
Institute, the Rensselaer Polytechnic Institute network sponsored by
the Electric Power Research Institute, and the Atmospheric
Deposition Network initiated by the U.S. Department of Agriculture.

viously[3] and research on this subject is increasing (Tamm and
Cowling[3]). The purpose of this report is to present the directions
these studies have taken in laboratories located within the U.S.A.
Research on possible secondary responses to acidic precipitation such
as altered susceptibility to nutritional deficiencies, diseases,
insects or other environmental factors (Tamm and Cowling[3]) are not
described in this review.

Currently, there are four major locations where experimental
research on direct effects of acidic precipitation on vegetation is
performed. The U.S. Department of Energy sponsors research at the
Brookhaven and Oak Ridge National Laboratories while the U.S.
Environmental Protection Agency sponsors research at the Corvallis
Environmental Research Laboratory (Lee and Weber[4]) and at the Boyce
Thompson Institute. Reference also will be made to publications from
other laboratories in the U.S.A. where this subject has been studied
in previous years. Field observations and studies which have not

Table 2. Precipitation Network Objectives and Associated Sampling Procedures.

Objective	Type of Sample[a]			Procedure for Collection[b]	
To determine:	Wet	Dry	Bulk	Weekly or Monthly	Event or Sequential
Geographical distribution and fate of pollutants emitted to the atmosphere	X	X		X	
Effect of changes in fuel usage and energy sources on distribution and fate of pollutants	X	X		X	
Effects on weathering and corrosion	X	X		X	
Effects on soils and nutrient supplies for forests and other ecosystems			X	X	
Effects on quality of surface and sub-surface waters and aquatic ecosystems			X	X	
Direct effects on growth, development, and yield of agricultural crops	X	X			X
Physical, chemical, meteorological processes of transport, transformations, and fallout of pollutants	X	X			X
Sources of components and precursors of acidity of precipitation	X	X			X

[a] Wet fallout is collected only during rain and snow events; dry fallout is collected during intervals between rain and snow events; bulk samples collect both wet and dry fallout.

[b] Event sampling is performed only during periods of rain or snow; sequential collection is performed by taking numerous samples within an event either on a time or volume basis.

proceeded to the point where publications or reports are available are not included. Terms used in this review article are defined in the Appendix to insure clarity.

THE PHYTOTOXIC COMPONENTS OF ACID PRECIPITATION

The major inorganic components of precipitation in the eastern U.S.A. (sulfate, nitrate, chloride, ammonium, calcium, magnesium, sodium, and potassium ions) either are benign or beneficial to plants at the concentrations occurring in rain and snow. Most investigators have assumed that the free hydrogen ion concentration of acidic precipitation is the component most likely to produce direct, harmful effects on vegetation and experimental studies have supported this assumption. In one series of tests, the addition of the inorganic ions commonly found in rainwater to acidic solutions did not alter the development of foliar symptoms as long as the pH was not affected (Jacobson and van Leuken[5]). However, responses other than the development of foliar symptoms might be altered by the concentration of other inorganic ions in precipitation (Evans and Bozzone[6]).

There has been some concern over the possible toxic effects of dissolved sulfur dioxide (Hill[7]) but, thus far, significant concentrations of dissolved sulfur dioxide have not been found in acidic precipitation collected in the eastern U.S.A. Injurious effects from this component are unlikely except perhaps at locations close to intense sources of sulfur dioxide emissions.

PLANT RESPONSE TO DIRECT CONTACT WITH ACIDIC PRECIPITATION

The most frequently reported response to simulated acidic precipitation is the formation of lesions or zones of dead tissue on the upper epidermis of leaves often near stomata, trichomes, and vascular tissues (Evans, Gmur, and Da Costa[8]; Jacobson and van Leuken[5]; Longnecker and Shriner, unpublished; Wood and Bormann[9]). Collapse and distortion of epidermal cells on upper leaf surfaces is followed by injury to palisade cells and ultimately both surfaces of the leaf are affected (Evans, Gmur, and Kelsch[10]). Erosion of cuticular surfaces also has been observed (Lang et al.[11]). In one study, gall formation due to abnormal cell proliferation and enlargement occurred in certain clones of hybrid poplar (Evans, Gmur and Da Costa[12]). Premature abscission of bean leaves also has been observed (Ferenbaugh[13]).

There is only one report of interference with a physiological process after contact with simulated precipitation. Sheridan and Rosenstreter[14] found marked reduction in phytosynthesis of mosses.

Changes in chemical composition of foliage produced by experimental exposures to acidic precipitation are now under investigation at Boyce Thompson Institute. Thus far, only reduced
chlorophyll content has been reported (Ferenbaugh[13]; Sheridan and
Rosenstreter[14]) although leaching of elements from leaves by rainfall is a well-known phenomenon (Lepp and Fairfax[15]; Tukey[16]).

No reports of beneficial effects on vegetation of direct
contact with acidic precipitation have appeared as yet. The likelihood of obtaining beneficial responses may increase when long
duration or repeated treatments are performed with plant species
resistant to injury. Certain elements have nutritional value after
absorption through leaves particularly with plants growing on
deficient soils.

Biological, chemical, environmental and cultural factors may
have a substantial influence on plant responses to acidic precipitation. Variation in response between different species and
cultivars is to be expected. The stage of foliar development most
susceptible to lesions after contact with simulated acidic precipitation is just prior to or during maximum leaf enlargement (Evans,
Gmur, and Kelsch[10]). Herbaceous species seem to be highly sensitive
to the formation of foliar lesions (Jacobson and van Leuken[5]) and
the foliage of deciduous trees is more susceptible than the needles
of coniferous trees (Evans, Gmur, and Kelsch[10]). Although there
have been no reports on the effects of climate and nutrition,
alterations in plant response due to these factors is quite likely.

Pubescence and the chemical composition both of the cuticle and
materials residing on the leaf surface may influence response to
acidic precipitation although no direct tests have been performed
as yet. It is likely that the toxicity of acidic precipitation is
affected by the wettability of the leaf and by reaction of components of precipitation with substances exuded from the interior of
leaves or deposited on surfaces in particulate matter.

Foliar response to acidic precipitation is markedly dependent
on the temporal and physical characteristics of the events. For
example, duration of exposure, frequency of exposure, duration
between rain events, and intensity of rainfall influence the
development of foliar symptoms (Jacobson and van Leuken[5]; Shriner[17]).
Effects of droplet size and variation in acidity during the event
also may be important although no tests of these two factors have
been reported. Environmental conditions during and after the event
could alter the response of plants by altering physiological
processes, the degree of contact with rainfall, the amount of liquid
remaining on leaf surfaces, or the rate of evaporation after
termination of the precipitation event.

Plant species in the eastern U.S.A. are exposed repeatedly to

elevated ozone concentrations during the growing season and to
elevated sulfur dioxide concentrations in specific localities.
Consequently, research on the possibility of interactive effects
between acidic precipitation and gaseous pollutants has been
initiated at the Oak Ridge National Laboratory.

THRESHOLD OF ACIDITY FOR PHYTOTOXIC EFFECTS

The "threshold" for the occurrence of an effect on a particular
organism or population of organisms is a convenient term that has
been widely-used both in scientific and non-scientific literature.
Unfortunately, "threshold" is difficult to define (see Appendix)
and it will vary with the type of effect measured, the techniques
used to measure the effect, the experimental conditions, and the
number of replicates used in the experiment. To avoid the complexity and imprecision of the threshold concept, it is desirable
to report dose-response relationships over a wide range of concentrations of toxicant and durations and frequencies of exposure.
This approach provides a more rigorous and specific basis for
characterizing the response of organisms to a toxic agent and is
valuable for determination of the existence of synergistic or
antagonistic effects of combinations of substances (Jacobson and
Colavito[18]). The typical form of these relationships is well known
(Munn et al.[19]; figures 2 and 3), but there has been only one
preliminary attempt to provide dose-response relationships for the
direct effects of simulated acidic precipitation on vegetation
(Jacobson and van Leuken[5]). Extension of this approach to a wide
range of plant species, conditions, and kinds of effects can provide
a rigorous basis for predicting plant response given the composition
of precipitation and the characteristics of the precipitation events.

The information now available only provides a rough guide to
the expected foliar response of susceptible plant species to direct
contact with acidic precipitation. Detectable foliar lesions
generally have been found at pH values between 3 and 4 in simulation
experiments (Table 3).

RESEARCH NEEDS CONCERNING DIRECT PHYTOTOXIC EFFECTS

Results obtained in simulation experiments performed to date
provide only a crude approximation of the response of field-grown
plants to acidic precipitation. Nevertheless, this approach is a
necessary and useful tool for providing clear, quantitative, cause
and effect relationships and a basis for the interpretation of field
observations. We must intensify our efforts to discover the full
range of effects of acidic precipitation and the factors which
influence them. There is a particular need to improve our attempts
to simulate actual rain events and field conditions.

Table 3. Maximum pH Value Producing Injury to Vegetation after
          Direct Contact with Simulated Acidic Precipitation

Effect	Receptor	pH Value	Reference
Foliar aberrations, decrease in growth	Bean	2.5	Ferenbaugh[13]
Foliar lesions, decrease in growth	Yellow birch	3.1	Wood and Bormann[9]
Foliar lesions	Bean, sunflower	3.1	Evans, Gmur and Da Costa[8]
Foliar lesions	Bean	3.1	Longnecker and Shriner[20] (Unpublished)
Foliar lesions	Hybrid poplar	3.4	Evans, Gmur and Da Costa[12]
Foliar lesions	Sunflower	3.4	Jacobson and van Leuken[5]
Reduction in dry weight	Bean	4.0	Shriner[17]

Little effort has been expended as yet to determine the
susceptibility of organs other than leaves and processes other than
vegetative growth. Interference with the fertilization process and
reproduction of plants (Evans and Bozzone[6]) may have a far greater
influence on the yield of cultivated crops and on seed production
and persistence of indigenous plant species than effects on foliage.
Exploration of effects on pollination and a determination of the
dose-response relationships for interference with this process should
be one important objective of future research on the direct effects
of acidic precipitation on vegetation.

APPENDIX

Defintion of Terms

Acidic Precipitation. Rain or snow containing a higher
concentration of free hydrogen ions than produced from the equilibration of pure water with atmospheric carbon dioxide at ambient
temperature and pressure (pH approximately 5.7).

Simulated Acidic Precipitation. A prepared solution composed of the major dissolved ionic components of ambient rain at concentrations approximating those occurring in a particular region. Investigators have assumed that the important components for phytotoxicity studies are the inorganic ions.

Direct Effects on Plants. Effects produced by contact of a toxicant with some portion of the plant structure, for example, foliage.

Phytotoxicity. A property of a specific physical, chemical or biological factor which produces abnormalities detectable with the naked eye or a light microscope or interferes with physiological functions of the plant.

Injury to Plants. Those alterations which are undesirable for plant growth, development, reproduction, yield, composition or competitiveness of a plant species or community.

Damage to Plants. A condition in which the economic, aesthetic, or ecologic value of plants are impaired.

Threshold. The lowest measurable combination of concentration of toxicant, duration of exposure, and frequency of occurrence at which a statistically significant effect is produced for a defined set of physical, chemical, and biological conditions. Threshold values will vary with the type of effect, measurement technique, experimental design, conditions, and number of replicates used in the study.

Dose-Response Relationship. A quantitative expression of the relationship between a specific effect and the time and concentration of exposure to a toxicant.

Beneficial Effects. Those changes that improve plant growth, development, reproduction, yield or competitiveness of a particular plant species or community.

REFERENCES

1. D. S. Ballantine, R. W. Beadle, and J. S. Jacobson, Research on the chemistry and ecological effects of atmospheric pollutants initiated by the Department of Energy. Presented at the: "Public Meeting on Acid Precipitation" sponsored by the New York State Assembly, Lake Placid, New York, May 4-5 (1978).

2. D. L. Dochinger, and T. A. Seliga, Proc. First Internat.
    Symposium on Acid Precipitation and the Forest Ecosystem.
    USDA Forest Service General Technical Report NE-23,
    Northeastern Forest Experiment Station, Upper Darby,
    Pennsylvania (1976).

3. C. O. Tamm, and E. B. Cowling, Acidic precipitation and forest
    vegetation, in: "Proc. First Internat. Symp. on Acid
    Precipitation and the Forest Ecosystem", USDA Forest
    Service, Gen. Tech. Report NE-23, Upper Darby,
    Pennsylvania, p. 854 (1976).

4. J. J. Lee, and D. E. Weber, A study of the effects of acid
    rain on model forest ecosystems, Paper 76-25.5, 69th
    Annual Meeting, Air Pollution Control Association,
    Portland, Oregon (1976).

5. J. S. Jacobson, and P. van Leuken, Effects of acidic rain on
    vegetation, Proc. Fourth Intern. Clean Air Congress,
    p. 124 (1977).

6. L. S. Evans, and D. M. Bozzone, Effects of buffered solutions
    and sulfate on vegetative and sexual development in
    gametophytes of Pteridium aquilinum, Amer. J. Bot.
    64:897 (1977).

7. D. J. Hill, Experimental study of the effect of sulphite on
    lichens with reference to atmospheric pollution,
    New Phytologist 70:831 (1971).

8. L. S. Evans, N. F. Gmur, and F. Da Costa, Leaf surface and
    histological perturbations of leaves of Phaseolus
    vulgaris and Helianthus after exposure to simulated acid
    rain. Amer. J. Bot. 64:903 (1977).

9. T. Wood, and F. H. Bormann, The effects of an artificial mist
    upon the growth of Betula allegheniensis Britt.,
    Environ. Pollut. 7:259 (1974).

10. L. S. Evans, N. F. Gmur, and J. J. Kelsch, Perturbations of
    upper leaf surface structures by simulated acid rain.
    Environ. and Experim. Bot. 17:145 (1977).

11. D. S. Lang, D. S. Shriner, and S. V. Krupa, Injury to
    vegetation incited by sulfuric acid aerosols and acidic
    rain, Paper 78-7.3, 71st Annual Meeting, Air Pollution
    Control Association, Houston, Texas (1978).

12. L. S. Evans, N. F. Gmur, and F. Da Costa, Foliar response of
    six clones of hybrid poplar to simulated acid rain.
    Phytopathology (in press).

13. R. W. Ferenbaugh, Effects of simulated acid rain on Phaseolus
     vulgaris L. (Fabaceae), Amer. J. Bot. 63:283 (1976).

14. R. P. Sheridan and R. Rosenstreter, The effect of hydrogen
     ion concentrations in simulated rain on the moss Tortula
     ruralis (Hedw.) Sm., The Bryologist 76:168 (1973).

15. N. W. Lepp, and J.A.W. Fairfax, The role of acid rain as a
     regulator of foliar nutrient uptake and loss, in:
     "Microbiology of Aerial Plant Surfaces", C. H. Dickinson
     and T. F. Preece, eds., p. 107, Academic Press,
     New York (1976).

16. H. B. Tukey, Jr., The leaching of substances from plants,
     Ann. Review of Plant Physiology 21:305 (1970).

17. D. S. Shriner, Atmospheric deposition: monitoring the phenomenon;
     studying the effects, in: "Handbook of Methodology for
     the Assessment of Air Pollutant Effects on Vegetation",
     S. Krupa, W. Heck, and S. Linzon, eds., Chap. 14b.
     Air Pollution Control Association, Pittsburg, Pennsylvania
     (in press).

18. J. S. Jacobson and L. J. Colavito, The combined effect of
     sulfur dioxide and ozone on bean and tobacco plants.
     Environ. Experim. Bot. 16:277 (1976).

19. R. E. Munn, M. L. Phillips, and H. P. Sanderson, Environmental
     effects of air pollution: implications for air quality
     criteria, air quality standards and emission standards.
     Science of the Total Environ. 8:53 (1977).

20. N. E. Longnecker and D. S. Shriner, Assessment of the effects
     of acid precipitation on Phaseolus vulgaris L. 'red
     kidney', Oak Ridge National Laboratory, Oak Ridge,
     Tennessee (unpublished manuscript).

ASSESSING THE CONTRIBUTION OF CROWN LEACHING TO THE ELEMENT

CONTENT OF RAINWATER BENEATH TREES

K.H. Lakhani and H.G. Miller

Monks Wood Experimental Station, Huntingdon and

The Macaulay Institute for Soil Research, Aberdeen

Rainwater that has passed over the surfaces of a tree shows a net gain in the loading of many chemical elements. This gain is partly through wash-down of elements trapped from the atmosphere by impaction or adsorption (dry deposition) and partly from elements derived from within the plant tissues (crown leaching).[1] Thus, while crown leaching represents part of the cycle of elements internal to the ecosystem, the wash-down of elements derived from the atmosphere represents an input to the site. Clearly it is important to be able to distinguish between internal redistribution and exogenous input, but there is no ready means for making this distinction.

Methods based on deriving the total nutrient balance (e.g. Likens et al.[2]) present problems resulting from the small size of the various fluxes in comparison to the size of the sinks, in particular that of the soil. On the other hand, attempts at direct estimation of the atmospheric input from the catch of air-borne particles as measured on artificial impactors (e.g. White and Turner[3]) involve problems of conversion to unit area of forest. In the absence of a better approach, Mayer and Ulrich[4] suggested that the gain shown by throughfall beneath beech during the leaf-less winter months could only come from atmospheric input. Accordingly, the proportional enrichment in winter could be used to separate input from crown leaching in summer throughfall. This however, is limited to deciduous species, and assumes that significant leaching only occurs through leaves, and that the leafless trees in winter have the same dry deposition "catching efficiency" as in summer.

More recently Miller et al. [5], working with Pinus nigra var.

maritima (Ait.) Melv., observed that the input of an element in throughfall plus stemflow (kg/ha/wk) is linearly related to the input in incident gross rainfall. They used the intercept of this regression as a measure of the enrichment of water beneath living trees in the absence of any input - i.e. release by the trees through crown leaching.

This regression approach makes the major assumption that the inputs of an element in rain and through dry deposition can be treated as a single variable. In this paper the underlying estimation problem is defined and a generally applicable estimation procedure that avoids the above assumption as developed by K.H. Lakhani. This technique is then applied to a two year segment of rainwater data from one of a new series of six experiments designed and managed by H.G. Miller and J.D. Miller of the Macaulay Institute. At these experiments gross rainfall is collected both in Nipher-shielded rain gauges (open gauges) and, following a suggestion by Miller et al.[5], in funnels surmounted by an inert wind-filter of polyethylene coated wire mesh (filter gauges)[7-10].

ESTIMATION TECHNIQUE

Suppose that rainwater is collected:
(i)   beneath the trees as throughfall,
(ii)   outside the forest in a standard open guage,
(iii) outside the forest in a filter gauge,
over n equal time periods extending over the total study period. Let X_{1i} denote the weight of a given element (kg/ha) in the rainwater collected as throughfall and stemflow during the ith time period, and let X_{2i} and X_{3i} denote the corresponding weights of the element in the open and filter gauges respectively. Thus for the n time periods there will be n triplets of observations X_{1i}, X_{2i}, X_{3i} (i = 1,2, ..., n).

Let W_i, D_i and L_i respectively denote the wet deposition, dry deposition and leaching of the element (kg/ha) during the ith time period. Then the X_{ji} values (j = 1,2,3, ; i = 1,2, ..., n) can be expressed in terms of these components and other effects.

The element content, X_{1i}, in the rainwater collected as throughfall and stemflow is essentially the total of wet deposition, dry deposition and leaching during the ith time period. There will, however, be some loss through foliar absorption, f_i, and some further loss, l_i, due to the rainfall for the ith time period failing to wash down all the dry deposition occurring during this time. On the other hand, there will be some gain, g_i, due to the rainfall of the ith time period succeeding in washing down some of the deposits not washed down in the rain of earlier periods. Further, the observed X_{1i} value will also be affected by chance variations making up the error term, e_i. Thus

$$X_{1i} = W_i + D_i + L_i - f_i - l_i + g_i + e_i \qquad (1)$$

The loss and gain terms, l_i and g_i, are unknown functions of a range of variables, but as these terms are time period boundary effects, they tend to cancel each other. In this formulation these boundary effects and the unknown foliar absorption term are absorbed into the error term, to obtain an overall error term $e_{1i} = e_i - l_i + g_i - f_i$, so that

$$X_{1i} = W_i + D_i + L_i + e_{1i} \qquad (2)$$

Next, the element content, X_{2i}, in the open gauge will be mainly due to wet deposition with some contamination due to dry deposition. It is reasonable to suppose that the dry deposition on the open gauge will be proportional to the dry deposition on the forest canopy, so that

$$X_{2i} = W_i + aD_i + e_{2i} \qquad (3)$$

where a is a positive constant of proportionality likely to be small and <1 and e_{2i} is assumed to be a random error term.

Finally, the element content, X_{3i}, in the filter gauge will also be a combination of wet deposition and dry deposition. However, the windfilter on this gauge will intercept some of the non-vertical rainfall which would otherwise not be captured, with the result that this gauge will tend to contain a greater amount of rainfall than the open gauge. Assuming that the rainfall catching efficiency of this filter gauge relative to the open gauge is constant, if most of the wet deposition is through normal rainfall, the amount of wet deposition in X_{3i} will be kW_i, where k is a constant likely to be >1. For the dry deposition, the wind-filter acting as an artificial tree may be assumed to have the catching efficiency b relative to the catching efficiency of the forest canopy, so that:

$$X_{3i} = kW_i + bD_i + e_{3i} \qquad (4)$$

As k is likely to be >1 and b>a, X_{3i} will be expected to be $>X_{2i}$.

Let V_i and U_i be the amounts of rainwater collected during the ith time period in the open and filter gauges. Then, the constant k can be estimated by $\hat{k} = \Sigma U_i / \Sigma V_i$. Assuming $\hat{k} \sim k$, and writing $b/\hat{k} = b'$ and $e_{3i}/\hat{k} = e'_{3i}$, we obtain the modified observation:

$$X'_{3i} = X_{3i}/\hat{k} = W_i + b'D_i + e'_{3i} \qquad (5)$$

To eliminate W_i from equations 2, 3 and 5 we define $Y_i = X_{1i} - X_{2i}$ and $X_i = X'_{3i} - X_{2i}$ to get

$$Y_i = (1 - a)D_i + L_i + e_{1i} - e_{2i} \qquad (6)$$

$$X_i = (b' - a)D_i + e'_{3i} - e_{2i} \qquad (7)$$

These equations show that if L_i is independent of D_i with the mean of L_i equal to M_L (this condition is less strict than the special case that L_i be constant), then Y_i is linearly related to X_i with the slope equal to $(1 - a)/(b' - a)$ and the intercept equal to M_L. The problem of estimating the parameters defining the structural relationship between two variables both of which are subject to random errors is intrinsically difficult and is discussed in detail by Kendall and Stuart[11]. The problem here appears even worse because the error term e_{2i} present in both equations makes the overall error terms $(e_{1i} - e_{2i})$ and $(e'_{3i} - e_{2i})$ correlated. Unlike the error terms e_{2i} and e_{3i}, the error term e_{1i} is a conglomeration in which because of the collosal spatial variation beneath a forest canopy[12] even the single term e_i is likely to be large relative to e_{2i} or e_{3i}. On the other hand, since k is likely to be greater than 1, e'_{3i} will tend to be less than e_{3i}. A practical approach therefore is to assume that e_{1i} is likely to be relatively large compared with e_{2i} or e'_{3i}, in which case the problem is reduced to the ordinary regression of Y_i on X_i. The parameters defining the relationship between Y_i and X_i are then readily estimated using the standard regression techniques.[13] In theory the magnitude of different errors can be controlled by appropriate sampling allocation.

APPLICATION OF METHOD

Two years rainwater data was taken from an experiment located at an altitude of 200m in Fetteresso forest[9] in North -east Scotland (Grid Ref. No. 8287). The crop is pole-stage Sitka spruce (_Picea sitchensis_ (Bong.) Carr.) that at the start of the measurement period was aged 27 years and of mean height 11m. Stocking was 2550 stems per ha, 31% of the basal area having been removed in a thinning one year previously. The site is exposed from the east and south-east to the sea, which at its nearest point is only 6km away, although the prevailing wind is south-westerly.

Six 200m^2 plots within the crop were each instrumented with four stemflow gauges and twelve throughfall guages, the construction of which has been described by Miller and Miller[7]. Four filter gauges, consisting of funnels surmounted by vertical cylinders of polyethylene coated wire mesh, were suspended above the forest canopy, and a standard Nipher shielded rain gauge[14] was mounted on top of a 6m pole in an immediately adjacent open area. Water

was collected from the forest at fourteen day intervals, and bulked
into periods of 28 days prior to analysis by atomic adsorption
spectroscopy for potassium and sodium. These elements are thought
to have very different patterns of crown leaching. Whereas leaching
of sodium varies little with season potassium shows a considerable
seasonal variation[5], to the extent that it has been suggested[6]
that separate estimates of leaching would have to be made for
different periods of the year.

Table 1 shows the rainfall (V_i and U_i) in the open and filter
gauges and the amount of sodium (kg/ha) in the rainwater collected
in the three types of gauges (the X_{ji} values) over twenty three
time periods of approximately 28 days each, extending from June
1974 to March 1976. The corresponding field observations for
potassium are shown in Table 2. As expected, the U_i values are
consistently larger than the corresponding V_i values. For both
sodium and potassium, the X_{3i} values are larger than the X_{2i} values
(see equations 2, 3 and 4). The modified values X'_{3i} (equation
5) are then obtained by dividing the X_{3i} values by k, which is
given by $\Sigma U_i / \Sigma V_i$ = 3430.50/1500.90 = 2.2856. As defined by
equations 6 and 7, the derived variables Y_i and X_i are then
obtained by subtracting X_{2i} from X_{1i} and X'_{3i} respectively.

Figure 1 shows the scatter diagram of the Y_i values for
sodium plotted against the corresponding X_i values. The fitted
least squares regression line is statistically significant
($p < 0.001$) and accounts for 70% of the total observed variation in
the Y_i values. The intercept is 0.597, which under the assumption
that L_i are independent of D_i values, provides the estimate $\hat{M}_L$ of
M_L, the mean amount leached (kg/ha/28 days). The estimated
standard error of $\hat{M}_L$ is 0.431. Thus, the estimate of sodium
leached per year is 7.8 kg/ha with a standard error of 5.6 kg/ha.

A similar scatter diagram of Y_i values for potassium plotted
against the corresponding X_i values showed a marked separation of
the observations from the two seasonal periods, June-November
and December-May, indicating differing extent of leaching during
these periods. The intercepts of the separate least squares
regression lines fitted to the observations of the two seasonal
periods gave the separate estimates 1.85 and 0.51 (kg/ha/28 days)
of the mean amounts of potassium leaching during these time
periods. However, it is possible for the pattern of the rainfall
to be different during the two time periods, requiring the
constant k in equation 4 to be estimated separately for the two
time periods. The observed rainfall totals in the open and filter
gauges during the June-November and December-May periods are
(877.7, 1853.4) and (623.2, 1577.1) mm providing the separate
estimates of k, 2.112 and 2.531, for these time periods. The X_{3i}
values for the two time periods were modified using these
estimates of K. Figure 2 shows the scatter of the resulting X_i,
Y_i values for the two time periods.

Table 1   Amounts of rainfall and sodium collected in different
            types of gauges during twenty-three time periods;
            Fetteresso forest, June 1974 - March 1976.

28 day time periods	Rainfall (mm)		Sodium (kg/ha)		
	Open	Filter	Throughfall + stemflow	Open	Filter
i	V_i	U_i	$X1_i$	$X2_i$	$X3_i$
1	73	105	2.2	1.1	14.9
2	85	145	2.8	3.0	13.4
3	49	80	2.3	2.1	15.7
4	95	195	7.3	5.7	58.1
5	54	91	6.5	3.6	51.0
6	103	242	16.0	10.4	134.0
7	68	175	9.6	5.8	87.1
8	22	87	4.9	1.6	39.2
9	107	325	17.4	10.3	127.5
10	40	123	5.9	3.2	54.3
11	59	113	7.6	5.3	54.1
12	64	118	5.7	1.4	18.7
13	26	62	1.6	1.0	19.8
14	66	153	6.4	0.7	27.3
15	71	135	2.9	0.6	10.9
16	45	91	0.8	0.2	5.2
17	111	291	1.7	0.1	10.2
18	58	147	2.4	0.5	23.3
19	68	179	5.4	2.8	55.9
20	50	147	4.1	3.6	32.4
21	57	134	4.0	1.8	34.9
22	80	176	14.8	5.1	149.3
23	50	117	10.8	6.8	84.5

Table 2   Records of potassium collected in different
           types of gauges during twenty-three time periods;
           Fetteresso forest, June 1974 – March 1976.

Potassium (kg/ha)			
Time periods of 28 days	Throughfall + stemflow	Open	Filter
i	X_{1_i}	X_{2_i}	X_{3_i}
A 1	1.58	0.01	0.91
2	1.99	0.51	1.09
3	1.90	0.15	0.54
4	3.37	0.48	2.05
5	1.91	0.36	1.47
6	3.43	0.82	3.49
B 7	1.02	0.03	1.29
8	0.53	0.01	0.60
9	1.40	0.05	0.16
10	1.08	0.14	1.63
11	1.05	0.53	1.15
12	0.74	0.13	0.53
13	0.53	0.13	0.42
C 14	2.05	0.03	0.40
15	2.81	0.07	0.20
16	1.69	0.09	0.27
17	2.55	0.07	0.33
18	1.29	0.07	0.75
19	2.43	0.08	2.28
D 20	0.80	0.22	1.29
21	1.06	0.04	0.92
22	2.51	0.01	7.52
23	1.78	0.59	5.58

A: June 1974 – November 1974;   B: December 1974 – May 1975

B: June 1975 – November 1975;   D: December 1975 – March 1976

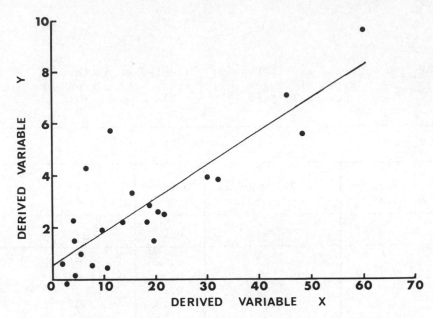

Figure 1.   Relationship between the derived variables Y and X for
            sodium; Fetteresso Forest, June 1974 – March 1976.
            Fitted line: Y = 0.597 + 0.127X

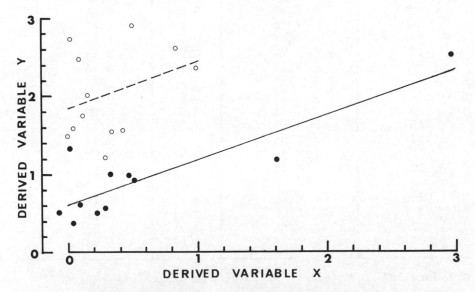

Figure 2.   Seasonal relationships between the derived variables
            Y and X for potassium; Fetteresso Forest, June 1974 –
            March 1976.
            June/November (open circles): Y = 1.836 + 0.588X
            December/May (closed circles): Y = 0.634 + 0.565X

From the separate regression analyses of the data for the
two time periods, the intercepts are estimated to be 1.84 and 0.63
with respective standard errors 0.23 and 0.12. Since the intercepts provide the estimates of potassium leaching (kg/ha) over 28
days the estimated annual rate of potassium leaching during June
to November is 23.93 ± 3.02 (kg/ha), and during December to May,
8.26 ± 1.50 (kg/ha). From these figures the overall annual
estimate is 16.10 ± 1.69 kg/ha/yr.

It is of interest to note that the constants a and b reflect
the dry deposition "catching efficiency" of the open and filter
gauges and are unlikely to show any great seasonal variation in a
study based on evergreen species. And since the estimates of k
for the two seasonal periods are of comparable magnitudes, the
slopes of the two regression lines in Figure 2 will be expected
to be comparable. For the June–November data, the observed range
of X_i is small, and possibly as a result, the fitted slope is
not significant. Nevertheless, the value of the slope is 0.588
which is remarkably close to the statistically significant
($p < 0.05$) slope of 0.565 for the December–May data. The following
calculations show that the intercept estimate of 1.84 for the
June–November data is of the right order, even though the fitted
regression line is not significant. From equation 6,
$\bar{Y} \approx (1-a)\bar{D} + \bar{L}$, and assuming a <1, $(1-a)\bar{D}$ will be$\geq$ 0, and so $\bar{Y}$ will
be expected to be$\geq \bar{L}$. Further, the June–November slope is likely
to be comparable to the December–May slope for which the 95%
confidence limits are 0.565 ± 0.247. The theoretical slope is
$(1-a)/(b'-a)$ and assuming the June–November slope to be less than
1, $b' > 1$. Then from equations 6 and 7, $Y_i - X_i \approx (1-b')D_i + L_i$,
where $(1-b')D_i \leq 0$, so that $\bar{Y} - \bar{X}$ is expected to be less than $\bar{L}$.
The observed value of $\bar{Y}$, which is the expected overestimate of $\bar{L}$,
is 2.02, while the observed value of $\bar{Y} - \bar{X}$, which is the expected
underestimate of $\bar{L}$, is 1.74. The intercept estimate of 1.84
falls within these narrow bounds.

DISCUSSION

Previously, Miller et al.[5] used the intercept of the regression
of X_{1i} against X_{2i} as an estimator of the extent of crown leaching.
Since $X_{1i} \sim W_i + D_i + L_i$ and $X_{2i} \sim W_i + aD_i$, this approach requires
not only that the error term be associated with X_{1i} only, but
also that D_i be proportional to W_i, so that they may be treated as
a single variable, and further that L_i be independent of W_i. It
is certainly true that D_i must be in part dependent on W_i, but it
seems unlikely that they can be regarded as showing a proprotional
relationship. Similarly, if crown leaching is an ion exchange
process, there must be a functional relationship between the ion
loading of rainwater and rate of crown leaching[15], in which case
L_i, although not dependent on W_i when expressed in terms of the same

ion, will be indirectly related to W_i in so far as this in turn is a reflection of the total ion loading.

In the method developed here, the presence of the additional variable X_{3i} makes it possible to eliminate W_i from the set of three equations. This enables the estimation of the extent of crown leaching on the assumption that it is independent of dry deposition. Strictly the problem belongs to the mathematically complex field of functional and structural relationships, but for practical purposes may be reduced to the simpler regression problem.

A single regression was required for sodium, an element that is not essential for growth. Potassium, however, which is not only essential for growth but is also highly mobile in the tree, required separate regressions for the summer and winter periods. Potassium contents in throughfall and stemflow are frequently high during times of maximum litter fall[5], when there is a large amount of easily leached dead and dying tissue on the trees. In this spruce crop, however, rates of litterfall were fairly equally distributed across the year and the operative factor appears to be the presence on the tree of young tissue from which potassium may be lost before the waxy cuticle is fully formed. A further factor, however, is that the two summers, particularly that of the first year, were marked by fairly heavy infestations of the green spruce aphid, Elatobium abietinum (Walker), and these must have been responsible for the loss of some of the potassium. However, as there was no similar variation in the leaching of sodium, it would seem that the aphids are unlikely to be solely responsible for the observed seasonal variation in potassium leaching.

The trial with results from Fetteresso suggested rates of crown leaching of sodium and potassium of around 8 and 16 kg of sodium and potassium per ha per year. These compare with values of 20 kg sodium and 6 kg potassium estimated using the previous method[5] for pine in close proximity to the sea, and with estimates from an inland site in Germany of 2 and 14 kg sodium and potassium for beech[4] and of 0.2 and 7 kg sodium and potassium for spruce[16].

The method proposed here is capable for further development. Presently K.H. Lakhani is developing ideas that take into account the fact that all the observed variables were subject to error, while H.G. Miller and J.D. Miller are examining experimentally the factors influencing the constant k and the effect of varying size and types of wind filter.

SUMMARY

The separation of crown leaching from the total input is of crucial importance in mass balance studies, but the problem is intrinsically difficult. The methods available are either impractical or based on assumptions which are unlikely to hold in practice.

Based on the type of field observations first suggested by Miller et al.[5] the problem is formulated in biomathematical terms and a robust estimation procedure is developed using mathematical/ statistical arguments.

The new technique is applied to sodium and potassium data arising from a two year segment of rain water data from an experiment carried out in Fetteresso forest, Scotland.

ACKNOWLEDGMENTS

The equipment used to collect rainwater in this study, including the filter gauge, was designed and maintained by J.D. Miller, who also supervised the experiments and all chemical analyses. The site at Fetteresso forest is used by permission of the Conservator, East (Scotland) Conservancy, Forestry Commission. Interest in rainwater nutrients was stimulated by the original work of Mr. R.A. Robertson. We are also grateful for the helpful comments on the manuscript made by Mr. M.D. Mountford and Mr. P. Rothery, both of the Institute of Terrestrial Ecology.

REFERENCES

1.   G. Stenlid, Handb. Pflphysiol. 4:615 (1958).

2.   G.E. Likens, F.H. Bormann, R.S. Pierce, J.S. Eaton and N.M. Johnson, "Biogeochemistry of a Forested Ecosystem", Springer-Verlag, New York (1977).

3.   E.J. White and F. Turner, J. Appl. Ecol. 7:441 (1970).

4.   R. Mayer and B. Ulrich, Oecol. Plant. 9:157 (1974).

5.   H.G. Miller, J.M. Cooper and J.D. Miller, J. Appl. Ecol. 13:233 (1976).

6.   H.G. Miller, paper presented to UNESCO-ITE Workshop to consider methods involved in studies of Acid Precipitation to Forests, Edinburgh (1977).

7.   J.D. Miller and H.G. Miller, Lab. Pract. 25:850 (1976).

8.   H.G. Miller and B.L. Williams, Rep. Forest Res., Lond.,
      143 (1972).

9.   H.G. Miller and B.L. Williams, Rep. Forest Res., Lond.,
      145 (1973).

10.   H.G. Miller and B.L. Williams, Rep. Forest Res., Lond.,
      51 (1977).

11.   M.G. Kendall and A. Stuart,"The Advanced Theory of Statistics",
      Vol.2, Griffin, London (1961).

12.   J.P. Kimmins, Ecology, 54:1008 (1973).

13.   N.R. Draper and H. Smith,"Applied Regression Analysis", John
      Wiley and Sons, New York (1966).

14.   F.E. Nipher, Proc. Am. Ass. Advmt. Sci., 27:103 (1878).

15.   G. Abrahamsen, K. Bjor, R. Horntvedt and B. Tveite,
      Fragrapport FR6/76, SNSF, Oslo-Ås,

16.   B. Ulrich, R. Mayer, P.K. Kanna, G. Seekamp and H.W.
      Fassbender, Verhandlungen der Gesellschaft für Okologie,
      Göttingen: 17 (1976).

INPUT TO SOIL, ESPECIALLY THE INFLUENCE OF VEGETATION IN

INTERCEPTING AND MODIFYING INPUTS - A REVIEW

R. Mayer and B. Ulrich

Institut für Bodenkunde und Waldernährung der
Universität Göttingen,
Federal Republic of Germany

Input to soil covered by vegetation would not require special attention if it would simply be the sum of atmospheric inputs, i.e. the sum of wet and dry deposition. But in contrary to this, atmospheric sources may contribute to soil input as well as vegetation itself. This may best be explained with the help of a schematic representation (Figure 1) which was essentially taken from Slinn.[1] The model applies only to terrestrial ecosystems. The designations for different inputs and fluxes shown in the figure are very much reflecting the methods used to measure them. Therefore discussion of the topic must include the discussion of methods.

There are two ways in which the canopy is modifying precipitation input to soil. One way is dry deposition. From an ecological point of view any vegetation canopy behaves like a filter or a sink for the fluxes of matter passing along its surface, the filter efficiency being very dependent upon the physical and chemical properties of the canopy. This has been recognized by many authors, like Eriksson,[2,3,4] Madgwick and Ovington,[5] Chamberlain,[6,7].

The other way of modifying soil input is leaching of substances previously taken up by the roots and translocated to the aboveground parts of the plant with the transpiration stream (Arens,[8] Tamm,[9] Stenlid,[10] Tukey and Morgan[11]).

Thus, when arriving at the soil surface the total precipitation input to soil is a mixture of substances coming from outside the ecosystem, i.e. from the atmosphere, and other substances which are merely completing an internal cycling. To any question

173

concerning the function of the soil as filter in the element flow,
its geochemical role, and the effect of acid precipitation or air
pollution to soil, a satisfactory answer can not be given unless
the share of leaching in the total soil input is known.

Many authors studying soil input with precipitation under a
vegetation cover were aware of this fact, and they tried to
attribute the modifying influence of the canopy to one or the
other of the two processes mentioned (Denaeyer-de Smet,[12]
Miller,[13] Grunert,[14] Carlisle et al.,[15] Attiwill,[16]
Ulrich et al.,[17]). The problem here is to measure the different
components of soil input, which would be the only way to quantify
the impact of a canopy on soil input. There are good methods to
show that leaching does occur, e.g. by using radioisotopes applied
to the roots or to the transpiration stream. But there is no method
to measure root uptake or leaching on an ecosystem level in a
quantitative manner.

The same is true for dry deposition. Attempts to measure dry
deposition to a canopy often yielded dry sedimentation (fallout)
plus filtering effect of a collection gauge because the difference
in the filter efficiency of a collector and the canopy was not
taken into account.

Up until now many authors are taking deposition with bulk
precipitation, measured above or apart from the canopy, that is
essentially wet deposition plus sedimentation, as the only atmospheric input to the soil, attributing the increase in the element
load of throughfall precipitation to leaching only. When the
problem was recognized, dry deposition data had to be estimated
on a very weak experimental base gained in laboratory experiments
since methods for the assessment under natural conditions were not
available. Junge[18] is giving the following global sulfur
budget (recalculated on a per hectare basis) for the land surface
of the earth:

Table 1: Annual sulfur budget (after Junge (1863).

INPUT TO SOIL	kg S $\cdot$ ha^{-1} $\cdot$ yr^{-1}
Precipitation deposit	4.7
Direct uptake by soil and plants	4.7
Release by weathering	1.7
OUTPUT FROM SOIL	
Release to the atmosphere	4.7
River runoff	6.4

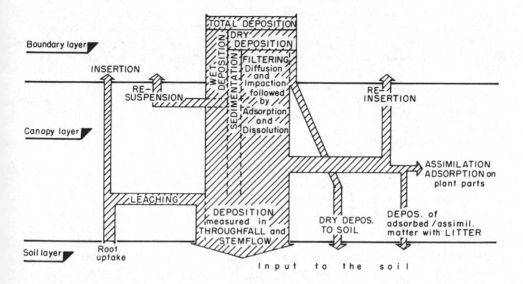

Figure 1

The only figures in this budget which are based at least on some representative measurements are for precipitation deposit (wet deposition) and river runoff. Dry deposition (filtering) is called "direct uptake by soil and plants", and is assumed to be in the same order of magnitude as wet deposition. This assumption is based upon some pot experiments with plants by Johansson,[19] Referring to the work of Eriksson [3,4] the release to the atmosphere is assumed to balance dry deposition. Release by weathering is then calculated as difference between river runoff and wet deposition.

It is obvious that this budget, even though it represents the best knowledge available in the late 1950's, cannot give a satisfactory picture of the sulfur budget. No account is taken in such global approaches of different land surface forms with specific deposition patterns. Furthermore, the assumption is not justified that the release to the atmosphere balances dry deposition, for a significant release of gaseous sulfur compounds (H_2S) from soil is observed only under specific conditions, like in tidal flats, swamps, and shallow lakes. If this is true, the flux data for weathering and dry deposition as well as the assumption of steady state for the soil must be doubted.

A better approach to the assessment of atmospheric deposition to soil under given environmental conditions is achieved when the mass balance is calculated for a small watershed where the precision of the flux measurements is such that dry deposition may be calculated from the overall budget. Such catchment studies extending to the element balance were done in New Zealand (Miller,[13] 1963) and in the U.S.A. (Johnson and Swank,[20] Likens and Bormann,[21] Elwood and Henderson,[22] Likens et al.[23]). For most elements the problem remains, though, the same for dry deposition must be calculated as difference between large fluxes. Furthermore, the catchment system to be balanced contains large storage compartments - like biomass, soil, bedrock, groundwater reservoir - and changes in the stores entering into the mass balance are difficult or impossible to measure.

This evidence suggests the need for a method to directly measure total or dry deposition with sufficient precision to judge the filtering effect of a vegetation canopy and its ability to modify soil input from the atmosphere. There were several promising attempts to measure dry deposition on different kinds of vegetation, e.g. a mixed deciduous forest (White and Turner[24]), and a coniferous forest (Nihlgard,[25] Wiman and Nihlgard[26]). Meteorologists have recognized the importance of the structure of a vegetation canopy, its physical, chemical and biological properties, in controlling the atmospheric input, especially dry deposition (cf. previous chapters of this volume). In their micrometeorological approach of the assessment of dry deposition (see previous session) they are using different parameters like deposition velocity, sink effectiveness etc., to account for the ability of the vegetation cover to modify atmospheric input. Unfortunately, none of these methods yield precise data on dry deposition over a time scale of months or years, or they are too complicated to be handled under most field research conditions.

Our own approach, chosen in the Solling Project, which led us to a conception of the filtering effect of a beech and a spruce canopy, is based on the assumption that total atmospheric input of many elements, including sulfur, to the soil may be measured by means of simple instrumentation beneath the forest canopy, close to the soil surface. This is possible because dry deposition to the canopy layer is being dissolved, provided the substance is soluble, by wet precipitation occurring after dry deposition, thus making it possible to collect the total soil input with traditional rain gauges. In addition, the following conditions, which will be discussed later, were given in the forest sites under consideration:

1. There is little or no dry deposition directly on the soil.

2. Adsorption and assimilation by leaves and subsequent deposition with litter fall is negligible for many elements. The same is true for the storage of these elements in the biomass.

3. Leaching of substances originating from the internal turnover is negligible for most elements, the same is true then for insertion of these elements.

Under these conditions the difference between the element flux measured beneath the canopy and wet deposition measured above or apart from the canopy equals the total dry deposition to the canopy layer, provided the sampling instrumentation is designed to avoid resuspension and reinsertion.

The conditions specified above are probably met in a great number of other ecosystems, so the same approach may be used. For the Solling forest sites they were verified by the following means:

- The non-existence of insignificance of dry deposition directly to the soil was found in lysimeter studies with the soil surface layer. No significant increase in element concentration could be observed in the percolating water compared to the precipitation input to soil except the known amount coming from litter decomposition. This applies also for sulfur, the element most susceptible for dry deposition. While bare soil shows considerable dry

deposition of sulfur (Nyborg et al.,[27] Kühn and Weller,[28])
the different behaviour of the forest soil may be
explained by little turbulent air movement above the
covered soil such that diffusion becomes rate limiting.

- The insignificance of adsorption and assimilation by
  leaves for most elements is revealed when element contents
  of litter are compared with the amounts transported with
  precipitation. Thus the annual flux of sulfur in precipitation under beech is about 50 kg/ha, but only 3 kg/ha in
  the litterfall. The major part of the latter flux, if not
  the total, is fed by root uptake of sulfur. The same must
  not be true for some nutrient elements, like N or P, as
  well as for many heavy metals that are strongly adsorbed
  by organic matter.

- The insignificance or lack of leaching is well demonstrated
  for most elements except K and Mn in a deciduous beech
  forest by comparison of soil input by precipitation during
  growing and dormant season, as it was done by Mayer and
  Ulrich 1974. Similar results were found for spruce in a
  greenhouse experiment. Several spruce trees 8 years old,
  grown in site specific soil were placed in a sealed growth
  chamber and kept under positive pressure of filtered clean
  air for a period of two weeks. During this period they
  were sprinkled 14 times with rain water. There was no
  significant increase in the amount of elements carried in
  the sprinkling water beneath the canopy, except in the
  case of K and Mn (cf. Fassbender,[29]).

A large number of soil input data, measured in throughfall
below a vegetation cover in ecosystems all over the world are found
in literature.

Many of these data may be used to calculate dry deposition
under various environmental conditions for different vegetation
types, provided similar criteria are applicable as in the case discussed above to exclude leaching.

The filtering effect of the canopy against air impurities
seems the most important process of dry deposition to forests, if
not the only one of significance. In table 2 data on dry deposition
on the canopy of a beech and a spruce forest are presented. The
data are calculated as follows: Filtering (part of dry deposition)
is taken as the difference between precipitation input above canopy
(wet deposition plus sedimentation) and below canopy (soil input)
for all elements except K and Mn. For K and Mn it is assumed that
filtering during May to November occurs at the same rate as during
the dormant season from December to April, and that no leaching
occurs during the leafless period. The filtering of K and Mn by

Table 2: Filtering effect of the canopies of spruce and beech

data in keq $\cdot$ ha^{-1} $\cdot$ a^{-1}

	H$^+$	Na$^+$	NH$_4^+$	K$^+$	Ca^{2+}	Mg^{2+}	Mn^{2+}	Cl$^-$	NO$_3^-$	SO$_4^{2-}$
spruce	2.28	0.43	0.23	0.23	1.18	0.25	0.05	0.64	0.48	4.04
beech	0.58	0.30	-	0.16	0.78	0.19	0.04	0.45	0.16	1.83
ratio beech/spruce	3.95	1.44	-	1.43	1.51	1.36	1.43	1.42	2.97	2.21
beech:ratio dry dep. to total dep.	0.39	0.47	-	0.64	0.57	0.52	0.74	0.47	0.24	0.56

last line: percentage of dry deposition to total deposition for beech

spruce canopy is calculated by multiplying the beech data with 1.43
(mean value of ratio spruce/beech for Na, Ca, Mg, Cl).

From the last line of Table 2 it may be seen that dry
deposition on a beech canopy is of almost the same size as wet
deposition. Exceptions are NH_4 and NO_3 which are taken up by the
leaves, filtering is therefore underestimated. For K and Mn
filtering is probably overestimated. The mechanism of dry
deposition seems therefore to be the same for all elements and
comparable to their appearance in rain. The beech canopy is a
sink for aerosols which are carried to the plant surfaces by air
movement (impaction).

Dry deposition to the spruce canopy is larger than to the
beech canopy. For aerosols (Na, Ca, Mg, Cl) the filtering by
spruce is about 150% of that of beech. NH_4 and NO_3 are less used
by the spruce needles, so the spruce data show that N is also
subjected to dry deposition. For SO_4 and H the filtering
effectivity of the spruce canopy is much higher than that of the
beech canopy. This indicates that, in the case of spruce, an
additional mechanism besides impaction of aerosols is acting. This
additional mechanism is probably the adsorption of SO_2 in the
water films adhering to the needle and bark surfaces most of the
time during winter. The SO_2 is oxidized to H_2SO_4 which is washed
off by the next rain. If for spruce the filtering of H^+ and SO_4
in the form of aerosols is calculated as above (multiplying beech
values with 1.43), the filtering by dissolution of SO_2 can be
calculated as the difference. The calculation yields 1.45 keq H^+
and 1.42 keq SO_4 per ha and year for the filtering by dissolution
of gaseous SO_2.

REFERENCES

1. W.G.N. Slinn. USDA Forest Service Gen. Techn. Rep. NE-23,
        p. 857, (1976).

2. E. Eriksson, Tellus 7:243, (1955).

3. E. Eriksson, Tellus 11:375, (1959).

4. E. Eriksson, Tellus 12:63, (1959).

5. H.A.I. Madgwick and J.D. Ovington, Forestry 32:1, (1959).

6. A.C. Chamberlain, Proc. Roy. Soc. Ser. A. 290:236, (1966).

7. A.C. Chamberlain in: "Vegetation and the Atmosphere," vol. 1,
        J.L. Monteith, ed., Academic Press, New York and London.
        (1975).

8. K. Arens, Jb. wiss. Bot. 80:248, (1934).

9. C.O. Tamm. Physiologia Plantarum 4:184, (1951).

10. G. Stenlid, in: "Hand. Pflanzenphysiologie," Bd. IV, p. 615,
        (1958).

11. H.B. Tukey and J.V. Morgan. Proc. 16th Int. Hortic. Congr.
        (Brussels) 4:153, (1962).

12. S. Denaeyer-de Smet. Bull. Soc. Roy. Bot. (Belgique) 94:285.
        (1962).

13. R.B. Miller. New Zealand J. Soil Sci. 6:388, (1963).

14. F. Grunert. Albrecht-Thaer-Archiv 8:435, (1964).

15. A. Carlisle., A.H.F. Brown and E.J. White. J. Ecol. 54:87,
        (1966).

16. P.M. Attiwill. Plant and Soil 24:390, (1966).

17. B. Ulrich., U. Steinhard and A. Müller-Suur. Gött. Boden-
        kundl. Ber. 29:133, (1973).

18. Ch.E. Junge. "Air Chemistry and Radioactivity", Academic
        Press, New York and London, (1963).

19. O. Johansson. Ann. Roy. Agr. Coll. (Sweden) 25:57, (1959).

20. Ph.L. Johnson and W.T. Swant. Ecology 54:70, (1973).

21. G.E. Likens and F.H. Bormann. Ecol. Studies 10:7, (1975).

22. J.W. Elwood and G.S. Henderson. Ecol. Studies 10:30, (1975).

23. G.E. Likens., F.H. Bormann., R.S. Pierce., J.S. Eaton and
        N.M. Johnson, in: "Biogeochemistry of a Forested
        Ecosystem," Springer Verlag, New York, 146 p. (1977).

24. E.J. White and F. Turner. J. Appl. Ecol. 7:441, (1970).

25. B. Nihlgard. Oikos 21:208, (1970).

26. B. Wiman and B. Nihlgard., Swedish Conif. For. Proj. Int.
        Rep. 49, (1977).

27. M. Nyborg and J. Creplin. USDA Forest Service Gen. Techn.
        Rep. NE-23, p. 767, (1976).

28. H. Kühn and H. Weller. Z. Pflanzenernährg. Bodenkd. 140:431 (1977).

29. H.W. Fassbender. Oecol. Plant. 12(3):263, (1977).

THE ACIDIFICATION OF SOILS

Byron W. Bache

Department of Soil Fertility

Macaulay Institute for Soil Research, Aberdeen,
AB9 2QJ, Scotland, U.K.

INTRODUCTION

Acidification is a natural process that occurs continuously in
soils through which water percolates. Precipitation acidified by
industrial pollutants does not constitute a special case but is
a source of acidity additional to a number of natural ones. Its
effects must therefore be assessed in the context of soil
acidification in general. The bulk of this paper describes the
principles of soil acidification and the factors that affect it.
Emphasis is given to the way in which the properties of soil
(related to what is generally understood by "soil type") modify the
acidifying effects. Some tentative conclusions are drawn about the
possibility of polluted precipitation acidifying soil at an
unacceptably rapid rate, and areas of uncertainty where more
research is required are identified.

GENERAL ASPECTS OF ACIDIFICATION

Parameters of Soil Acidity and their Measurement

Three different parameters are involved in defining soil
acidity: the <u>total</u> acidity, the <u>degree</u> of activity, and the manner
in which the degree of acidity varies when total acidity is varied,
i.e. the <u>buffer capacity</u>.

Total acidity, A_t, is given by the amount of base required to
bring the soil to a pre-determined pH value under standardized   •

conditions. It is normally determined by reacting soil with 0.25 M $BaCl_2$ solution controlled at pH 8.0 with a triethanolamine-hydrochloride buffer system,[1] and can be expressed in mili-equivalents per kilogram of soil (meg/kg). Exchangeable acidity, A_e, is part of the total acidity, and can be estimated by leaching or extracting soil with an unbuffered concentrated salt solution (1.0 mol/l KCl or NH_4Cl), and titrating the extract with base. This fraction consists mainly of exchangeable aluminum and hydronium ions. The balance is the "residual acidity", A_r, so that $A_t = A_e + A_r$. Residual acidity constitutes the major part of the total acidity and it exists in soil mainly in un-dissociated forms.

The total acidity is dissociated to a small extent, and this dissociation produces free hydronium ions (H_3O^+) that determine the degree of acidity. Degree of acidity is expressed on a pH scale, where pH = $-\log_a H^+$. A pH measurement can be made with a pH meter, but the meaning of pH is rather uncertain in an electrically-charged heterogeneous system such as soil. When reporting pH values it is essential to record the soil: solution ratio and the nature of the equilibrating solution because the value may vary considerably with these factors. The pH reading approximates closely to that of the soil solution when using a low solution: soil ratio (e.g. 1:1) with a similar electrolyte concentration to the soil solution. However, as natural electrolyte concentrations of soil solutions vary widely, it is better for comparative purposes to use a standard electrolyte solution such as 10^{-2}M calcium chloride[2].

The acid-base buffer capacity is the amount of acid or base required to change the pH of the soil by one unit, i.e. b = $\Delta A/\Delta pH$ = $\Delta B/\Delta pH$. Its value can be read directly from a pH titration curve, or estimated by approximate methods such as those used for lime requirements of soils[3]. The buffer capacity may vary with pH, depending on the different reactions involved in the acid-base equilibria over different pH ranges, and it may also be time-dependent, because some reactions, such as the dissolution of aluminum from minerals, may be very slow.

Acid Inputs to the Soil Environment

There are a variety of sources of hydronium ions that are responsible for soil acidification. The following are more important.

Carbonated water: $CO_2 + 2H_2O \rightleftharpoons H_3O^+ + HCO_3^-$. The pH of the water depends on the substances dissolved in it. Water that is otherwise pure, but is in equilibrium with carbon dioxide in the atmosphere (0.03%) has pH 5.6, but in topsoils the CO_2 concentration is higher because of the respiration of the soil flora and fauna. The CO_2 concentration of soil air varies widely and it may occasionally exceed 1 per cent, giving pH 4.9 and a dissolved CO_2 concentration > 0.4 mM. The total CO_2 produced by topsoils is of the order of 10^4 kg/ha per annum[4]. The acidity produced depends very much on local circumstances, but is of the order of 1 kEq/ha per annum. (Ulrich[8] estimates 2keq/ha per annum for a beech forest.)

Mineral acid from nitrification. Ammonium ions are produced naturally in soils during the decomposition of plant residues and humus by the general heterotrophic microflora[5]. They are also added to agricultural land in considerable quantities as fertilizers. Specialized autotrophic bacteria oxidize ammonium ions to nitric acid, the overall reaction being: $NH_4^+ + 2O_2 + H_2O \rightarrow 2H_3O^+ + NO_3^-$. The theoretical amount of acid produced by the nitrification of 100 kg N added as an ammomium salt is 13.3 keq. This is equivalent to 286 kg Ca and would require 714 kg of lime ($CaCO_3$) to neutralize it. Ammonium sulphate is the most acidifying fertilizer. It has been found in a number of field experiments that 100 kg N as ammonium sulphate releases about 210 kg Ca into the drainage water, 25% below the theoretical amount. Arable soils without fertilizer additions nitrify between 20 and 100 kg/ha N annually[4]: the acidity produced here seems to be about half of the theoretical amount, i.e. 1.5 - 7 keq, and may therefore cause the leaching of 30 - 140 kg/ha Ca.

Organic acids from the decomposition of plant residues. A variety of organic acids are produced by the microbial decomposition of dead plant remains, or released directly from decaying vegetation. Some are rapidly metabolised in the soil, but others such as phenolic acids from conifer and heath litter are more stable[6]. Some have strong chelating properties and so may be more effective in removing metal cations from soil than simple organic or mineral acids. These acids are an important factor in the formation of the acid podzolic soils with low base status typical of the northern moist boreal zone[7]. The acidity produced in this way varies with local conditions, and is difficult to quantify, but Ulrich[8] estimates 5 keq/ha per annum under a beech forest in W. Germany if all the nitrogen is taken up as NH_4^+, but that zero acidity is produced if nitrogen is taken up as NO_3^-.

Oxidation of pyrite. Pyrite, (FeS_2), may occur in chemically reducing environments, and is found in some argillaceous sediments, but more importantly in recently formed estuarine muds. On exposure to the air it oxidizes to ferric sulphate and sulphuric acid, forming the specialized group of acid sulphate soils[9]. This process is unlikely to contribute much acidity to mature free-draining soils, but it is possible that aeration of previously anaerobic soil horizons by deep cultivation may release some sulphuric acid in this way.

Acids deposited in precipitation. Rainfall may contain small amounts of nitric acid, especially during thunderstorms, but the amounts are insignificant, apart from the acids produced by industrial pollution. 1000 mm of rainfall at pH 5.6 will add 0.025 keq/ha acidity to the land surface, while at pH 4.0 it will add 1.0 keq/ha.

General Effects of Acidification

By definition, acidification increases the total acidity of the soil and reduces pH. A number of associated effects are also observed, of which the following are the more important[10].

Loss of base cations, Ca^{2+}, Mg^{2+}, K^+, Na^+. Calcium is the main cation lost, because it is usually the dominant cation in soils. It is leached with the co-ions bicarbonate, nitrate, sulphate, chloride or organic anions depending on circumstances. The leached calcium may arise from:
(i) solution of free carbonates: $H_3O^+ + CaCO_3 \rightarrow Ca^{2+} + HCO_3^- + H_2O$. If present, free carbonate provides a tremendous reserve of buffering, and pH is unlikely to fall below 6.5 - 7 until all free carbonate is dissolved.
(ii) displacement of Ca from weak acid exchange sites, such as occur particularly in humus, but also in poorly-ordered alumino silicates and hydrous oxides:

$$(R.O)_2Ca + 2H^+ \rightarrow 2ROH + Ca^{2+} \qquad \dots\dots\dots(1)$$

(iii) displacement of Ca from permanent-charge exchange sites. It seems unlikely that much direct exchange of H_3O^+ for Ca^{2+} occurs here, but rather exchange of aluminum ions (see below).

Base cation losses are very dependent on soil pH. Gasser summarized work from field experiments on agricultural land, and showed that average lime losses drop from 900 to 100 kg/ha $CaCO_3$ in geometric progression as pH drops from 8 to 5[11]. The losses at a given pH depend on fertilizer additions, soil type and the amount of percolating water, so that the maximum loss may be twice the minimum loss.

Reduction in cation-exchange capacity, CEC. The charges on the weak-acid exchange sites referred to in (ii) above is strongly pH-dependent, the acids being dissociated at high pH and undissociated at low pH. Thus CEC, and the associated ability of the soil to store reserves of nutrient cations, is reduced as pH drops[10, 12]. In peat soils and organic horizons, the process may continue until humic acids are undissociated, at pH 3. Figure 1 illustrates the variability in CEC as a function of pH.

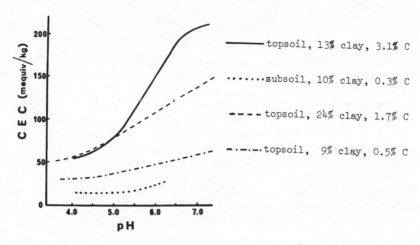

Figure 1.   Effect of soil reaction (pH in 10^{-2} M $CaCl_2$) on CEC, for some contrasting soils adjusted to different reactions in the field.

Mobilization of aluminum ions.   Solution and precipitation of aluminum occurs between pH 3.5 and 5.5, and provides the main buffer mechanism over this pH range in mineral soils. The reaction scheme in a pure Al system[13] (omitting co-ordinated water molecules) can be simplified to:

$$Al^{3+} \leftrightharpoons Al(OH)^{2+} \leftrightharpoons Al_n OH_m^{(3n-m)+} \leftrightharpoons Al(OH) \leftrightharpoons Al(OH)_4^- \quad (2)$$

mono-nuclear ions	poly-nuclear ions	solid

Approx. 3.5   5   6.5   8
  pH:

In soils, solid-phase aluminum occurs in the lattice structure of minerals, in inter-layer sites of expanding clay minerals, and in poorly-ordered minerals (allophane and hydrous oxides) of variable composition. As surface pH drops, hydronium ions are adsorbed at the surface giving positively charged sites[14]:

$$\begin{array}{c} -O \\ \diagdown \\ -O \diagup \end{array} Al-OH \ + \ H_3O^+ \rightarrow \begin{array}{c} -O \\ \diagdown \\ -O \diagup \end{array} Al^+ \ + \ 2H_2O \qquad \ldots\ldots\ldots (3)$$

but eventually discrete soluble aluminium ions are produced, initially poly-nuclear hydroxy ions, but eventually the simple mono-nuclear ions $[Al(H_2O)_6]^{3+}$. These then occupy permanent-charge exchange sites, displacing more Ca^2. Aluminium ions then become an important feature in soil solution. Thus an acid mineral soil is an "aluminium soil"[4]. The exchangeable hydroxy-Al ions and the hydroxy-Al surfaces provide effective buffering so that the pH of acid mineral soils rarely falls below 4[10,14].

Mineral degradation. Extreme acidity, following sustained inputs of large amounts of mineral acid may cause degradation of primary and secondary soil minerals.

Ancillary chemical effects. Changes in surface properties of minerals and the composition of soil solutions affect a number of adsorption and solubility reactions in soils[4, 10].

Biological activity. Reduced pH, and the presence in solution and on mineral surfaces of toxic metal cations such as aluminium and manganese have a profound effect on a variety of organisms. This is discussed in detail in this volume and elsewhere[15,16].

DETAILED PROCESSES OF ACIDIFICATION

The complex nature of soils ensures that the processes involved in acidification are complicated. Quantitative adsorption of the added acid followed by the leaching of an equivalent amount of base cation is unlikely to occur. There is little exact information for situations of practical interest, and at present we have to make reasonable deductions from known principles. The principles that seem to be most relevant are those of cation exchange processes and water infiltration.

Cation Exchange Equilibria

Cation exchange is a consequence of the negative electric charges on the colloidal clay and humus particles of the soil matrix. Permanent charge is generated by partial isomorphous substitution within the lattices of clay-size layer silicates, and in particular the substitution of Al III for Si IV in the tetrahedral sheet and of Mg II or Fe II for Al III in the

octahedral sheet. The consequent deficiency in positive valency
results in a crystal with an excess negative charge. The variable
charge is generated mainly by the pH-dependent dissociation of
hydroxyl groups, which may occur in silanol groups, located at
surfaces of alumino-silicate gels or the edges of layer silicate
crystals, or in phenols, or in carboxylic acids. Where
appreciable amounts of organic matter are present, such as in
surface soils, the major part of the variable charge arises from
proton dissociation from carboxyls and phenols. Thus, the
negative charge depends not only on the nature and amount of
silicate clay minerals, but also on the pH and organic matter
content. Electrical neutrality is preserved by the adsorption
of positively-charged counter ions. These are mainly Ca^{2+}, but
also K^+, Na^+, NH_4^+, Mg^{2+}, Mn^{2+} and Al^{3+} depending on circumstances
There is very little truly exchangeable hydronium, meaning
diffusible H_3O^+ electrostatically attracted to charged surfaces,
except at pH <4[10]. Most surface hydrogen is undissociated proton
of weak acids.

When soil particles are immersed in a solution, the adsorbed
cations form an ionic atmosphere around the charged particles,
which also incorporates the cations in solution, and cation-
exchange equilibria are thus established between the solution
and the charged surfaces. The distribution of each cation between
the surface and the outer solution depends on (a) the charge, size
and hydration state of the cations; (b) the surface charge density
of the soil particles; (c) the ionic strength of the solution; and
(d) any specific interactions that may occur, such as a size-
preference for K^+ by some minerals or organic-complexing of some
cations.

For a given ion pair, the distribution between the surface
and solution can be quantified by a distribution coefficient.
There are many formulations for cation-exchange coefficients[17].
The most generally satisfactory is probably the "corrected rational
selectivity coefficient", because it can be related to the thermo-
dynamic exchange constant by including the activity coefficients
of the ions in the adsorbed phase.

One of the most important reactions for soil acidity studies
is the exchange of Al^{3+} for Ca^{2+} and Mg^{2+}, the dominant base
cations[18,19]. Ca^{2+} and Mg^{2+} will be considered as one ionic
species in what follows, because they behave similarly, although
not identically, in cation exchange reactions; where Ca is
referred to, (Ca,Mg) is understood [20,21].

The exchange reaction between Al and Ca can be written:

$$3 \text{ Soil Ca} + 2Al^{3+} \rightleftharpoons 3Ca^{2+} + 2 \text{ Soil Al} \qquad \ldots\ldots\ldots (4)$$

for which the selectivity coefficient, normally expressed in
equivalent fractions, can be simplified to:

$$K_{Ca}^{Al} = \frac{(q_{Al})^{2q_o}}{(q_{Ca})^3} \cdot \frac{(a_{Ca})^3}{(a_{Al})^2} \qquad \ldots\ldots (5)$$

where q is the amount (meq/kg) of adsorbed exchangeable ions,
$q_o = q_{Ca} + q_{Al}$, approximately the cation-exchange capacity of the
soil, and a indicates the activity of the ions in solution. The
base saturation and pH of acid mineral subsoils has been shown to
be controlled by this reaction [19], but in surface soils and
humic horizons where organic matter-hydronium interactions are
also involved the dissociation-exchange phenomena are more
complex and $Ca^{2+} \rightleftharpoons Al^{3+}$ exchange is only one of a number of possible
reactions[22]. Poly-nuclear hydroxy-aluminium ions are also
involved, probably in association with organic matter[23].

A similar formulation for hydronium-calcium exchange cannot
be made, because most of the hydrogen is undissociated. Solution
hydronium is in equilibrium with a variety of acid-base pairs,
principally the weak acids responsible for pH-dependent charge,
but also the hydroxy-aluminium groups illustrated in equations (2)
and (3). However, for a generalized acid-calcium dissociation
reaction such as equation (1) an empirical equilibrium quotient
can be formulated, as follows

$$Q = \frac{(q_{ROH})^2}{q_o \cdot q_{Ca}} \cdot \frac{a_{Ca}}{(a_H)^2} \qquad \ldots\ldots (6)$$

where ROH signified any proton donor.

It is probably most convenient to quantify ROH as the total
acidity determined at pH 8.0 as described earlier; q_o is again the
cation exchange capacity but here it includes all the variable
charge and should be measured at pH 8 in the same buffer system,
i.e. $BaCl_2$-triethanolamine hydrochloride[24]. These quantities
could also be estimated from a pH titration curve using calcium
hydroxide in a background electrolyte. (These equations would
not apply to soils containing free carbonate.)

Equation (6) can be rearranged to enable the base saturation
of the soil to be expressed in terms of the other variables, i.e.

$$\frac{q_{Ca}}{q_o} = \frac{(q_{ROH})^2}{(q_o)^2} \cdot \frac{1}{Q} \cdot \frac{a_{Ca}}{(a_H)^2} \qquad \ldots\ldots (7)$$

An alternative cation exchange formulation for acidity based on the Gapon equation should be mentioned, because it relates the calcium saturation directly to the total acidity rather than to its square:

$$Q_G = \frac{q_{ROH}}{q_{Ca}} \cdot \frac{(a_{Ca})^{\frac{1}{2}}}{a_H} \qquad \cdots\cdots\cdots (8)$$

$$\therefore \frac{q_{Ca}}{q_o} = \frac{q_{ROH}}{q_o} \cdot \frac{1}{Q_G} \cdot \frac{(a_{Ca})^{\frac{1}{2}}}{a_H} \qquad \cdots\cdots\cdots (9)$$

The quotients Q and Q_G are likely to vary with the pH at which they are measured, because different reactions may be involved. Equations (7) and (9) are useful for helping to define buffer capacity but they are also important because they identify the solution component of soil acidity as the activity ratio $a_{Ca}/(a_H)^2$, rather than the hydronium activity alone[2]. This activity ratio can be transformed to give a function related to pH:

$$-\tfrac{1}{2}\log\,[a_{Ca}/(a_H)^2] = pH - \tfrac{1}{2}pCa \qquad \cdots\cdots\cdots(10)$$

Therefore Ca^{2+} and Mg^{2+} concentration must be measured as well as pH. The function $pH-\tfrac{1}{2}p(Ca,Mg)$ is the correct solution measurement for indicating the acidity level, or the acid–base status, of a soil. The composition of percolating solutions should also be presented in terms of H_3O^+ and $(Ca,Mg)^{2+}$ rather than as pH alone, if their reaction with soil is to be considered. This expression also removes much of the disturbing effects of variable salt (anion) concentrations on soil pH measurements because both pH and $\tfrac{1}{2}pCa$ vary together, giving an approximately constant $pH-\tfrac{1}{2}pCa$ value[21].

The function $pH-\tfrac{1}{2}pCa$ was introduced by Schofield[21]. He related it to a chemical potential (μ) and called it the lime potential of soil. It could equally well be referred to as the acidity potential, which would be more informative in the present context, except that its numerical value decreases as acidity increases, as with pH. Because this seems somewhat illogical it is preferable to retain the term "lime potential". For systems with the same $(Ca,Mg)^{2+}$ concentration, lime potential is directly proportional to pH.

The measurement of lime potential is straightforward for neutral or slightly acid agricultural soils which have moderately

high salt contents in the soil solution, when it is given by adding
1.13 (a value which includes the activity coefficient) to the pH
measured in 0.01 M $CaCl_2$ solution. More strongly acid soils are
likely to be well-leached and contain much lower salt contents in
soil solution, so that using 0.01 M $CaCl_2$ will give less accurate
results. In these cases the interpolation of the equilibrium
(Ca,Mg) concentration from a simple adsorption isotherm may be
necessary[21]. The acidity potential can then be calculated from
this equilibrium value and the pH measured at this (Ca,Mg)
concentration. This procedure is more rapid and reliable than
determining the composition of displaced soil solution, which is
always experimentally tedious, and may give inaccurate results
when electrolyte concentration is low[25].

There have been many comparisons between pH, or lime potential
and the degree of base saturation of soils[18,19,26]. To the present
time, no relationships of general applicability appear to have
come of this work, most likely because the reactions occurring are
somewhat different in soils of different clay mineral and organic
matter composition.

Disturbance of Exchange Equilibria by Acid Solutions

Direction of change. A simple comparison between the lime
potential of the soil and that of rainfall will show the direction
of possible change. When the difference is large, a pH measurement
may be all that is needed, but for critical studies accurate lime
potential measurements are necessary.

Extent of change: equilibrium considerations. An acid input
disturbs the equilibrium

$$\text{Soil Ca} + 2H^+ \rightleftharpoons \text{Soil } H_2 + Ca^{2+} \qquad \ldots\ldots\ldots (11)$$

so that the reaction proceeds to the right, and surface calcium is
desorbed into solution. When this is eventually leached out in
drainage water the new lime potential of the solution is lower
than it was initially. It does not follow that all the added
hydronium must be adsorbed, and release an equivalent amount of
calcium: the extent of reaction, or the position of the new
equilibrium, is governed by the buffer capacity of the soil. This
is a general term that includes three factors implicit in the
exchange equations referred to above.

(i) The lime potential, or pH, of the soil. A soil with a high
degree of Ca saturation will lose more Ca as a result of a given
change in lime potential than one with low Ca saturation. This
follows from the shape of the calcium adsorption isotherm, which

is illustrated in Figure 2. Various lines of evidence support
this, such as the data on calcium losses in drainage waters from
agricultural land referred to above[11], and work relating pH to
base saturation[26].

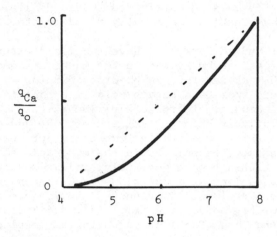

Figure 2.   Degree of Ca^{2+} saturation of the cation exchange complex
            of soil as a function of pH (diagrammatic).

(ii) The cation exchange capacity (CEC) of soil is the main factor
determining buffer capacity, and it is here that soil type has most
influence on rate of acidification. The permanent charge component
of CEC resides mainly in the crystalline alumino-silicate clay
minerals, but magnitude varies with the mineral species. Thus
vermiculite has CEC $\simeq$ 1.5 meq/g. montmorillonite $\simeq$ 1.0 meq/g, illite
$\simeq$ 0.2 meq/g and kaolinite $\simeq$ 0.04 meq/g[10]. Variable, or pH-
dependent charge, in non-crystalline alumino-silicates (allophane)
varies from zero to pH 5 to about 0.2 meq/g at pH 7[29], and that in
humus varies from 0.1 meq/g at pH 3 to about 2 meq/g at pH 7[30,10].
The CEC of a whole soil will be a composite of the materials that
it contains. Few soils have exclusively any one type of clay
mineral; exceptions are vertisols (montmorillonite), ultisols
(kaolinite and hydrous oxides) and andosols (allophane). In cool
temperate regions most soils have a mixture of minerals, so that
in general one can say that the higher the contents of clay and
humus, the higher will be the CEC of the soil.

(iii) The relative bonding strength of the soil materials for
hydrogen as against base cations will affect the buffer capacity
to some extent. This is reflected in the equilibrium coefficient
Q. For instance, important differences have been shown in the
exchange isotherms for bases against hydronium for the three clay
minerals kaolinite, illite and montmorillonite[27]. One can also
expect differences between clays (permanent-charge CEC) and the
variable charge materials, especially humus. These differences
may not be very obvious in cultivated surface soils, where humus
and various clay minerals are usually mixed together.

Wilklander and co-workers[27] have published extensive data on
the release of cations by hydronium for model systems in the
laboratory. They showed that as pH dropped below 4, Ca^{2+} is
displaced by acid from Ca-saturated clay and soil in much less
than equivalent amount. In these experiments the acid concentration was relatively high (10^{-3}M) and without added electrolyte.
The addition of salts predictably lowered pH but resulted in less
Ca^{2+} being exchanged, especially in the more acid samples.
Continuous ion exchange gave generally similar conclusions on the
removal of base cations as a function of pH, but showed differences
between different cations. Continuous ion exchange favoured the
desorption of weakly bound ions (e.g. Na^+, Mg^{2+}) and conserved the
more firmly bound ions (Ca^{2+}) compared to the results from batch
experiments[27].

Bergseth[28] studied cation release as a function of HCl
concentration from humic surface horizons of brown earth and podzol
soils, in laboratory experiments equilibrated for 18 hours. Soil
of high pH and high exchangeable cation contents released large
amounts of cations with small acid additions, accompanied by little
change in solution pH. In some cases, the $(Ca,Mg)^{2+}$ release was
more than equivalent to the acid added, but this dropped with
higher acid additions. Soils with low exchangeable cations and pH
released small amounts, but the soil pH dropped considerably, i.e.
very little of the added acid was adsorbed, showing the strong
binding of small amounts of base cations to the soil.

Two further points need to be made about buffer capacity and
soil composition. Peats, or soil horizons composed exclusively of
organic matter show the strongest buffering at pH>4, but below pH
4 the carboxy-groups are mostly undissociated and may therefore
provide rather less-effective buffering. On the other hand, a
mineral soil that contains some clay as well as sand, silt and
humus, will show the strongest buffering around pH 4 because of the
large reserve of aluminium in the clay that can come into solution.
The soil may then become strongly acid in terms of Al^{3+} rather than
H_3O^+.

Water Movement in Soils

The effect of a reacting solution must depend on the pattern of
infiltration and flow through the soil. In general terms, this is
governed mainly by the structure of the soil in each layer, and by
the rate of application of the water.

Infiltration has been extensively studied by soil physicists,
and detailed equations are available to describe it for simple
situations where the initial moisture content and structure are
uniform with depth and the wetting front develops a constant shape.
Even if infiltration occurs according to this simple pattern, it
may displace the native soil water directly, or mix with it over
a diffuse zone of dispersion i.e. miscible displacement[31]. Thus
the solution that reacts with the soil may have composition of the
influent solution, or a composition between that of the native
soil solution and the influent solution. The transport and
dispersion of reactive solutes through homogeneous soil columns,
and through different layers of otherwise homogeneous material,
has recently been analysed both theoretically and experimentally[32].
Whatever the detailed pattern of dispersion, it seems that the
whole of the influent solution is likely to react with soil
surfaces at fairly low infiltration rates. This is not likely
to be so if the flow is rapid, resulting from high infiltration
rates.

In field situations, where many soils show a marked degree of
heterogeneity, simple infiltration models do not apply. In the
first place, there may be surface run off, or lateral flow within
the soil profile. Then for vertical gravitational movement some
degree of channelled flow is likely to be the norm, where much of
the influent water travels down a few larger pores or channels.
This can only react with the main soil mass by lateral diffusion.
Recent work using nitrate and chloride solutions as tracers showed
that applied solutions flow much deeper into soil than predicted
by miscible displacement theory[33]. Phillips and co-workers compared water movement in soils of different texture and different
initial moisture contents, using chloride as the tracer. They
showed that up to 80 per cent of the applied water appears to
equilibrate with soil water in soils that are initially relatively
dry and poorly structured, but that a smaller proportion (as low
as 20 per cent in some cases) does so in well structured clays
that are initially relatively wet. Some hours after the
application of water, most of the water moving at lower depths was
doing so in isolated pores, thus permitting drainage to occur even
though the water content of the soil as a whole was below the
upper limit of available water[33]. Channelled flow is particularly
likely to occur where there are layers of low permeability, either
at the surface or within the soil profile, thus causing water to

pond and build up a hydrostatic pressure. The resistance of
large pores is low and they become the main drainage channels.

The flow characteristics of water through soil therefore
indicates that in practical field situations, the whole of the
influent solution is unlikely to react with the soil. The proportion that will do so may vary from very little to almost 100
per cent depending in a complex manner on the detailed
circumstances in a given situation.

Detailed Analysis of Acidification Dynamics

The soil acidification problem has obvious analogies with
chromatographic transport. Transport theories have been applied
to soils, and the problems revolve around the conditions under
which the basic differential equations can be applied. The
relevant conservation of matter equation, for variation with
time T and depth X is:

$$\frac{\partial (S + \theta C)}{\partial T} = \theta \cdot D_s \cdot \frac{\partial^2 C}{\partial X^2} - V \cdot \frac{\partial C}{\partial X} + Q$$

where S is the amount of ion sorbed, C is its concentration in
solution, θ is the volumetric moisture content, D_s is the
dispersion coefficient, V is the velocity of flow of the
reacting solution and Q is a production term to account for the
appearance or disappearance of the reacting ion by an
irreversible mechanism.

This equation was examined for soils by Reiniger and Bolt[34],
with particular reference to the effects of attainment of
equilibrium, the shape of the adsorption isotherm, and dispersion.
It has also been applied to transport through layered soils[32].
Analytical solutions are available for relatively simple
systems[32,35], but simulation models are more versatile and have
recently been described in some detail[36]. Modification of the
simulation models to give the changes in cation balance of the
soil and the composition of the percolating solution holds out the
best hope of theoretical treatment, provided reliable data for
the inputs can be obtained.

EXPERIMENTAL STUDIES IN RELATION TO SOIL TYPE

There are a few direct experimental studies in progress and some
preliminary reports have appeared in the literature.

Overrein[37] measured calcium leached over a 40 day period by
500 mm/month of "rainwater" acidified to pH from 5.0 to 2.0 with

sulphuric acid, from 40 cm deep columns of four soils varying in
texture from fine sand to clay loam, in lysimeters. There was a
sharp increase in Ca loss as pH dropped below 4, and differences
between the soils became apparent at pH <3. No chemical data on
the soils before and after leaching were given. Abrahamsen et
al.[38] describe field experiments on forests on five podzolic soils.
irrigated during the summer months with artificial rain at 50 mm/
month containing about 0.13 mM $(Ca,Mg)^{2+}$ and acidified with
sulphuric acid between 6 and 2. Preliminary results in one of the
soils showed a significant difference in base saturation of the
soil after 2 years of pH 3 water. Lamm et al.[39] found that
calcium loss and the acidity of the leachate from a sandy soil in
40 cm lysimeters, caused by adding 100 kg/ha sulphuric acid was
more than for a non-acid control, but less than for added NPK.
This is an important comparison because it puts calcium loss by
acidification in the context of the wider sphere of soil management. The calcium loss from the soil was less than equivalent to
the acid added.

     One must be careful in interpreting these experiments. There
is no question that high concentrations of acid acidify soils
(equation (11) above) but the strategy of simulating a moderate
acid input over a long period by using a high acid input over a
short period[37,38,40], is not valid. The effects of acid rain
are not cumulative, because artificial lowering of pH or lime
potential may initiate a process that otherwise may not have
occurred at all. The situation must be judged in terms of cation
exchange equilibria, and must consider not only the hydronium
ions in precipitation, but also the bases that neutralize some
of the acidity[41], and the resulting lime potential of the water.
Even though the precipitation is acid, acidification of soil
cannot occur unless the lime potential of the precipitation is
lower than that of the soil.

     Some of these principles can be illustrated by comparing the
pH (or lime potential) of precipitation with that of soil
horizons, using the data in Tables 1 and 2. The composition of
the precipitation in a relatively unpolluted area of eastern
Scotland (Table 1) shows that although the pH of total deposition
is lower than that of open-pan rainfall, the relatively greater
concentration of cations in the total deposition gives a higher
lime potential so therefore a less-acidifying solution. Comparing
the lime potential values with those of two soil profiles on open
moorland a few miles away (Table 2) shows that it is higher than
that of the freely-drained humus-iron podzol (Site 1) but less
than that of the brown forest soil (Site 2). One might therefore
not expect acidification at the first site, but expect it at site
2. The position is more complicated, however, because site 2 is
a water-receiving site with a subsoil that has imperfect drainage;
it would seem that in these circumstances some rainfall will run

Table 1.   Acidity of precipitation at Fetteresso forest,
Kincardineshire, Scotland.

	Wet deposition			Wet + dry deposition		
	(Ca,Mg) mM	pH	pH$-\frac{1}{2}$pCa	(Ca,Mg) mM	pH	pH$-\frac{1}{2}$pCa
1973–74	0.036	4.15	1.92	0.352	3.81	2.06
1974–75	0.032	4.40	2.14	0.168	4.04	2.13
1975–76	0.045	4.15	1.96	0.208	4.10	2.24

Figures are means of measurements at two-week intervals. pH and
pCa were calculated from mean molar concentrations and then logged.
(Data by permission of Dr. H. G. Miller, Macaulay Institute.)

off the surface rather than leach the profile, and therefore have
less acidifying effect on the subsoil, although it may affect sites
lower down the slope.

Another feature of these results is the difference between
horizons in pH (lime potential) and the Ca-Al balance of the
exchangeable cations. This illustrates the different H-(Ca,Mg)-Al
exchange properties of the different material in the soil horizons.
In the podzol (site 1) the B horizons have presumably come to
equilibrium with the leachates from the humic layers over about
10,000 years of soil formation. They are de-calcified to a greater
extent than the humic horizons, as shown by the degree of base
saturation and the Ca-Al balance, but at the same time they have a
higher pH than the surface horizons.

CONCLUSIONS

The interaction of acid precipitation with soils is a complex
problem. Mainly, it involves cation exchange processes, which
have been well researched by soil scientists. However, their
application to this problem needs further investigation of the
buffering effect of the soil profile as a whole, particularly as a
function in time, and the importance of aluminum ions in exchange
reactions. These topics are not well understood. In general, the
effect of soil type on acidification is related to the pH (or lime
potential) of the soil and to its buffer capacity[42].

Table 2. Acidity of two semi-natural soil profiles, Mount Shade, Kincardineshire, Scotland. (Vegetation dominated by calluna, soils formed from granitic till.)

Site 1, a shedding site
(Humus iron podzol)

Horizon	Solution composition		Exchangeable cations (meq/kg)	
	pH	pH$-\frac{1}{2}$pCa	$(Ca,Mg)^{2+}$	Al^{3+}
F	3.13	1.72	26	5
H/A$_1$	3.06	1.59	11	14
A$_2$	3.24	1.85	1	8
B$_{21}$	3.74	2.33	0.6	28
B$_3$	4.48	3.15	0.4	7

Site 2, a receiving site
(Brown forest soil)

Horizon	Solution composition		Exchangeable cations (meq/kg)	
	pH	pH$-\frac{1}{2}$pCa	$(Ca,Mg)^{2+}$	Al^{3+}
A	4.48	3.03	6	5
B$_2$	4.34	2.95	2	7
B$_3$/C	4.38	3.03	1	5

Soil type also affects the pattern of water movement through soils. This is a more intractable problem. Although it is being actively investigated by soil physicists, it is difficult to quantify without doing detailed experiments in the field. In the present context it is probably sufficient to realize that variations in the pattern of flow may affect the extent of the reaction of acidified precipitation with soil.

REFERENCES

1. M. Peech, in: "Methods of Soil Analysis", C. A. Black, ed., Amer. Soc. Agronomy, Madison (1965).

2. B. W. Bache, Soil Reaction, in: "Encyclopedia of Soil Science", Dowden, Hutchinson and Ross, Stroudsburg, Pa. (1978), (in press).

3. A. Mehlich, S. S. Bowling and A. L. Hatfield, Comm. Soil Sci. Plant Analysis 7:253 (1976).

4. E. W. Russell, "Soil Conditions and Plant Growth", 10th ed., Longman, London (1973).

5. M. Alexander, in: "Soil Nitrogen", W. V. Bartholomew and F. E. Clark, ed., Amer. Soc. Agronomy, Madison (1965).

6. H. Shindo and S. Kuwatsuka, Soil Sci. Plant Nutrition 22:23 (1976).

7. Ph. Duchaufour, "Precis de Pedologie", Masson, Paris (1960).

8. B. Ulrich, this symposium.

9. N. van Breemen, in: "Proc. Int. Symp. Acid Sulphate Soils", Wageningen (1972).

10. For a general account see N. T. Coleman and G. W. Thomas, in: "Soil Acidity and Liming", R. W. Pearson and F. Adams, Amer. Soc. Agronomy, Madison (1967).

11. J. K. R. Gasser, Experimental Husbandry 25:86 (1973).

12. B. W. Bache, J. Sci. Food Agric. 27:273 (1976).

13. C. F. Baes and R. E. Mesmer, "The Hydrolysis of Cations", Wiley-Interscience, N.Y. (1976).

14. M. L. Jackson, Soil Sci. Soc. Amer. Proc. 27:1 (1963).

15. W. A. Jackson, in: "Soil Acidity and Liming", Amer. Soc.
        Agronomy, Madison (1967).

16. C. O. Tamm, Ambio 5:235 (1976).

17. F. Helfferich, "Ion Exchange", McGraw-Hill, New York (1962);
        G. Bolt, J. Agric. Sci., Neth., 15:81 (1967).

18. J. S. Clark, Canada J. Soil Sci. 46:94 (1966);
        R. C. Turner and J. S. Clark, Soil Sci. 99:194 (1965).

19. B. W. Bache, J. Soil Sci. 25:320 (1974).

20. P. H. Beckett, Soil Sci. 100:118 (1965).

21. R. K. Schofield and A. W. Taylor, Soil Sci. Soc. Amer. Proc.
        19:164 (1955); B. W. Bache, J. Soil Sci. 21:28 (1970).

22. P. Schachtschabel and M. Renger, Z. PflErnahr. Dung. Bodenk.
        112:238 (1966); P. B. Hoyt, Canada J. Soil Sci. 57:221
        (1977).

23. B. W. Bache and G. S. Sharp, J. Soil Sci. 27:167 (1976).

24. C. L. Bascomb, J. Sci. Fd. Agric. 15:821 (1964).

25. J. V. Lagerwerff, Soil Sci. Soc. Amer. Proc. 28:502 (1964).

26. G. R. Webster and M. E. Harward, Soil Sci. Amer. Proc. 23:446
        (1959); R. C. Turner and J. S. Clark, "Soil Chemistry
        and Fertility", Int. Soc. Soil Sci., Aberdeen (1966);
        D. L. Blosser and H. Jenny, Soil Sci. Soc. Amer. Proc.
        35:1017 (1971).

27. L. Wiklander and A. Andersson, Geoderma 7:159 (1972);
        L. Wiklander, Geoderma 14:93 (1975); L. Wiklander and
        S. K. Ghosh, Acta Agric. Scand. 20:105 (1970), 27:280
        (1977).

28. H. Bergseth, Acta. Agric. Scand. 25:225 (1975).

29. K. Wada, in: "Minerals in Soil Environment", Soil Sci. Soc.
        Amer., Madison (1977).

30. J. P. Andre, Ann. Agron. 27:17 (1976).

31. D. R. Nielsen and J. W. Biggar, Soil Sci. Soc. Amer. Proc.
        25:1 (1961); 26:125 (1962); U. Zimmermann, K. O. Munnich
        and W. Roether (1967), in: "Isotope Techniques in the
        Hydrologic Cycle", Amer. Geophys. Union, Washington, D.C.,
        M. Collis-George; Water Resources Res. 13:395 (1977).

32. H. M. Selim, and R. S. Mansell, Water Resources Res. 12:528
        (1976); H. M. Selim, J. M. Davidson and P. S. C. Rao,
        Soil Sci. Soc. Amer. Proc. 41:3 (1977).

33. V. L. Quisenberry, and R. E. Phillips, Soil Sci. Soc. Amer.
        Proc. 40:484 (1976). D. D. Tyler, and G. W. Thomas,
        J. Environ. Quality 6:63 (1977). G. W. Thomas,
        R. E. Phillips and V. L. Quisenberry, J. Soil Sci. 29:32
        (1978).

34. P. Reiniger, and G. H. Bolt, J. Agric. Sci., Neth., 20:301
        (1972).

35. D. E. Elrick, P. H. Groenevelt and T. J. M. Blom, in:
        "Heat and Mass Transfer in the Biosphere", Part 1,
        D. A. de Vries and N. H. Afgan, eds., Wiley, New York,
        pp.537, (1975).

36. M. J. Frissel and P. Reiniger, "Simulation of accumulation and
        leaching in soils", Centre for Agricultural Publishing,
        Wageningen (1974).

37. L. N. Overrein, Ambio 1:145 (1972).

38. G. Abrahamsen, R. Hornvedt and B. Tveite, Research Report 2
        (1975); G. Abrahamsen, K. Bjar and O. Teigen, Research
        Report 4 (1976), SNSF Project, Norway.

39. C. O. Lamm, G. Wiklander and B. Popovic, 1st Int. Symp. Acid
        Precipitation and Forest Ecosystem, Columbus, Ohio
        (1975).

40. N. Malmer, Ambio 5:231 (1976).

41. R. Mayer, and B. Ulrich, Water, Air and Soil Pollution 7:409
        (1977).

42. B. W. Bache, this symposium.

EFFECT OF LOW pH ON THE CHEMICAL STRUCTURE AND REACTIONS

OF HUMIC SUBSTANCES

Morris Schnitzer

Chemistry and Biology Research Institute

Agriculture Canada, Ottawa, Ontario, K1A 0C6

INTRODUCTION

Humic substances are probably the most widely distributed organic carbon-containing compounds on the earth's surface, occurring in soils, fresh waters and in the sea. After many years of stagnation, interest in humic substances has been increasing in recent years. Soil and water scientists, geochemists, environmentalists, biologists and chemists have come to realize that humic substances participate in, and often control, many of the reactions that take place in soils and waters. Also, the relatively recent commercial availability of such sophisticated and powerful analytical instruments as the gas chromatographic-mass spectrometric-computer system and advanced Nuclear Magnetic Resonance and Electron Spin Resonance Spectrometers have made possible significant advances in our knowledge of the chemical structure and reactions of humic substances.

SOIL ORGANIC MATTER

The organic matter in soils consists of a mixture of C.B.R.I. Contribution No. 1022 plant and animal residues in various stages of decomposition, of substances synthesized chemically and biologically from the breakdown products, and of microorganisms and small animals and their decomposing remains. The importance of soil organic matter C in terms of the C cycle as a major source of CO_2 and as a C reservoir sensitive to changes in climate and in atmospheric CO_2 concentrations has been emphasized by recent estimates of Bohn[1,2] According to the latter (Table 1), the mass of soil organic C (30.0×10^{14} kg) more than equals those of other surface C reservoirs combined (20.8×10^{14} kg). The decay

of soil organic matter provides the largest of CO_2 input into the
atmosphere. It is true that deeper C deposits in the form of
marine organic detritus, coal, natural gas and petroleum, deep sea
solute C and C in sediments are much larger, but these are
physically separated from active interchange with surface C
reservoirs.[1]

For the sake of simplicity, soil organic matter is usually
partitioned into non-humic and humic substances. Nonhumic
substances include those with still-recognizable physical and
chemical characteristics, such as carbohydrates, proteins,
peptides, amino acids, fats, waxes, alkanes and low-molecular
weight organic acids. Most of these compounds are attacked
relatively readily by microorganisms in the soil and have a short
survival rate.

The major portion of the organic matter in most soils and
waters, however, consists of humic substances. These are
amorphous, dark colored, hydrophilic, acidic, partly aromatic,
chemically complex organic substances that range in molecular
weights from a few hundred to several thousand.

Based on their solubility in alkali and acid, humic
substances are partitioned into three main fractions; (i) humic
acid (HA), which is soluble in dilute alkali but is precipitated
by acidification of the alkaline extract; (ii) fulvic acid (FA),
which is that humic fraction which remains in solution when the
alkaline extract is acidified, that is, it is soluble in both
dilute alkali and dilute acid, and (iii) humin which is that humic
fraction that cannot be extracted from the soil or sediment by
dilute base and acid. (Figure 1). From analytical data published
in the literature[2] it becomes apparent that structurally the three
humic fractions are similar, but differ in molecular weight,
ultimate analysis, and functional group content, with FA having a
lower molecular weight, but higher content of O-containing
functional groups (CO_2H, OH, C=O) per unit weight than the other
two humic fractions. Important characteristics exhibited by all
humic fractions are resistance to microbial degradation, and
ability to form stable water-soluble and water-insoluble complexes
with metal ions and hydrous oxides and to interact with clay
minerals.

It has been estimated that between 70-80% of the organic
matter in most inorganic soils consists of humic materials. The
remainder is composed mainly of polysaccharides and protein-like
materials. Thus humic substances are the major components of
soil organic matter and I shall confine this discussion to these
materials.

During the first part of my talk I shall describe analytical

Table 1. Distribution of C over the earth's surface[1].
Soil organic C

Soil organic C	30.0×10^{14} kg
Atmospheric CO_2	7.0×10^{14} kg
Biomass C	4.8×10^{14} kg
Fresh Water C	2.5×10^{14} kg
Marine C	6.5×10^{14} kg

$\left.\begin{array}{l} \\ \\ \\ \end{array}\right\} 20.8 \times 10^{14}$ kg

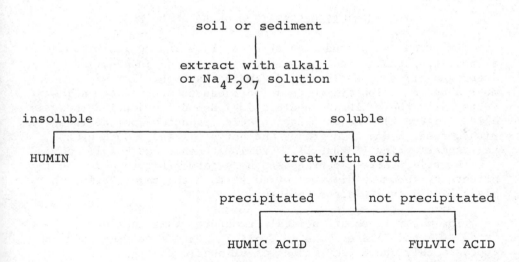

Figure 1. Extraction and fractionation of humic substances

characteristics and current concepts of chemical structure of
HA's and FA's and how these are affected by low pH. The second
part of my talk will focus on metal-and mineral-humic interactions.

ANALYTICAL CHARACTERISTICS OF HUMIC SUBSTANCES

The elementary composition and functional group content of a
"model" HA and FA are shown in Table 2. These data are the means
of numerous analyses done on HA's and FA's extracted from soils
from widely differing origins[3].

A more detailed analysis of the data shows that: a) the
"model" HA contains approximately 10% more C but 10% less O than
does the "model" FA; b) there is little difference between the
two materials in H, N and S contents; c) the total acidity and
CO_2H content of the FA are appreciably higher than those of the
HA; d) both materials contain per unit weight approximately the
same numbers of phenolic OH, total C=O and OCH_3 groups, but the
FA is richer in alcoholic OH groups than the HA. The fact that
the E_4/E_6 ratio of the "model" FA is much higher than that of the
"model" HA indicates that the FA has a lower particle or molecular
weight than the HA.

THE CHEMICAL DEGRADATION OF HA'S AND FA'S

The oxidative degradation of HA's, FA's and humins produces
aliphatic, phenolic and benzenecarboxylic acids in addition to
freeing n-alkanes, n-fatty acids and dialkyl phthalates[3]. The
most abundant aliphatic degradation products are n-fatty acids
(especially the n-C16 and n-C18 acids) and di- and tri-carboxylic
acids (Figure 2). Major phenolic acids produced include those
with between 1 and 3 OH groups and between 1 and 5 CO_2H groups
per aromatic ring (Figure 3). Prominent benzenecarboxylic acids
are the tri-, tetra-, penta-, and hexa-forms (Figure 4). The
structures shown in Figures 2-4 constitute the major "building
blocks" of HA's, FA's and humins.

The major types of chemical structures that make up the
"model" HA and FA are shown in Table 3. The "model" HA contains
approximately equal proportions of aliphatic and phenolic
structures, but a greater percentage of benzenecarboxylic
structures or structures producing benzenecarboxylic acids on
oxidation. By contrast, the "model" FA contains more phenolic
than aliphatic and benzenecarboxylic structures. Both the "model"
HA and FA contain approximately equal proportions of aliphatic
structures and have similar aromaticities.

Thus, the "models" HA and FA have similar chemical structures,

Table 2. Analysis of "model" HA and FA.

Element %	HA	FA
C	56.2	45.7
H	4.7	5.4
N	3.2	2.1
S	0.8	1.9
O	35.5	44.8

Functional groups
(meq/g)

	HA	FA
Total acidity	6.7	10.3
CO_2H	3.6	8.2
Phenolic OH	3.9	3.0
Alcoholic OH	2.6	6.1
Quinonoid C=O Ketonic C=O	2.9	2.7
OCH_3	0.6	0.8
E_4/E_6	4.8	9.6

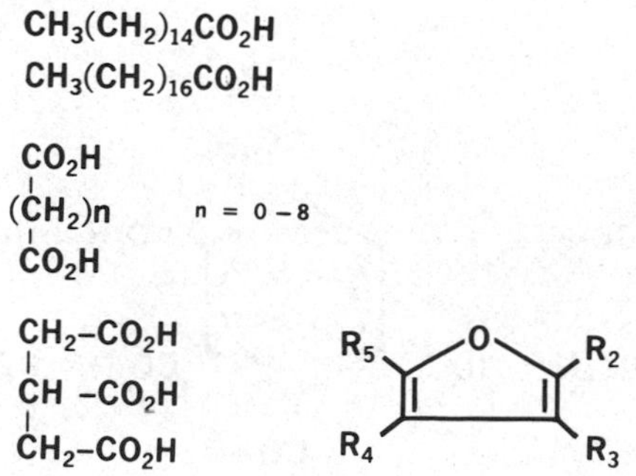

$$CH_3(CH_2)_{14}CO_2H$$
$$CH_3(CH_2)_{16}CO_2H$$

$$CO_2H$$
$$(CH_2)n \qquad n = 0-8$$
$$CO_2H$$

$$CH_2-CO_2H$$
$$CH-CO_2H$$
$$CH_2-CO_2H$$

Figure 2. Major aliphatic degradation products.

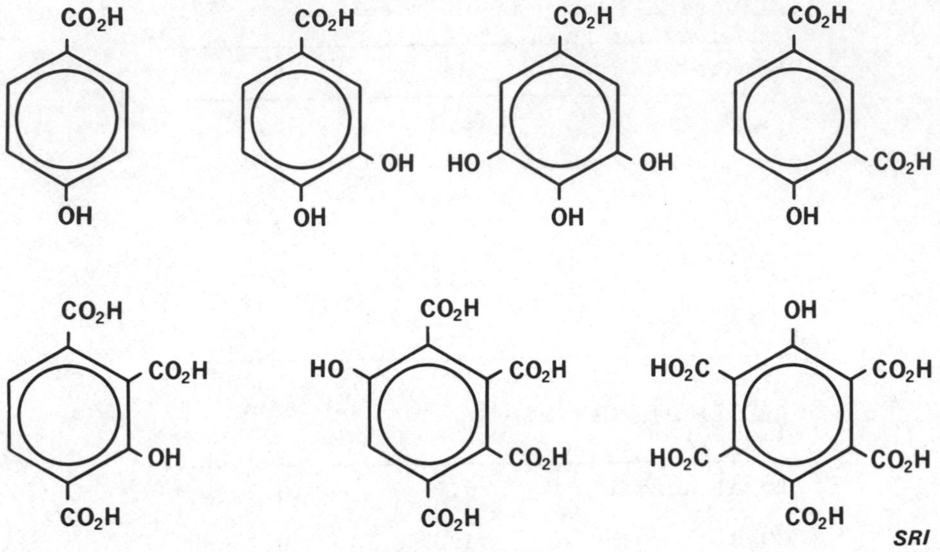

Figure 3. Major phenolic degradation products.

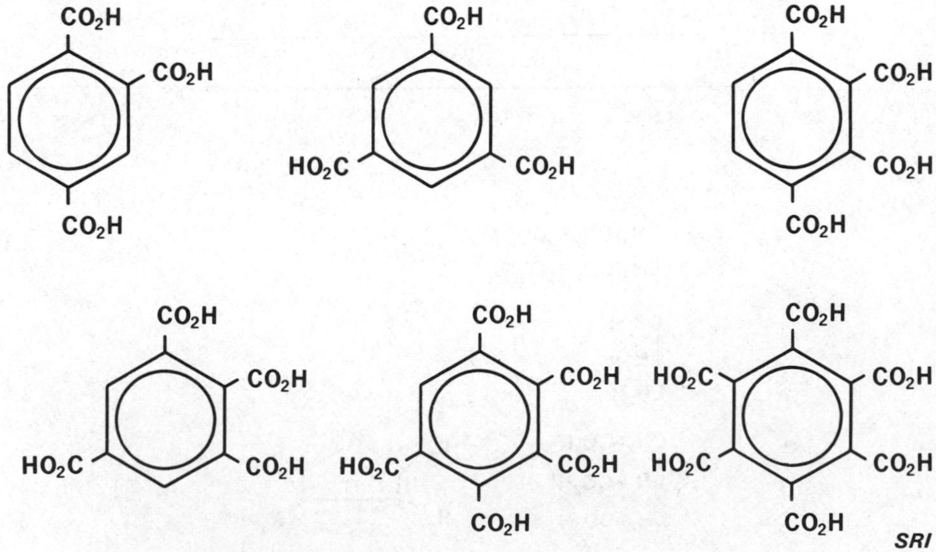

Figure 4. Major benzenecarboxylic degradation products.

except that the FA is richer in phenolic but poorer in benzene-carboxylic structures than the HA.

THE CHEMICAL STRUCTURE OF HUMIC SUBSTANCES

From experimental data that we have obtained in our laboratory over a period of many years, it appears that up to 50% of the aliphatic structures in HA's and FA's consists of n-fatty acids esterified to phenolic OH groups. The remaining aliphatics are made up of more "loosely" held fatty acids and alkanes that seem to be physically adsorbed on the humic materials, and which are not structural humic components, and possibly of aliphatic chains joining aromatic rings. As shown by a wide variety of chemical degradation experiments, the major HA and FA degradation products are phenolic and benzenecarboxylic acids. These could have originated from more complex aromatic structures or could have occurred in the initial humic materials in essentially the same forms in which they were isolated, but held together by relatively weak bonding. If the latter hypothesis is correct, then the phenolic and benzenecarboxylic acids would be the "building blocks" of humic materials and future research should be directed toward finding out how the "building blocks" fit together and what type of structural arrangement is produced. If, on the other hand, the degradation products originate from more complex chemical structures, further research should be concerned with developing milder degradation methods that would permit the isolation and identification of larger fragments of the total structure.

Present indications are that humic substances are not single molecules but rather associations or aggregates of molecules of microbiological, polyphenolic, lignin and condensed lignin origins. These are the benzenecarboxylic and phenolic acids referred to earlier as "building blocks". In HA's and humins the "building blocks" appear to be more complex and stable than in FA's[3]. But in all cases the forces holding the "building blocks" together are similar, consisting mainly of low-energy bonds. The importance of intra- and inter-molecular hydrogen bonds in holding together constituent molecules within an aggregate and linking aggregates with each other has recently been highlighted by Wershaw et al.[4]

X-ray analysis, electron microscopy and viscometry[5] all point to a relatively "open", flexible structure perforated by voids of varying dimensions that can trap or fix organic and inorganic substances that fit into the voids, provided that the charges are complimentary, and that can also interact with these substances on its surface. A chemical structure that is in harmony with many of the requirements that I have so far discussed is shown in Figure 5. We have isolated each of the components of the structure from HA's and FA's before and after chemical degradation. Bonding

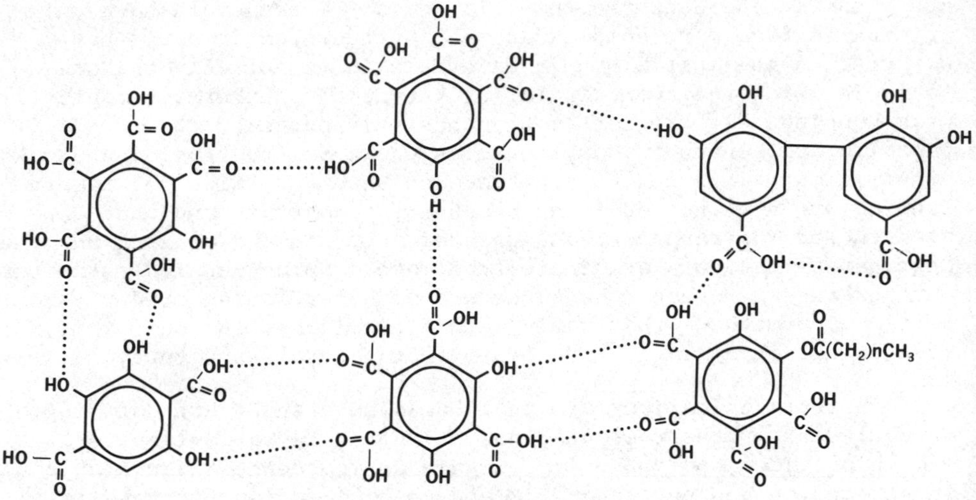

Figure 5. Partial structure for FA.

between the "building blocks" is by hydrogen-bonds, which makes the structure flexible, permits the "building blocks" to aggregate and disperse reversibly, depending on the pH, ionic strength, the presence of metal ions, etc., and allows the humic substance to interact with inorganic and organic soil constituents either via oxygen-containing functional groups on the large external and internal surfaces, or by trapping them in internal voids which appear to be less hydrophilic than the external surfaces.

Effects of Low pH on a Number of FA Characteristics

(a) Effects observed under the Scanning Electron Microscope (SEM). The effect of varying the pH on the shape and particle arrangement of FA is shown in Figure 6. At pH 2 (Figure 6a), the FA consists of elongated fibers and bundles of fibres. The fiber thickness ranges from 0.1 to 0.4 μm. The short, thick protrusions with rounded heads are 0.25 - 0.45 μm long and about 0.15 μm thick. At pH 4 (Figure 6b) the fibers tend to become thinner, and this tendency becomes more pronounced at pH 6 (Figure 6c), when a greater proportion of the FA occurs in bundles of closely knit fibers. At pH 7 (Figure 6d) we observe a fine network of tightly meshed fibers which tend to coalesce. At pH 8 (Figure 6e), the FA particles form a sheet like structure of varying thickness. At pH 9 (Figure 6f), the plastic-like sheets tend to thicken whereas at pH 10 (Figure 6g) the particles appear to be fine-grained, with a high degree of homogeneity.

Thus, the micrographs show that pH has a marked effect on the trimensional arrangement of FA particles. We observe a gradual transition from a more random structure at low pH, to a more organized and oriented one at higher pH. Simultaneously, the particles become smaller as the pH increases. It is not possible to do the same type of experiment with HA because it is not water-soluble at pH < 6.5.

(b) Effect on viscosity. Figure 7 shows the relationship between η red and pH at a constant FA concentration of 0.6 g/100 ml. As the pH is lowered, aggregation or association of FA particles occurs, so that the viscosity increases; the FA particles are practically uncharged under these conditions, and electrostatic repulsion is not an important factor.

(c) Effects on ESR parameters. Effects of pH on ESR parameters of Fa and HA in aqueous solutions are listed in Table 4 and ESR spectra are shown in Figure 8. Regardless of pH, all ESR spectra consist of single, symmetrical lines devoid of hyperfine splitting.

Spin concentrations in both FA and HA increase with rise in pH. This is also true for g-values but line widths for FA's are narrower at high relative to low pH.

Table 3. Major chemical structures in "model" HA and FA[3].

Major Products	HA (%)	FA (%)
Aliphatic	24.0	22.2
Phenolic	20.3	30.2
Benzenecarboxylic	32.0	23.0
Total	76.3	75.4
Ratio $\dfrac{benzenecarboxylic}{phenolic}$	1.6	0.8
Aromaticity	69	71

Table 4. ESR parameters for FA and Ha solutions at various pHs[5].

Material	pH	Spins/g(X 10^{-17})	Line width(G)	g-values
FA	2.0	1.44	2.5	2.0038
FA	5.0	1.50	3.0	2.0037
FA	7.0	1.56	3.0	2.0039
FA	8.0	1.51	2.5	2.0040
FA	9.0	1.69	2.0	2.0044
FA	10.6	2.19	2.0	2.0044
FA	11.6	2.56	2.1	2.0044
FA	12.5	12.06	2.2	2.0045
HA	6.0	23.51	3.3	2.0031
HA	8.0	26.24	3.5	2.0035
HA	9.3	35.00	3.5	2.0035
HA	10.3	37.00	3.6	2.0039
HA	11.1	37.43	3.5	2.0041

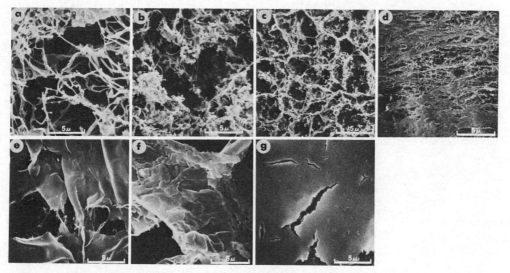

Figure 6. Scanning electron micrographs of FA at various pH levels:
a = pH 2; b = pH 4 c = pH 6; d = pH 7; e = pH 8; f = pH 9; g = pH 10.

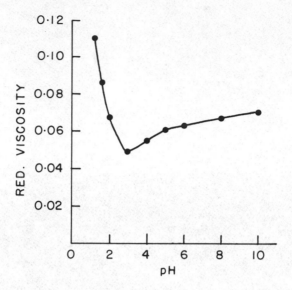

Figure 7.   Effect of pH on η red of FA.

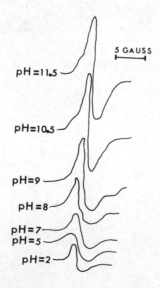

Figure 8.   ESR spectra of FA at different pH levels.

Most of the g-values are in the 2.0038-2.0042 range, characteristic of semiquinones or substituted semiquinones. The latter are stabilized as semiquinone ions at higher pH. The semiquinones are formed by the chemical or biological dehydrogenation of phenolic hydroxyl groups. Broadening of line widths at more acid pH is most likely due to increased molecular association with the aid of hydrogen bonds and/or dipolar spin-spin interactions between electron spins.

The aggregation of HA and FA particles at low pH can be explained by hydrogen-bonding, van der Waal's interactions, and interactions between π electrons of adjacent molecules, as well as by homolytic reactions between free radicals. As the pH increases, these forces become weaker, and because of increasing ionization of carboxylic acid and phenolic hydroxyl groups, particles separate and begin to repel each other electrostatically, so that the molecular arrangements become smaller but better oriented.

Thus, the HA and FA behave like flexible, linear, synthetic polyelectrolytes. Aggregation at low pH is the result of intermolecular attraction, whereas dispersion at high pH derives from intermolecular repulsion.

(d) Extraction of organic matter from soils by dilute acid. In a previous investigation[6] we extracted with 0.1N HCl a soil sample taken from a Podzol Bh horizon (15-20 cm below the surface). The pH of the extract was 1.9. Following removal of most of the Cl⁻ the extract contained 2.6% ash. One extraction removed 40% of the organic matter in the initial soil sample.

Elementary compositions of organic matter extracted with 0.5 N NaOH and 0.1N HCl (Table 5) were very similar. Both contained approximately 50% C and 45% O, typical of FA's. Both extracts were soluble in alkali and acid. Oxygen-containing functional groups in the two extracts were also of the same order of magnitude (Table 5). These data convey the impression that the two extractants removed very similar materials from the soil sample.

Infrared spectra of the two extracts were practically identical and so was their separation on sephadex G 50 and G 25 gels (Figure 9).

Thus, from soils containing appreciable concentrations of FA (such as Podzols), dilute acids can solubilize and remove substantial amounts of low molecular weight humic materials, that is, FA's.

On the other hand, dilute inorganic acids are incapable of extracting more than traces of HA's from both organic and inorganic soils. For all practical purposes, HA's are insoluble in dilute inorganic acids.

Table 5. Chemical characteristics of purified organic matter
extracted from a Podzol Bh horizon soil by 0.5N NaOH
and by 0.1N HCl[6].

	Elementary composition, % (dry, ash-free)	
	0.5N NaOH extract	0.1N HCl extract
C	49.50	48.61
H	3.60	4.22
N	0.75	0.47
S	0.25	0.89
O	45.90	45.67
	Oxygen-containing functional groups meq/g (dry, ash-free)	
Total acidity	12.4	11.5
Carboxyl	9.1	8.7
Phenolic hydroxyl	3.3	2.8
Alcoholic hydroxyl	3.6	3.0
Carbonyl	3.1	3.6

METAL-ORGANIC INTERACTIONS

It may be appropriate to refer at this point to the major types
of reactions that are known to occur between FA and metal ions
(Figure 10).

According to reaction (1) one CO_2H group reacts with one metal
ion to form an organic salt or monodentate complex. Equation (2)
describes a reaction in which one CO_2H group and one adjacent OH
group react simultaneously with the metal ion to form a bidentate
complex or chelate. According to equation (3), two adjacent CO_2H
groups interact simultaneously with the metal ion to also form a
bidentate chelate. Chelation according to equations (2) and (3)
has been discussed in considerable detail by Gamble and
Schnitzer.[9]

Measurements of metal-HA and metal-FA stability constants have
been done in recent years in a number of laboratories[7]. There is

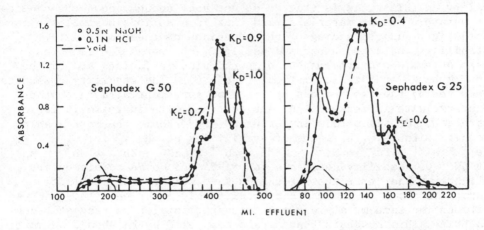

Figure 9. Gel filtration of the organic matter extracts.

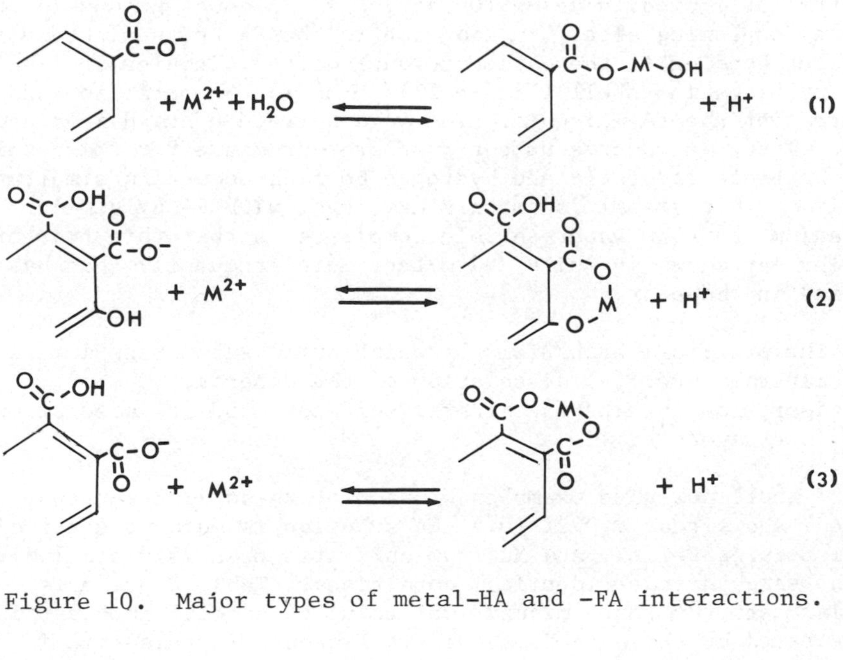

Figure 10. Major types of metal–HA and –FA interactions.

still considerable uncertainty about how to define the problem, which method to use and how useful such constants are once they have been determined. The main reason for the current state of uncertainty in this field is that we do not know enough about the chemical structure of humic materials, and this is a serious obstacle to an intelligent understanding of the reactions of these materials. Stability constants, expressed as log K, for metal-FA complexes were determined by the method of continuous variations and by the ion exchange equilibrium method[8] (Table 6). The stability constants measured by the two methods are in good agreement with each other, increase with increase in pH, but decrease with increase in ionic strength. Of all metals investigated, Fe^{3+} forms the most stable complex with FA. The order of stabilities at pH 3.0 is: $Fe^{3+} > Al^{3+} > Cu^{2+} > Ni^{2+} > Co^{2+} > Pb^{2+} \simeq Ca^{2+} > Zn^{2+} > Mn^{2+} > Mg^{2+}$. At pH 5.0, stability constants for Ni-FA and Co-FA complexes are slightly higher than that of the Cu-FA complex. The stability constants shown in Table 4 are considerably lower than those formed between the same metal ions and synthetic complexing reagents such as EDTA. This suggests that metals complexed by FA should be more readily available to plant roots, microbes and small animals than when sequestered by EDTA and similar reagents.

In a recent study on water-soluble Mn^{2+}-FA complexes by ESR spectrometry[9], it was noted that Mn^{2+} was complexed to FA by electrostatic binding as $Mn(OH_2)^{2+}$, with FA donor groups in outer sphere complexing sites, in addition to the FA being linked via hydrogen bonding to the cation through water molecules in its primary hydration shell (Figure 11). Mn^+ is the metal ion and R the rest of the FA molecule. The bond between O and H is a hydrogen bond. Thus, we observe an outer sphere structure for the complex, with both electrostatic and hydrogen bonding occurring simultaneously. Other transition metals may react with FA by similar mechanisms to form water-soluble complexes so that this type of complex may occur in soils and waters more frequently than has been assumed in the past.

The principal mechanisms by which humic substances interact with minerals are: (a) dissolution of the mineral, (b) adsorption on external mineral surfaces, and (c) adsorption in clay interlayers[10].

I shall now give examples of each of these interactions. Table 7 shows that a 0.2% (w/v) FA solution is much more efficient in dissolving Fe, Al, and Mg from chlorites than is distilled water at the same pH under identical conditions. Table 8 presents similar data for three micas. One point of special interest is illustrated by the data in these two Tables: minerals rich in Fe are more susceptible to attack by FA than are those free of Fe or containing little Fe. The strong complexing of Fe by humic materials appears to have an adverse effect on the structural stability of Fe-

Table 6. Stability constants of metal-FA complexes[8] (CV = method
          of continuous variations; IE = ion-exchange equilibrium
          method).

| | Log K | | | |
| Metal | pH 3.0 | | pH 5.0 | |
	CV	IE	CV	IE
Cu^{2+}	3.3	3.3	4.0	4.0
Ni^{2+}	3.1	3.2	4.2	4.2
Co^{2+}	2.9	2.8	4.2	4.1
Pb^{2+}	2.6	2.7	4.1	4.0
Ca^{2+}	2.6	2.7	3.4	3.3
Zn^{2+}	2.4	2.2	3.7	3.6
Mn^{2+}	2.1	2.2	3.7	3.7
Mg^{2+}	1.9	1.9	2.2	2.1
Fe^{3+}	6.1*	-	-	-
Al^{3+}	3.7**	3.7**	-	-

*determined at pH 1.70
**determined at pH 2.35

Table 7. Dissolution of metals (Fe, Al, Mg) from chorites by
          0.2% FA solution and by H_2O after 360 h of shaking at
          pH 2.5[10].

| Type of Mineral | % of sample dissolved by | |
	H_2O	0.2% FA
Leuchtenbergite (Fe-poor)	2	4
Thuringite (Fe-rich)	6	26

Table 8. Dissolution of micas by 0.2% FA (pH 2.5)[10].

Mineral	Element dissolved	% dissolved after 710 h
Biotite	Fe	11.5
(Fe-rich)	Al	14.5
	Mg	17.0
	K	18.5
	Si	14.0
Phlogopite	Al	8.0
(Fe-poor)	Mg	8.5
	K	8.0
	Si	9.0
Muscovite	Al	2.3
(Fe-poor)	K	1.8
	Si	0.6

Table 9. Effect of pH on adsorption of FA by 20 mg of Kaolinite
          (<1μ and 2-5μ), muscovite and sepiolite.

pH	<1μ Kaolinite	5-2μ Kaolinite	Muscovite	Sepiolite
		mg FA		
2.5	1.46	0.74	0.98	2.61
3.5	0.84	0.56	0.69	2.00
4.5	0.60	0.48	0.61	2.02
5.5	0.62	0.38	0.64	1.66
6.5	0.68	0.44	0.54	1.46

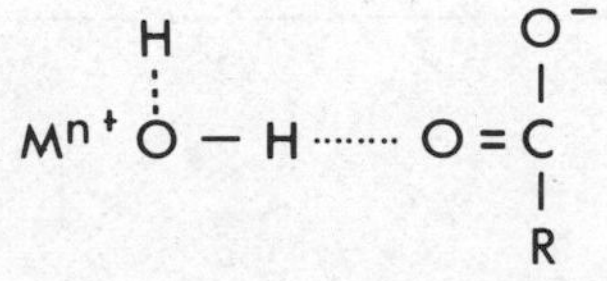

$$M^{n+} \; \overset{\text{H}}{\underset{}{\text{O}}} - H \cdots\cdots O = \overset{O^-}{\underset{R}{C}}$$

Figure 11. Outer-sphere metal-FA complex.

rich minerals so that along with Fe, other major constituent
elements such as Mg, Al, K and Si are also released.

    The extent of adsorption of humic materials on external
mineral surfaces depends on the surface area, geometry and chemistry
of the surface, the pH of the system and the water content. The data
in Table 9 illustrate these effects. Note the relatively high
adsorption of FA by sepiolite, which has a channel-like structure
formed by the joining of edges of slender and elongated talc-like
layers.

    The interlayer adsorption of FA by Na-montmorillonite is pH-
dependent, being greatest at low pH, but no longer occurring at
pH > 5.0 (Figure 12). The main reaction mechanism governing the
interlayer adsorption of low-molecular weight humic materials
appears to be the ability of relatively undissociated humic materials
to displace water from clay interlayers, and so approach the dominant
cation. Thus, under acidic conditions, substantial amounts of Fa
will be adsorbed in the interlayers of expanding clay minerals.
In the long run, we may expect increasing degradation of most
minerals that are normally found in soils under conditions of even
moderate acidity.

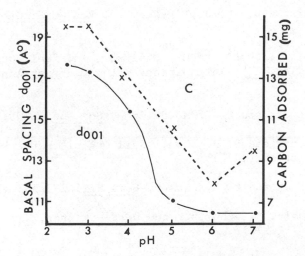

Figure 12.   Effect of pH on d_{001} and adsorption of FA (%C · 2).

SUMMARY

Conditions of moderate acidity tend to aggregate HA particles but lead to the dissolution and mobilization of FA's. In soils such as Podzols, where FA's are the major organic matter components, losses of organic matter brought about in this manner can be substantial. A soil pH of between 2 and 3 will favour the adsorption of FA on mineral surfaces. Also, acid conditions will increase the interlayer adsorption of FA by expanding clay minerals. As these conditions persist for a long time both protons and solubilized FA will attack and degrade minerals and so initiate fundamental changes in the inorganic and organic composition of soils. The net result will be a less favourable environment for biological activities and a concomitant reduction in soil fertility.

REFERENCES

1. H. L. Bohn, Soil Sci. Soc. Am. J. 40:468 (1976).

2. M. Schnitzer and S. U. Khan, "Humic substances in the Environment", Marcel Dekker, New York (1972).

3. M. Schnitzer, in: "Soil Organic Matter Studies, II", International Atomic Energy Agency, Vienna (1977).

4. R. L. Wershaw and D. J. Pinckney, J. Res. U.S. Geol. Survey 5:571 (1977).

5. N. Senesi, Y. Chen, and M. Schnitzer, in: "Soil Organic Matter Studies, II", International Atomic Energy Agency, Vienna (1977).

6. M. Schnitzer, and S. I. M. Skinner, Soil Sci. 105:392 (1968).

7. M. Schnitzer and S. U. Khan, "Soil Organic Matter", Elsevier, Amsterdam (1978).

8. M. Schnitzer and E. H. Hansen, Soil Sci. 109:333 (1970).

9. D. S. Gamble, M. Schnitzer and D. S. Skinner, Can. J. Soil Sci. 57:47 (1977).

10. M. Schnitzer and H. Kodama, in: "Minerals in Soil Environments", Soil Science Society of America, Madison, Wisconsin (1977); M. Schnitzer and H. Kodama, Geoderma 15:381 (1976).

PODZOLIZATION: MECHANISM AND POSSIBLE EFFECTS OF ACID PRECIPITATION

Leif Petersen

Royal Veterinary and Agricultural University, Chemistry Department, 1871 Copenhagen V, Denmark

INTRODUCTION

Large areas in the cool humid regions of North America, Europe and Asia are covered by podzol soils. These soils are normally confined to sandy parent materials where the precipitation is high enough to cause a severe leaching. Their natural vegetation is most often coniferous forest or heath.

A well developed podzol has a number of characteristic features including an acid reaction, a low content of most plant nutrients, a mor type humus, and a pronounced differentiation of the profile into eluvial (A) and illuvial (B) horizons. The eluvial horizon has suffered a loss of aluminum and iron compounds, and it is strongly bleached. The metals have moved downwards with leaching water and have, at least partly, been deposited together with organic matter in the illuvial horizon. Due to this deposition the illuvial horizon is dominated by dark or reddish colours. The material below the illuvial horizon is normally considered to be fairly unchanged parent material.

The soil processes leading to the formation of podzols have been studied by soil scientists for more than a hundred years but, nevertheless, many aspects of the podzolization processes are still incompletely understood. The horizon sequence shows clearly that a downward translocation of iron, aluminum and organic matter has taken place within the profile. In order to explain these translocations it is necessary to understand how the compounds are mobilized in eluvial horizon and how they are immobilized in the illuvial horizon.

Many studies have been concerned primarily with the mobilization of iron and aluminium in the eluvial horizon. However, in
order to arrive at a complete understanding of the podzolization
processes it is necessary also to consider the reactions by which
the organic matter becomes mobile and the reactions causing deposition of the metals as well as the organic matter in the illuvial
horizon.

CONDITIONS FOR PODZOLIZATION

As mentioned above podzols occur mainly on sandy parent materials in cool humid climates where the leaching is high. Podzols
may also be found in warm humid climates, even in the tropics, but
here they are also confined to sandy parent materials. Podzols are
found only under conditions of high leaching and on sandy parent
materials because podzolization requires an acid soil reaction. The
high leaching causes a removal of basic compounds such as carbonate
and, to a large extent, of mono- and divalent metal ions from the
soil. The low buffer capacity of sandy soils accelerates acidification. Furthermore, sandy soils have low contents of iron and aluminium compounds and, as will be shown below, this also makes these
soils liable to podzolization.

The high leaching leads to a low content of plant nutrients
in the soil and this in connection with the acid reaction causes
rather poor conditions for the microorganism in the soil. One of
the consequences of this is that the decomposition of the plant
residues is retarded and a mor type humus is formed. This is usually found as an almost purely organic layer (O) on top of the
mineral soil.

TRANSLOCATION OF ORGANIC
MATTER, IRON AND ALUMINIUM COMPOUNDS

Several mechanisms have been proposed in order to explain the
translocation of organic matter, iron and aluminium during podzolization. Translocation of the metals as simple ions is generally
ruled out because of the low solubilities of iron(III) and aluminium hydroxide. Calculations[1] based on the solubility products of
iron(III) and aluminium hydroxide, and taking into account the
formation of soluble hydroxy-complexes, have shown, however, that
the solubility of aluminium hydroxide is high enough to account
for a significant equilibrium concentration of dissolved aluminium
at pH 3-3.5. pH values in this range are often found in the upper
horizon of podzols. Since iron(III) hydroxide is less soluble than
aluminium hydroxide, the equilibrium concentration of dissolved iron
is much lower than that of aluminium and hardly large enough to

account for the amount of iron that may be translocated during
podzolization. The solubility of iron may be increased if reducing
conditions prevail in the soil and this may play a role in some
soils but hardly in all podzols.

Some authors have explained podzolization in terms of translocations of electrically charged iron(III) hydroxide, aluminium
hydroxide and humus sols[2].

At present there seems to be a fairly general agreement that
podzolization is caused by organic compounds which are capable of
forming soluble compounds with the metals. Several experiments
have shown that aqeuous extracts of various plant and soil materials contain organic compounds capable of dissolving iron and aluminium and maintaining these elements in solution under conditions
where they would otherwise be precipitated as hydroxides[3-13].

The dissolution of iron and aluminium is assumed to proceed
through formation of complex compounds between the organic matter
and the metals but little is known about the nature of the active
organic compounds. Coulson, Davies and Lewis[5] identified a number
of so-called pólyphenols in beech litter and soil extracts and
found that these compounds could reduce and mobilize substantial
amounts of iron from precipitated ferric hydroxide and from diato-
maceous earth, aluminium oxide and sand columns impregnated with
ferric chloride. The reduction of iron took place even under aerobic conditions. Bloomfield[14] and Davies[15] were of the opinion that
polyphenols play a major role in mobilizing iron but attention has
also been paid to many other compounds.

Kaurichev and Nozdrunova[5,16] and Kaurichev, Ivanova and
Nozdrunova[17] attributed the major iron-dissolving effect of aqueous
soil extracts to simple organic acids such as formic, oxalic,
citric, succinic, malic, lactic and various amino acids. Jacquin
and Bruckert[18] identified a number of simple aliphatic and phenolic
acids in extracts from oak and pine litter, and Bruckert[10] found
that an aqueous extract of a mor-layer contained simple aliphatic
acids as well as polymeric unidentified compounds. Among the simple
aliphatic acids, citric, malic, oxalic, malonic, succinic and lactic
acids were identified and determined quantitatively. Although these
acids apparently accounted for only a relatively small fraction of
the organic matter of the extracts, they were reported to account
for 75% of the capacity of the extract for complexing iron.

Nykvist[19] found significant amounts of simple organic acids
in extracts from oak, beech, ash, alder, birch, and pine litter.
However, in Nykvist's experiments these acids proved highly unstable in the presence of oxygen and organisms originating from
the litter, and he concluded that they would decompose inside the

litter and be of minor importance for the mobilization of iron and
other elements in the soil.

The processes causing deposition of the compounds in the illuvial horizon are not well understood. If the compounds active in
the mobilization of the metals are simple organic acids it is rather difficult to imagine a mechanism by which they accumulate in a
specific soil horizon. The deposition has often been explained as
being due to biological decomposition of the organic molecules. It
has been demonstrated that bacteria are able to precipitate ferric
hydroxide from ferric citrate solutions[20-22], and Crawford[23] found
that the illuvial horizon of some podzols contained a large number
of such bacteria. On the other hand, McKenzie, Whiteside and
Erickson[24] were unable to confirm this, in fact they found that the
number of citrate oxidizing bacteria was substantially lower in
illuvial horizons than in eluvial horizons of the same podzols. A
more elaborate deposition mechanism involving decarboxylation of
the organic acids followed by biological decomposition and transformation into humus compounds has been suggested by Bruckert[10].

A deposition caused by biological decomposition would not be
limited to simple organic acids, it could also affect other organic compounds active in the complexation of the metals. It has been
mentioned above that some authors regard polyphenols as the compounds active in the complexation and mobilization of iron and aluminium. The polyphenols do not constitute a well-defined group of
compounds but include a variety of organic compounds from simple
polyhydroxybenzenes to complex polymeric phenolic compounds. Their
chemical structure is related to that of the tannins and they have
a number of properties in common with the humus compounds. Many
authors, including Davies[15], regard polyphenols as precursors of the
true humus compounds.

It has been found in many experiments[25-28] that addition of
iron(III) or aluminium salts to solutions containing dispersed soil
organic matter may cause a mutual precipitation of the metals and
the organic matter. Bloomfield[29] found that addition of ferric hydroxide to plant extracts caused a decline in the content of dissolved organic matter and iron in the extracts. These results indicate
that precipitation could be due to a large uptake of iron and/or
aluminium by the organic compounds in the illuvial horizon. This
would be consistent with a number of properties of podzols. As
shown by Petersen[1] the ratio between organic carbon and iron + aluminium decreases sharply with depth in podzols. Several authors[28,
30-33] have demonstrated that the organic matter of podzol illuvial
horizons may be extracted almost completely by a variety of acids,
complex-forming compounds, bases and cation exchange resins. These
reagents are all capable of removing metals such as iron and aluminium from the organic matter as soluble or insoluble compounds. On

the other hand, the only reagent that can extract a substantial
part of the organic matter from O and A horizons is concentrated
sodium hydroxide. These results may be interpreted to confirm that
the organic matter is deposited and maintained insoluble in the
illuvial horizon by iron and aluminium, as also suggested by
Ponomareva[28], Schnitzer[34] and Duchaufour[35].

MODEL EXPERIMENT ON PODZOLIZATION

In order to obtain some information regarding the ability of
water-soluble soil organic matter to react with certain metal ions,
an experiment was carried out as described below.

A soil extract was prepared by a prolonged treatment of soil
from the eluvial horizon of a podzol with distilled water. Overall
aerobic conditions were maintained during the extráction period by
passing atmospheric air through the soil-water mixtures. The composition of the soil extract is shown in Table 1.

Table 1. Composition of Soil Extract from Podzol Eluvial Horizon
(mg per 1)

C	148
Na	3.8
K	5.9
Ca	0.65
Mg	0.24
Fe	1.74
Al	0.83

Five cation exchanger columns(heigth approx. 100 mm, diameter
4,5 mm) were prepared from a Dowex 50 W x 8 (50-100 mesh) resin and
saturated with hydrogen, sodium, calcium, aluminium and iron(III)
ions respectively. The columns were leached with ten 5 ml portions
of distilled water, ten 5 ml portions of the soil extract mentioned
above, and ten 5 ml portions of distilled water,in that order. The
flow rate was adjusted so that it took 6-8 hours for each 5 ml portion to pass through the column. Each 5 ml portion was collected
separately and analysed for carbon and the metal present in the
column as described elsewhere[1].

The carbon contents of the leachates are shown in Table 2. By
comparison with Table 1 it is seen that there is a considerable
reduction in the carbon content of the soil extract that has passed the aluminium saturated ion exchanger column.A smaller, but
still marked, reduction is noted for the soil extract that has passed the column saturated with iron(III). As compared with these
reductions the decreases in the carbon content caused by the hyd-

Table 2. Carbon Contents of Leachates from Ion Exchanger Columns
(mg per l)

Liquid applied	Portion No.	Column saturated with				
		H	Na	Ca	Al	Fe
	1	3	3	4	3	1
	2	6	3	6	2	2
	3	5	6	10	6	2
	4	6	4	7	2	1
Distilled	5	2	3	7	2	2
water	6	1	3	2	2	1
	7	2	6	0	0	1
	8	1	3	3	2	1
	9	2	5	3	1	1
	10	2	7	7	1	1
	1	84	111	88	54	49
	2	135	161	140	120	116
	3	148	155	132	109	123
A_o-A_1	4	117	165	130	108	120
soil	5	137	154	141	98	128
extract	6	140	157	124	99	129
	7	129	157	131	113	116
	8	130	153	133	100	128
	9	146	156	140	111	126
	10	149	159	130	122	123
	1	67	70	68	54	68
	2	4	8	9	33	17
	3	9	11	9	13	17
	4	7	2	3	10	5
Distilled	5	4	11	7	14	7
water	6	5	5	6	4	6
	7	5	4	6	5	5
	8	1	3	6	8	4
	9	3	1	2	7	5
	10	3	5	3	5	4

rogen and calcium saturated resins, and the increase caused by the
sodium saturated resin, are rather small. When evaluating the decreases and increases it should be taken into account that the
accuracy obtained by the method used for determination of organic
carbon is rather low. However, the decreases caused by the aluminium and iron saturated resins are highly significant (P < 0.001).
The decrease caused by the calcium saturated resin is significant

to the five percent level, while the effects of the hydrogen and sodium saturated resins are not significant to this level.

The results shown in Table 2 are confirmed by the observations made during the leachings. The original soil extract had a brown colour and this remained unaffected by the passage of the hydrogen, sodium and calcium saturated columns. No or only a negligible deposition of organic matter was observed in these columns (Fig. 1). By contrast the colour intensity of the soil extract was strongly reduced by passage of the aluminium saturated column and a strong deposition of organic matter took place in the column. This caused the column to change its colour, which originally was faint yellow and identical to that of the hydrogen, sodium and calcium saturated columns, to dark brown, almost black (Fig. 1). A deposition of organic matter was also observed in the column saturated with iron(III) but it had no pronounced effect on the colour of the column, because the ferric ions caused the resin to become dark brown. Although the iron saturated column caused a marked decrease in the carbon content of the soil extract it had no visible effect on its colour, in fact measurements of the optical density showed an increase in the colour intensity. This may be due to formation of coloured iron-organic matter complexes.

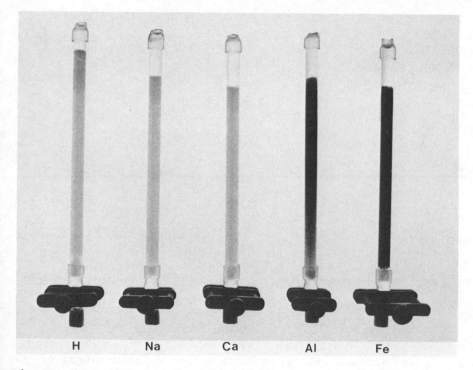

Fig. 1. Ion exchanger columns after final leaching with water.

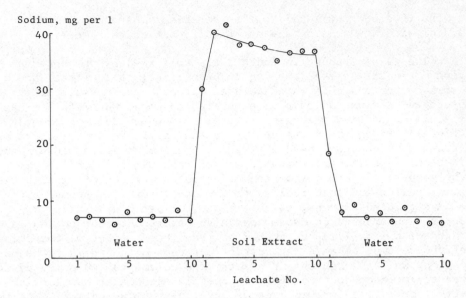

Fig. 2.   Sodium contents of leachates from sodium saturated ion exchanger column.

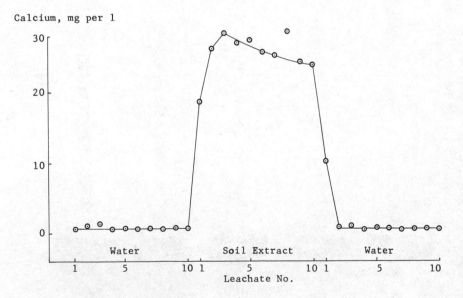

Fig. 3.   Calcium contents of leachates from calcium saturated ion exchanger column.

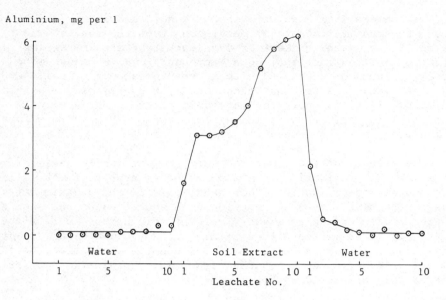

Fig. 4. Aluminium contents of leachates from aluminium saturated ion exchanger column.

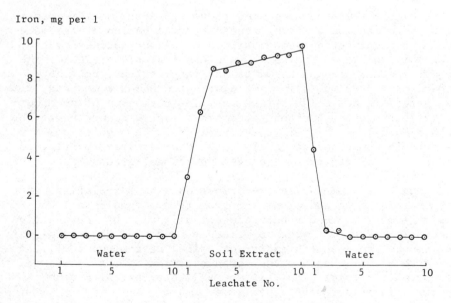

Fig. 5. Iron contents of leachates from ion exchanger column saturated with ferric iron.

It appears from Figs. 2-5 that the soil extract is able to
remove substantial amounts of metal ions from the resins. The
changes in the metal content of the leachates with increasing
amount of soil extract applied are probably due to change in pH
For all columns pH decreased with increasing amount of soil extract applied. This was most pronounced for the aluminium saturated column (Fig. 4) where the marked increase in the aluminium
content of the soil extract leachates was closely correlated with
an increase in the hydrogen ion concentration.

The results of the experiment confirm that water-soluble organic compounds originating from a podzol are capable of mobilizing
iron and aluminium. On the other hand, part of the organic matter
looses its solubility upon reaction with sufficient amounts of the
two metals. According to Table 2, more organic carbon is retained
in the column saturated with aluminium than in that saturated with
iron(III). This may indicate that aluminium is more effective in
precipitating organic matter than iron. Additional evidence for
this fact is presented by Petersen[1].

It is also clear from the experiment that sodium and calcium
ions have no or only a negligible capability to cause precipitation and deposition of the organic compounds.

THE PODZOLIZATION PROCESS

From the evidence presented above it seems justified to explain podzolization as being caused by water-soluble organic compounds produced during decomposition of plant residues in the
upper soil horizons. The nature and chemical properties of the organic compounds are not known to any large extent, but they seem
to have many properties in common with the humus compounds, although
the latter are not soluble in water. The formation of such compounds in the upper horizons of podzols may be connected to the
low biological activity in these horizons. The high solubility of
the organic compounds is likely to be due, at least partly, to a
high content of negatively charged functional groups.

When these compounds move downwards with percolating water
they dissolve iron and aluminium from the inorganic soil constituents. The metals cause a partial neutralization of the negative
charge carried by the organic molecules and when the charge has
been reduced to a sufficiently low value the compounds will no
longer be soluble and a mutual precipitation of the metals and the
organic matter takes place. It seems likely that aluminium is more
important in this respect than iron.

In the initial stages of the development of a podzol sufficient

iron and aluminium will be available to cause deposition of the
organic compounds at shallow depth. However, as the production of
the active organic compounds continues,all available iron and aluminium will be removed from the upper soil layers, and the organic
compounds must be leached to greater depth in order to accumulate
sufficient iron and aluminium to become precipitated. Some iron
and aluminium may be dissolved from metal-organic compounds precipitated earlier. In the long run this causes the illuvial horizon
to descend gradually. Since the organic compounds are not identical, they probably require different amounts of metals to become
precipitated. Those requiring the smallest amounts will precipitate
first, and hence the upper part of the illuvial horizon becomes
comparatively rich in organic matter. The organic compounds requiring larger amounts of metals proceed to deeper layers of the illuvial horizon and here the organic matter-metal ratio becomes smaller.
This in agreement with the usual distribution of organic matter
and metals in the illuvial horizon. Some organic compounds may not
become precipitated at all, but may be leached completely out of
the soil profile.

As mentioned above podzolization requires an acid reaction.
One of the reasons may be that the active soluble organic compounds
are not formed in neutral soils. The fact that clayey soils are unlikely to become podzolized,even if they have low pH-values, may
be explained in terms of the larger content of iron and aluminium
compounds in these soils as compared with sandy soils. The high
metal content will cause precipitation of the organic compounds in
the upper soil layers where they are formed, and hence no translocation will take place. This is in agreement with results obtained
by Duchaufour and Souchier[36] according to which podzolization is
inversely related to the content of iron and clay in the soil.

POSSIBLE EFFECTS OF ACID PRECIPITATION ON PODZOLIZATION

According to Duchaufour[35] the time required for development
of a podzol may vary from a few hundred years under optimum conditions to several thousand years where the conditions for podzolization are less appropriate. However, even the shorter period is long
as compared with the period in which observations on the occurence
and effects of acid precipitation have been carried out.Hence, it
is not to be expected that it is possible as yet to make direct
observations on the effect of acid precipitation on podzol development.

Since acidification of the soil is a necessary prerequisite
for podzolization one would expect that a content of sulphuric
acid or other strong acids in the precipitation would shorten the
period required for the soil to become so acid that podzolization

will take place. Several investigations have shown that acid precipitation causes an increased leaching of divalent cations from
the soil and a corresponding decrease in the base saturation percentage[37-40]. The magnitude of this effect depends,of course, on the
amount of acid received by the soil. To estimate the relative importance of the strong acids it is further necessary to compare
with the amounts of weaker acids produced by natural processes in
the soil.

The effect on the soil reaction depends on the buffer capacity
of the soil. If this is sufficiently large the effect may be insignificant, and it may take a long time before any effect on the soil
pH is noted. The low buffer capacity of soils liable to become podzolized make these rather sensitive to the application of strong
acids, and an accelerated decrease in the soil pH may occur within
a rather short period.

When the soils become so acid that a mor-layer develops the
buffer capacity may increase somewhat,and in addition the acids
will be less effective in replacing adsorbed cations[41]. Furthermore,
mono- and divalent cations, which are always present in natural precipitation together with the strong acids, reduce the acidifying
effect of the precipitation on the soil,because these cations com-
pete with the hydrogen ions for the exchange sites[41].

Whether strong acids will have an effect on the translocation
processes taking place during podzolization is unknown. A rough
comparison of the amounts of sulphur commonly found in the precipitation[42-45] with the contrations of organic acids in the soil
solution of upper horizons of podzols[1] indicates that the former
are small compared with the latter, except possibly in heavily polluted areas. Sulphate and nitrate ions are probably unable to sub-
stitute for organic compounds in the translocation processes discussed above.

By lysimeter experiments with a podzol soil Abrahamsen et al.[40]
have found that the aluminium content of the leachate increases
with increasing amount of sulphuric acid in simulated rain applied
to the lysimeters.According to the podzolization mechanism described
in the preceding section such an increased leaching of aluminium
from the profile would favour the mobilization of organic compounds
but hamper their deposition in the illuvial horizon. Hence, the
results could be an increased production of water-soluble organic
compounds in the upper horizons and a leaching of these to greater
depth before precipitation takes place. Actually, this means that
podzolization is amplified.

However, it should be stressed that insufficient evidence is
available to substantiate these assumptions, and that the addition

of strong acids to podzols could cause unknown changes in the
chemical and biological properties of the soils, which in turn could
lead to different results.

REFERENCES

1. L. Petersen, "Podzols and Podzolization", DSR Forlag, Copenhagen,
    (1976).

2. S. Mattson, Soil Sci. 36: 149 (1933); S. Mattson, and Y.
    Gustafsson, Lantbrukshogsk. Ann. 1: 33 (1934); S. Mattson, and
    Y. Gustafsson, Lantbrukshogsk. Ann. 2: 1 (1935); S. Mattson,
    and I. Nilsson, Lantbrukshogsk. Ann. 2: 115 (1935); S. Mattson,
    and Y. Gustafsson, Lantbrukshogsk. Ann. 4: 1 (1937); S. Mattson,
    and Y. Gustafsson, Soil Sci. 43: 453 (1937).

3. C. Bloomfield, Nature 170: 540 (1952); C. Bloomfield, J. Soil
    Sci. 4: 5 (1953); C. Bloomfield, J. Soil Sci. 4: 17 (1953);
    C. Bloomfield, J. Soil Sci. 5: 39 (1954); C. Bloomfield, J.
    Soil Sci. 5: 46 (1954); C. Bloomfield, J. Soil Sci. 5: 50 (1954);
    C. Bloomfield, Trans. Intern. Congr. Soil Sci. 6th Paris B: 427
    (1956); C. Bloomfield, Rep. Rothamsted Exp. Stn., 73 (1966).

4. W.A. DeLong and M. Schnitzer, Soil Sci. Soc. Am. Proc. 19: 360
    (1955); M. Schnitzer, and W.A. DeLong, Soil Sci. Soc. Am. Proc.
    19: 363 (1955).

5. C.B. Coulson, R.I. Davies and D.A. Lewis. J. Soil Sci. 11: 20
    (1960); C.B. Coulson, R.I. Davies, and D.A. Lewis, J. Soil Sci.
    11: 30 (1960).

6. I.S. Kaurichev, and E.M. Nozdrunova, Soviet Soil Sci. 1057
    (1961).

7. F.L. Himes, R. Tejeira, and M.H.B. Hayes, Soil Sci. Soc. Am.
    Proc. 27: 516 (1963).

8. J.W. Muir, R.I. Morrison, C.J. Bown, and J. Logan, J. Soil Sci.
    15: 220 (1964); J.W. Muir, J. Logan, and C.J. Bown, J. Soil Sci.
    15: 226 (1964).

9. H.G.C. King, and C. Bloomfield, J. Soil Sci. 19: 67 (1968).

10. S. Bruckert, Ann. Agron. 21: 421 (1970); S. Bruckert, Ann.
    Agron. 21, 725 (1970).

11. R. C. Ellis, J. Soil Sci., 22:8 (1971).

12. J. F. Dormaar, J. Soil Sci., 22:350 (1971).

13. A. Saas, and M. Matteoli, Add. Trans. Intern. Symp. Humus et
     Planta V Prague (1971).

14. C. Bloomfield, in: "Experimental Pedology", 257, Butterworths,
     London, (1965); C. Bloomfield, Welsh Soils Discussion Group
     Report No. 11:112 (1970).

15. R. I. Davies, Soil Sci., 111:80 (1971).

16. I. S. Kaurichev, and E. M. Nozdrunova, Pochvovedenie No. 12:
     30 (1960).

17. I. S. Kaurichev, T. N. Ivanova, and E. M. Nozdrunova,
     Pochvovedenie No. 3:27 (1963).

18. F. Jacquin, and S. Bruckert, Comp. Rend. 260:4556 (1965).

19. N. Nykvist, Studia Forestalia Suecica No. 3: 1 (1963),
     N. Nykvist, Trans. Intern. Congr. Soil Sci. 8th Bucharest
     V:1129 (1964).

20. E. G. Harder, U.S. Geol. Surv. Profess. Papers 113:1 (1919).

21. G. S. Mudge, Soil Sci., 23:467 (1927).

22. M. Alexander, "Introduction to Soil Microbiology", John Wiley
     and Sons Inc., New York (1961).

23. D. V. Crawford, Trans. Intern. Congr. Soil Sci. 6th Paris C:197
     (1956).

24. L. J. McKenzie, E. P. Whiteside, and A. E. Erickson, Soil Sci.
     Soc. Am. Proc. 24:300 (1960).

25. O. Ashan, Z. prakt. Geol. 15:56 (1907).

26. A. E. Martin, and R. Reeve, J. Soil Sci., 11:369 (1960);
     A. E. Martin, J. Soil Sci., 11:382 (1960).

27. J. R. Wright, and M. Schnitzer, Soil Sci. Soc. Am. Proc. 27:171
     (1963); M. Schnitzer and S. I. M. Skinner, Soil Sci. 98:
     197 (1964).

28. V. V. Ponomareva, "Theory of Podzolization", "Nauka",
     Moskva-Lenningrad, 1964, Israel Program for Scientific
     Translations, Jerusalem, (1969).

29. C. Bloomfield, J. Soil Sci., 6:284 (1955).

30. M. Schnitzer, and J. R. Wright, Can. J. Soil Sci. 37:89 (1957);
     M. Schnitzer, J. R. Wright, and J. G. Desjardins,
     Can. J. Soil Sci., 38:49 (1958); M. Levesque, and
     M. Schnitzer, Can. J. Soil Sci., 47:76 (1967).

31. F. Jacquin, G. Juste, and P. Dureau, Comp. Rend Acad. Agr.
     France, 51:1190 (1965).

32. J. A. McKeague, Can. J. Soil Sci. 48:27 (1968).

33. G. Hubert, and A. Gonzalez, Can. J. Soil Sci., 50:281 (1970).

34. M. Schnitzer, Soil Sci. Soc. Am. Proc. 33:75 (1969).

35. P. Duchaufour, Comp. Rend. 259:3307 (1964); P. Duchaufour,
     "Osnovy pochvovedeniya evolyutsiya pochv", Paris (1965),
     Progress Moscow (1970); P. Duchaufour, Anal. Edafol.
     Agrobiol., 26:241 (1967).

36. P. Duchaufour, and B. Souchier, Geoderma, 20:15 (1978).

37. S. Odén, "Nederbördens och luftens försurning – dess orsaker,
     förlopp och verkan i olika miljöer", Statens
     Naturvetenskapliga Forskningsråd, Stockholm, (1968).

38. L. N. Overrein, Ambio, 1:145 (1972).

39. O. Haugbotn, Meldinger fra Norges Landbrukshøgskole 55(8):
     (1976).

40. G. Abrahamsen, K. Bjor, R. Horntvedt, and B. Tveite, in:
     "Impact of Acid Precipitation on Forest and Freshwater
     Ecosystems in Norway", Research Report No. 6, 37, SNNS,
     Oslo (1976).

41. L. Wiklander, Grundförbättring 26:155 (1973/74); L. Wiklander,
     Geoderma 14:93 (1975).

42. E. Eriksson, Tellus, 12:63 (1960).

43. G. E. Likens and F. H. Bormann, Science, 184:1176 (1974).

44. B. Bolin, and C. Persson, Tellus, 27:281 (1975).

45. H. Dovland, E. Joranger, and A. Semb, in: "Impact of Acid
     Precipitation on Forest and Freshwater Ecosystems in
     Norway", Research Report No. 6, 15, SNNS, Oslo, (1976).

INTERACTION BETWEEN CATIONS AND ANIONS INFLUENCING ADSORPTION AND LEACHING

Lambert Wiklander, Department of Soil Science,

The Swedish University of Agricultural Sciences

Uppsala, Sweden

INTRODUCTION

The soil has amphoteric properties exhibited in the capacity of adsorption and exchange of cations and anions. The behaviour of cations has been extensively studied regarding the influence of the mineral composition, the organic matter, the moisture content, the adsorption energy of ions and the complementary ion effect. A great deal of knowledge about the anion behaviour is also available but far less than for cations.

A possible interaction between cations and anions of soluble salts affecting the adsorption and mobility of the ions has, however, attracted only little interest. Formation and exchange adsorption of complex ions as $CuCl^+$, $CuOH^+$, $ZnCl^+$, $ZnOH^+$, $Al(OH)_n^{(3-n)+}$ etc. are known and also the increase of the cation exchange capacity (CEC) caused by adsorbed phosphate. But this knowledge has not stimulated further research on the influence of a possible interaction between cations and anions on their behaviour in soils and the significance of soil properties in this connection.

In exchange reactions cations and anions have generally been treated as independent entities influencing each other only by Coulumbic forces as expressed by the ionic strength, activity coeffients and chemical potentials. Thus, in cation exchange expressions the anion in usually neglected and in anion exchange, conversely, the cation:

$R-Na + K^+ = R-K + Na^+$ and

$R-Cl + NO_3^- = R-NO_3 + Cl^-$,

where R stands for the ion exchanger. Likewise in dynamic processes, i.e. in leaching, soluble ions have usually been assumed to react independently.

A closer examination of a possible interaction between cations and anions in soils leads to the conclusion, however, that polyvalent anions will decrease the solubility, increase the adsorption, and reduce the leaching of cations in a way not explicable by the general interionic forces in solutions; this anion effect varying with the soil properties.

This paper gives a short report on experiments conducted to elucidate these problems. The soils used are widely different in their properties:

(1) Cambisols, cultivated, loam-loamy clay, $(Fe,Al)_2O_3/SiO_2$ low;

(2) Humo-orthic podsol, A_2 (bleached) and B_2 (rusty) horizons, sand, $(Fe,Al)_2O_3/SiO_2$ of B_2 medium;

(3) Latosol from Kenya, lateritic clay, $(Fe,Al)_2O_3/SiO_2$ high.

THE MECHANISM OF ANION ADSORPTION

The cation exchange process is fairly well understood and will not be treated here. On the other hand, the mechanism of anion adsorption and exchange is more complex, especially for polyvalent ions as phosphate.

Polyvalent anions can be bound to basic groups of a soil particle by one single bond leaving the secondary and even the third ionization step able to act as acidic group as shown below:

$RFeOH + KHSO_4 = RFeOH_2^+-SO_4^{2-}-K^+$,

the ions K^+ and SO_4^{2-} being exchangeable.

$RFeOH + KH_2PO_4 = RFeOH_2^+-HPO_4^{2-}-K^+$,

HPO_4^{2-} being firmly adsorbed but exchangeable and K^+ easily exchangeable. Phosphate and sulphate are more firmly bound by substitution of OH or OH_2 attached to surface-located Fe and Al of the soil particles:

$$RFeOH + KH_2PO_4 = RFePO_4HK + H_2O$$

$$RFeOH_2 + KH_2PO_4 = RFePO_4H_2K + H_2O,$$

where R signifies soil particle. Russell et al.[1] have proposed the existence of three types of exposed OH groups of goethite, common iron mineral in soils. Based on the structural positions and bondings of OH, the conclusion can be drawn that the acid – base properties of the OH groups are different and by that the affinity for phosphate ions.

The adsorbed phosphate creates new cation exchanging groups. There are other reactions of phosphate increasing the density of negative charges but they will not be discussed here.

Soil properties of importance for the adsorption of anions are mineral composition, texture and pH. Soils rich in hydrated oxides of Fe and Al have high capacity for binding anions (Figure 1). Removal of hydrated oxides by treatment with dithionite and NaH-oxalate, strongly reduces the phosphate adsorption (Figure 2). Also clay soils with low Me_2O_3/SiO_2 can retain considerable amounts of SO_4^{2-}, which is of significance regarding the atmospheric deposition of sulphate (Figure 3).

As a consequence of the increase of acid groups, adsorption of polyvalent anions leads to decreased solubility of cations of added salts as well as to increased CEC of the soil. Further, this anion effect will be noticeable in a reduced migration velocity and a decreased leaching of cations due to electric attraction by the adsorbed polyvalent anions. The anion effect increases with the anion adsorption capacity of the soil and thus with the ratio of $Al_2O_3 + Fe_2O_3$ to SiO_2.

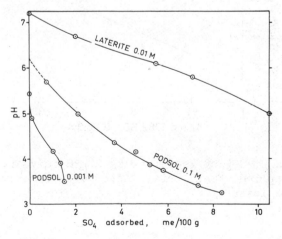

Figure 1. The influence of pH, salt concentration and soil properties on retention of SO_4^{2-} in laterite (Me_2O_3/SiO_2 high) and podsol B (Me_2O_3/SiO_2 medium). Me = Fe,Al.

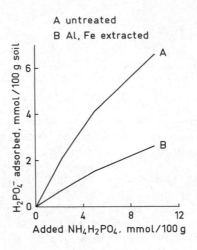

Figure 2. Adsorption of $H_2PO_4^-$ by latosol from Kenya, untreated and treated with dithionite and oxalate, pH 3.2, to remove hydrated oxides of Al and Fe, then K-saturated.

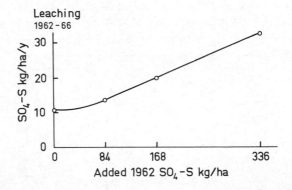

Figure 3. Loss of SO_4 by leaching, kg SO_4-S/ha/year, as related to the amounts applied. Field trial. After 4 years only 14, 22 and 27 of added S are lost.

THE EFFECT OF ANIONS ON THE ADSORPTION OF CATIONS IN EQUILIBRIUM
SYSTEMS

Na^+, K^+, Mg^{2+} and Ca^{2+}, in the mole proportions of $1:1:0.5:1$,
combined with either Cl^-, NO_3^-, SO_4^{2-}, $H_2PO_4^-$ or HPO_4^{2-}, were added
to different types of soil. The salts were added as solutions in
amounts usual in fertilization but also higher. After shaking until
equilibrium, the solution was separated by centrifugation or membrane filtration and analysed.

Without any exception the adsorption of Na, K, Mg and Ca varied
with the type of salt anion and increased in the order:

$$Cl^- \sim NO_3^- < SO_4^{2-} < H_2PO_4^- < HPO_4^{2-},$$

Figures 4, 5.

The solubility of the added salts was highest for chlorides and
nitrates, somewhat less for sulphates and least for phosphates. The
results were similar for single salts with Na and NH_4 as for combined salts with Ca, Mg, K and Na. Lateritic soil, B horizon of podsol and cultivated soils of cambisol type behaved similarly; the
anion effect varying quantitatively with the soil type.

Removal of hydrated oxides in the latosol (Figure 2) previous
to the salt addition, strongly reduced the differences in retention
of Ca, Mg, K, Na between the H_2PO_4, SO_4 and NO_3 systems (Table 1)
caused by decreased adsorption of H_2PO_4 and of SO_4.

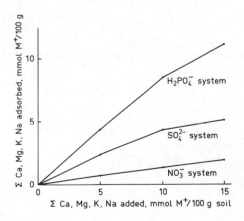

Figure 4. Adsorption of Ca, Mg, K and Na by latosol, equilibrated
with solutions of NO_3^-, SO_4^{2-} and $H_2PO_4^-$. Soil : solution = 1:4. Soil
pH 5.8.

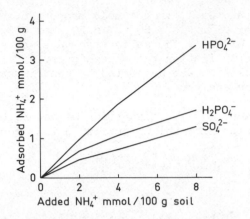

Figure 5. Adsorption of NH_4 by cambisol (loam) as influenced by the anions SO_4^{2-}, $H_2PO_4^-$ and HPO_4^{2-}. Soil : solution = 1:10.

Experiments with Na salts and podsol yielded results similar to those with NH_4. The adsorption of Na as well as of NH_4 was noticeabl higher in the HPO_4^{2-} system than in the $H_2PO_4^-$ system despite the twice higher phosphate concentration in the latter. This indicates that part of Na was held as $-OPO_3Na_2$ and part as $-OPO_3HNa$, the same being true of NH_4. The higher pH in the HPO_4 systems than in the H_2PO_4 systems is a contributing factor for the difference in cation bonding capacity.

Table 1. Adsorption of Ca, Mg, K, Na from solutions of NO_3^-, SO_4^{2-} or $H_2PO_4^-$ by latosol before and after removal of hydrated oxides. Rel. values, NO_3 = 1. (Cf. Figures 2, 4)

Solutions	NO_3^-	SO_4^{2-}	$H_2PO_4^-$
Before treatment	1	2.00	4.54
After treatment	1	1.09	1.18

THE EFFECT OF ANIONS ON THE CATION EXCHANGE CAPACITY (CEC)

CEC was determined by shaking and centrifuging 10 g soil with 4x100 ml solutions of either KCl, K_2SO_4 or KH_2PO_4/K_2HPO_4 (0.2 M K^+, pH 7.0), 4 washings with water and alcohol and replacement of exchangeable K^+ by shaking with 4x100 ml 0.5 M NH_4OAc (pH 7.0). Determinations made in duplicate.

Table 2 and Figure 6 show that the capacity of a soil to bind cations in exchangeable form depends not only on the soil properties but also on the anion of the salt used for saturation. A significant increase in K^+ retention is shown not only by phosphate but also by sulphate, chloride taken as a basis. The results agree well with the adsorption data and evidence the theoretical conclusions about the interaction between adsorbed polyvalent anions and base cations from the soil solution.

The difference between CEC in $H_2PO_4^-$ system and Cl^- system increases with the contents of sesquioxides and by that with the capacity of adsorbing phosphate: Cambisol < Podsol < Latosol, as estimated by the HCl-soluble H_3PO_4 in Table 2.

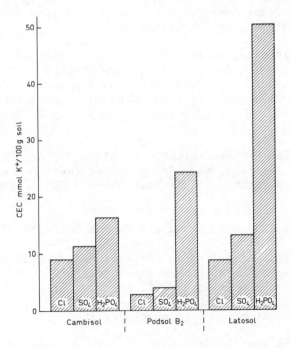

Figure 6. The retention of K^+ at pH 7.0 (CEC) by: Cambisol (cultivated loam); Podsol B_2 (sand); Latosol (clay). Saturation by treatment with KCl, K_2SO_4 or KH_2PO_4/K_2HPO_4.

Table 2. Retained K^+ after treatment with KCl, K_2SO_4 or $KH_2PO_4/$ K_2HPO_4 at pH 7.0. Adsorbed H_3PO_4 dissolved by the K replacement (4x100 ml NH_4OAc) and H_3PO_4 soluble in hot 2 M HCl (2 h on steam bath)

Soil	KCl	K_2SO_4	KH_2PO_4	Difference KH_2PO_4-KCl	Dissolved by	
					NH_4OAc	HCl
	mmol K^+/100 g soil				mmol H_3PO_4/ 100 g soil	
Cambisol	9.0	11.3	16.3	7.3	3.5	9.3
Podsol B_2	2.8	4.0	24.4	21.6	2.8	20.0
Latosol	8.8	13.2	50.5	41.7	9.4	33.2

The phosphate adsorbed by the KH_2PO_4/K_2HPO_4-treatment is bound partly in exchangeable forms as $RFeOH_2^+$-PO_4^-HK and partly in difficultly exchangeable form as $RFePO_4HK$. The stability of the phosphate-induced CEC increase depends therefore on the type of bonding and on the anion exchange occurring in soils.

Table 2 gives the amounts of adsorbed phosphate that were replaced in the three soils by treatment with NH_4OAc (pH 7.0) in connection with the CEC determination. Such a phosphate desorption will, of course, decrease the CEC of the phosphated soil, CEC being very high for the latosol and even for the podsol B_2 despite the sandy texture.

A comparison between the HCl-soluble H_3PO_4 and the difference in the K^+ retention between the KH_2PO_4 and KCl systems indicates that part of the exchangeable K^+ was held as $-OPO_3HK$ and part as $-OPO_3K_2$.

THE INFLUENCE OF SALT ANIONS ON THE LEACHING OF CATIONS

The influence of anions on adsorption and solubility of cations in equilibrium systems as demonstrated indicates that a similar effect is likely to appear in dynamic systems. If so this will result in a retarded migration and a decreased leaching of cations added to soils when the salt anion is adsorbed by the soil.

Experiments for investigation of this problem have been performed by leaching of salts added as solution onto the top of isotropic water-saturated soil columns, area 7.1 cm^2. The leaching device enabled a slow and constant rate of infiltration of distilled water, pH 7, and an automatic sampling of the effluent, each sample 3.6 cm^3, number of successive samples up to 55. Rate of percolation was 0.042 cm^3 min^{-1} cm^{-2}. The added salts amounted, in the experiments reported here, to 1.8 mmole per 100 g soil, the cations being Ca^{2+}, Mg^{2+}, K$^+$ and Na$^+$ and the anions Cl$^-$, NO$_3^-$, SO$_4^{2-}$ or H$_2$PO$_4^-$.

Figure 7 shows the leaching of Cl$^-$, SO$_4^{2-}$ and H$_2$PO$_4^-$. The Cl curve is almost symmetrical with the break-through in the ninth effluent fraction. The break-through of the SO$_4$ leaching appears in the 16th fraction and the distribution is asymmetrical with a pronounced tailing. This behaviour is a natural consequence of a stronger retardation and a slower migration of SO$_4^{2-}$ than of Cl$^-$. Not less than 29 % SO$_4$ was retained but of Cl only 4.9 %. The affinity for H$_2$PO$_4$ is still higher, resulting in a total retention in the upper part of the column and no leaching whatsoever.

The leaching of cations was studied in 25 cm columns of podsol B$_2$, 80 g soil, the added salts consisting of either NaCl, NaNO$_3$, Na$_2$SO$_4$, or Na$_2$HPO$_4$. Systems of either Cl, SO$_4$, or H$_2$PO$_4$ combined with equivalent amounts of Ca, Mg, K and Na were also prepared. In the SO$_4$ system Cu^{2+} was used instead of Ca^{2+} to avoid precipitation of CaSO$_4$. Untreated B$_2$ contained 3.9 % dithionite-soluble Al + Fe as Al$_2$O$_3$,Fe$_2$O$_3$, pH being 4.2 and base saturation 10.5 %.

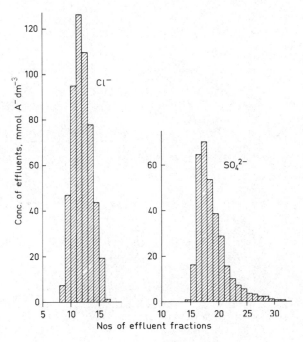

Figure 7. Leaching of Cl$^-$, SO$_4^{2-}$ and H$_2$PO$_4^-$, added as Na salts, in 25 cm column of podsol B$_2$, pH 4.2. 4.9 % Cl, 29 % SO$_4$ and 100 % H$_2$PO$_4$ retained (H$_2$PO$_4$ therefore not included).

Figure 8 shows the large difference in Na leaching due to the
anion. In the Cl system the break-through of Na appears in the 8th
fraction and the distribution is only slightly asymmetric in contrast to the conditions in the H_2PO_4 system. The leaching patterns
of Na and Cl are almost identical, indicating a simple mechanical
transport by the percolating solution. In the H_2PO_4 system, on the
other hand, adsorbed $H_2PO_4^-$ attracts Na^+ forming groups of $RFePO_4^-HNa^+$
thereby retarding the migration velocity, which appears in the break-
through point, and causing the pronounced tailing as well as the increased retention of Na.

Comparison with systems of single salts shows that the Na adsorption is strongly reduced by the presence of the more firmly
bound Ca, Mg and K. Thus, 66 % of added Na was retained in a system
with the single salt NaH_2PO_4 but only 8.4 % in the Ca,Mg,K,Na-H_2PO_4
system. Also in the Cl and SO_4 systems similar but smaller differences were found.

Based on the data from the NaCl system it can be calculated for
the NaH_2PO_4 system that 35 % of the Na retained is bound by inherent
acid groups of the soil and 65 % by adsorbed phosphate ions as
$RFePO_4HNa$ and $RFePO_4Na_2$.

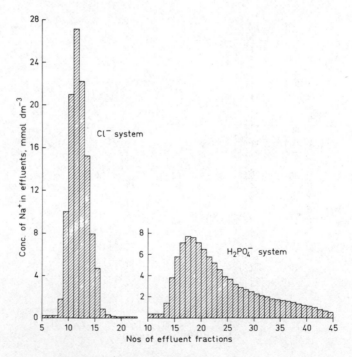

Figure 8. Leaching of Na^+ in podsol B_2, Na together with Ca, Mg, K
added as NaCl, Na_2SO_4 or NaH_2PO_4. 0.25 % Na retained in Cl system,
2.5 % in SO_4 system (not included) and 8.4 % in H_2PO_4 system.

The leaching of K in the same systems as for Na is given in
Figure 9. K was leached in the Cl and SO_4 systems but not at all in
the H_2PO_4 system. K^+ is much more firmly bound than Na^+, which explains the differences in leaching between the two ions. The retention of K in the systems of Cl, SO_4 and H_2PO_4 is 42, 44 and 100 %
of added respectively. Actually the figure for the SO_4 system is
somewhat decreased as Ca was replaced with Cu in the added salt and
Cu^{2+} is more firmly adsorbed than Ca^{2+} resulting in a higher leaching of K. The point of break-through and the strong tailing of the
K leaching in the Cl and SO_4 systems are in accordance with the
firm bonding of K.

That the phosphate ion, added in soluble salts, after adsorption attracts the accompanying cations is evidenced by the fact that
the total leaching of Ca, Mg, K, Na was 23 % of added in the H_2PO_4
system as compared with 62 % in the Cl system. No K, Mg and Ca was
leached in H_2PO_4 system but as much as 91 % of the weakly bound Na.
In the Cl system 65 % of the leached Cl was accompanied by Ca, Mg,
K, Na and 35 %.was lost as HCl.

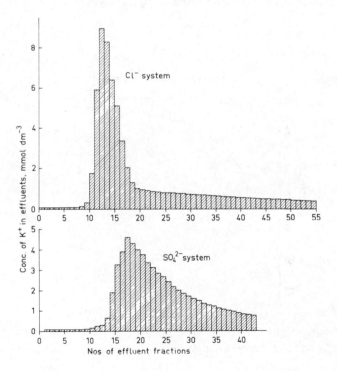

Figure 9. Leaching of K^+ in podsol B_2, the same systems as in
Figure 8. 42% K retained in Cl system, 44% in SO_4 system and
100% in H_2PO_4 system (not included).

In the Cu, Mg, K, Na-SO$_4$ system no Cu was leached. The distribution of Cu exchangeable by NH$_4$Cl is shown in Figure 10. The curve
indicates that the added Cu^{2+} is firmly adsorbed and retained in
the upper part of the soil column.

THE COEXISTENCE OF ADSORBED PHOSPHATE AND RETAINED CATIONS

It can be concluded from the experimental data that there
should be a positive correlation between the location of adsorbed
phosphate in the soil column and that of retained cations as K, Mg
and Ca. For an investigation of this problem two soil columns were
filled with A$_2$ and B$_2$ material from Häggbygget humo-orthic podsol
as shown in Figures 11 and 12. B$_2$ has a high capacity for phosphate
adsorption whereas A$_2$ has a very low capacity.

Ca,Mg,K,Na-H$_2$PO$_4$ (proportions of base cations on mole basis being 0.5:0.5:1:1) was added onto the top of the columns and leached
with distilled water until 50 effluent fractions were collected for
analysis. After finished leaching the soil columns were cut every
1 cm followed by extraction with AL-solution, using 3.5 g soil in
100 ml solution, shaking 1.5 h and centrifuging.

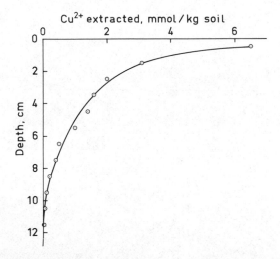

Figure 10. Distribution of Cu^{2+} soluble in NH$_4$Cl in podsol B$_2$,
Cu,Mg,K,Na-SO$_4$ added and leached with water. 21 % of added extracted.

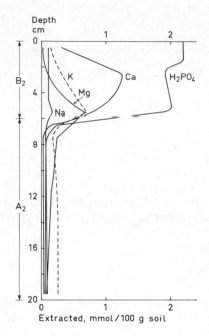

Figure 11. Distribution of
H_2PO_4, Ca, Mg, K and Na retained in soil column built of
6 cm B_2 and 14 cm A_2 from podsol. The salts added onto the
top of soil and leached until
50 effluent fractions collected.
Subsequent extraction by AL-solution: NH_4 lactate (0.1 M) +
HOAc (0.4 M), pH 3.75.

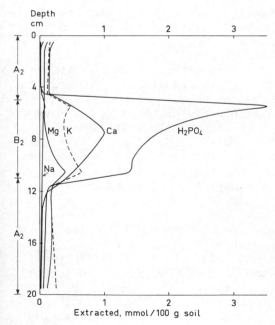

Figure 12. The same as in
Figure 11 but the soil column
consisting of 5 cm A_0 + 6 cm
B_2 + 9 cm A_0.

The distribution of extracted ions (Figures 11, 12) shows clearly

(1) that the phosphate adsorption is located at the B_2 horizon,

(2) that the phosphate ion has migrated through A_2 down to B_2 (Figure 12),

(3) that the major part of the retained base cations accompany H_2PO_4, and

(4) that the co-ordination of H_2PO_4 and cations is of the order: Ca > Mg,K >> Na on equvalent basis. Comparison with the retention of these cations after leaching Cl systems of B_2 shows that about 60 % of the adsorbed Ca, Mg, K, Na were held by $H_2PO_4^-$ and HPO_4^{2-} groups and about 40 % of inherent acid groups. The increased cation-holding capacity by the adsorbed phosphate corresponds to about 470 kg Ca per ha and 10 cm depth.

CONCLUSIONS

The increased retention and reduced leaching of salt cations by adsorbed polyvalent anions are processes of great significance in soil formation, geochemical circulation of nutrients, fertilization in agriculture and horticulture and also in pollution of soils and waters. Application of polyvalent anions, as phosphate, in fertilizers or other compounds to soils implies a gradual change of the physico-chemical properties of soils. This effect is more marked in soils with high capacity for anion adsorption, that is, in soils rich in hydrated oxides of Al and Fe.

Phosphate firmly bound or fixed in nonexchangeable form has low plant availability and, added as fertilizers, low response in crop yield. On the other hand, it increases the CEC of soils, favours the retention of cation nutrients and reduces the salt leaching and possible pollution of waters. It is also likely that adsorption of phosphate has some bearing on the availability of associated cation nutrients. K and NH_4 in combination with NO_3 or H_2PO_4 are used as fertilizers. If nitrates are applied, the concentration of K and NH_4 in the soil solution will be higher than if phosphate is applied, which may affect the uptake by plants.

Another factor of importance is the hampered migration of cations caused by phosphate. The major part of the nutrients is taken up by plant roots directly from the soil solution and this process is connected with ion exchange and diffusion of nutrients in the soil solution and on the particle surfaces. The contact area between the active parts of plant roots and the surface of soil particles, and by that the contact with soil solution, is very limited, not exceeding a few per cent of the total soil surface. The nutrient uptake, therefore, induces diffusion processes, the rate of which will be restrained by adsorbed polyvalent anions, especially phosphate.

It is likely that phosphate exerts a depression on the cation availability to plants, caused by the mentioned effects, when comparison is made with corresponding nitrate fertilizers. A similar but weaker influence of sulphate is probable.

For a control of water pollution by fertilizers it would be advisable, based on the present results, to use more phosphate and less nitrate. From previous lysimeter experiments[2] the conclusion can be drawn that a partial replacement of nitrate in fertilizers by ammonium (NH_4^+, NH_3) will lower the leaching of nitrogen.

The experiments show that there are possibilities to increase the cation adsorbing capacity of soils and to reduce the leaching of cations by using more phosphate as fertilizers, especially in soils rich in hydrated oxides of Fe and Al.

SUMMARY

The capacity of soils to adsorb and retain anions increases with decreasing pH, with the salt concentration and with the contents of hydrated oxides of Fe and Al. The retention of anions common in atmospheric salts and fertilizers is of the order:

$$Cl^- \simeq NO_3^- < SO_4^{2-} < H_2PO_4^-.$$

Experiments with equilibrium and dynamic systems of soils and various salts have shown that polyvalent anions of soluble salts added to soils increase the adsorption and decrease the leaching of the salt cations. Thus, the solubility and retention of nutrient cations in soils are determined not only by the bonding energy of the cations but also by the kind of the associated anion. The effectiveness of the anions studied proved to be in the order:

$$Cl^- \sim NO_3^- < SO_4^{2-} < H_2PO_4^- < HPO_4^{2-}.$$

The polyvalent anions are bound by hydrated oxides of Al and Fe or by other minerals leaving free, to a varying extent, one or two negative charges attracting and adsorbing salt cations. This anion effect increases with the ratio of $Al_2O_3 + Fe_2O_3$ to SiO_2 of the fine material of the soil.

For the experiments different soil types were used: Cambisols, ortho-humic podsol and latosol. Salt cations used: Na^+, K^+, Mg^{2+}, Ca^{2+} and Cu^{2+}.

The difference in CEC between phosphate and chloride systems
increased for the soils as follows:

Cambisol < Podsol B_2 < Latosol.

Of Ca,Mg,K,Na-H_2PO_4 added to podsol B_2 only Na was leached.
The total leaching in the H_2PO_4 system was 23 % of added cations
whereas the corresponding figure from the Cl system was 62 %.
Analysis of soil columns showed a distribution of retained H_2PO_4
similar to that of the retained Ca, Mg and K, indicating the
adsorbed phosphate as the active groups for holding salt cations.

REFERENCES

1. J. D. Russell, R. L. Parfitt, A. R. Fraser, and V.C. Farmer,
    Nature 248: 220 (1974).

2. K. Vahtras, and L. Wiklander, Acta Agric. Scand. 27: 165 (1977).

PRODUCTION AND CONSUMPTION OF HYDROGEN IONS IN THE ECOSPHERE

B. Ulrich

Institut für Bodenkunde und Waldernährung der

Universität Göttingen, Buesgenweg 2, D-3400 Goettingen

INTRODUCTION

Soil acidity is not caused only by the anthropogenic acid rain of today. There are other sources of H ions leading to soil acidification as shown by the existence of acid soils under almost undisturbed conditions in tropical rain forests. On the other hand, further influences of man exerted on ecosystems besides acid rain may also lead to soil acidification, e.g. cropping, grazing, fertilizing etc.. In most parts of Europe, including Scandianavia these influences of man have been quite strong and have persisted since 1,000 B.C.. Under the present conditions of ecosystem management in Norway, Rosenquist,[1] estimates that acid rain is of minor importance for the acidification of river waters. He assumes that the buffering capacity of the organic humus layer has changed considerably during the last century due to forestry, agriculture and changes in animal husbandry. He concludes that the acid run-off in water courses is mainly due to ion exchange processes in the catchment area.

In this paper an attempt is made to develop a general scheme for estimating man-made as well as natural sources of H ions in terrestrial ecosystems. It is quite clear that the organic matter is a strong potential source of protons by forming OH-groups able to dissociate. But if organic acids are completely decomposed again, no protons are left over. The turnover of protons is therefore bound to the dissipation of the following processes in space and time: formation and decomposition of organic matter, uptake and mineralization of chemoelements.

The data needed for exemplifying the general scheme are taken from our own studies of a beech (_Fagus silvatica_) and a spruce (_Picea abies_) forest ecosystem in the Solling mountains in Central Europe. These data are used in a model to demonstrate potential effects under specified conditions.

HYDROGEN ION INPUT BY RAIN IN THE PREINDUSTRIAL TIME

If we assume that the H^+ concentration in rain water is determined by, and is in equilibrium with, the CO_2 in the atmosphere, the pH value of the rain water may be calculated from equation (1) to (3)

(1) $$ CO_{2(g)} + H_2O \rightleftharpoons H_2CO_3 \quad ; \quad pK = 1.46 $$

(2) $$ H_2CO_3 \rightleftharpoons H^+ + HCO_3^- \quad ; \quad pK = 6.35 $$

(3) $$ \left[H^+ \right] = \left[HCO_3^- \right] $$

For a partial pressure of CO_2 = 0.0003 bar as in the atmosphere one gets pH = 5.65. An amount of rain of 1000 mm with pH 5.65 transports 0.022 keg H^+/ha. This figure gives an indication of the theoretical preindustrial hydrogen ion input into ecosystems from the atmosphere.

EFFECTS OF ROOT AND DECOMPOSER RESPIRATION

Whereas the assimilated carbon is taken almost completely from the atmospheric CO_2 pool directly, the CO_2 produced by root and decomposer respiration appears mostly in the soil. The CO_2 partial pressure in the soil air therefore exceeds atmospheric CO_2 pressure with values of 0.002 to 0.007 bar and in some extreme cases with 0.1 bar. Since the water entering the soil is in equilibrium with the CO_2 pressure in the atmosphere, it is undersaturated with respect to the CO_2 pressure existing in the soil. CO_2 is therefore dissolved in soil water according to equation (1). Below pH 5.4 to 4.9 (corresponding to CO_2 pressures of 0.001 and 0.01 bar, respectively) this process is stopped and the opposite reaction starts: CO_2 is set free from the solution. pH values of 5 to 5.5 are therefore typical for (subsurface) soils where carbonic acid is the only acid formed.

In calcareous soils the rate of dissolution of $CaCO_3$ is fast enough to keep the system close to the equilibrium described in equation (4)

(4) $CaCO_3 + H_2O = CO_{2(g)} \rightleftharpoons Ca^{++} + 2HCO_3^-$;   pK = 5.83

Combination of equation (1) and (2) shows that in the case of root
and decomposer produced CO_2, 1 mole H^+ from carbonic acid is consumed per mole Ca for the transformation of carbonate into bicarbonate.

Using the pK values given, the Ca^{++} concentration in the equilibrium solution of calcite can be calculated for different CO_2
partial pressures (see table 1) (system: $CaCO_3 - H_2O - CO_{2(g)})$:

Table 1: CONCENTRATION OF Ca^{++} IN THE SYSTEM $CaCO_3 - H_2O - CO_2$

CO_2 partial pressure in mbar	0.3	1	3	10	100
Ca^{++} concentration in mmol/l	0.5	0.7	1.0	1.5	3.4

As an example the following system may be considered:

calcareous soil: rate of seepage   =   200 mm per year;
CO_2 partial pressure in soil at-   =   3 mbar; equilibrium
mosphere between solution and
calcite.

For these conditions there will be 2 kmol H^+ per ha and year
consumed for the dissolution of 2 kmol $CaCO_3$. At a bulk density
of 1.5 this corresponds to 0.014 % $CaCO_3$ per ha and dm soil depth.
The carbonic acid needed for this dissolution process stems almost
completely from root and decomposer respiration; the carbonic acid
content in precipitation water in equilibrium with the atmospheric
CO_2 would allow only the dissolution of 2% of the amount of $CaCO_3$
mentioned (22 eq H^+ + 22 eq HCO_3^- per ha and year; assumption: no
source for CO_2 in soil). With decreasing pH values the solubility
of carbonic acid decreases. At pH below 5.5, depending on CO_2
partial pressure, HCO_3^- is present in soil solutions only in
insignificant amounts. In these soils CO_2 is given off to the
atmosphere without forming carbonic acid. The proton production
by root and decomposer respiration CO_2 is therefore restricted to
calcareous soils and the dissolution and leaching of $CaCO_3$. The
result in the ecosphere has been that in the postglacial period
the carbonate has been leached relatively rapidly from the root
zone. The process stopped at the lower boundary of the root zone.

Table 2: ION UPTAKE IN THE ABOVE SOIL PART OF A BEECH (FAGUS SILVATICA) STAND IN SOLLING DISTRICT (FRG).   data in keq · ha^{-1} · a^{-1}

line		Na^+	K^+	Ca^{2+}	Mg^{2+}	Mn^{2+}	Al^{3+}	cation sum	Cl^-	SO_4^{2-}	$H_2PO_4^-$	anion sum	N	balance* A	balance* B
1	litter	.03	.41	.81	.13	.19	.06	1.63	.02	.20	.13	.35	3.5	-2.22	+4.78
2	crown leaching		.49			.09		.58				0		+.58	+.58
3	phytomass increment	.004	.17	.39	.14	.12	.01	.83	.003	.085	.068	.16	.93	- .26	+1.60
4	sum = uptake	.03	1.07	1.20	.27	.40	.07	3.04	.02	.29	.20	.51	4.43	-1.90	+6.96

* balance A: N uptake as NO_3^-:   cation sum - (anion sum + N)

  balance B: N uptake as NH_4^+:   (cation sum + N) - anion sum

EFFECTS OF ION UPTAKE

In table 2 data concerning ion uptake of a beech stand in the soiling area, FRG, are shown. The uptake is calculated as the sum of the transport rates in litter, crown leaching and phytomass increment. The main problem in calculating uptake rates is the distinction between crown leaching and filtering. The sum of both rates is calculated as the difference between the monthly flow rate of elements below canopy (throughfall + stemflow) and above canopy (open land precipitation). In this paper we consider this difference as filtering (part of dry deposition) for all elements except K and Mn. For K and Mn it is assumed that filtering during May to November occurs at the same rate as during the leafless period December to April, and that no leaching occurs during the leafless period. The data presented in table 2 are calculated on the basis of these assumptions. They represent mean values of a 8 year measuring period (1969 to 1976). In ion uptake salts pass over from the soil compartment to the plant (root) compartment. In both compartments the condition of electroneutrality must be satisfied during uptake. Since nitrogen is the element taken up in highest amounts either as cation (NH_4^+) or anion (NO_3^-) the salt (cation/anion) balance depends very much upon the form of N uptake. Two extreme cases are considered:

Hypothesis A: N uptake solely as NO_3^-

Hypothesis B: N uptake solely as NH_4^+

Table 2 represents also the corresponding salt balances (A and B). Further information used in the following regards the selectivity of ion uptake. For the beech stand under consideration (Prenzel,[2]) calculated the mass flow coefficients (MFC) given in table 3. The mass flow coefficient is defined as the ratio between total uptake and mass flow uptake, where mass flow uptake means the uptake calculated from transpiration rate and element concentration in soil solution.

Table 3: MASS FLOW COEFFICIENTS

Al	Cl	Na	S	Fe	Mg	Mn	Ca	K	N	P
.076	.086	.34	.77	1.5	1.7	2.2	2.9	8.3	11	120
	discrimination				selective uptake					

Hypothesis A: It is assumed that nitrogen is taken up solely as nitrate. In this case (see line 4 of table 2) the anion uptake (4.94 keq) exceeds the cation uptake (3.04 keq) by 1.9 keq. This anion surplus has to be balanced by an H^+ consumption in soil. On the basis of the findings that Na is discriminated in ion uptake and accumulated in the root zone (Ulrich,[3]), the following reaction taking place at root surface may be postulated (p = localized in plant, s = localized in soil):

$$(5) \quad CO_{2(g)}^{(p)} + H_2O + NaNO_3^{(s)} \longrightarrow NaHCO_3^{(s)} + HNO_3^{(p)}$$

This equation describes the production of H^+ ions by root respiration and their (partial) consumption to balance nitrate uptake. The salt balance in the soil is maintained by the formation of (dissolved) sodiumbicarbonate. In acid soils the soil bound reaction continues:

$$(6) \quad NaHCO_3 + H^+ \longrightarrow Na^+ + CO_{2(g)} + H_2O$$

The consumption of H^+ as demonstrated by equation (6) should result in a pH increase in the rhizosphere; this effect can be demonstrated by growing plants in unbuffered nutrient solutions pH < 5.

It is possible that the selective uptake of all cations except Na and Al is also related to nitrate uptake. The most important cation in this respect is K, which is taken up selectively in amounts allowing substantial crown leaching. Equation (5) thus describes only that fraction of nitrate uptake which is not balanced by uptake of cations.

Hypothesis B: It is assumed that nitrogen is taken up solely as ammonium. In this case the cation uptake (7.47 keq) exceeds the anion uptake (0.51 keq) by 6.96 keq. This cation surplus may also be connected with the CO_2 produced by root respiration. NH_4^+ is selectively taken up, using HCO_3^- for balancing in the plant compartment, whereas CL^- is not taken up and balanced by H^+ ions in the soil compartment:

$$(7) \quad CO_{2(g)}^{(p)} + H_2O + NH_4Cl^{(s)} \longrightarrow NH_4HCO_3^{(p)} + HCl^{(s)}$$

Equation (7) is supported by the findings that Cl is discriminated in ion uptake. The ability of plants to lower the pH of a nutrient medium below the range covered by carbonic acid is well established. Considerable differences between species exist.

EFFECTS OF MINERALIZATION

Besides ion uptake the mineralization has to be considered also. In case of the anions the complete oxidation leads to strong acids:

(8) $R - SH + H_2O \longrightarrow R - OH + H_2S \rightsquigarrow H_2SO_4$

(9) $R - NH_2 + H_2O \longrightarrow R - OH + NH_3 \rightsquigarrow HNO_3$

(10) $R - O - PO(OH)_2 + H_2O \longrightarrow R - OH + H_3PO_4$

On the other hand the mineralization of organic salts leads to the formation of strong bases (Na_2O, K_2O, CaO, MgO...):

(11) $-C \overset{\displaystyle O}{\underset{\displaystyle ONa}{\big\langle}} + H_2O \rightsquigarrow C \overset{\displaystyle O}{\underset{\displaystyle OH}{\big\langle}} + NaOH$

In case of S and N the mineralization may stop with H_2S and NH_3. H_2S is a weak acid (pK_1 = 7.05, pK_2 = 13.0), which forms with metals like Fe sulfides

(12) $H_2S + FeCl_2 \longrightarrow FeS + 2HCl$

The weak acid H_2S is transformed into the strong acid HCl. If nitrification is incomplete, there will be a H ion consumption according to

$NH_3 + H^+ \longrightarrow NH_4^+$

being equivalent to the H ion production during NH_4 uptake.

COUPLING OF ION UPTAKE AND MINERALIZATION BY THE HYDROGEN BALANCE

Under undistrubed conditions it is most common that ion uptake and mineralization proceed simultaneously and within the same soil space. This is valid for uncropped systems (no export of organic matter) and sites which are characterized by mull as humusform: uptake and mineralization take place mainly within the A horizon. Since the temperature dependency for both effects is similar, they will balance each other to some extent also with respect to time. In soils with a pronounced rooting below the A horizon some dissipation takes place, since mineralization prevails in the upper soil (O and A) horizons, whereas ion uptake goes on also in lower soil horizons. This is especially true for nitrate. This means that

according to hypothesis A there is, under undisturbed conditions,
a tendency for H ion consumption in deeper soil layers and for H
ion production in upper soil layers (either by NH_4 uptake or by
nitrification). This is in agreement with the common depth function
of pH in soils.

As can be seen from the data in table 2 the H ion turnover
quantitatively strongly depends upon the nitrification process.
With degradation of the biological soil status the nitrification
is slowed down. Baum,[4] found in mineralization studies the
following relations between the NH_4/NO_3 ratio and the humus form:

humusform:	mull	moder	raw humus
NH_4/NO_3 ratio:	~0.1	~ 1	~ 10

Hypothesis A applies therefore mainly to mull soil, hypothesis B
to raw humus soils like podzols. In raw humus soils the processes
of mineralization and uptake are spatial separated. The mineralization takes place in the F layer of the top organic O horizon and
is connected with a H ion consumption. The uptake takes place in
the horizons following below (H layer, A_e horizon) and is connected
with a H ion production. In these horizons the pH values may reach
values below 3. The soil acidification can proceed as deep in the
soil as NH_4 is transported. It seems that the inhibition of nitrification and the spatial dissipation of formation and uptake of
NH_4 are of importance to the podzolization process. The soil
acidification by difficult decomposable litter may therefore only
partly be caused by soluble organic acids (phenols).

ACCUMULATION OF ORGANIC MATTER IN SOIL

With amounts between 50,000 and 150,000 kg/ha soils contain
considerable amounts of organic matter which has been withdrawn
from the decomposition process. This organic matter has produced
hydrogen ions by dissociation of OH groups according to its cation
exchange capacity.

As an example a soil may be considered with an amount of
100,000 kg organic matter, the carbon content of organic matter =
50%, C/N ratio = 15, CEC = 2 eq/kg organic matter. During the
accumulation of this organic matter 200 keq H^+ have been produced
by dissociation of OH groups. In case of hypothesis A (nitrate
uptake) this amount of H^+ is balanced by the withdrawal of 238 keq N
from mineralization and nitrification, since with the nitrification
an equivalent amount of H^+ would have been produced. Accumulation
of soil organic matter with a narrow C/N ratio has therefore no
acidifying influence on the soil; the contrary may even happen to
limited degree.

Hypothesis B is bound to nitrification inhibition and formation of raw humus on top of the mineral soil. The withdrawal of N from mineralization (C/N = 30) means a decrease in H ion consumption according to equation (12) of 120 keq; this amount has to be considered as H ion production. On the other hand the CEC_e of the organic matter accumulated is much smaller, $\sim$ 20 keq/ha, due to low pH. Accumulation of raw humus is therefore connected with a substantial H ion production (120 + 20 = 140 keq/ha); if the accumulation occurs within 30 years, the mean annual H ion production is 4 keq $\cdot$ ha^{-1} $\cdot$ a^{-1}. This figure may be compared with 0.02 keq $\cdot$ ha^{-1} $\cdot$ a^{-1} reaching soil by rain in the preindustrial time or with 2 keq $\cdot$ ha^{-1} $\cdot$ a^{-1} produced by root and decomposer respiration in calcareous soils.

The periodic build-up of organic top layers is very common in managed coniferous forests in Central Europe. In a spruce or pine plantation the organic top layer develops, reaching steady state (equilibrium between rate of litter fall and of mineralization) after 30 to 60 years stand age. Due to drastic changes in the microclimate of the decomposition layer, clear cutting is often connected with a rapid mineralization of great parts of the organic top material; at least part of the N mineralized underlies nitrification. Thus the system swings periodically from H ion production due to accumulation of organic matter according to hypothesis B, to H ion production due to nitrification after clear cutting. This periodical change in H ion production within the life cycle of managed forest ecosystems should substantially contribute to soil acidification. The old assumption that planting of spruce forests has an acidifying influence on soil seems therefore to have a realistic background. These effects are minimized if during the whole stand life and especially during the regeneration phase care is taken to maintain climate and biological status of the soil. This can only be achieved by allowing a ground vegetation layer to develop.

EFFECTS OF BIOMASS EXPORT FROM THE ECOSYSTEM

As pointed out there would be no net change if rate and location of ion uptake are equal to rate and location of mineralization. In a managed forest the phytomass increment (table 2 line 3) is exported from the ecosystem thus introducing a net change equal to the cation/anion balance of the phytomass increment. It can be seen from table 2, that in case of hypothesis A the hydrogen ion balance in the soil is shifted to a surplus of H^+ consumption over H^+ production of 0.26 keq $\cdot$ ha^{-1} $\cdot$ a^{-1}. In case of hypothesis B the opposite happens, the H^+ consumption is lowered by 1.6 kval $\cdot$ ha^{-1} $\cdot$ a^{-1}. This means that only on already acid soils with a low degree of nitrification wood export increases soil acidity.

It seems important to realize that this kind of feedback leads to an exponential increase in soil acidification, once the mechanism starts to work. A steady rate will be reached if nitrification is stopped completely. The mechanism should be acting in most European coniferous forests, as indicated by the humus forms moder and raw humus.

In many regions of Europe the litter has been exported from forests during centuries to increase agricultural yields. This should have stopped nitrification rapidly, thus initiating a strong soil acidification. Using the data of table 2 as an example the maximal net H^+ production would be for phytomass increment 1.6, for litter 4.78, in total 6.4 keq $\cdot$ ha^{-1} $\cdot$ a^{-1}. Since the biomass production goes down due to nitrogen deficiency these values are not reached in reality. The strong acid character of soils subjected to litter export is a well known fact.

ACCUMULATION AND USE OF FOSSIL FUELS

During the development of the earth, both processes of ion uptake during biomass formation and ion mineralization during biomass decomposition have been discoupled to a considerable extent, leading to the accumulation of fossil biomass.

By using fossil biomass as fuel in the time being a further discoupling is introduced: During the burning process the elements forming acids are transformed into gases (SO_2, NO_x, Cl) whereas the elements forming bases (Na, K, Ca, Mg...) remain in the ash (the role of NH_3 will be considered later). This discoupling is the major source of the acidic rain. Rain acidity is thus finally also a consequence of mineralization of biomass.

The gases emanating during burning processes reach the ecosystem again either by wet or dry deposition.

WET DEPOSITION: RAIN INPUT

Table 4 shows the annual rain input (mean of 6 years) in the Solling district, FRG. The mean precipitation is 985 $1/m^2$ ± 217 (standard deviation between years). Statistical parameters for these data are given in Ulrich et al.[5]

Potential acid formers are sulfate, nitrate and chloride. Calculations show that only 3% of the sulfate come from sea water, but 81% of the chloride. The natural background of nitrate may be in the order of 15% (= 1 kg N $\cdot$ ha^{-1} $\cdot$ a^{-1}). If the remaining is considered as acid, this would correspond to 1400 eq H_2SO_4, 430 eq HNO_3 and 90 eq HCl. SO_2 is thus the main source (73%) of rain acidity,

NO_x contributes 22%, HCl is insignificant (5%). The amount of H ions in rain (791 eq $\cdot$ ha^{-1} $\cdot$ a^{-1}) is smaller than the amount of acid forming anions (1920 eq).

Table 4: ANNUAL RAIN INPUT IN SOLLING DISTRICT, FRG DATA

IN eq $\cdot$ ha^{-1} $\cdot$ a^{-1}

cations:

H^+	Na^+	NH_4^+	K^+	Ca^{2+}	Mg^{2+}	Fe^{2+}	Mn^{2+}	Al^{3+}	$\sum i^+$
791	345	776	89.1	549	170	29.7	12.4	125	2887

anions:

NO_3^-	Cl^-	P^-	SO_4^{2-}	$Norg^-$		$\sum i^+$
505	477	17.9	1442	353		2795

This indicates that a substantial neutralization (58% of the acids formed) takes place in air before the rain drops arrive at the ecosystem. This neutralization is caused by NH_3, by lime dust (Ca, Mg) and soil dust (Al, Fe). The sum of these cations (1650 eq) exceeds the amount needed for neutralization (1130 eq), indicating that some other minor sources exist for these elements in order to dissolve in rain water (e.g. see water spray or in combination with organic bound nitrogen which is due mainly to contamination of the sampling vessels).

Even after the neutralization of 58% of the acids in air the rain input of H ions today (800 eq $\cdot$ ha^{-1} $\cdot$ a^{-1}) exceeds the natural one (22 eq) by a factor of 40. The weighted mean pH of the rain today is 4.07 compared with 5.65 calculated for preindustrial times.

DRY DEPOSITION: PLANT FILTERING

The filtering effect of the canopy against air impurities seems the most important process of dry deposition to forests, if not the only one of significance. In table 5 data on dry deposition to the canopy of a beech and a spruce forest are presented. The data are calculated as mentioned in paragraph 4 and described in the paper of Mayer and Ulrich (this publication). The dry deposition seems to be due to the impaction of aerosols. For this reason dry deposition includes the same ions and in the same ratio as wet

Table 5:   FILTERING EFFECT OF THE CANOPIES OF SPRUCE AND BEECH

data in keq $\cdot$ ha^{-1} $\cdot$ a^{-1}

cations:

	H^+	Na^+	NH_4^+	K^+	Ca^{2+}	Mg^{2+}	Mn^{2+}
spruce	2.28	.43	.23	.23	1.18	.25	.05
beech	.58	.30	0	.16	.78	.19	.04

anions:

	NO_3^-	Cl^-	SO_4^{2-}
spruce	.64	.48	4.04
beech	.45	.16	1.83

deposition. There is one exception: the spruce canopy filters out H^+ and SO_4 by dissolution of SO_2 in the water films adhering to the needle and bark surfaces during great parts of the winter. This effect accounts for 1.45 keq $\cdot$ ha^{-1} $\cdot$ a^{-1} H and SO_4.

The effect of SO_2 and NO_x on rain acidity becomes evident by comparing the following data:

	pH	H^+ input keq$\cdot$ha$^{-1}\cdot$a^{-1}	relation
preindustrial rain	5.65	.022	1
rain today (Solling)	4.07[+]	.814[+]	37
rain under beech canopy	3.79[+]	1.39[+]	63
rain under spruce canopy	3.38[+]	3.09[+]	140

+ weighted mean 1969 - 1976

EFFECT OF TOP ORGANIC LAYER (O HORIZON)

After the passage through the canopy the rain water has to pass through the top organic layer (O horizon) before reaching mineral soil. If we consider the amount of organic matter stored in that layer as being independent of time (equilibrium between annual rates of litter production and decomposition), there is still the possibility of H ion production or consumption by mineralization and uptake. Furthermore organic matter carrying acidic groups (fulvic acids) can be transferred in dissolved form from the top organic layer to the mineral soil.

At the Solling sites the amount of carbon dissolved in the percolation water leaving the top organic layer was determined for a period of one year. The weighted mean carbon concentration is ≈ 200 kg C $\cdot$ ha^{-1} $\cdot$ a^{-1} which corresponds to $\approx 10\%$ of the rate of leaf litter production. Thus a substantial amount of organic matter reaches the mineral soil surface as fulvic acid dissolved in water. According to Schnitzer (this volume), fulvic acids carry 10 meq acidic groups per g; with this value the transport rate of acidic groups from 0 to A horizon can be calculated as ≈ 2 keq $\cdot$ ha$^{-1} \cdot$ a^{-1}.

Information about the net effect of all the processes mentioned may be obtained by comparing the rain input of H$^+$ into the top organic layer and the percolation output of H$^+$ from it. Pertinent data are presented in table 6.

Table 6: H ION INPUT TO AND OUTPUT FROM TOP ORGANIC LAYER (DATA IN keq$\cdot$ha$^{-1}\cdot$a^{-1}, 95% CONFIDENCE LIMITS IN PARENTHESES)

	n yrs	beech canopy drip	beech stem flow	spruce
input I	8	.852 (.74 - .97)	.538 (.38 - .70)	3.09 (2.9 - 3.3)
output 0	6	1.27 (.67 - 1.86)	.538	2.69 (2.34 - 3.03)
I - 0	6	- .420 (-.86 - .02	0	+ .355 (.14 - .57)

In the case of beech the interaction of stem flow with top organic layer has not bee assessed and is considered to be zero. In the case of spruce, stemflow is negligible. Table 6 shows that under beech there is some H ion production, under spruce some H ion consumption, but both effects are small.

This experimental result shows that the H ion input into the top organic layer represents the determining function. All other processes are connected in such a way that the amount of H ions passing through that layer is not changed. There are indications that, in addition to the processes mentioned, the transfer of Al ions from mineral soil to the top organic layer via roots, also plays an important role.

If this interpretation is true, (i.e., if the H ion input forces all processes going on in the top organic layer), then it makes no sense to ask how great the contribution of that layer to the H ion input into mineral soil is. Looking for causality it is obvious that the present H ion input into mineral soil is due to wet and dry deposition of SO_2 and NO_x. This is the actual mechanism operating. The effects of organic matter accumulation in soil or of biomass export are involved in the whole process chain and should contribute to the final effects in the long run, but they do not regulate the system in its present transient state.

LEACHING OF ORGANIC MATTER FROM SOILS

The leaching of acidic organic matter from soils is known for peat. Streams leaving peat are dark coloured and acid. By mixing with water of higher pH and reacting with oxygen, the phenolic substances are partly oxidized, partly polymerized and eventually enter the sediment.

From mineral soils a leaching of organic acids into stream water is possible only if the mineral soil is covered with an organic top layer and itself either watersaturated or frozen. Under these conditions no infiltration of rain water or melted snow into the mineral soil is possible. The water will therefore percolate down through the organic top layer and carry soluble substances with it. This process, according to Rosenquist,[1] plays a significant role by increasing acidity in river waters in Scandinavia but not in Western and Central Europe.

CONSUMPTION OF H IONS BY SILICATE WEATHERING

The fate of H ions in the weathering of silicate minerals may be demonstrated with orthoclas as an example:

(13) $KAlSi_3O_8 + 4 H^+ \longrightarrow K^+ + Al^{3+} + 3SiO_2 + 2H_2O$

Since silica acid is finally converted to SiO_2 (Opal), the H ion consumption corresponds to the sum of charges carried by the cations dissolved (Na^+, K^+, Ca^{2+}, Mg^{2+}, Al^{3+} ...). This weathering reaction runs through many stages. One important intermediate stage is the clay minerals, which are formed by acid hydrolysis of silicates:

(14) $KAlSi_3O_8 + H^+ + \frac{1}{2} H_2O \longrightarrow K^+ + \frac{1}{2} Al_2Si_2O_5(OH)_4 + 2SiO_2$

Equation (14) shows the weathering of orthoclas to kaolinite. Comparison of equation (13) and (14) shows that the main H ion consumption occurs during the weathering of clay minerals. If the formula of kaolinite is rewritten as $AlOOH \cdot SiO_2 \cdot \frac{1}{2} H_2O$ it becomes apparent that the H ion consumption by the weathering of clay silicates may be simply described by the following equation

(15) $AlOOH + 3 H^+ \longrightarrow Al^{3+} + 2 H_2O$

This equation shows that an understanding of the buffering of H ions by acid soils requires detailed knowledge about the chemistry of aluminum in the pH range 3 to 6.

BUFFER RANGES OF MINERAL SOILS

With respect to the chemical state of a soil, the following pH ranges may be distinguished (CEC_e = effective cation exchange capacity determined with N NH_4Cl; CEC_t = total CEC determined with $BaCl_2$ pH 8; X_i^S = equivalent fraction of ion i in the exchanger phase; X_i^L = equivalent fraction of ion i in the solution phase):

pH range

6.5 – 8.3   buffering of H^+ by carbonate:

neutral $CaCO_3 + H_2CO_3 \longrightarrow Ca^{2+} + 2HCO_3^-$

buffer capacity: 1% $CaCO_3 \approx 150$ kmol $H^+ \cdot ha^{-1} \cdot dm^{-1}$

pH depending upon partial pressure of CO_2 in soil air:

$pH - 0.5$ pCa $= 4.9 + 0.5$ pCo_2

$CEC_e = CEC_t$

$X_{Al}^S = 0$; $X_{Ca}^L > 0.5$

salt concentration in soil solution high (HCO_3^-)

leaching of Ca

soil fabric is stable

5 – 6.5	beginning of aluminum buffering
moderately	pH correlated with X_{Ca}^S and/or X_{Al}^S
acid	$CEC_e/CEC_{t_S} > 0.7$
	$0.2 < X_{Al}^S < 0.5$
	buffer capacity: $\simeq$ CEC : 1% clay $\simeq$ 6 kmol $H^+ \cdot ha^{-1} \cdot dr$
	salt concentration in soil solution low
	leaching of Ca
	soil fabric is relatively unstable
3 – 5	aluminum buffering:
strongly	$AlOOH \cdot H_2O + xH^+ \longrightarrow Al(OH)_{3-x}^{x+} + xH_2O$
acid	buffer capacity: 1% clay $\simeq$ 100–150 kmol $H^+ \cdot ha^{-1} \cdot d$
	pH determined by hydrolysis of Al
	$0.15 < CEC_e/CEC_t < 0.7$
	$0.5 < X_{Al}^S < 0.9$
	leaching of Al and Mn
	soil fabric is stable
< 3	iron buffering:
extremely	$Fe(OH)_3 + xH^+ \longrightarrow Fe(OH)_{3-x}^{x+} + xH_2O$
acid	growth inhibition on mineral soils

Common pH values of unfertilized forest soils on carbonate free parent material in Central Europe vary between 3 (top soil, A_eh horizon) and 4.5 (B_v horizon), that is in the range of strongly acid soils characterized by aluminum buffering.

SOME ASPECTS OF ALUMINUM CHEMISTRY IN ACID SOILS

As Bache et al.[6] have shown, soil solutions of strongly acid soils contain different Al ion species including polynuclear species The Al ion species which can be expected according to literature data are compiled in Table 7 together with their mass action constan (pK values) for the dissociation reaction. For the hydroxy aluminum species additional corrected values are listed; these values are obtained by comparison with experimental data (for detail see Nair,[7]).

In tab. 7 the relative concentrations of the different mono-nuclear and polynuclear hydroxy-Al ion species are given. The values have been calculated with a computer programme described

by Nair and Prenzel,[8]. The partitioning between the different
ion species is not only controlled by pH, but also by the total Al
concentration also.

As equilibration studies with natural occurring gibbsites
revealed, a considerable time is necessary for establishing equilibrium between gibbsite and its constituent ions. However, during
the equilibrium process between the polynuclear species $Al_7(OH)_{17}^{4+}$
and $Al_{13}(OH)_{34}^{5+}$[51]) which starts after a few days of the initiation
of dissolution, the $pAl^{3+} + pOH^-$ values of the solution are close to
the solubility product of gibbsite.

Similar observations have been made with soils and their equilibrium solutions. With an equilibration period of 24 h there was no
agreement between experimentally determined polynuclear Al and the
values calculated on the bases of the pK values. Agreement was much
better after an equilibrium period of 20 days. Soil solutions collected in the field by aid of alundum tension lysimeter plates were
also in disagreement between experimental and theoretical values
for polynuclear Al. This indicates that at percolation rates of
500 mm per year the Al species in soil solution may not be in equilibrium. Since gibbsite is precipitated only after reaching final
equilibrium between polynuclear Al ion species, the condition of
nonequilibrium found in soils may be the reason for the lack of
gibbsite formation in spite of the possibility according to the solubility product principle.

With increasing concentration of Al^{3+}, $AlSO_4^+$ comes into play.
In the upper part of the Solling soils, where the pH values are low
$(< 4)_+$ and the Al concentrations in soil solution relatively high,
$AlSO_4^+$ can make up to 20% of the Al ions in solution, the rest being
mainly Al^{3+}.

As with the Al ion species in solution there is still uncertainty concerning the Al compounds existing in solid phase. As regards
the mass, only hydroxides and sulfates of Al have to be considered.
In most acid soil gibbsite is missing despite the fact that at pH
values above 4.1 to 4.2 the composition of the soil solution corresponds to its solubility product. As pointed out this may be due to
nonequilibrium with respect to polynuclear Al ion species. Instead
the formation of gibbsite, polynuclear hydroxo aquo complexes of Al
build up in the interlayer space of clay minerals like smectite and
vermiculite.

By extraction with 0.5 N NaOH at 60° C an attempt was made to
differentiate between "free" and "silicate bound" Al. In the beech
soil in the depth interval 0 - 50 cm. 700 kmol $Al \cdot ha^{-1}$ corresponding
to 9% of the Al bound in the illite clay minerals have been extracted. This free Al consists mainly of polymeric interlattice
Al. Subsequent X-ray analysis revealed that only around 2/3 of the

interlattice Al has been extracted. The charge neutralized by the
extractable interlattice Al, calculated as difference between CEC_t
($BaCl_2$ pH 8) and CEC_e (unbuffered N NH_4Cl), is equal to 370 $keq \cdot ha^{-1}$.
From these data the positive charge of the interlayer Al can be
calculated as $\sim$0.5 per Al atom. Thus interlayer Al maybe approximated by the formula $Al_6(OH)_{15}^{3+}$.

   With respect to sulfate bound Al, Singh et al.,[16] postulated
the formation of crystalline basaluminite $Al_4(OH)_{10}SO_4$
(pK_{sp} = 117.31) by reaction of sulfate ions with the hydroxy
aluminum interlayer. From the composition of the equilibrium soil
solution Van Breemen[17] postulated the existence of amorphous $AlOHSO_4$
(pK_{sp} = 17.23). Solubility products found in the Solling soils and
many other acid soils in Central Europe (Fassbender et al.,[18]) are
in agreement with the finding of Van Breemen and contradic the
existence of basaluminite at least on the more acid side of the pH
range. If polynuclear Al species are taken into account the
solubility product of the $AlOHSO_4$ as postulated by Van Breemen is
in the range of 18.5.

   In fig. 1 a hypothetical reaction scheme for the buffering of
H ions in acid soils by liberation of aluminum from clay minerals
is given.

   At the upper limit of the pH range (pH 5) the Al concentration
is low (pAl 6 to 7) and, according to tab. 7, the dominant Al species
in solution is $Al(OH)_3$. The disintegration of clay(a) is therefore forumulated as hydrolysis of a heterovalent Si-O-Al bond at a
clay mineral surface leading to the liberation of $Al(OH)_3$. This
Al species reaches equilibrium with polynuclear species like
$Al_7(OH)_{17}^{4+}$ consuming H ions(b). The polynuclear species are
assumed to be in adsorption equilibrium with surfaces. In the
interlayer space they are transformed to hydroxo Al polymers without
substantial change in charge(c). This reaction chain starts below
pH 6.5 and continues to about pH 4. Exchangeable Al^{3+} appears as
soon as the pH and pAl values in the double layer allow (cf. tab. 7).
By lowering pH below $\sim$4 the hydroxo Al polymers in the interlayer
space are becoming disintegrated, the reaction(d) is formulated as
$Al(OH)_2^+$ being formed. In reality this reaction seems much more
complicated and may often involve the formation of organic Al complexes with organic ligands (fulvic acids). $Al(OH)_2^+$ is according
to fig. 1 only of minor importance, but in the presence of HSO_4^-
it may be responsible for the formation of $AlOHSO_4$ complexes(e)
which are adsorbed at surfaces and form a solid phase. The final
ion species of the buffering reactions are $AlSO_4^+$ and Al^{3+}, depend-
in upon SO_4 concentration. The ion species leached with seepage
is Al^{3+}. The whole reaction chain is driven by H ion input to soil;
the different reactions are separated in time and space. The first
reaction, the hydrolysis of the clay mineral lattice, is only

indirectly influenced by pH, its rate is not proportional to H^+ input. This explains why the buffering system reacts relatively poorly and does not keep the pH at the upper limit of the buffer range (pH 5) in spite of the great buffer capacity.

There are some analytical results in favour of this hypothetical reaction scheme:

- As Fiedler et al.,[19] already showed, in the strongly acid soils of Central Europe the hydroxo Al polymers in the inter-layer space have their minimum in the A horizon. This applies also to the Solling soils.

- By measuring the adsorption isotherms of SO_4^{2-} in the Solling soils Meiwes et al.,[20] found that there is no sulfate ion adsorption in the 0 - 10 cm layer and that the adsorption increases with increasing soil depth and pH. The same is true for the Al-sulfate stores (determined by Na_2CO_3 extraction).

In agreement with these results the uppermost soil layers are already reacting at the lower limit of the buffer range at pH 3. The interlayer Al accumulated in times of lower H ion input and higher pH are, at present, dissolved as shown in the reaction scheme. Even for the formation of $AlOHSO_4$ the pH is too low in the top soil.

INPUT – OUTPUT BALANCE OF THE SOIL

The input – output balance of the soil should show the net effects of the processes going on and allows conclusions regarding the rates. If the sum of all outputs (0) is subtracted from the sum of all inputs (I), the difference indicates changes in soil storage. The soil can behave either as sink (I > 0) or as source (0 > I) or as inert matrix (0 = I). For the Solling ecosystem I is equal to the sum of the annual flow rates of wet and dry deposition; 0 is equal to the sum of percolation output and phyto-mass increment (Mayer et al.,[21]). Pertinent data from the Solling are presented in table 7. As the ratio (I - 0)/I shows, larger effects are observed in case of H, Ca (spruce only), Mg, Mn, Al and SO_4. The soil acts as a sink for H, Ca and SO_4, and as a source for Mg, Mn and Al. The source function of the soil with respect to Mg and Mn is part of the overall reaction described in equation (13): during silicate weathering Mg and Mn have been converted to other binding forms which are now "dissolved". Mg seems to be leached from the exchangeable pool; its loss due to soil acidifica-tion is already of relevance to forest growth in Central Europe (Altherr et al.,[22]; Ulrich,[23]). Mn is dissolved according to equation (16).

(16) $\frac{1}{2} Mn^{4+}Mn^{2+}(O, OH)_2 + 3H^+ + e^- \longrightarrow Mn^{2+} + 2H_2O$

    As can been seen from table 8, the exchange and dissolution of Mg and Mn by H^+ with subsequent leaching from the soil are much smaller in size than the dissolution and leaching of Al. The disintegration of clay minerals is thus the main process consuming H ions in acid soil.

Table 7: Al ION SPECIES IN AQUEOUS SOLUTIONS AND THEIR
DISSOCIATION CONSTANTS

species	charge per Al atom	pK	reference	corrected pK (see text)
Al^{3+}	+ 3			
$Al(OH)^{2+}$	+ 2	5.02	Frink et al.,[9]	
$Al_2(OH)_2^{4+}$	+ 2	7.55	Kenttämaa,[10]	
$Al(OH)_2^+$	+ 1	9.70	Turner,[11]	
$Al(OH)_3^o$	± 0	13.02	Dalal,[12]	
$Al_6(OH)_{15}^{3+}$	+ 0.5	47.00	Sillen et al.,[13] 2M $NaClO_4$ 40° C	44.17
$Al_7(OH)_{17}^{4+}$	+ 0.57	48.80	Sillen et al.,[13] 3 M $NaClO_4$ 25° C	45.18
$Al_{13}(OH)_{34}^{5+}$	+ 0.38	97.60	"	91.05
$Al_{13}(OH)_{32}^{7+}$	+ 0.54	104.50	Aveston,[15] 1M $NaClO_4$ 25° C	98.65
$ALSO_4^+$		3.20	Sillen et al.,[13]	

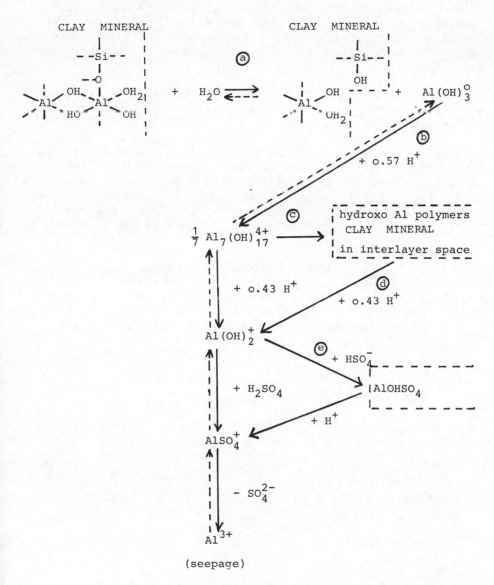

Figure 1
A hypothetical reaction scheme for the buffering of H ions in
acid soils

The sink function of the soil with respect to H and SO_4 is a consequence of the buffering reactions described already.[4] The rate of sulfate accumulation therefore corresponds to the net rate of $AlOHSO_4$ formation (buildup), if other changes in soil store can be excluded.

The sink function of the spruce soil for Ca may indicate that as a consequence of SO_2 and NO_x emissions the Ca input by wet and dry deposition has increased.

From the input – output differences (I – O values) presented in table 8 the equivalent sum of ions produced by H ion consumption can be calculated as sum of Mg + Mn + Al + SO_4. This sum exceeds the input – output difference of H ions by a factor of 2 (spruce) to 3 (beech). This is a consequence of the earlier accumulation of intermediate reaction products in the soil as demonstrated for Al (interlayer Al, $AlOHSO_4$).

Earlier leaching losses of Al can be estimated from a balance of the total Al storage in the upper 50 cm of the beech soils as given by Heinrichs et al.,[24]. By comparison with the sediment (loess) from which this layer was developed a loss of 350 kmol Al/ha can be estimated for the postglacial period.

From the data presented in this paper the consumption of H ions due to the reaction

$$(17) \quad AlOOH \quad H_2O = xH^+ \longrightarrow Al(OH)_{3-x}^{x+} = H_2O$$

can be estimated for the postglacial period as follows (example: soil under beech):

	keq
Leaching of 370 kmol Al^{3+}	1110
Buildup of exchangeable Al^{3+}	360
Buildup of interlattice $Al^{0.5+}$	500
Buildup of $AlOHSO_4$	40
Sum	2010

If the rates of leaching of Al (1.6 keq $\cdot$ ha^{-1} $\cdot$ a^{-1}) and formation of $AlOHSO_4$ (.95 keq) of table 7 are used, it would need 700 years to obtain the leaching loss and 42 years to arrive at the sulfate storage in soil.

The soil under study (for description see Benecke et al.,[25]) is for different reasons not very suited for a balance of this kind, therefore, only the order of magnitude of the data should be considered. With this restriction the data show that the stronger

TABLE 8:   INPUT/OUTPUT BALANCE OF A BEECH AND A SPRUCE ECOSYSTEM IN
THE SOLLING DISTRICT
I = WET + DRY DEPOSITION (cf. table 5);
O = PERCOLATION OUTPUT + PHYTOMASS INCREMENT (cf. table 2)
All values given as $keq \cdot ha^{-1} \cdot a^{-1}$

	n Jahre	H^+	Na^+	K^+	Ca^{2+}	Mg^{2+}	Mn^{2+}	Al^{3+}	Cl^-	SO_4^{2-}
beech										
I	8	1.39	.63	.24	1.33	.35	.05	.20	.93	3.31
O	5	.53	.63	.26	1.21	.45	.36	1.81	1.06	2.36
I-O		+.86	+ 0	-.02	+.12	-.10	-.31	-1.61	-.13	+.95
(I-O)/I		.62	0	.08	.09	.29	6.2	8.0	.14	.29
spruce										
I	8	3.09	.76	.31	1.73	.41	.06	.33	1.12	5.52
O	2	.31	.70	.29	.95	.61	.43	3.01	.97	2.79
I-O		+2.78	+.06	+.02	+.78	-.20	-.37	-2.68	+.15	+2.73
(I-O)/I		.90	.08	.06	.45	.49	6.2	8.1	.13	.49

acidification of this soil has probably started with the human use
and misuse of the primary forest around 1000 to 2000 years ago.
The acidification may be a consequence of biomass export including
litter, partly by burning to get potash, and the resulting changes
in humus form and ion uptake. This acidification was probably not
a continuous process but related to human interferences of the kind
mentioned, with the effect levelling out if the ecosystem was not
subjected to new interferences.

The buildup of Al sulfates in soil is a very young process and
started with the emission of SO_2 due to coal burning in the last
century. The present rate is the maximum rate reached during the
last 10 to 20 years.

POSSIBILITIES OF SOIL PROTECTION

For managed ecosystems a protection of the soil against the H
ions produced either as consequence of biomass export or deposited
by wet and dry deposition is possible and bears no big problems
besides that of the costs. Liming is a well established practice
in agriculture and is the appropriate measure to consume the H ions
produced or deposited. It is interesting to note that in the
Federal Republic of Germany the mean application of lime per ha of
arable land corresponds to the amount of H ions deposited; the
increase of lime use in the last 25 years may very much reflect
the increase in H ion deposition. The protection of arable land
is thus finally paid for by the food consumer. This seems to be
in order, since this consumer is, on the other hand, the producer
of SO_2 and NO_x (by heating houses, using electrical current and
driving cars).

With a few exceptions (e.g. N in Sweden) forests are not yet
fertilized. This has to change if the acid rain continues and
whole tree crop systems are introduced. There exist already
experiences concerning the fate of lime in forest soils (Ulrich,[26])
and recommendations regarding liming to compensate H ion input
(Ulrich,[27]).

FINAL REMARKS AND CONCLUSIONS

The effect of air contaminations on ecosystems is very difficult
to assess. Air pollution is so widespread that an area not influenced (i.e. a zero plot) does not exist. This is true for
Western and Central Europe. As part of human activities air contaminations build up at the same rate as human societies are changing. Their effects are therefore below the perception ability of
human beings. Only in retrospect, by comparing the situation
today with the situation some decades ago, is perception possible.

As a consequence, the classical research approach in applied science, the factorial experiment, is either not applicable or restricted to very narrow limited partial problems.

The research approach followed in this paper may be demonstrated by two examples:

1. We define for some chemoelements the excess of throughfall rate above precipitation rate as dry deposition. It is very difficult if not impossible to prove this directly. What remains is the possibility to disprove other possible explanations. This has been done experimentally as shown in the paper of Mayer. This way of drawing conclusions meets with displeasure. For those using this approach it is important to realize that any other possible explanation not disproved makes the conclusion meaningless.

2. Another approach used in this paper is the application of a very fundamental theorem (principle of electroneutrality) to separate processes such as uptake and mineralization. This approach requires completeness with respect to the processes considered. Completeness means in this sense that all chemoelements of quantitative importance have to be considered. Those not considered are assumed to be quantitatively within the error limits of the cation and anion sum. If this basic assumption is accepted, this approach yields hypotheses which have to be proved or disproved experimentally. As pointed out in the paper experimental evidence is often already available from older investigations.

The result presented can be summarized in two statements:

1. Human use and misuse of forest ecosystems has caused soil acidification since prehistoric times.

2. SO_2 and NO_x emissions cause an H ion input by wet and dry deposition to the soil which leads to substantial changes in the chemical soil state especially in the uppermost soil layers.

The logical consequence of statement 1 is that forest use of today may also cause soil acidification. This should be kept in mind and investigated especially if whole tree crop systems and clear cutting are practised.

Statement 2 leaves the question open whether or not there are already consequences of ecological significance. This question cannot be answered by looking solely on H ions and their effect. With acid rain more than 20 kg N $\cdot$ ha^{-1} $\cdot$ a^{-1} are put into the ecosystems in Central Europe. Without doubt the N input must be also of ecological significance. The question therefore is "what

is the balance of the combined effects in various ecosystems?" In
Central Europe the N input seems to have increased forest growth on
many sites. The evidence for this is indirect. N fertilization
experiments started immediately after World War II have led to
substantial growth increases. As a result, a lot of fertilizer
experiments have been laid out in the 1960's, but almost none of
these trials showed the expected growth increases.

Another way of arguing rests on the input - output balance
of N. The N stored in the annual increment of the stands is
completely covered by the excess of imput over : output. Furthermore,
in the beech ecosystem the soil storage of N is increasing. This
indicates that the conditions for the decomposers are positively
changing.

On the other hand, it seems from observations of the humus form
that the opposite is true for mull soils developed on limestone or
basalt.

According to Röhrig et al.[28] difficulties observed during
the natural regeneration of beech stands are caused by the high
acidity of the uppermost mineral soil layer and thus caused by
acid rain.

REFERENCES

1. I. Th. Rosenquist, Sur jord-surt vann, Ingenioerforlaget
      A/S, Oslo (1977).

2. J. Prenzel, Simulationsmodelle von Waldökosystemen: Wie
      wird der Teilprozeß Mineralstoffaufnahme gesteurt? in:
      "Verh. Ges. f. Ökologie" Elsevier Verlag, Göttingen, p. 43
      (1977).

3. B. Ulrich, Kationenaustausch-Gleichgewichte im Boden,
      Z. Pflanzenernährn. Düng. Bodenkd. 113:141 (1966).

4. U. Baum, Stickstoff-Mineralisation und Stickstoff-Fraktionen
      von Humusformen unterschiedlicher Wald-Ökosysteme,
      Göttinger Bodenkdl. Ber. 38, 1-96 (1975).

5. B. Ulrich, R. Mayer and P. K. Khanna, Fracht an chemischen
      Elementen im Niederschlagswasser im Solling, Z. Pflanzener-
      nährn. Düng. Bodenkd. 142: 601 (1978).

6. B. W. Bache and G. S. Sharp, Soluble polymeric hydroxy-aluminium
      in acid soils, J. Soil Sci. 27:167 (1976).

7. V. D. Nair, Aluminium species in soil solutions, Dissertation, Universität Göttingen, (1978).

8. V. D. Nair and J. Prenzel, Calculations of equilibrium concentrations, Z. Pflanzenernährn. Düng. Bodenkd. 141: 741 (1978).

9. C. R. Frink and M. Peech. Soil Sci. Soc. Amer. Proc. 26:346 (1962).

10. J. Kenttämaa, Acad. Sci. Fenn. Ann., Ser. A. Sec. II No. 67 (1955).

11. R. C. Turner, Soil Sci. 106:291 (1968).

12. R. C. Dalal, Soil Sci. 119:127 (1975)

13. L. G. Sillen and A. E. Martell, "Stability constants of metal ion complexes," London (1964).

14. L. G. Sillen and A. E. Martell, "Stability constants of metal ion complexes," London (1971).

15. J. Aveston: J. Chem. Soc. 4438 (1965).

16. S. S. Singh and J. E. Brydon, Activity of aluminium hydroxy sulfate and the stability of hydroxy aluminium interlayers in montmorillonite, Can. J. Soil Sci. 50:219 (1970).

17. N. Van Breemen, Soil forming processes in acid sulphate soils, in: "Acid Sulphate Soils," Proc. of the Int. Symp. Wageningen, Vol. 1, p. 66 (1973).

18. H. W. Fassbender and E. Matzner, Zur Bildung von basischen Aluminiumsulfaten im Boden, Mitt. Dtsch. Bodenkdl. Ges. 25:175 (1977).

19. H. J. Fiedler and S. Lentschig, Die Bedeutung der "freien Oxide" für die Systematik der Mittelgebirgsbraunerden, Chemie der Erde, 26 (1967).

20. K.-J. Meiwes and P. K. Khanna, Adsorptions- and Desorptionsverhalten von sauren Braunerden für Sulfat und deren Zusammenhang zum Sulfat-Input durch Immission, Mitt. Dtsch. Bodenkdl. Ges. 25:169 (1977).

21. R. Mayer and B. Ulrich, Conclusions on the filtering action of forests from ecosystem analysis, Oecol. Plant. 9:157 (1974).

22.   E. Altherr and F. H. Evers, Allg. Forst- u. Jagdztng.,
       145:121 (1974).

23.   B. Ulrich, Die Umweltbeeinflussung des Nahrstoffhaushaltes
       eines bodensauren Buchenwalds, Forstwiss. Centralbl.
       94:280 (1975).

24.   H. Heinrichs and R. Mayer, Distribution and cycling of
       major and trace elements in two Central European Forest
       Ecosystems, J. Environ. Qual. 6:402 (1977).

25.   P. Benecke and R. Mayer, Aspects of soil water behaviour
       as related to beech and spruce stands, Ecological Studies
       2:154 (1971).

26.   B. Ulrich, Die Reaktionen von Calciumcarbonat bei der
       Einarbeitung von Kalkmergel in stark versauerte Walböden
       mit Auflagehumus, Allg. Forst Jagdztg. 141:5 (1970).

27.   B. Ulrich, Grundsätzliches zur Forstdüngung, Der Forst- und
       Holzwirt 26:433 (1971).

28.   E. Röhrig, H. Bartels, H.-A. Gussone and B. Ulrich, Untersuch-
       ungen zur naturlichen Verjungung der Buche (Fagus
       silvatica). Forstwissenschaftl. Centralbl. 97: 121(1978).

THE EFFECTS OF SOIL ACIDITY ON NUTRIENT AVAILABILITY AND PLANT RESPONSE

I. H. Rorison

Unit of Comparative Plant Ecology (NERC),

Department of Botany, University of Sheffield, UK

INTRODUCTION

Historical

More than two hundred years of study has led to our present knowledge of the physical and chemical soil factors that influence plant distribution. In selecting some of the more recent developments in acidic soil-plant relationships, I have assumed a back-ground knowledge of the subject as presented in recent reviews[1,2,3]. The results, derived from mainly agronomic and ecological studies, are currently applied to problems of the reclamation and management of infertile soils[4], of acid sulphate soils[5], of industrial spoil[6], and, most recently, of habitats affected by acid precipitation[7].

There is as yet little direct experimental evidence concerning the effect of acid precipitation on vegetation but it seems likely that it will make the greatest impact through long-term deposition in soils with relatively low buffering capacities.

The extent of effects on the plant canopy and of resulting changes in the soil is the subject of other contributions to this volume.* In this paper the response of herbaceous plants to any given set of acidic conditions and their own influence on these conditions is considered.

*e.g. H. B. Tukey Jnr.

Current approaches

A change in emphasis in the study of plant-soil relationships has emerged over the last ten years.   Previously the major preoccupation was with the relatively static approach in which plant distribution was related to environmental factors as they occurred at one point in time.   The environmental factors were seen as the dominant and the plant the passive component.   Now, as with so many disciplines, the approach is more dynamic.

Emphasis is on the plant and its adaptation to environmental factors, the potential for these adaptations to evolve and the strategies involved.   It is realised that the presence of vegetat-ion has profound influences on the nature of the environmental 'sieve'[8], not only on the changing state of the soil environment but also on the invading individual progeny.   Plant demographers, preoccupied with the numbers of births and deaths in a population, have so far had little time for the physiological processes which occur during the life-span of plants.

Experimental ecologists, especially those concerned with plant-soil relationships, have tended to examine the responses of individuals of one or a few species of plants to individual environ-mental factors.   Results have been considered in isolation from the study of the population or of the evolutionary component.   Such people would argue that it is as important to understand the physiology of a plant, i.e., how it functions in relation to its immediate environment, as it is to know that it has evolved this function or may evolve further in the future.

Both approaches have merit and a great stimulus can be given to our attempts to understand the functioning of plants by the coming together of demographers and experimentalists as occurred in Wageningen last year[9].

To be most effective hypotheses must be based on relevant studies of the physiological mechanisms controlling plant distri-bution.   It is also important to know not only the current state of the growth medium at the time an organism is introduced to it but also the possible extent of changes which could take place during the establishment and subsequent life cycle of the organism.

Concomitantly a knowledge of the tolerance of plant species to a wide enough range of levels of major variables is desirable in order to select plants for a particular site or in order to ameliorate the site before introducing the plants[10].   Tolerance also may vary with age which again underlines the need for a time-course study.   The physiological pathways by which tolerance and susceptibility operate are beginning to emerge from such studies. Having identified the crucial pathways, their evolution and possibilities for improvement can then be examined with profit.

THE CONSTRAINTS OF ACIDIC SOILS

Acidic soils are typically deficient in essential nutrients and usually contain an imbalance of both essential and non-essential elements.   The chemical factors involved are as listed by Hewitt[11](Table 1) and refer mainly to well-drained soils. Under wetter, reducing conditions there is a tendency for poly-valent cations to assume a lower valency.   This is particularly marked in acid sulphate soils[12] which are defined as "All materials and soils in which as a result of processes of soil formation sulphuric acids either will be produced, are being produced or have been produced in amounts that have lasting effects on main soil characteristics."

Table 1.   Factors associated with acidic soils.   After Hewitt[11], 1952.

---

1. Direct effect - injury by hydrogen ions

2. Direct effect due to low pH

    (a) Physiologically impaired absorption of Ca, Mg and P.
    (b) Increased solubility and toxicity of Al, Mn, Fe, etc.
    (c) Reduced availability of P - Al x P interaction
    (d) Reduced availability of Mo

3. Low base status

    (a) Ca deficiency
    (b) Deficiencies of Mg, K and possibly Na

4. Abnormal biotic factors

    (a) Impaired nitrogen cycle and fixation
    (b) Impaired mycorrhizal activity
    (c) Increased attack by soil pathogens

5. Accumulation of soil organic acids or other toxic compounds due to unfavourable oxidation-reduction conditions

---

When dry they have pHs as low as 2.0 and polyvalent ions such as Fe^{3+}, Al^{3+} and Mn^{4+} appear in solution in high concentrations. When waterlogged, H_2S is produced and pHs rise towards neutrality leaving Fe^{2+} and Mn^{2+} in solution but resulting in the precipitation of Al^{3+} and Fe^{3+},

The effective draining and leaching of these soils takes ten or more years to accomplish and isolated additions of lime are of no lasting value[13]. Similar problems occur with the reclamation of pyritic waste[14]. On a smaller scale the centre of soil crumbs may remain as wet, reduced microsites in an otherwise well-drained profile[15].

Plant selection

Unlike the agronomist concerned with optimal economic yield, the ecologist is usually interested in survival under agronomically adverse conditions. The antidote for acidic soils to the agronomist is chemical and possibly physical amelioration: liming to reduce toxicity and to release certain essential elements followed by the addition of more essential elements to achieve a suitable balance.

To the ecologist amelioration may be neither economic nor desirable and the introduction of species adapted to a particular environment may be all that is required. A successful choice demands a knowledge of the physiological response of the plants to be introduced, during both the establishment[16] and the mature phases of their life cycle and is helped considerably by knowing something of the potential for adaptation which lies in every population. One of the most exciting realisations of recent years has been that the evolution of tolerant races is rapid and forms a very useful basis for the solution of problems of heavy metal toxicity and other stress conditions[17].

For example, Walley and co-workers[18] sowed 1000 seeds of a normal population of _Agrostis tenuis_ on a 12:1 copperwaste/compost mixture. After six months' growth ca. 30 reasonably healthy plants survived. Some individuals showed almost complete tolerance after only one generation of selection, indicating that there is genetic variability for tolerance even in non-tolerant populations. The intensity of selection can be varied by using different proportions of mine waste and potting compost. The result is obviously specific to the material under test but the technique can be used on a range of materials.

Having outlined areas of current interest we may note advances in knowledge of some individual soil factors and then consider evidence for mechanisms of adaptation exhibited by plants.

Individual factors

pH remains the most informative preliminary measurement to be made when investigating soil acidity problems.* Definitive experimental evidence on the impact of H ions and OH ions on plant growth is still awaited.

Calcium and nitrogen are two elements whose form affects pH. The lack of Ca as $CaCO_3$ is a feature of acidic soils and is reflected in the response of calcifuge plants; many species[19] and populations[20] having a low requirement.

Nitrogen is commonly found to be in the NH_4-N form in acidic soils and as NO_3-N elsewhere. However, there are reports of heterotrophic nitrification taking place[21] in very acidic soils and of the absence of nitrification both in waterlogged and in cold soils[22].

It is well established that NH_4-N is toxic to calcicole species under acidic conditions and that the presence of NO_3-N at circumneutral pHs is toxic to calcifuges[23].** It should also be noted that calcifuges are not only tolerant of NH_4-N but can also utilize any NO_3-N produced under acidic conditions. The important work of Havill, Lee and Stewart[24] confirmed that, among all the herbaceous species tested, only certain members of the Ericaceae had insignificant nitrate reductase activity. Thus the enzyme system necessary to utilize NO_3-N is present and inducible in most calcifuge species (Table 2).

Phosphorus is normally considered to be immobile in soil where it occurs mainly as a sparingly soluble iron, aluminium and/or calcium phosphate according to pH. In soils underlying undisturbed

Table 2. Mean NRAs of plants from acidic, calcareous and wasteland sites. Activities for Ericaceous species were <0.1.
After: Havill, Lee and Stewart (1974)

Mean nitrate reductase activity (NRA)

	Control μ mol hr^{-1} g FW^{-1}	Introduced	% increase
Calcifuges	0.4	2.1	525
Calcicoles	1.1	3.1	281
Ruderals	4.1	5.2	128

*Bache, this volume   **Alexander, ibid.

vegetation, however, as much as half the total phosphorus may occur in organic form[25]. Because of this the effective mobility of P in the soil profile may be greater than indicated by the few milli- metres around the root surface which is all that is depleted under conditions imposed by many laboratory experiments (Fig.1)[26]. In addition to diffusion gradients at the root surfaces, and mycor- rhizal associations which facilitate the movement of P, it can also be moved through the soil profile in short-lived fungal and bacterial[27] tissues whose breakdown releases it into a competitive pool. In acidic soils its uptake and utilization by many plants is inhibited in ways which have been extensively reviewed[2].

Aluminium remains the polyvalent cation most likely to disrupt plant growth in acidic soils both directly and by its effect on P uptake[2,28]. It is not only metallic but also exhibits both ionic and covalent bonding[29]. It may be displaced by iron at pHs <3.0 but in the range 3.5-4.5 it is a major toxin.

In well-aerated soils of pH less than 3.5, ferric iron is likely to be toxic to plants; above pH 3.5 its solubility is very low and toxicity unlikely. Iron as Fe^{2+} is soluble and liable to reach toxic levels in marshlands and other waterlogged soils with pHs as high as 6.0. Many calcifuges, marsh plants and plants from waterlogged habitats have been shown to be relatively resistant to iron toxicity. Conversely, calcicole species tend to be very susceptible, and the rapidity with which they recover from iron deficiency in the labor- atory suggests that they have a low requirement for optimum growth (see also p.13). This is an obvious advantage in their usual habit- ats on circumneutral soils in which the availability of iron is very low.

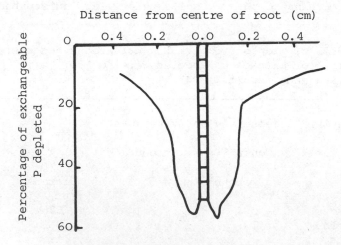

Fig.1 A concentration profile of phosphate in soil adjacent to a root of rape (Brassica napus) after eight days. The diameter of the root cylinder is indicated by the hatched column (redrawn from Bhat & Nye 1973[26]).

Manganese is another polyvalent cation whose solubility in the reduced state Mn^{2+} makes it potentially toxic in waterlogged soils over a wide range of pH.   Its toxicity is ameliorated in acidic soils by the presence of several elements, notably silica, iron and aluminium[30].   Calcicoles are particularly sensitive to manganese toxicity which providentially is ameliorated by calcium.

Sulphate ions are neither strongly adsorbed on soil particles (unless on iron or aluminium hydroxide films, see Wiklander, this volume) nor absorbed by plants in large amounts[31].   As a result a concentration gradient can build up at the root surface[32] which could lead to an enhancement of acidification particularly if the same roots are releasing H-ions.   An upset due to an ionic imbalance is unlikely because plants so far tested exhibit a wide tolerance of the presence of sulphate ions[33].   Their immediate requirement of sulphur for growth is considered to be small and is met in natural communities from the breakdown of organic material. With the purification of inorganic fertilizers some sulphur deficiencies have been reported in crops, in which case small aerial depositions of S containing salts may not be disadvantageous[34].

Interactions: Nitrogen source and aluminium

Single factors may have a dominant influence in a habitat but it is unlikely that they will operate in isolation.

Both aluminium and NH_4-N are major chemical components of acidic soils but until recently it was not known whether their combined effects are more or less toxic than their individual effects. Now the growth response of a number of species to acidic media which contain both a range of Al levels (O-2 X 10^{-4} M) and nitrogen as either NH_4-N or NO_3-N (2-8 x 10^{-4} M) has been measured[35].   Experiments were carried out in a controlled environment which included constant flow culture solutions maintained at pH 4.2 $\pm$ 0.2.

The hypotheses tested (Fig.2) were that: (1) species may be indifferent to both N source and to variation in the level of Al supplied; (2) they may be drastically affected by the form of N supplied, thus masking any possible effect of Al; (3) adverse effects of N source might be ameliorated in the presence of Al; (4) no adverse effects of N source but a classic response to Al toxicity.   All species tested so far have shown one or more of these basic responses in accordance with their ecological distribution.

The calcifuge species Deschampsia flexuosa was largely unaffected by NH_4-N or Al.   The calcicole species Scabiosa columbaria was strongly inhibited by NH_4-N both in the presence and absence of Al, and by Al in the presence of NO_3-N.

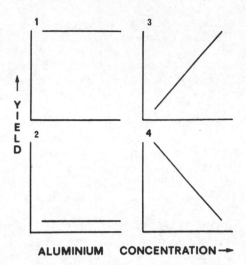

ALUMINIUM   CONCENTRATION⟶

Fig.2 Hypothesis: the four basic plant responses to nitrogen source and aluminium (see text).

Results for populations of the widely distributed species Holcus lanatus showed classic patterns of aluminium toxicity in the presence of NO_3-N and varying degrees of amelioration in the presence of NH_4-N.

Consideration of the mechanisms involved is still at a preliminary stage but initial chemical analysis of root and shoot material shows one common pattern exemplified by results for a population from an acidic soil with a pH of ca. 4.5 (Table 3).

In the absence of added aluminium, the aluminium content of both roots and shoots of the NH_4 and NO_3 plants is very low.   In solutions containing 2×10^{-4} M Al the content of both roots and shoots increases and the roots contain much more than the shoots.

The NO_3-N plants contain much more Al than those grown with NH_4-N.   In particular the roots of the NO_3 plants have a very high aluminium concentration.

The internal phosphorus concentration of the plants is not noticeably depressed in the presence of 2×10^{-4} M Al solution. Notably there is a high P concentration in the roots of the NO_3-Al_3 which corresponds with the high Al concentration in that treatment. The high Al-P concentration in the roots could be due to adsorption and to precipitation in intercellular spaces.

Table 3.   The aluminium and phosphorus concentrations of roots and shoots of <u>Holcus lanatus</u> grown from seed for five weeks in complete nutrient solution[23].   Nitrogen was supplied as either NH_4-N or NO_3-N and aluminium as $Al_2(SO_4)_3$ 16 H_2O. > indicates significant differences at the 0.1% level.

Holcus lanatus

		Root		Shoot	
[Al]		NH_4	NO_3	NH_4	NO_3
	Al_0	0.08	0.18	0.05	0.05
	Al_3	3.2	>15.1	0.18	> 0.45
[P]					
	Al_0	4.3	7.0	4.9	5.6
	Al_3	7.7	>15.3	6.3	< 3.9

mg/g dry wt

The idea of immobilization in the root is supported by the significantly depressed concentration of P in the shoots of the NO_3-Al_3 treatment.   Therefore, the presence of NH_4-N ensures a low Al concentration in the roots and no depression of P translocation to the shoots.

Several questions remain unanswered:

Precisely where and how is the Al intake to the plant regulated? and what role does NH_4-N play?

CONSTRAINTS OF PLANTS

Plant effects on soil pH

There has been growing interest in the last ten years in the influence that higher plants, their litter, and their associated rhizosphere can have on ion uptake[31].   The role of soil buffering capacity and of microorganisms, both free-living and mycorrhizal, is the subject of other contributions to this volume.*

*Alexander, Tukey and others.

This section is limited to a consideration of ion uptake and of the internal metabolic activity in the plant which results in pH changes at the root surface. This involves the release of H^+ and HCO_3^- and of organic exudates, some of which may chelate metals in the rhizosphere. Whatever the nature of these substances it is vital that electrochemical neutrality be maintained both in the plant and in the growth medium, irrespective of pH[36].

Currently the main source of interest centres on plant response to nitrogen source which itself is determined inter alia by soil pH. Raven and Smith[37] have given a clear account of present knowledge concerning the way intracellular pH is regulated in plants as a result of the assimilation and transport of nitrogen. It is as a part of this pH regulation that hydrogen and hydroxyl ions are released by roots, thus affecting the rhizosphere pH.

The assimilation of each NH_4^+ and NO_3^- produces approximately one H^+ and one OH^- respectively. Any H^+ or OH^- produced in excess of that required to maintain cytoplasmic pH must be neutralized. This is achieved by a mixture of biochemical and long and short distance transport processes which enable cells remote from a large sink of H^+ or OH^- to produce protein without unfavourable pH changes. For example, NH_4-N is assimilated into organic-N in the roots. As a result the shoots are supplied with a mixture of amino acids, amides and organic acids which can be incorporated with neutral photosynthate into cell material without damaging pH changes. The only way a build-up of acidity can be prevented in the root cell cytoplasm is for the excess H^+ that the process generates to be released into the soil solution (Fig.3).

When NO_3 is the source of N taken into the plant its reduction in the root leads to the production of OH^-. This is partly released into the soil solution and partly neutralized by a 'biochemical pH stat'[38] which produces strong organic acids.

Both these systems operate also when NO_3^- is assimilated in the shoot. Excess organic acid anions generated during pH regulation are removed either by conversion to an insoluble form such as calcium oxalate or transported to the roots in soluble form (Fig.4). There OH^- is regenerated following organate breakdown and is released into the soil solution.

OH^- may be released in combination with CO_2 as HCO_3^- which also has the effect of raising soil pH close to the surface. CO_2 which is respired by the root does not normally have an acidifying effect locally because under aerobic conditions it will diffuse rapidly away through the air-filled pore space[31]. There is less likely to be pH stress within a plant if both NH_4^+ and NO_3^- nitrogen is taken up simultaneously in the same cells[39]. The deposition of acid rain is normally likely to shift the balance towards uptake of NH_4^+ due

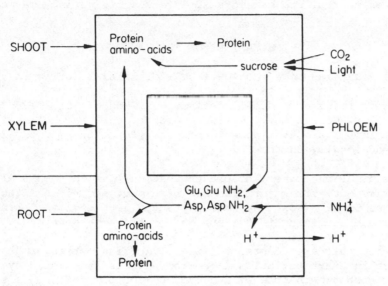

Fig. 3   pH regulation during ammonium assimilation in roots.[37]

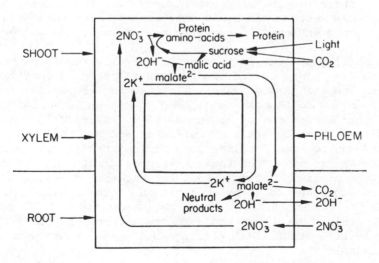

Fig. 4   pH regulation during nitrate assimilation in shoots: malate
         transport to the root.[37]

to inhibition of nitrification. If, however, the rain contains a
predominance of dilute nitric acid a reasonable balance may be
retained.

MECHANISMS OF TOLERANCE

(a) Plant responses to low levels of mineral nutrients

Acidic soils are usually deficient in most major elements,
i.e. N, P, K, Ca and Mg, whereas several of the trace elements
(and aluminium) are available in potentially toxic amounts, i.e.
Fe, Mn and locally Zn and Cu.

The question arises: do plants that survive in such soils have
low requirements for major elements or a very efficient method of
uptake, or both? (How they are adapted to high levels of trace
elements and aluminium is the subject of the next section.)

Plants that are adapted to mineral nutrient stresses tend to
have certain characteristics in common. These are listed by Grime[40]
as restricted growth form with a lack of sharply defined seasonal
growth pattern, slow relative growth rate and the ability to con-
serve absorbed mineral nutrients rather than to maximise the
quantity captured. Some are also able to survive lower levels of
availability of such elements as phosphorus and, in the lowest
deficiency ranges, to maintain as high or higher relative growth
rates as species from fertile habitats (see Figs. 5a and b in Roriso
1969).[10]   There remains some uncertainty over the general application of these results and it is important to note the range of
nutrient levels employed in experiments whose results do not agree.

The ability of species to survive very low levels of nutrient
supply implies a form of efficiency both in terms of specific
absorption and utilization rates. It also introduced the likelihood that when such plants are exposed to higher levels of the same
element their uptake can be increased to an injuriously high level.
This has been shown to occur by Jefferies and Willis[19] and many
others and can be seen from the data of Ingestad[41]. In Table 4 it
is of interest to note that toxic levels of calcium can be detected
in calcifuge species, not only when the external concentration of
calcium is increased but also when the pH of the growth medium is
raised to ca. 7.0 and nitrogen supplied in the form of NO_3-N instead
of NH_4-N. This suggests an upset in the internal pH of the plant
which would also occur if a calcifuge plant was transplanted into
calcareous soil.

Another stress response mechanism is suggested for Fe 'efficien
calcicole genotypes of tomato by Brown[42].   He reports that when irc

Table 4.   The influence of pH and nitrogen source on the concentration of calcium found in root and shoot fractions of the calcifuge Deschampsia flexuosa and the calcicole Scabiosa columbaria[23].

	N source		Calcium contents mg. g^{-1} DW ± SD		
			pH		
Species			7.2	5.8	4.2
Scabiosa columbaria	NO_3	Shoot	10.3 ± 0.4	18.0 ± 1.0	11.6 ± 0.2
		Root	4.5 ± 0.1	5.1 ± 0.1	4.4 ± 0.4
Deschampsia flexuosa	NO_3	Shoot	14.8 ± 0.8	4.8 ± 1.1	3.8 ± 0.3
		Root	52.0 ± 4.2	2.9 ± 0.1	3.2 ± 0.1
	NH_4	Shoot	2.8 ± 0.1	2.7 ± 0.1	2.3 ± 0.1
		Root	7.7 ± 0.4	6.0 ± 0.6	2.0 ± 0.1

becomes deficient in the growth medium H-ions are released from the plant roots.   This is followed by the release of a 'reductant' substance once the pH in the surrounding nutrient solution drops to ca. 4.5[43].   When a plant with this mechanism is transferred to an acidic soil it would be liable to suffer from excess uptake of iron.

     (b) Plant response to polyvalent cations at the whole plant level

Most polyvalent cations are potentially toxic and current interest centres: (1) on whether plants can evolve tolerance to one or more element and (2) on the mechanisms of tolerance.

A few cases of multiple tolerance have been reported and one of the most recent confirmed that, with respect to a population of Agrostis stolonifera, simultaneous tolerance to copper and zinc is regulated by two specific and different types of binding site[44].

Movement of most ions into the plant may be regulated in a number of ways.   Control sites may be both in the root and in the shoot of the plant, at the wall of individual cells or within the cytoplasmic tissue[2,3].

It is in some ways easier and certainly more conventional to examine the state of leaf tissue but it must be remembered that pattern in the root may or may not be similar[10].   A simple diagram illustrates the types of response commonly found (Fig.5).

In the first instance the ion moves readily through the whole plant system accumulating the ion to a uniformly high concentration so long as there is a minimum external level on which to draw.   If such plants are perennial they may accumulate in woody tissues and if deciduous they will shed material annually back into the external system[3].

In the second case the amount of the ion moving into the shoot is regulated so that internal concentration reflects external concentration.

Thirdly, there are plants in which translocation is kept to a minimum until, at a certain external concentration metabolism is disturbed and the concentration of the potentially toxic ion in the shoot increases rapidly and to the detriment of the plant.

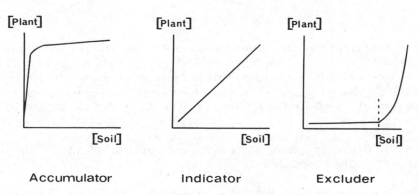

| Accumulator | Indicator | Excluder |

Fig.5   Three ways in which the response of plants to increasing soil concentration is reflected in their internal plant concentration[45].

(c) Plant response to polyvalent cations at the cellular
     level

It is suggested that control of movement of polyvalent cations is achieved by their precipitation, adsorption or chelation in the plant system (Fig. 6).

The initial control zone lies in the root and most experimental evidence links it with adsorption in and around the cell wall. More possibilities arise at the plasmalemma and tonoplast boundaries of the protoplasm. These include the existence of specific, metal-resistant enzymes and special inter-boundary carrier systems that operate from cell wall to vacuole.

Experimental evidence of carrier systems is lacking but there are indications of differential metal tolerance among key enzyme systems. Fig.7 shows the effect of heavy metals on the nitrate reductase activity (NRA) of a zinc-tolerant population of Silene cucubalus. Other enzymes such as peroxidases have been shown to be less sensitive than NRA.

However, many enzyme systems have yet to be tested; measurements of enzyme activities are sparse and results of in vivo and in vitro studies often differ. With further development of techniques and more extensive comparisons between tolerant and non-tolerant plants, results could be of considerable value to our understanding of plant adaptation to acidic conditions.

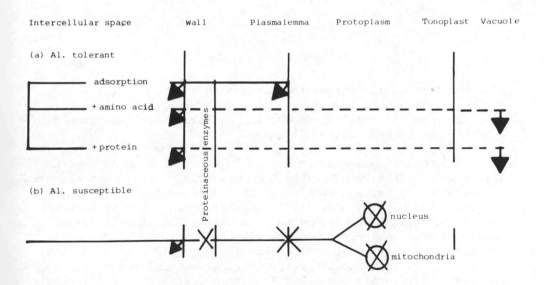

Fig.6 Possible sites of reaction in and around the cortical cells of roots of (a) tolerant and (b) non-tolerant plants[2].

▼ - sites of sequestration   X - sites of metabolic disruption

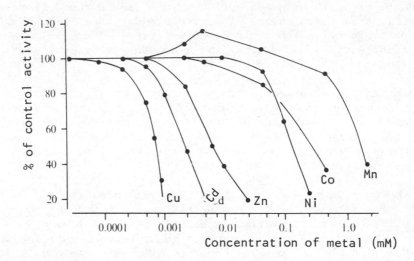

Fig.7 The effects of some heavy metals (all as sulphates in solution) added to the incubation medium on the activity of nitrate reductase (after Mathys, 1975)[46].

Another possibility is the production of metabolites which could in some way sequester metals in tolerant plants. These could take the form of metallo-proteins and tricarboxylic acids and be produced intra- and extra-cellularly.

The complexing mechanism may be highly efficient in dealing with a specific heavy metal which is potentially toxic but essential to the plant in trace amounts. It need not cause a deficiency because there will be some local competition in the root between systems requiring the element and the complexing agents. Some complexed metal may also move through the plant and be dissociated by enzymatic activity at sites where it is required (e.g. in the leaf). Synthesis of necessary proteinaceous complexes can be rapid. The raw materials are available in the cell wall and in the plasma-lemma. Synthesis might be triggered off by a rise in the external concentration of the element to be 'filtered' and continue until that concentration rose to an uncontrollable level (Fig.8). At this point the filter system would no longer be effective and tox-icity would occur due to an excess of the particular cation at metabolically sensitive sites. The problem of isolating and identi-fying specific complexes without denaturing them remains[2].

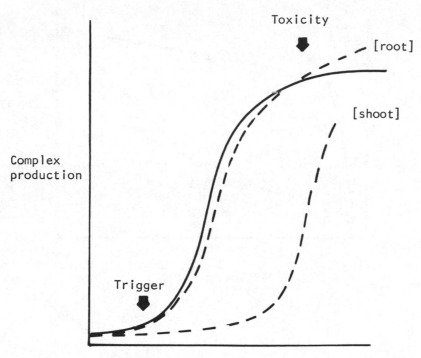

Fig. 8 Exclusion mechanisms[2]. A theoretical outline of the
relationship between complex formation (—) and the levels of the
cation (---) in the roots and shoots of an excluder plant.
Supporting evidence is provided by the data of Wu, Thurman and
Bradshaw[44], redrawn in Figs. 9a and b.

CONCLUSIONS

     Our understanding of the ways in which soil acidity affects
plant growth is advancing on several fronts.   We know that acidic
soils are generally characterised by having relatively low levels
of essential major nutrients and high levels of many polyvalent
cations including elements required by plants in trace amounts.

     This is paralleled by the adaptation of the indigenous flora
to utilize low levels of the major elements and to exclude any
excess of polyvalent ions.   Where the respective levels of supply
are gradually reversed, as when pH rises, the plants may become
stunted due to an excess uptake of major elements and a deficiency
of trace elements.

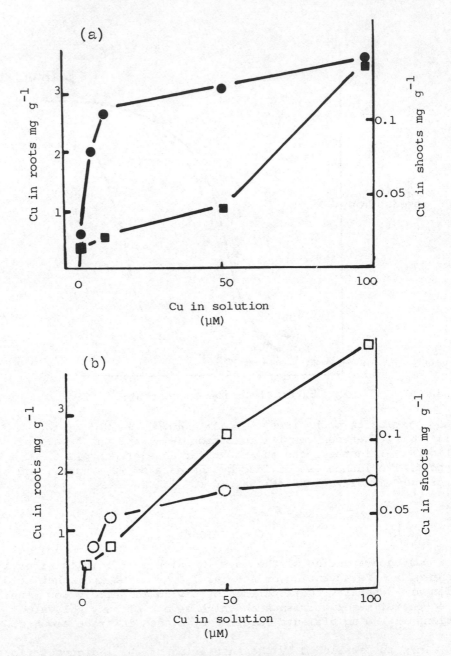

Fig.9 The copper content of leaves (■□) and roots (● O) of
(a) copper-tolerant and (b) non-tolerant clones of Agrostis
stolonifera after 8 days exposure to copper concentrations over
the range 1-100 μM (after Wu, Thurman and Bradshaw 1975)[44].

Physiological evidence continues to support the idea of specific ion tolerance although the precise mechanisms which enable a plant to tolerate the presence of one or more toxic elements have still to be identified, i.e. Fig.6 is a skeleton which still needs modifying and clothing.

Even when regulation is determined at the cellular level it must still be related to the response of the whole plant system. So there is much work to be done - remembering that the plant in its natural environment is responding to varying rather than steady-state conditions and that the variations occur both in time and space. This applies to both climatic and chemical factors associated with acidic soils.

It also applies to biotic factors which include not only natural diseases and predators (Cowling, this volume) but also the range of man's activities.

Determination of the processes underlying adaptation to acidic conditions needs to include consideration of the plant's efficiency in utilizing available nutrients during its life span. There may be different needs at different times.

Fortunately, together with increasing knowledge of the effects and interactions of environmental factors, there is a developing awareness of the value of utilizing genetic variability as a source of adaptive material. This should lead not only to a reduction of practical problems concerning the selection of tolerant plants but also to more refined experimental tests of plant response.

The rapid development of interest in isoenzymes[47] - multiple molecular forms of enzymes - has awakened the hope that their identification may be used as a measure of genetic variation[48]. The debate continues as to whether such molecular variation has adaptive significance or is just evolutionary 'noise'.

Mention of what happens at the cellular level may have seemed academic in relation to whole-plant reactions and especially to gross problems of reclamation and management. But threads which improve our understanding are being drawn together through the disciplines of ecology, biochemistry and genetics. Understanding of the role of nitrogen in particular has increased over the last ten years. Not least, the realization that evolution of tolerance is widespread and potentially rapid[44] has revealed a new and exciting bank of research material.

REFERENCES

1. H. Ellenberg, Hdb. PflPhysiol. IV:638 (1958).

2. I. H. Rorison, in: "Acid Sulphate Soils", Vol. 1, H. Dost, ed.,
    Wageningen, 223 (1973).

3. W. Ernst, in: "Effects of Air Pollution on Plants", T. A.
    Mansfield, ed., Cambridge University Press, 115 (1976).

4. M. J. Wright, ed., "Plant Adaptation to Mineral Stress in
    Problem Soils", Cornell University (1977).

5. H. Dost, ed., "Acid Sulphate Soils", Vols. 1 and 2,
    Wageningen (1973).

6. M. J. Chadwick and G. T. Goodman, eds., "The Ecology of
    Resource Degradation and Renewal", Blackwell Scientific
    Publications, Oxford (1975).

7. L. S. Dochinger and T. A. Selida, eds., "Proceedings of the
    First International Symposium on Acid Precipitation
    and the Forest Ecosystem". (USDA For. Serv. Gen. Tech.
    Rep. NE-23),(1976).

8. J. L. Harper, "Population Biology of Plants", Academic Press,
    London (1977).

9. J. W. Woldendorp, ed., "A Synthesis of Demographic and
    Experimental Approaches to the Functioning of Plants",
    Wageningen (1978).

10. I. H. Rorison, ed., "Ecological Aspects of the Mineral Nutrition
    of Plants", Blackwell Scientific Publications, Oxford, 155
    (1969).

11. E. J. Hewitt, Trans. Int. Soc. Soil Sci. Jt. Meet., Dublin 1:
    107 (1952).

12. L. J. Pons, in: "Acid Sulphate Soils", Vol. 1, H. Dost, ed.,
    Wageningen, 3 (1973).

13. J. K. Coulter, in: "Acid Sulphate Soils", Vol. 1, H. Dost,
    ed., Wageningen, 255 (1973).

14. R. P. Gemmell, and T. M. Roberts, personal communication.

15. A. M. Smith, Soil Biol. Biochem. 8:293 (1976).

16. I. H. Rorison, J. Ecol. 55:725 (1967).

17. M. O. Humphreys and A. D. Bradshaw, in: "Plant Adaptation to Mineral Stress in Problem Soils", M. J. Wright, ed., Cornell University, 95 (1977).

18. K. Walley, M. S. I. Khan and A. D. Bradshaw, Heredity 32:309 (1974).

19. R. L. Jefferies and A. J. Willis, J. Ecol. 52:691 (1964).

20. R. W. Snaydon and A. D. Bradshaw, J. Appl. Ecol. 6:185 (1969).

21. M. Runge, Oecol. Plant. 9:219 (1974).

22. F. N. Ponnamperuma, in: "Plant Adaptation to Mineral Stress in Problem Soils", M. J. Wright, ed., Cornell University, 341 (1977).

23. A. Gigon and I. H. Rorison, J. Ecol. 60:93 (1972).

24. D. C. Havill, J. A. Lee and G. R. Stewart, New Phytol. 73: 1221 (1974).

25. B. J. Halm, J. W. B. Stewart and R. L. Halstead, "Isotopes and Radiation in Soil-Plant Relationships including Forestry", IAEA, Vienna, 571 (1972).

26. K. K. S. Bhat and P. H. Nye, Plant and Soil 38:161 (1973).

27. R. J. Hannapel, W. H. Fuller and R. H. Fox, Soil Sci. 97:421 (1964).

28. C. D. Foy, in: "The Plant Root and its Environment", E. W. Carson, ed., Univ. Press of Va, 601 (1974).

29. E. O. McLean, Commun. in Soil Science and Plant Analysis 7:619 (1976).

30. D. E. Peaslee and C. R. Frink, Proc. Soil Sci. Soc. Amer. 33: 569 (1969).

31. P. H. Nye and P. B. Tinker, "Solute Movement in the Soil-Root System", Blackwell Scientific Publications, Oxford (1977).

32. S. A. Barber, J. M. Walker and E. H. Vasey, J. Agric. Food Chem. 11:204 (1963).

33. E. G. Bollard, in: "Sand and Water Culture Methods used in the Study of Plant Nutrition", E. J. Hewitt, ed., Commonwealth Agric. Bureaux Tech. Comm. No. 22, 198 (1966).

34.   L. H. P. Jones, D. W. Cowling and D. R. Lockyer, Soil Sci.
      114:104 (1972).

35.   I. H. Rorison, J. Sci. Fd. Agric. 26:1426 (1975).

36.   E. A. Kirkby, in: "Ecological Aspects of the Mineral Nutrition
      of Plants", I. H. Rorison, ed., Blackwell Scientific
      Publications, Oxford, 215 (1969).

37.   J. A. Raven and F. A. Smith, New Phytol. 76:415 (1976).

38.   D. D. Davies, Symp. Soc. Exp. Biol. 27:513 (1973).

39.   D. Merkel, Z. Pflanzenernahr Bodenk. 134:236 (1973).

40.   J. P. Grime, Amer. Natur. 111:1169 (1977).

41.   T. Ingestad, Physiol. Plant. 38:29 (1976).

42.   J. C. Brown, in: "Plant Adaptation to Mineral Stress in Problem
      Soils", M. J. Wright, eds., Cornell University, 83 (1977).

43.   J. C. Brown, and W. E. Jones, Physiol. Plant 38:273 (1976).

44.   L. Wu, D. A. Thurman and A. D. Bradshaw, New Phytol. 75:225
      (1975).

45.   A. J. M. Baker, Ph.D. Thesis, University of London (1974).

46.   W. Mathys, Physiol. Plant. 33:161 (1975).

47.   R. M. Cox and D. A. Thurman, New Phytol. 80:17 (1978).

48.   K. Salander, in: "Molecular Evolution", F. J. Ayala, ed.,
      Massachusetts, 21 (1976).

EFFECTS OF ARTIFICIAL ACID RAIN ON THE GROWTH AND NUTRIENT STATUS OF TREES

Bjørn Tveite and Gunnar Abrahamsen

Norwegian Forest Research Institute, 1432 Ass-NLH, Norway

This report is SNSF-contribution FA 29/78

ABSTRACT

This paper gives results of growth measurements in four field irrigation experiments with artificial acid rain within southern Norway. Foliar analyses were carried out in three of the experiments. Height and diameter growth were stimulated by increased "rain" acidity in a Scots pine sapling stand: The reason for this is probably increased nitrogen uptake from the soil. A beneficial effect of sulphur application either alone, or in combination with increased nitrogen uptake is also possible. In the other experiments no treatment effects on height or diameter growth were found.

INTRODUCTION

The possible effect of acid rain on tree growth has been studied by comparing the tree-ring development in areas with soils considered to differ in sensitivity to acidification, and in areas with different deposition of acid air pollutants. Jonsson,[1] concluded from a previous Swedish study that acidification could not be excluded as a possible cause of poorer growth development on soils considered to be susceptible. However, in studies in eastern North America and Norway no clear effects have been found (Cogbill,[2]; Abrahamsen et al.,[3,4]). Due to a large random variation in tree ring analyses, the low precision of the estimated differences should be considered.

The growth of trees exposed to artificial acid rain has also been studied in laboratory and field experiments. Wood and

305

Bormann,[5] found no effect on the growth of yellow-birch seedlings
exposed to mist with pH[3] and above. Mist with pH 2.3 reduced the
growth of the seedlings. However, a similar study with eastern
white pine demonstrated highest growth at the most acid mist level
(pH 2.3) (Wood & Bormann,[6]). The increased growth was explained
as a nitrogren fertilizer effect as nitrate constituted 24% of the
anions (in equivalents), in the acidified mist. As large reductions
in the exchangeable K, Mg and Ca of the soil accompanied the
acidification, the authors "felt" that the increased productivity
was a short term phenomenon.

Field experiments, applying sulphuric acid to pine and spruce
forest are carried out in Scandinavia. Tamm,[7] and Tamm
et al.[8] found no effect upon growth of Scots pine by applying up
to 150 kg H_2SO_4 ha^{-1} yr^{-1} over a 5 year period. In our experiments,
however, a certain stimulation of the height growth was found by
the application of rain with pH 4 and 3 to a lodgepole pine forest
(Abrahamsen et al.,[3]). Stimulated height growth was also
found on transplants of Norway spruce in a forest nursery, after
the application of pH 4 and pH 3 "rain".

The present paper gives new results from the Norwegian field
experiments with artificial acid rain and liming to pine and spruce
forest.

DESCRIPTION OF THE FIELD EXPERIMENTS

The field experiments are located in two different areas of
southern Norway, area A about 40 km north of Oslo, and area B about
180 km south west of Oslo. Three experiments are established in
area A and two in area B. In this report only two experiments
from area A will be mentioned.

Vegetation and soil characteritstics of each field experiment
are given in Table 1. All experiments were located on flat plains
of glacifluvial sediments deposited above the marine limit. The
deposits are dominated by sand and are exceedingly thick - in area
A approximately 60 m. The soil in area B contains no silt, whereas
in area A the content is approximately 20%.

All experiments include treatment with 25 or 50 mm month^{-1} of
artificial rain with different pH, applied during the frost-free
period of the year. Usually five waterings are carried out each
year. The artificial rain is produced by mixing sulphuric acid
with groundwater. The chemical characteristics of the groundwater
are shown in Table 2.

The experimental design, treatments with "rain" and lime,
number of replications, plot sizes and commencement of "rain"

Table 1.   Vegetation and soil characteristics for the field experiments
with artificail acid rain.

Experi-ment	Tree species	Tree height in 1975 m	Dominating ground cover species	Soil profile	Soil horizon	Soil chemical properties			
						Loss on ignition %	pH (H$_2$O)	Base saturation %	N, % of oven dry material
A - 1	*Pinus contorta*	3	*Deschampsia flexuosa*	Entisol	O B	28 4	4.4 4.7	25 7	0.60 0.07
A - 2	*Picea abies*	4	*D. flexuosa Pleurozium schreberi*	Entisol	O B	63 2	4.3 5.4	21 6	0.90 0.07
B - 1	*Pinus sylvestris*	1.5	*V.vitis-idaea D. flexuosa Calluna vulgaris*	Spodosol	O B	87 5	3.7 5.2	15 2	1.6 0.10
B - 2	*Pinus sylvestris*	17	*V.myrtillus V.vitis-idaea C.vulgaris*	Spodosol	O B	84 5	3.9 4.7	11 2	1.2 0.07

Table 2. Groundwater characteristics.

Experiment	Depth to ground water table m	pH	Concentration $^{mg}/l$					
			K	Na	Ca	Mg	Fe	NO_3-N
A - 1	6 - 7	5.6±0.1	0.56	2.00	1.65	0.57	0.96	0.08
A - 2	2 - 4	6.1±0.2	0.65	2.58	3.69	0.71	0.38	0.08
B - 1, B - 2	5 - 6	6.0±0.2	0.32	1.51	1.16	0.33	–	0.22

Table 3. Treatments, experimental design, number of replicates and plot size of the field experiments.

Experiment A-1: Free randomization, 3 replicates, plot size 3 by 5 m. Watering commenced August 11, 1972.

Lime applied as $CaCO_3$ CaO kg ha^{-1}	No watering	Watering					
		25 mm month^{-1}			50 mm month^{-1}		
		pH 5.6	pH 4	pH 3	pH 5.6	pH 4	pH 3
No lime	x	x	x	x	x	x	x
1500 kg				x			
3000 "		x	x	x			
6000 "				x			

Experiment A-2: Free randomization, 3 replicates, plot size 150 m^2 Watering commenced June 28, 1973.   (circle).

Experiment B-2: Randomized blocks, 10 replicates, plot size 25 by 25 m. Watering commenced August 18, 1975.

	No watering	Watering 50 mm month^{-1}			
		pH 6	pH 4	pH 3	pH 2.5
No lime	x	x	x	x	x

Experiment B-1: Free randomization, 3 replicates, plot size 75 m^2 Watering commenced August 14, 1974.   (circle).

Lime applied as $CaCO_3$ CaO kg ha^{-1}	No watering	Watering 50 mm month^{-1}				
		pH 6	pH 4	pH 3	pH 2.5	pH 2
No lime	x	x	x	x	x	x
500 kg	x	x		x		x
1500 "	x	x	x	x	x	x
4500 "	x	x		x		x

application are given in Table 3.

   More details of the field experiments are given by Abrahamsen et al.,[9].

MEASUREMENTS AND ANALYSIS

   In three sapling stands (exp. A-1, A-2 and B-1) height or height growth was measured on all trees at the end of each growth period. Height growth before commencement of the experimentation was also measured for possible use in covariance analysis. In all experiments, girth at breast height (1.3 m above ground), or at 0.5 m (in exp. B-1) was measured at the end of each year, using steel tapes. Apart from exp. B-2, the girth was not measured at the start of the experiment, mainly because of the small tree dimensions. In exp. B-2 girth increment was also measured in the growth period before the treatments started.

   The nutrient status has been observed by needle analyses in the three sapling experiments. Pooled samples of needles from 5-8 trees per experimental plot were taken in October-December every year. The needles were separated into current and previous year needles. The samples were collected at the third or fourth branch whorl from the top of the tree. Nutrient analyses are based on standard methods as described by Ogner et al.,[11].

   The data were subjected to analyses of variance of covariance. Two of the experiments have a factorial design, but not fully balanced. Balanced parts of these experiments were first analysed to trace possible interactions and to study main effects. Data from nonwatered plots were not used in the present evaluation.

TREE GROWTH

   The result of the growth measurements are summarized in Figure 1 and Table 4-7.

   The Scots pine sapling stand (B-1) has obviously increased height growth by the application of "rain" with pH 2, pH 2.5 and pH 3, independently of the lime levels used. There were no treatment effects in the first growth season following the start of experimentation. The increased growth at pH 3 was not significant in 1977 (P=0.07 for the contrast pH 6 - pH 3). A positive effect of "rain" applied at pH 2 (P=0.005 for the contrast pH 6 - pH 2) is also apparent for the girth increment in 1977, but the evidence is not as clear as that for height growth. No consistent effect of liming is apparent.

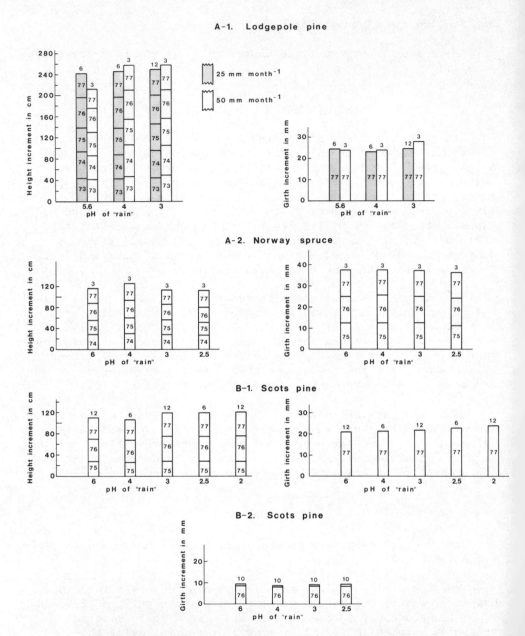

Figure 1.   Height and girth increments in the different experiments.
Numbers within columns: growth year, e.g. 77 = 1977.
Numbers at top of columns: plot number behind each treatment mean.   Treatment means in exp. A-1 and B-1 are based
on all lime levels.

Table 4. Height (cm) and girth (mm) increments of lodgepole pine in
          experiment A-1. Watered plots only.

Height increment for the total
experimental period 1973-1977,   Girth increment in 1977.
adjusted by analysis of covari-   s=2.4 mm, CV=9.7%
ance. s=15.4 cm, CV=6.3%

CaO	"Rain"	pH of "rain"			CaO	"Rain"	pH of "rain"		
kg ha^{-1}	mm month^{-1}	5.6	4.0	3.0	kg ha^{-1}	mm month^{-1}	5.6	4.0	3.0
0	25	247.2	244.4	239.8	0	25	25.1	23.0	25.6
0	50	212.4	257.0	258.2	0	50	23.8	23.7	28.0
1500	25	-	-	236.1	1500	25	-	-	23.8
3000	25	237.8	247.0	269.0	3000	25	24.1	23.7	24.9
6000	25	-	-	250.5	6000	25	-	-	25.0

Table 5. Height (cm) and girth (mm) increments of Norway spruce in
          experiment A-2. Watered plots only.

Height growth for the total ex-   Girth increment for the period
perimental period 1974-1977.   1975-1977. s=2.7 mm, CV=7.4%
s=11.0 cm, CV=9.6%

pH of "rain"				pH of "rain"			
6.0	4.0	3.0	2.5	6.0	4.0	3.0	2.5
116.0	123.4	112.1	109.5	37.6	37.2	36.8	35.7

Table 6. Height (cm) and girth (mm) increments of Scots pine in
          experiment B-1. Watered plots only.

Height increment for the total experi-
mental period 1975-1977, adjusted by   Girth increment in 1977.
analysis of covariance.   s=2.2 mm, CV=9.8%
s=9.0 cm, CV=7.7%

CaO	pH of "rain"					CaO	pH of "rain"				
kg ha^{-1}	6.0	4.0	3.0	2.5	2.0	kg ha^{-1}	6.0	4.0	3.0	2.5	2.0
0	107.3	102.2	122.8	120.4	120.7	0	25.3	21.3	20.5	23.6	23.9
500	108.9	-	123.2	-	124.6	500	19.1	-	22.5	-	23.0
1500	113.3	111.7	112.4	121.1	119.8	1500	19.9	21.1	21.6	21.7	22.4
4500	113.0	-	121.8	-	123.0	4500	18.9	-	22.0	-	24.5

Table 7. Girth increment (mm) of Scots pine in experiment B-2 for
          the total experimental period 1976-1977, adjusted by
          analysis of covariance. s=1.3 mm, CV=14.1%.

pH of "rain"	6.0	4.0	3.0	2.5
Increment	9.2	8.5	8.9	9.3

There was no significant treatment effects in the Norway
spruce sapling stand (A-2). A somewhat lower height growth at
pH 2.5 is indicated for the total experimental period, but in the
last year (1977) the height growth is the same for plots treated
with groundwater (pH 6) or "rain" at pH 2.5.

The girth increment in the old Scots pine stand (B-2) is
generally low. The extremely small increments for 1977 compared
to 1976 were probably caused by bark swelling or shrinkage.
"Rain" at pH 2.5 gave the highest increment in 1977, but the
contrast pH 6 - pH 2.5 is not significant.

The lodgepole pine experiment (A-1) is the oldest established
in 1972. Compared with the older experiments smaller quantities
of acid are applied in this experiment. Most of the plots receive
only 25 mm month^{-1} of "rain", and pH 3 is the most acid treatment.
After 5 years of treatment no negative growth effects of the acid
applications are apparent, nor have any effects of liming been
found. The only significant effect is a lower height growth after
treatment with groundwater at 50 mm month^{-1}. No similar effect
was found by applying 25 mm month^{-1} of groundwater. Addition of
water should have a beneficial effect on the actual site,
especially as summer precipitation in the years 1974-1977 has been
subnormal. We therefore now ascribe the apparent negative effect
of groundwater application to random variation. This is in
contrast to earlier evaluation (Abrahamsen et al.,[3]) where a
certain stimulation of the height growth by applying 50 mm
month^{-1} of rain with pH 4 and pH 3, was reported.

NUTRIENT STATUS OF THE TREES

The discussion of nutrient status is restricted to the current
year's needles. The data are summarized in Table 8, which shows
the total range of nutrient levels for all treatments and years
within each experiment. Apart from nitrogen and perhaps sulphur,
the nutrient levels indicate no severe departure from optimum
values (Ingestad,[12]; Zottl,[13]). We assume that lodgepole

pine behaves like Scots pine. Magnesium, phosphorus and potassium concentrations are maybe somewhat suboptimal in the pines (A-1 and B-1), but we will limit further discussion to nitrogen and sulphur (Figure 2).

In Scots pine (B-1) a certain rise in the nitrogen level was found after the most acid treatment in the same autumn as the experiment started. This effect was accentuated the following year (1975) when a definite increase in nitrogen concentration was found with increased "rain" acidity. The following year (1976) no such effects were found, but somewhat lower nitrogen levels are indicated after application of "rain" at pH 3 (P = 0.01) for contrast pH 6 – pH 3). Sulphur was only analysed in 1975 and 1976. The sulphur concentration has increased after treatment with "rain" of pH 2.5 and pH 2. There are no indications of interactions between liming and "rain" acidity. There was an indication in 1976, that liming had a depressing effect on nitrogen levels (P=0.05 for the contrast between no lime and the highest lime level).

In Norway spruce (A-2) a rise in the nitrogen level was found at pH 2.5 in the autumn of 1974, i.e. one year after commencing the treatments. Otherwise no treatment effects on nitrogen levels were found. Sulphur contents were significantly affected by treatment with "rain" of pH 2.5 in 1975 and 1976 (P 0.001 for the contrast pH 6 – pH 2.5).

No treatment effects on nitrogen or sulphur levels were found in lodgepole pine (A-1) in any year.

Table 8.   The range in concentration of elements in the current year's needles from the field experiments.

Experiment	Concentration, % of dry weight						
	K	Ca	Mg	Mn	N	P	S
A - 1 Lodgepole pine	0.431	0.097	0.054	0.032	1.03	0.110	0.063
	0.627	0.267	0.090	0.097	1.43	0.155	0.086
A - 2 Norway spruce	0.572	0.340	0.087	0.115	1.00	0.161	0.074
	0.743	0.491	0.125	0.147	1.23	0.228	0.130
B - 1 Scots pine	0.422	0.142	0.100	0.021	1.03	0.113	0.069
	0.675	0.320	0.143	0.043	1.56	0.175	0.135

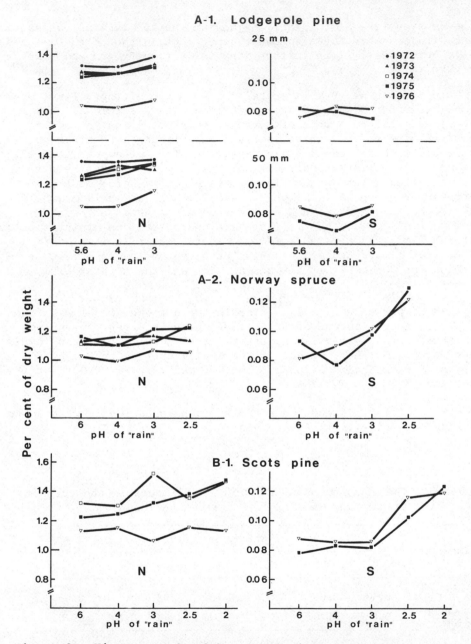

Figure 2.   Nitrogen and sulphur levels in current year needles.

In some recent studies the S:N ratio of foliage has been used to evaluate the sulphur status and possible deficiency in forest trees (Kelly and Lambert,[14]; Lambert et al.,[15]; Malcolm and Garforth,[16]). Using the organic fractions only, this ratio is, for several tree species, found to be about 0.030, which is approximately the S:N ratio in foliar proteins. Using total nitrogen and total sulphur, a S:N ratio lower than 0.030 should indicate possible sulphur deficiency, while higher ratios should indicate free sulphate in the needles. Using this criterion, there are no clear indications of sulphur deficiency in plots treated with non-acidified groundwater (Table 9). Significant year to year variations were found in the pine experiments, the year 1976 giving the highest S:N ratios. "Rain" of pH 2.5 (or pH 2) has consistently resulted in higher S:N ratios.

Turner et al.,[17] proposed to use foliage sulphate concentrations to evaluate the sulphur status. Sulphate has not yet been determined in our foliar analyses.

Table 9.   The S:N ratio for the different experiments and years. Treatment means in experiments A-1 and B-1 are based on all lime levels. Watered plots only.

Experiment	Water mm month^{-1}	Year	pH of "rain"				
			6	4	3	2.5	2
A - 1 Lodgepole pine	25	1975	0.029	0.027	0.025	–	–
	25	1976	0.032	0.035	0.034	–	–
	50	1975	0.027	0.023	0.027	–	–
	50	1976	0.036	0.033	0.032	–	–
A - 2 Norway spruce	50	1975	0.035	0.030	0.035	0.047	–
	50	1976	0.034	0.040	0.042	0.050	–
B - 1 Scots pine	50	1975	0.028	0.030	0.027	0.033	0.037
	50	1976	0.034	0.033	0.036	0.044	0.047

DISCUSSION

The stimulated height growth, and to some extent diameter growth
in the Scots pine sapling stand (B-1), were probably caused by
increased nitrogen uptake from the soil. The failing correlation
between growth response in 1977 and the nitrogen levels in the
needles collected in the autumn of 1976 might be explained by
dilution effects (cfr. e.g. Fagerstrom and Logm,[16]. This
explanation assumes that needle biomass increased in 1975 and 1976
in the most acid treatments due to higher nitrogen concentrations
in the needles. The increased needle biomass in turn caused the
stimulation of height and diameter growth in 1977.

A beneficial effect of sulphur application alone, or in
combination with increased nitrogen uptake, cannot be excluded.
The sulphur content of the needles was slightly suboptimal.
However, the increased sulphur concentration in the Norway spruce
needles (A-2) was not accompanied by increased growth. The S:N
ration also indicates sufficient supply of sulphur in the plots
treated with non-acidified groundwater.

In a previous paper, increased leaching of metal cations
from the foliage and soil, along with reduced base saturation in
the top soil by increased "rain" acidity was reported (Abrahamsen
et al.,[3]). As also observed by Wood and Bormann,[6] and Tamm et
al.,[17] these effects have not influenced short-term productivity
of the trees. At present, effects of artificial acidification
appear to be dominated by increased nitrogen or sulphur uptake.
The increased nitrogen and sulphur uptake have, however, only been
demonstrated in experiments with very high quantities of acid. No
results indicate similar effects of natural acid precipitation.

Acid precipitation involves deposition of relatively large
amounts of nitrogen. The annual deposition of inorganic nitrogen
in southern Norway is of the order 7-15 kg ha^{-1}. The sulphur
deposition is even slightly larger. This deposition of nitrogen
and possibly sulphur, is likely to increase short-term forest
productivity. The long-term effect of an increased loss of metal
cations, is hard to predict. These nutrients are generally in
surplus at present and little is known about the effect of acid
precipitation on the mobilization of the same nutrients through
weathering.

REFERENCES

1. B. Johnson. Soil acidification by atmospheric pollution and
        forest growth. Water, Air and Soil Pollut.7:497 (1977)

2. C.V. Cogbill. The effect of acid precipitation on tree
    growth in eastern North America. Water, Air and Soil
    Pollut. 8:89 (1977).

3. G. Abrahamsen., K. Bjor., R. Horntvedt and B. Tveite.
    Effects on acid precipitation on coniferous forests.
    "Impact of acid precipitation on forest and freshwater
    ecosystems in Norway." F.H. Brække, ed., SNSF-project
    FR 6/76, Oslo-Ås. p. 36 (1976).

4. G. Abrahamsen., R. Horntvedt and B. Tveite. Impacts of acid
    precipitation on coniferous forest ecosystem. Water,
    Air and Soil Pollut. 8:57 (1977). Also published as
    SNSF-project FR 2/75, Oslo-Ås, 15 pp. (1975).

5. T. Wood and F.H. Bormann. The effects of an artificial acid
    mist upon the growth of Betula alleghaniensis, Britt.
    Envir. Poll. 7:259 (1974).

6. T. Wood and F.H. Bormann. Short-term effects of a simulated
    acid rain upon the growth and nutrient relations of
    Pinus strobus, L. Water, Air and Soil Pollut. 7:479
    (1977).

7. C.O. Tamm. Acid precipitation: biological effects in soil
    and on forest vegetation. Ambio 5:235 (1976).

8. C.O. Tamm., G. Wiklander and B. Popovic. Effects of
    application of sulphuric acid to poor pine forests.
    Water, Air and Soil Pollut. 8:75 (1977).

9. G. Abrahamsen., K. Bjor and O. Teigen. Field experiments
    with simulated acid rain in forest ecosystems. 1. Soil
    and vegetation characteristics, experimental design and
    equipment. SNSF-project FR 4/76, Oslo-Ås, 15 pp. (1976).

10. G. Ogner., A. Haugen., M. Opem., G. Sjøtveit and B. Sørlie.
    Kjemisk analyseprogram ved Norsk institutt for
    skogforskning. (The chemical analysis program at The
    Norwegian Forest Research Institute.). Meddr. Norsk.
    inst. skogforsk. 32:207 (1975).

11. G. Ogner., A. Haugen., M. Opem., G. Sjøtveit and B. Sørlie.
    Kjemisk analyseprogram ved Norsk instiutt for
    skogforskning. Supplement 1. (The chemical analysis
    program at The Norwegian Forest Research Institute.
    Supplement 1.). Meddr. Norsk. inst. skogforsk. 33:85
    (1977).

12. T. Ingestad. Macroelement nutrition of pine, spruce and
     birch seedlings in nutrient solutions. Meddn. St.
     SkogforskInst. Stockh. 51(7): 1 (1962).

13. H.W. Zöttl. Diagnosis of nutritional disturbances of
     forest stands. in: FAO/IUFRO international symposium
     on forest fertilization, p. 75 Paris 3-7 Dec. (1973).

14. J. Kelly and M.J. Lambert. The relationship between sulphur
     and nitrogen in the foliage of Pinus radiata. Pl. Soil
     37: 395 (1972).

15. M.J. Lambert., J. Turner and D.W. Edwards. Effects of
     sulphur deficiency in forests. Voluntary paper, XVI
     IUFRO World Congress, Oslo, 5 pp. (1976).

16. D.C. Malcolm and M.F. Garforth. The sulphur-nitrogen ratio
     of conifer foliage in relation to atmospheric
     pollution with sulphur dioxide. Pl. Soil 47:89 (1977).

17. T. Fagerström and U. Lohm. Growth in Scots pine (Pinus
     silverstris L.). Mechanism of response to nitrogen.
     Oecologia 26:305 (1977).

18. J. Turner., M.J. Lambert and S.P. Gessel. Use of foliage
     sulphate concentrations to predict response to urea
     application by Douglas-fir. Can.J. For. Res.
     7:476 (1977).

FOREST ECOSYSTEM RESPONSES TO ACID DEPOSITION - HYDROGEN ION BUDGET AND NITROGEN/TREE GROWTH MODEL APPROACHES

Folke Andersson, Torbjörn Fagerström, and S. Ingvar Nilsson

Swedish Coniferous Forest Project, Department of Ecology and Environmental Research, Swedish University of Agricultural Sciences, S-750 07 Uppsala, Sweden

INTRODUCTION

In order to assess the nature and the extent of acid deposition in the environment an ecosystem approach is often recommended. What do we then mean with ecosystem approach? - It can be defined as studies on an areal basis of basic ecosystem processes such as primary production, decomposition, mineralization and leaching and the integration of these processes as appearing in the biogeochemical cycling of elements. Preferably, there ought to be a coupling between trophic levels or processes representing different levels in order to describe and analyze the time behaviour of different functional parts of the ecosystem.

In the literature today several examples are available on the effects of acid deposition representing partial as well as whole ecosystem approaches. The latter are mostly represented by watershed studies. A watershed integration over a large area is often obtained representing more than one type of ecosystem. Seldom do these studies refer to a delimited and uniform ecosystem. In the following an ecosystem approach will be presented by comparing hydrogen budgets of three forests. Furthermore, a nitrogen dependent tree growth model will also be discussed in order to integrate plant and soil responses.

The aim with this paper is merely to present two types of frameworks within which an assessment of effects of acid deposition on whole systems may be done. Only to a minor extent will an assessment be done as it is considered that we today still lack sufficient and consistent data for a proper analysis of effects of acid deposition on terrestrial systems as a whole.

MATERIAL

The data presented below are derived from ecosystem investigations
in Sweden, Western Germany, and the U.S. The Swedish data have
been obtained from a detailed analysis of a poor Scots pine
(*Pinus sylvestris*) forest, the main investigation objective of the
Swedish Coniferous Forest Project (SWECON) at Ivantjärnsheden,
Jädraos, central Sweden.[1] The investigated forest is situated
185 m above the sea level. The mean annual precipitation is
607 mm. Scots pine is the dominating tree species and the ground
vegetation consists of *Calluna vulgaris* and *Vaccinium vitis-idaea*
with feather mosses and lichens, mainly *Cladonia* species. The
soil profile is an iron podsol, vey shallow in depth. The F/H-
layer is only a few centimeters thick, the A_2-layer being of the
same thickness. The B-horizon occupies approximately 30-40 cm.
The site is extremely poor in nutrients. The humus layer has a
pH_{KCl} = 3.4. The biogeochemical cycling in this 120-140 year old
pine forest has been summarized by Bringmark.[2] The data of the
nitrogen budget have also been used in the tree growth model discussed below.

The German data are derived from detailed ecosystem investigations in Solling, central Germany. It refers to a mature beech
forest, (*Fagus sylvatica*) situated 500 m above the sea level. The
annual precipitation is 1000 mm. The soil is an acid moder with
loess as the main constituent. Data for the budget construction
are mainly taken from Mayer[3] and Mayer and Ulrich.[4]

The U.S. data refer to an investigation by Sollins et al.[5]
in a forest dominated by 450 year old Douglas fir (*Pseudotsuga
menziesii*). Annual precipitation is 2300 mm and the mean temperatures for January and July are $0^{\circ}C$ and $+19^{\circ}C$ respectively. The
soil has app. 20 cm of weakly developed A_1-horizon overlaying
50-80 cm of a weakly developed B_1-B_2-B_3 sequence. The textural
fraction < 2 mm contains 30-50 per cent clay. Soil pH is approximately 6.1 in both the A_1- and the B-horizons.

A HYDROGEN ION BUDGET APPROACH

Procedure

Acidification in soil is usually regarded as an increase of
hydrogen ions. It is tempting to try to evaluate actual flows of
H^{+} ions as well as potential changes. It must however be borne in
mind that a hydrogen ion budget is a result of calculations of the
flows of other main ions taking part in hydrological, biological
and chemical soil processes. To some extent it is an artificial
budget compared with e.g. the mineral element budgets. The idea

has been adopted from Sollins et al.[5]

A hydrogen ion budget is established by calculating mass balances of mineral cations (M) and anions (A) according to the following formulas:

$$\Sigma M^+ = Na^+ + K^+ + Ca^{2+} + Mg^{2+} + NH_4^+ \qquad eq \cdot ha^{-1} \cdot yr^{-1}$$

$$\Sigma A^- = Cl^- + SO_4^{2-} + NO_3^- + H_2PO_4^- \qquad eq \cdot ha^{-1} \cdot yr^{-1}$$

$\Sigma M^+ - \Sigma A^- < 0$ and $\Sigma M^+ - \Sigma A^- > 0$ indicates a net flow of H^+ and HCO_3^- respectively.

The hydrogen budgets will be used to compare three types of forests under different deposition regimes, especially regarding:

-    the relative importance of acid deposition as compared to hydrogen ion flows connected to internal soil processes such as root uptake and litter mineralization and

-    to state a maximum value for the neutralization of hydrogen ions caused by the weathering soil minerals and finally

-    to state possible H^+ losses from the system.

The following structure has been chosen (Figure 1).

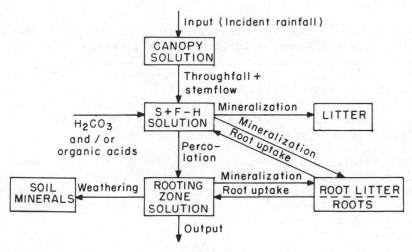

Figure 1. Basic model structure for calculation of H^+ budget.

Some of the flows can be measured, others might be calculated adopting various assumptions. Flows in wet deposition and soil water can be measured and calculated from analytical data, whereas more or less indirect estimations must be used for root uptake and litter mineralization. Root uptake is calculated according to the following formula:

root uptake = increment in plant biomass + litter formation
              + through fall + stem flow - incident rain fall

Figures for litter mineralization should preferably be taken from litter bag studies including those on litter components such as needle litter, branches and roots. Weathering is calculated to make the whole budget balance.

In the specific examples presented below, "hydrogen ion" flows are calculated directly in "Input", "Throughfall + Stemflow", "Percolation + Output".

Root uptake is calculated according to the formula given above, where both "increment in plant biomass" and "litter formation" include fine roots.

As only data concerning the distribution of fine roots have been available from the Solling site, the proportion between fine root uptake and total tree uptake obtained from Jädraos has been used to get a tentative figure for the total uptake process in the Solling beech stand.

The mineral element transfers and thereby the "hydrogen ion" transfers connected with the mineralization procedure have been obtained from:

- actual weight loss and mineral content values for surface and root litter (Jädraos - Staaf and Berg[6], Bringmark[2]),
- a combination of net output data from "humus lysimeters" and assumptions of a complete annual turnover of fine roots (Solling - Mayer[3], this paper) and
- difference calculations including uptake, litter fall and hydrologic flows under the assumption of steady state conditions (H.J. Andrews-Sollins et al.[5]).

No H^+ increments have been allowed in the solution compartments. Balance has been achieved by assuming a transfer of H_2CO_3 and/or organic acids into the humus solution. The mineral soil solution compartment is balanced with H_2CO_3 flows and the net neutralization connected with soil mineral weathering.

RESULTS

The H^+ budgets are presented in figures 2-4 and Table 1.

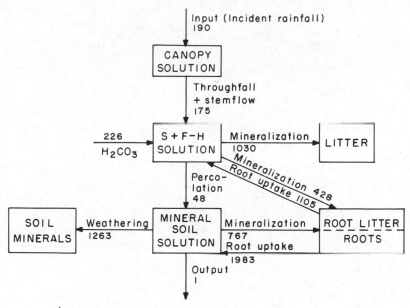

Figure 2. H$^+$ flows in a mature Scots pine forest, Jädraås, central
Sweden. Flows given in eq · ha^{-1} · yr^{-1} .

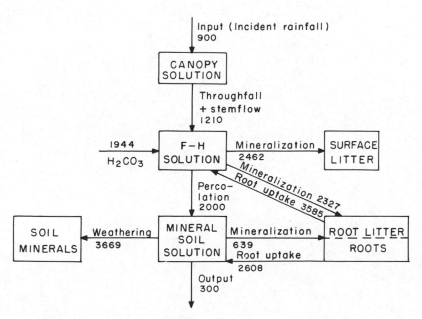

Figure 3. H$^+$ flows in a mature beech forest. Solling, West Germany.
Flows given in eq · ha^{-1} · yr^{-1} .

Table 1. Summary of cation and anion flows in a poor Scots pine forest (Jädraos, Central Sweden), and a beech forest (Solling, Central West Germany). Unit: Eq·ha^{-1}·yr^{-1}. ΣM^+ indicates equivalent sum of cations, H^+ excluded. ΣA^- represents equivalent sum of anions, HCO_3^- excluded.

	ΣM^+	Ca^{2+}	NH_4^+	(in % of ΣM^+)	ΣA^-	NO_3^-	(in % of ΣA^-)	H^+	SO_4^-
				Jädraos – Scots pine					
Input	248	42	141	57	506	114	23	190	355
*Thrf + stfl.	329	88	98	30	562	98	17	175	392
Output	502	115	5	1	482	5	1	1	300
Mineralization	2551	699	1494	59	325	0?	0	2226	203
Root uptake	3848	854	2119	55	760	238	31	3088	340
				Solling – beech					
Input	2680	690	1180	44	2610	510	20	900	1530
*Thrf + stfl.	4290	1300	1230	29	4230	530	13	1210	2760
Output	3626	830	287	8	2798	143	5	300	1710
Mineralization	12841	2088	8218	64	7463	5896	79	5378	895
Root uptake	15294	4282	8541	56	9101	3823	42	6193	4391

*Thrf = throughfall
 stfl = stemflow

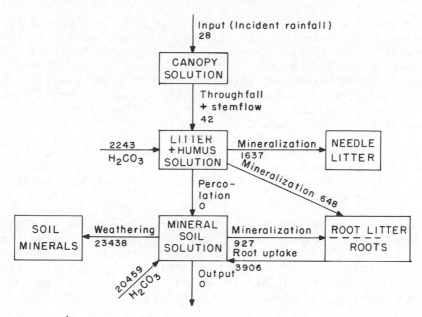

Figure 4. H^+ flows in an old-growth Douglas fir forest. Flows given in $eq \cdot ha^{-1} \cdot yr^{-1}$.

The hydrological flows are small as compared to the biological and the chemical ones. Both root uptake and litter mineralization account for hydrogen ion flows well above the atmospheric deposition, which is particularly striking at the pine site.

One of the main differences between H.J. Andrews and the other two sites is seen in the H^+/HCO_3^- ratio in the soil solution. At Jädraos and Solling H^+ is clearly dominating and HCO_3^- is only of minor importance, whereas the opposite case is found at H.J. Andrews, where the impact of atmospheric deposition should be almost negligible.

A tentative way for calculating the buffer capacity of the whole ecosystem for a given amount of deposition would be to form the ratio between "Output" and "Weathering". The ratios found are thus 0, 0.0008 and 0.08 for H.J. Andrews, Jädraos and Solling respectively. It should be noted that the figure from Solling should probably have been much closer to that of H.J. Andrews provided that the atmospheric deposition had been the same.

As nitrogen is the most important mineral element involved in the root uptake and the litter mineralization processes, the NH_4^+/NO_3^- balance determines the hydrogen ion flows to a very large extent, the nitrification process acting in an acidifying direction (Reuss[7]).

The importance of biological hydrogen ion flows can be further illustrated, by estimating the indirect hydrogen ion input caused by immobilization by tree growth or caused by tree harvests. There are examples which show, that this accumulated "input" divided with the stand age at harvesting corresponds to roughly speaking, the annual input from the atmospheric deposition, which means, that it cannot be neglected when an evaluation of different acidification sources is to be made, (cf. Rosenqvist[8]).

Conflicting evidence is available however. Tveite[9], has stated that the importance of tree harvest is definitely much less than the atmospheric deposition, as far as the southern parts of Norway are concerned.

According to estimations by Odén[10] the importance of biological hydrogen flows has been enhanced, due to an increased deposition of ammonium ions. These "extra" ions cause soil acidification either directly through root uptake or indirectly through nitrification. According to Odén, the deposition of NH_4^+ would account for 40% of the total atmospheric load.

CONCLUDING REMARKS

Restrictions in the budget approach:

-   indirect calculations (root uptake, mineralization) play a crucial role

-   throughfall and stemflow include not only physiological leaching but substantial fractions of the mineral element flow should emanate from dry deposition.

-   the weathering figures are maximum values and they are calculated. An increased soil acidification should have both short term and long term effects corresponding to a decreased base saturation and increased weathering rate respectively. Which of the two effects that is prevailing at a certain point in time must be deduced from actual measurements and experiments

Input-output calculations from water-sheds sometimes reveal a surprisingly good equivalence between "hydrogen ion" input and the output of calcium, magnesium and aluminium, (cf. reference 11). Similar results have been obtained for water-shed studies in the Hubbard Brook area. The following could be quoted from Johnson et al.[12]: "The fact that the rate of chemical denudation is directly proportional to the amount of water, or more correctly, the amount of hydrogen ion that is flushed through the system, suggests that the entire weathering reaction obeys a simple mass action rule with relatively fast kinetics".

The general validity of this conclusion could be questioned however, partly because studies as those referred to above, are rather few in number, partly because there are field experimental data, showing a net accumulation of "hydrogen ions" in the soil[13],[14]. It should be admitted however, that in the field experiment referred to, the H^+ input is fairly high (1020 eq·ha^{-1}· yr^{-1} or 2040 eq·ha^{-1}·yr^{-1}).

As a final evaluation the main advantage with a budget approach is, provided that the data basis is accurate enough, that the relative importance of acid deposition, as compared to the internal hydrogen ion turnover can be stated, at least to the correct order of magnitude. Data concerning leaching and weathering provide further information for a more proper evaluation and formulation of hypothesis to be tested in field trials and/or experimentally.

There are very few localities from which proper data bases are available at present. In order to make the hydrogen ion budget a more powerful research tool, more mineral cycling studies comprising cation-anion budgets of the whole forest ecosystem are badly needed.

A NITROGEN DEPENDENT TREE GROWTH MODEL APPROACH

Background

Considering the indirect effects on the forest ecosystem of acid deposition a decrease of tree growth in a regional scale has not yet been convincingly demonstrated. There are obvious reasons for this fact as the forest or any other terrestrial ecosystem is of a complex nature with buffering mechanisms in vegetation and soil and also embracing processes operating in different time scales. Further, addition of components counteracting negative changes occurs, especially nitrogen.

It is a well known fact that plant growth in temperate climate usually is limited by nitrogen. It has also been found that there exists a relationship between forest yield (forest site index) and the acid-base status of the soil. It has also been demonstrated a relationship between the acid base status and the nitrogen content of the humus layer. It is most likely that adverse effects will include not only the biological soil processes involved in the transformation of nitrogen, but also chemical changes of soil processes. In this way a decrease in available nitrogen limiting production in a regional scale.

Our possibility to combine and synthesize influences and processes acting in different time scales is very limited without
using some kind of an integrating tool. A powerful tool to use in
this context is a mathematical model, which allows testing of the
assumptions and hypotheses involved in the plant/soil system. This
is especially valid to probelms regarding large time domains. At
the same time it must be stressed that it becomes more difficult
to validate modelling results as the time domain increases.

A modelling exercise

Possible effects of acid deposition on soil biological and
chemical processes have been summarized by Tamm[15] and Malmer[16]
respectively. Tamm (*op. cit.*) emphasizes the need for a synthetic
approach in interpreting the effects on the plant/soil system.
This could be done in following changes in primary production
(tree growth) coupled to changes in the turnover of nitrogen. The
same author also states that we "need a theory of site fertility
that reveals the causal relationships between the various soil processes influencing plant growth." An outline to such a theory was
also presented. For some of the processes involved experimental
evidences valid for short time perspectives (Tamm[15], Booth et al.[17])
are available indicating effects on e.g. decomposition, organism
activity. On this basis a modelling exercise was carried out using
basic data on tree biomass and nitrogen cycling of the mature Scots
pine forest described previously in this paper[18]. The aim of the
exercise was:

-       to identify those processes in the nitrogen cycle where a
        change in rate could give a maximum effect on tree growth, and

-       to identify the time behaviour on the tree growth response.

The effects of acid deposition on forest growth have been
tested by assuming that parts of the nitrogen cycle is effected.
The hypothetical mechanisms involved are summarized in figure 5.
Figure 6 shows the compartments and the fluxes of nitrogen in the
forests, rate constant of flows are also included. The time domain
considered as relevant to the problem correspond to a rotation period

The processes in the nitrogen cycle have been described by
assuming that all flows are linear and donor dependent, i.e. a constant fraction leaves a component per unit time. A set of constants
corresponds then to a unique steady state. Such a steady state has
been calculated using information from the Scots pine forest described earlier. The growth of stem wood is considered as a function
of the total needle biomass and the rate of production of needle
biomass is determined by its nitrogen concentration. The theory
applied to growth and turnover of nitrogen in needles follows
reference 19.

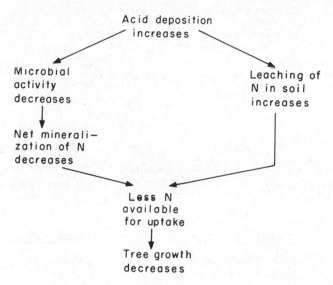

Figure 5.   Hypothetical mechanism behind decreased fertility
            of a pine forest, caused by acid deposition.

The effects of acidification have been studied by sensitivity
analysis changing the rate constants either for an individual process or processes in combination according to different assumptions
(table 2).

RESULTS

The perturbation of an individual process as N deposition from
atmosphere, gave either of the following two answers:

internal process (mineralization of N in litter or N in fine
root litter, turnover time of fine roots).

All response variables except one reacts with some years time
delay, passes the maximum after 1-20 years and reach the
starting value after a long time (> 150 years). The only compartment which deviates from this behaviour is the nitrogen
pool in needle litter as a consequence of the changed turnover
rate. This compartment reaches a new steady state.

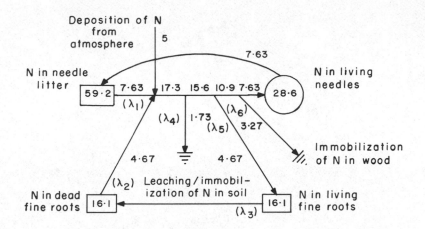

Constant	Verbal statement	Numerical value
λ_1	Rate-constant for mineral-ization of N in needle litter	0.13 yr^{-1}
λ_2	Rate-constant for mineral-ization of N in fine root litter	0.29 yr^{-1}
λ_3	Rate-constant for death of fine roots	0.29 yr^{-1}
λ_4	Fraction of mobile N in soil that is annually leached/immobilized	0.10
λ_5	Fraction of N taken up by pine that is trans-located to fine roots	0.30
λ_6	Fraction of N taken up by pine that is trans-located to wood	0.30

Figure 6.   Schematic representation of the most important pools and flows of nitrogen in a Scots pine forest. "Needle litter" means the total amount of dead needles in the litter. Quantitative data pertaining to N turnover. Pool sizes are given in kg N $\cdot$ ha^{-1}; fluxes are given in kg N $\cdot$ ha^{-1} $\cdot$ yr^{-1}.

Table 2.   Sensitivity table.   Figures for response variables pertain to percentage deviation from unaffected stand.

RESPONSE VARIABLE

		Annual wood production	Annual leaching/ mineralization of N in soil	Annual N deposition with needle death	N in needle litter	N in dead fine roots	N in live fine roots
a	Rate of mineralization of needle litter (-50%)	-20	-22	-22	+96	-18	-18
b	Rate of mineralization of fine roots (-50%)	-12	-14	-13	-7	+100	-9
c	Average life span of fine roots (+50%)	+6	+6	+6	+3	+2	-33
d	Fraction of mobile N in soil that is annually leached/ immobilized (+50%)	-16	+50	-17	-17	-17	-17
e	Annual deposition of N with precipitation (+50%)	+50	+50	+50	+50	+50	+50
f	a + b	-36	-36	-40	+95	+96	-25
g	a + b + d + e	-23	+95	-26	-26	+143	+146	-16

PERTURBED QUANTITY

-   Exchange with the surroundings (N deposition from atmosphere, leaching of N, immobilization of N in plant biomass).

    All variables respond after some years time delay with a slow change (> 150 years) towards a new steady state.

    A great number of alternatives are possible for simultaneous perturbation of two or more processes in combination. The effects can co-operate or counteract each other depending on sign and strength. Figure 7 shows the effects of a simultaneous decrease in mineralization rate of litter nitrogen (needle litter and root litter) and increase in leaching and atmospheric input of nitrogen, all 50%. The net effect of the time behaviour of these disturbances was quite different from the behaviour attained for individual processes. As an example the effect on annual wood production is close to zero during a period up to 30 years after the perturbation. It must be stressed that the assumed perturbation of 50% of all four processes and the assumption that the processes react instantaneously as a result of an increased acidification is arbitrary. Other combinations could have eliminated the answer. However, the result

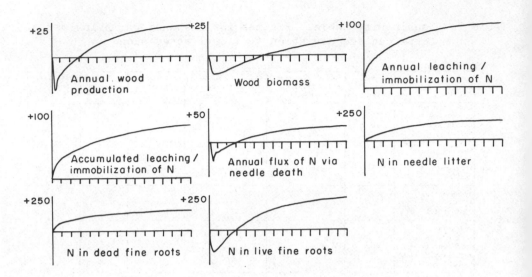

Figure 7.   Time curves of different response variables subsequent to
             a 50% simultaneous change in the rate of mineralization of
             nitrogen in litter (-), leaching of nitrogen (+), depos-
             ition of nitrogen from atmosphere (+), expressed as per-
             centage deviation from unaffected stand.

indicates that measurement of a single variable for example year-
rings does not necessarily reveal an ongoing longterm change of
site fertility.

      The sensitivity calculation of each variable was done by
forming the quotient starting value/maximum response value.   This
allows only a comparison within classes of "internal processes"
and "N-exchange with the surroundings" respectively.   The following
main conclusions can be drawn (table 2):

-     The system as a rule dampens the perturbation.

-     Within the class of internal processes the system in most
      sensitive to the perturbation of the mineralization of
      nitrogen in needle litter.

-     Within the class of "nitrogen exchange with the surroundings"
      the system is sensitive to change in input of nitrogen from
      the atmosphere.

-   There is no evidence that a certain response variable (i.e.
    year-ring width) in each situation is the best one in order to
    reveal an ongoing change of the site fertility in a rapid way
    with great response. The choice of variable ought to be
    governed by the hypothesis raised for the causal mechanisms
    for the effects of the acid precipitation and the costs for
    different methods of measurements. According to table 2, if
    the hypothesis g was accepted it would be four times more
    effective to measure leaching and immobilization of nitrogen
    in the soil profile compared to measurement of year-ring width
    (95% compared to 23%).

FINAL REMARKS

The hydrogen ion budget and the nitrogen dependent tree growth
model so far can be regarded as blunt weapons. The work must be
continued with further testing of hypotheses and more realistic
values on perturbations. Further experimental work is needed. However, the approaches described offer a framework for an analysis of
complex problems. In order to assess the extent of soil acidification a more proper definition of soil sensitivity to acid deposition
ought to be established in both chemical and biological terms.
Present Swedish efforts are directed towards experimental investigations of acid-basic properties as well as nitrogen turnover characteristics of a range of forest soils in order to obtain better
support for a sensitivity classification. On the same time work
on tree growth response to a change in nitrogen availability is
carried out.

ACKNOWLEDGEMENT

This review is based on work carried out within the Swedish
Coniferous Forest Project with support from the Swedish National
Science Research Council, Swedish Environmental Protection Board
and the Swedish Council for Forestry and Agriculture. Background
data has been delivered by a great number of collaborators. Special
consideration is here given to T. Fagerström, P. Sollins and
C.O. Tamm. In this paper S.I. Nilsson has compiled and analyzed
the hydrogen ion budgets.

REFERENCES

1. S. Brakenheim. Technical Report, Swedish Coniferous
         Forest Project 15 (1978).

2. L. Bringmark. Silva Fennica 11:201 (1977).

3. R. Mayer. Göttinger Bodenkundliche Berichte 19:1 (1971).

4. R. Mayer and B. Ulrich. USDA Forest Service General
    Technical Report NE-23:737 (1976).

5. P. Sollins, C.C. Grier, K. Cromack, Jr., F. Glenn and R.
    Fogel. The internal nutrient cycle of an old-growth
    Douglas-fir stand in Western Oregon, Ecological
    Monographs (in press).

6. H. Staff and B. Berg. Silva Fennica 11:210 (1977).

7. J.O. Reuss. USDA Forest Service General Technical Report
    NE-23:791 (1976).

8. I.T. Rosenqvist. Sur Jord - Surt Vann, Oslo. (1977).

9. B. Tveite. Report, SNSF-project (1977).

10. S. Odén. USDA Forest Service General Technical Report
    NE-23:1 (1976).

11. E.T. Gjessing, A. Henriksen, M. Johannessen and R.F. Wright.
    SNSF-project, Research Report 6:65 (1976).

12. N.M. Johnson, G.E. Likens, F.H. Bormann and R.S. Pierce.
    Geochimica et Chosmochimica Acta 32:531 (1968).

13. C.O. Tamm, Svenska Skogsvorsdförbundets Tidskrift 75:189 (1977)

14. I. Nilsson. (this conference).

15. C.O. Tamm. Ambio 5:235 (1976).

16. N. Malmer. Ambio 5:231 (1976).

17. E. Bååth, B. Berg, U. Lohm, B. Lundgren, H. Lundkvist,
    T. Rosswall, B. Söderström and A. Wirén. (this conference

18. T. Fagerström. Internal Report, Swedish Coniferous Forest
    Project 51: (1977).

19. T. Fagerström and U. Lohm. Oecologia 26:305 (1977).

PREDICTING POTENTIAL IMPACTS OF ACID RAIN ON ELEMENTAL CYCLING

IN A SOUTHERN APPALACHIAN DECIDUOUS FOREST AT COWEETA

Bruce Haines[1] and Jack Waide[2]

Department of Botany,[1] University of Georgia
Athens, Georgia 30602, U.S.A.

Department of Zoology,[2] Clemson University
Clemson, South Carolina 29631, U.S.A.

INTRODUCTION

Acidification of rain fall in Europe[1,2] and in the eastern
United States has been reported.[3,4,5] Reports of damage to
individual organisms and to aquatic ecosystems, though few, are
growing in number. The ability of hydrogen ions to displace
biologically essential cations from exchange sites in soil and
from plant tissues is especially significant. Chronic exposure
of ecosystems to acid rain has a potential to disrupt the cycling
of plant nutrients upon which continued plant productivity in part
depends. Understanding the magnitude and consequences of this
disruptive effect is critical to the development of alternative
ecosystem management strategies. The problem is to predict ecosystem responses to acid rain so that costs and benefits of
alternative management plans can be evaluated. Investigations
presently underway are designed to provide preliminary predictions
for southeastern deciduous forests. Work is being conducted on
experimental watersheds at the Coweeta Hydrologic Laboratory,
operated by the U.S. Forest Service, Franklin, North Carolina,
USA.[6] Here we briefly describe the main features of our investigations for the interests of others planning related investigations in other ecosystems.

The plant leaf, the plant root, and the soil seem likely to
be important components determining ecosystem responses to
differing regimes of acid precipitation. Leaves are the most
physiologically active of the plant organs exposed directly to
acid rain. The buffering capacity of the soil is likely to

335

control nutrient losses from systems by leaching, and also to
control the chemical environment in which both nutrient uptake by
roots and nutrient release by decomposed organisms occurs. The
nutrient processing responses of leaves, roots, and soil to differing regimes of acid precipitation are being experimentally
characterized. Recharge of soil exchange sites following cessation of acid rain inputs is also being characterized. These
nutrient-processing responses will eventually be incorporated
into existing ecosystem-level models in order to simulate nutrient
cycling and production processes under acid rain regimes differing
in intensity and in duration. System behavior following cessation
of acid rain inputs can also be simulated. This combined experi-
mental-simulation approach can provide preliminary answers to
potentially critical questions concerning responses of terrestrial
ecosystems to acid rain.

EXPERIMENTAL INVESTIGATIONS

    Work in progress is designed to test the following null hypot-
heses: 1) losses of ions from leaves of dominant tree species
are independent of rainfall pH over the range from pH 5.5 to 2.5;
2) if rainfall pH significantly affects quantities of elements
leached per unit leaf area, leaching rates are identical for dominant tree species; 3) kinetics of ion uptake by roots of dominant
tree species are independent of hydrogen ion activity over the
range from pH 5.5 to 3.5; 4) if kinetics of ion uptake by roots
are affected by pH, roots of all dominant species respond in the
same manner; 5) nutrient losses from forest soil are independent
of the pH of extraction solutions over the range of pH 5.5 to 3.5;
and 6) if significant nutrient losses occur, the curves characterizing nutrient recharge of soil exchange sites are simply the
reciporcals of the cumulative leaching curves. In other words,
there is no hysteresis.

    Dominant tree species collectively comprising more than 50%
of the basal area on the watershed[7] are Carya glabra, Quercus
coccinea, Quercus prinus, Cornus florida, Liriodendron tulipifera,
and Acer rubra.   Because of low transplant success, Carya
illinoiensis and Quercus rubra have been substituted for the first
two species.

    Leaves of seedlings of dominant species are being leached
with simulated acid rain for 1 hour per week at the average rainfall rate for the watershed of 0.89 cm/hr.[8] Simulated acid rain
mimics rain chemistry reported from Coweeta,[9] and is acidified to
pH 5.5, 4.5, 3.5 and 2.5 with reagent acids in ratios of SO_4:NO_3:
Cl of 10:7:1 following the acid rain chemistry reported from
New York State.[10]

    Soil leaching studies have been performed on aliquots of a

statistically representative composite sample from the forested
watershed. Leaching solutions contained a salt component and an
acid component, each at 3 levels. Salt concentrations were based
on elemental contents of soil solutions sampled with zero tension
lysimeters[11] beneath the litter layer in the watershed. Salt
levels 10 and 100 times greater were also used in case the artificial acid rain increased the elemental contents of leaf and
litter leachates by large factors, thereby increasing concentrations of counter ions in the soil solution. These salt solutions
were acidified just as were the leaf leaching solutions, but only
to pH 5.5, 4.5 and 3.5, resulting in 9 different pH-salt combinations. To condense time, aliquots of sieved soil, corresponding
to the amount in a core 1 cm in cross-sectional area and 10 cm
deep were equilibrated with the average rainfall of 215 cm per
year for 20 hours on a reciprocating shaker. Samples were centri-
fuged and the supernate decanted for chemical analysis. New
annual increments of artificial acid rain were then equilibrated
with the soil. The process was repeated 10 times on duplicate
samples.

Quantification of effects of H^+ on elemental uptake by tree
roots has proven to be the most difficult part of the investigation, with attempts being made using isolated roots[12] and growing
seedlings potted in sand and irrigated with experimental solutions.

Solution samples from leaf leaching, soil leaching, and root
uptake experiments are analyzed for Na, K, Mg, Ca, NH_4, SO_4, NO_3,
and Cl. Data will be subjected to appropriate analysis of
variance procedures to test the 6 null hypotheses. Where acid
treatments significantly affect movements of NH_4, NO_3, K and Ca,
these responses will be incorporated into the ecosystem simulation
models. Modeling efforts for other elements have not yet been
initiated.

MODELING AND SIMULATION

Current understanding of element cycling processes in southern
Appalachian deciduous forests have been synthesized into compart-
mental models for N[13] and for K and Ca.[14] Refinements of all models
are currently being developed. These elemental cycling models will
be combined in the future to produce an integrated model accounting
for both production and elemental cycling, again with major emphasis
on the elements K, Ca and N.

Figure 1 illustrates the types of predictions for which these
simulation models have been used to date. The nitrogen model has
been used to predict consequences of various types of forest
harvesting on the nitrogen cycle and on the long-term sustainable
productivity of southeastern forests receiving increasing demands
for wood products.[15,16] Simulations were performed for deciduous

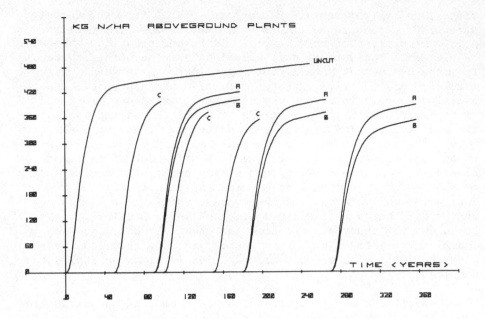

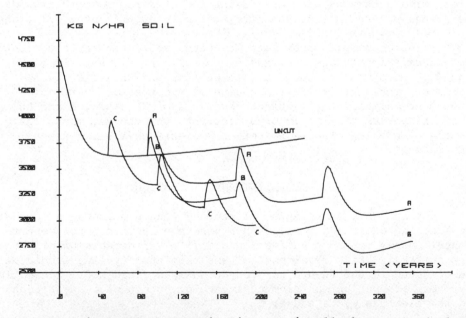

Figure 1.   Simulated changes in nitrogen in all above ground plants
and total nitrogen in soil of a southern Appalachian deciduous fores
subject to three types of forest harvesting: A. merchantable timber
removal, 90 yr rotation length; B. complete tree harvest, 90 yr
rotation length; C. merchantable timber removal, 50 yr rotation leng

forests based on data from Coweeta and also for a loblolly pine
stand in the Duke forest based on published literature data.
Various types and frequencies of cuts were simulated for both
forest types. Such simulation results (Fig. 1) clearly suggest
that increasing the amount of forest biomass removed in harvesting
operations or shortening the period between successive harvests,
may prove especially damaging to the ability of southeastern
forests to sustain high levels of productivity over long time
periods, as well as to the functioning of the forest ecosystem
itself.

Integration of our experimental work and the continually evolving
ecosystem models will make possible such predictive simulations
of potential long-term responses of southeastern forests to acid
rain regimes differing in intensity and in duration, as well as
of ecosystem recovery following cessation of low pH rainfall. It
is clear that acid rainfall is an ecosystem level problem and
that a full understanding of its potential impact can best be
gained by appropriate combination of experimental and modeling
approaches such as those described here.

REFERENCES

1. E. Barrett, and G. Brodin, Tellus 7:251 (1955).

2. C. Brosset, Ambio 2:2 (1973).

3. C. V. Cogbill and G. E. Likens, Water Resources Research
        10:1133 (1974).

4. G. E. Likens and F. H. Bormann, Science 184:1176 (1974).

5. G. E. Likens, Chem. and Eng. News, Nov. 22, p.29 (1976).

6. W. T. Swank and J. E. Douglass, Publication No. 117 de l'Assoc.
        Internationale des Sciences Hydrologigues Symposium de
        Tokyo (Decembre 1975), 445.

7. F. P. Day, Jr., Ecology 55:1066 (1974).

8. W. T. Swank, personal communication.

9. W. T. Swank and F. S. Henderson, Water Resources Research
        12:541 (1976).

10. C. V. Cogbill and G. E. Likens, Water Resources Research
        10:1133 (1974).

11. G. R. Best, Ph.D. diss. Univ. of Ga., Athens (1976), and
        G. R. Best, unpublished.

12.   E. Epstein, "Mineral Nutrition of Plants: Principles and
           Perspectives", John Wiley and Sons, Inc., New York (1972).

13.   J. E. Mitchell, J. B. Waide, and R. L. Todd, A preliminary
           compartment model of the nitrogen cycle in a deciduous
           forest ecosystem, in: ""Mineral Cycling in Southeastern
           Ecosystems", F. G. Howell, J. B. Gentry, and M. H. Smith,
           eds., U.S.E.R.D.A. Conf., 740513, p.41.

14.   J. B. Waide and W. T. Swank, Comparative studies of nutrient
           cycling in watershed ecosystems in the southern
           Appalachians. Paper presented at Symp. on the Metabolism
           of Terrestrial and Freshwater Ecosystems, A.I.B.S.
           Meeting, June 1974, Temple, Ariz. (1974).

15.   J. B. Waide, and W. T. Swank, Proc. Soc. Am. For. 1975:404
           (1975).

16.   J. B. Waide and W. T. Swank, Simulation of potential effects
           of forest utilization on the nitrogen cycle in different
           southeastern ecosystems, in: "Watershed Research in
           Eastern North America", D. L. Correll, ed., Vol. II,
           Chesapeake Bay Center for Environmental Studies,
           Edgewater, Maryland, p.767 (1977).

EFFECTS OF ARTIFICIAL ACID RAIN AND LIMING ON SOIL ORGANISMS AND THE DECOMPOSITION OF ORGANIC MATTER

Gunnar Abrahamsen[1], Jon Hovland[2], Sigmund Hågvar[1]

Norwegian Forest Research Institute[1],
1432 Aas-NLH, Norway

Department of Microbiology[2],
Agricultural University of Norway
1432 Aas-NLH, Norway

INTRODUCTION

Litter decomposition is an important part of nutrient cycling in natural ecosystems. A major part of the nutrients taken up by the vegetation is returned to the soil through litter fall. Decomposition begins when living tissue dies, even if it is still attached to the plant. In the boreal forest, decomposition is often slow and an accumulation of organic matter on the soil is common. This layer is inhabited by enormous numbers of soil organisms: fungi, bacteria and invertebrate animals. These organisms take part in the decomposition of organic matter; as the nutrients are moving within complicated food webs the importance of these various organisms is hard to predict.

The effect of acid precipitation on the decomposition processes and participating organisms is difficult to forecast. However, potential effects have been suggested (Royal Ministry for Foreign Affairs & Royal Ministry of Agriculture[1]; Malmer[2]; Tamm[3,4]).

The present paper summarises Norwegian studies on the effect of artificial acid rain on decomposer organisms and decomposition processes in coniferous forests. Some of the studies include the effects of liming.

This report is SNSF-contribution FA 28/78.

DESCRIPTION OF THE FIELD EXPERIMENTS

Field experiments have been laid out to examine the effect of
artificial acid rain and lime upon the following: tree-growth,
ground cover vegetation and the chemical and biological properties
of the soil. The experimental sites are located in two different
areas in Southern Norway, area A about 40 km north of Oslo and area
B about 180 km south-west of Oslo. Three experiments are established
in area A and two in area B. In this report only one of the
experiments in area B will be mentioned.

Vegetation and soil characteristics of each field experiment
are given in Table 1. All experiments were located on flat plains
of glacifluvial sediments deposited above the marine limit. The
deposits are exceedingly thick - in area A about 60 m, where they
are dominated by sand. The soil in area B contains no silt, whereas
in area A the content is approximately 20%.

All experiments include treatments with 25 or 50 mm month^{-1} of
artificial rain (groundwater), applied during the frost-free period
of the year. Usually five waterings are carried out each year. The
different experiments include treatment with "rain" of different
pH. The pH of the "rain" is adjusted by means of sulphuric acid.
The waterings commenced on the following dates: A-1 September 1972,
A-2 June 1973, A-3 June 1974 and B-1 August 1974. Crushed limestone
has been applied in three experiments in a semi-factorial design.
Further description of the experiments is given by Abrahamsen[5].

The effects of the different treatments on soil acidity are
given in Table 2. Soil pH has not yet been measured in experiment
A-3.

EFFECTS ON DECOMPOSERS

Fungi

Studies on the growth of some fungi isolated from decomposing
material have been carried out. The material from which the fungi
were isolated was withered, yellowish-brown needles of lodgepole
pine (*Pinus contorta*) and needles of Norway spruce (*Picea abies*).

In field experiment A-1 the lodgepole pine needles were
collected aseptically from the trees (Ishac & Hovland[6]). To avoid
accidental contamination with epiphytic microorganisms, the needles
were surface sterilised by treatment with a 1% solution of acetyl-
pyridine chloride followed by washing with sterile water. The
needles were then cut into two pieces, placed on Hagem's agar
(Modess[7]), and incubated at 15 and 25°C. Isolations were made from

Table 1. Vegetation and soil characteristics for the field experiments with artificial acid rain.

Experiment	Tree species	Tree height in 1975 m	Dominating ground cover species	Soil profile	Soil horizon	Soil chemical properties: Loss on ignition %	pH (H_2O)	Base saturation %	N, % of oven dry material
A - 1	*Pinus contorta*	3	*Deschampsia flexuosa*	Entisol	0	28	4.4	25	0.60
					B	4	4.7	7	0.07
A - 2	*Picea abies*	4	*D. flexuosa*	Entisol	0	63	4.3	21	0.90
			Pleurozium schreberi		B	2	5.4	6	0.07
A - 3	*Picea abies*	0.2	*D. flexuosa*	Entisol	0	51	3.9	15	0.8
					B	3	4.7	3	0.07
	Pinus sylvestris	0.2	*Vaccinium myrtillus*						
	Betula verrucosa	-	*V.vitis-idaea*						
B - 1	*Pinus sylvestris*	1.5	*V.vitis-idaea*	Spodosol	0	87	3.7	15	1.6
					B	5	5.2	2	0.10
			D. flexuosa						
			Calluna vulgaris						

Table 2.　Effects of acidification and liming ($CaCO_3$) on soil $pH_{(H_2O)}$.　Soil samples were collected in November 1975.

Experiment	CaO kg ha^{-1}	"Rain" mm month^{-1}	Not watered	pH of "rain"				
				6	4	3	2.5	2.0
	0	25	4.7	4.8	4.5	4.5		
	0	50		4.6	4.5	4.6		
A-1	1500	25				6.4		
	3000	25		6.9	7.2	6.8		
	6000	25				7.2		
A-2	0	50	4.1	4.1	4.0	3.9	3.7	
	0	50	4.1	4.1	4.1	4.2	4.1	3.9
	500	50	4.4	4.3		4.2		3.9
B-1	1500	50	4.4	4.4	4.2	5.0	4.2	4.3
	4500	50	4.6	5.0		4.7		4.2

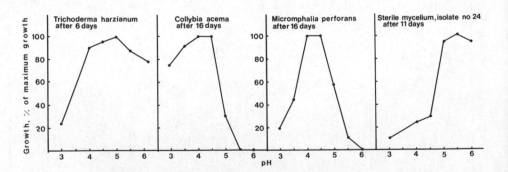

Fig. 1.　Effect of pH on the growth of fungi isolated from decaying needles of lodgepole pine and Norway spruce.

suitable colonies and transferred to slants of Hagem's agar. Strains
of the isolated fungi were tested for their ability to decompose
cellulose. The four most active cellulose decomposing fungi
isolated: *Trichoderma harzianum, Coniothyrium* sp., *Cladosporium
macrocarpum*, and a sterile, white mycelium (isolate no. 24), were
grown at different pH values. The following medium was used:
glucose 10 g, asparagine 1.1 g, KH_2PO_4 1 g, $MgSO_4 \cdot 7H_2O$ 0.5 g,
yeast extract 0.5 g, Fe-citrate $\cdot 5H_2O$ 6 mg, thiamine 50 µg,
succinic acid 7.8 g, distilled water to 1 liter. The pH of the
medium was adjusted with 0.1 M NaOH. Incubation time was dependent
on the growth rate of the fungi, growth was measured as dry weight.

The spruce needles were collected from the litter layer in a
raw humus (Hovland & Olsen, in prep.). In this case a special
interest was taken in litter-decomposing basidiomycetes, isolations
were also made as tissue cultures from fruit-bodies. This work is
still in progress. The effect of pH on the growth of the basidiomycetes was tested on the medium of Lindeberg and Lindeberg[8], as
modified by Guttormsen[9]: Glucose 20 g, $(NH_4)_2$-tartrate 3 g, K_2HPO_4
1 g, $MgSO_4 \cdot 7H_2O$ 0.5 g, Fe-citrate $\cdot 5H_2O$ 5.25 mg, $ZnSO_4$ 4.6 mg,
$MnSO_4 \cdot 4H_2O$ 5 mg, $CaSO_4 \cdot 2H_2O$ 200 mg, Hoagland's A-Z solution
1 ml, biotin 1 µg, thiamine 50 µg, succinic acid 7.8 g, agar (Difco),
15 g. The medium was adjusted to the appropriate pH with 1 N NaOH
and made up to 1:1 with distilled water. Growth was measured as
radial growth in petri dishes.

Of the fungi isolated from pine needles, *T. harzianum* was the
one least affected by low pH (Fig. 1), while the most acid tolerant
basidiomycete was *Collybia acema* which appeared to grow almost as
well at pH 3 as at pH 4.5. The other basidiomycetes had a higher
pH-optimum as exemplified by *Micromphalia perforans*.

Invertebrates

Zoological studies included enchytraeids (Oligochaeta),
different groups of mites, and collembols (Hågvar and Abrahamsen[10];
Hågvar[11]). Enchytraeids were extracted from the soil by the wet-
funnel method (O'Connor[12]) and the arthropods by means of a high
temperature gradient apparatus (Macfayden[13]).

Different experiments were carried out to study the effect of
soil acidification on these animals.

Results from the field experiments. Soil samples from the
field experiments were used to estimate the abundance of each group
of animals and the population densities of the different species.
The field experiments were primarily designed for studies on tree-
growth and soil chemical properties. This meant the number of
replications had to be restricted to three or four. As the distri-

bution of soil animals is very patchy the results are likewise
encumbered with large, random variation.

The abundance of enchytraeids, mites and collembols was
examined in experiments A-1, A-2 and B-1. The enchytraeids were
always identified to species. The species composition of mites and
collembols was only examined in A-1 and A-2 which are the experiments
that have run for the longest period. In experiment A-1 the control-
watered and limed plots were studied with respect to species
composition of collembols and mites, while the acidification effect
was studied in A-2.

Figure 2 shows the abundance of collembols, mites and
Cognettia sphagnetorum, which is the dominating enchytraeid species
in this soil, (experiments A-1 and A-2). The abundance of collembols
increased significantly with increasing acidity of the rain
($P < 0.01$). The reduced abundance of collembols at the limed plots
($P < 0.05$) corresponds to the effect of acidification. The total
number of mites was apparently not influenced by the "rain" acidity.
However, the effect of lime, although not significant, indicates
also that the density of mites is highest in the more acid condition

The abundance of *C. sphagnetorum* was apparently increased by a
slight acidification ($P < 0.1$), but this species was almost eliminated from the most acidified plots. The lime effect ($P < 0.1$)
demonstrates that the species apparently prefers rather acid conditions. Another species, *Mesenchytraeus pelicencis*, behaved the same
way. Only one species, *Enchytronia parva*, increased in abundance
with liming ($P < 0.001$).

The observed acidification effects on the total abundance of
collembols (Fig. 2), were mainly caused by three dominant species
viz. *Tullbergia krausbaueri* ($P < 0.005$), *Willemia anophthalma*
($P < 0.05$) and *Anurida pygmaea* (not significant) (Fig. 3). All
species appeared to increase in abundance as a result of acidification and to be reduced in abundance by liming although only
Willemia anophthalma significantly ($P < 0.001$). Only one abundant
species *Isotomiella minor*, was apparently not influenced by the
acidification. For the other less-abundant species no statistically
significant effects could be found either in the acidified plots or
in the limed plots. The only species which might be reduced in
abundance by acidification is *Isotoma notabilis*.

Among the mites, the only significant effect of the acidification was the increased abundance of Brachychtoniidae in the 3-6 cm
soil layer ($P < 0.01$) (Fig. 4). In line with this there are indications that liming decreased the abundance of this family. Also the
two oribatids *Nothrus silvestris* ($P < 0.01$) and *Nanhermannia* sp.
($P < 0.05$) were reduced in abundance by liming. It should be noted
that the predatory mites, viz. Gamasina among Mesostigmata, tend to
be reduced in abundance at the highest acidity level.

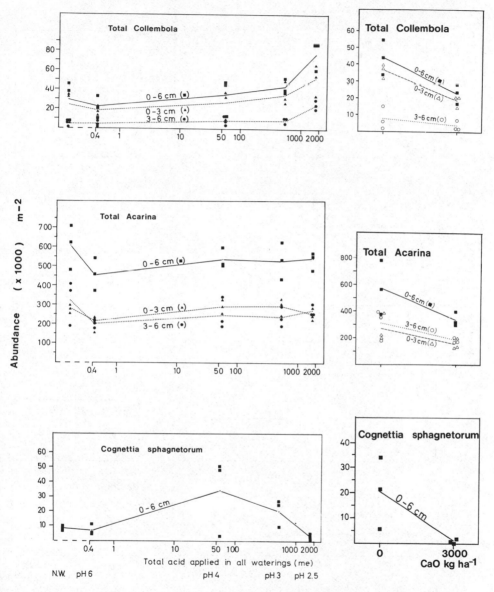

Fig. 2. The effect of acidification and liming on the abundance of collembols, mites, and *Cognettia sphagnetorum* (Enchytraeidae). Results from field experiments. N.W. = not watered.

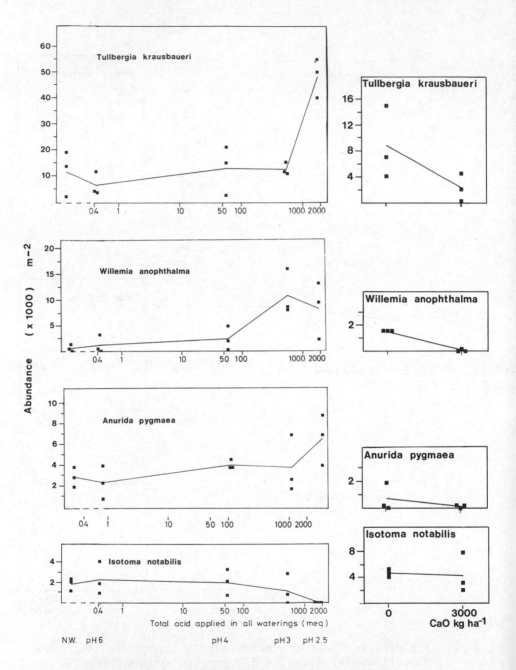

Fig. 3.   The effect of acidification and liming on the population
          densities of several dominant collembole species(0-6 cm).
          Results from field experiments.
          N.W. = not watered.

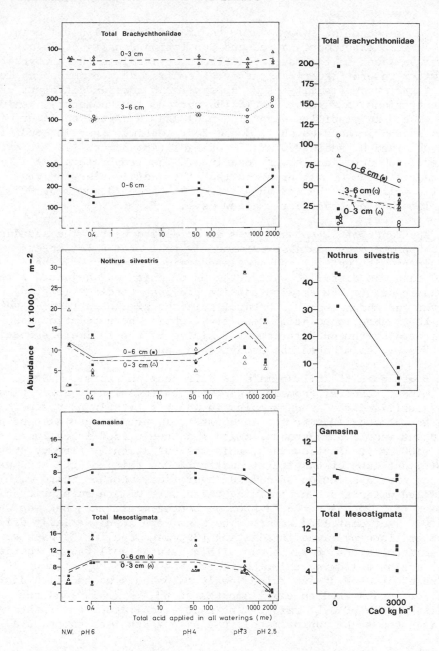

Fig. 4.   The effect of acidification and liming on the abundance
of selected mites: Brachychthoniidae, *Nothrus silvestris*
(both Oribatei), Gamasina among Mesostigmata, and total
Mesostigmata.   Results from field experiments.
N.W. = not watered.

Tolerance of *C. sphagnetorum* towards dilute sulphuric acid.
In the field experiments a reduced population density of *C. sphagnetorum* was found in the most acidified plots. Since the enchytraeids
are water tolerating animals, this effect may be reexamined by looking at the susceptibility of *C. sphagnetorum* towards different
concentrations of sulphuric acid (Hågvar and Abrahamsen[10]). Animals
were placed in distilled water (pH 5.6) and in distilled water with
added sulphuric acid to pH 4, 3 and 2.5. Optimal osmotic conditions
were obtained by adding 20 mM l^{-1} of NaCl (Abrahamsen[14]). At each
pH level 160 individuals were examined. The temperature was
maintained at 20°C. It appeared that the individuals survived
equally well at pH 5.6 and pH 4 (Fig. 5). At pH 3 and 2.5 100%
mortality was observed after a few days.

Abundance of *C. sphagnetorum* in acidified peat. The abundance
of *C. sphagnetorum* was also examined in peat from a lysimeter
experiment with natural and simulated rain (Braekke[15]). The simulated rain was distilled water, the pH of which was adjusted with
sulphuric acid. The results confirm previous findings (Fig. 6)
(Hagvar and Abrahamsen[10]). Natural rain and simulated rain at pH4
give the highest population densities, while the population was
eradicated by the pH 2-treatment. Liming of the lysimeters receiving
pH 4 "rain", eliminated the enchytraeid fauna.

Preference of soil animals to different pH regimes. Some
preliminary results, from an experiment on invasion of soil animals
into sterilised raw humus samples in which a certain pH regime had
been established (Hågvar and Abrahamsen, in prep.), can also be given.
Raw humus was lightly ground, mixed and then placed in cylindrical
bags (100 cm^3), of nylon mesh, this was sufficiently fine to prevent
significant loss of particulate matter. One third of the bags were
treated with a solution of calcium hydroxide, another third with
distilled water and the remaining third with dilute sulphuric acid
at a pH of 2.25. After 8 days of such treatment, each bag was washed
with 0.7 l of distilled water. The bags were then partially dried
and sterilised by gamma irradiation (3.2 mega rad). The bags were
placed into layers of raw humus with a natural soil fauna. During
the 5 months following this, the limed bags were subjected to 10 mm
of "rain" at a pH of 5.3 twice weekly. The bags washed with distilled
water were subjected to equal amounts of pH 4.3 "rain" and the
acidified bags pH 3.5 "rain". The neutral salt content of the rain
was adjusted to be approximately the same as that of the natural rain.

The results strongly support previous findings: *Tullbergia
krausbaueri* and *Anurida pygmaea* are collembols that increase in
abundance with acidification (Fig. 7). *Isotoma notabilis*, on the
other hand, was significantly less abundant in the acidified bags
than in the limed bags. *Isotomiella minor* was not influenced by the
treatments. The enchytraeid *Cognettia sphagnetorum* was significantly
more abundant in the pH 4.3 and pH 3.5 treated bags than in the limed
bags. The mite material has not yet been examined.

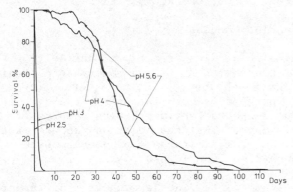

Fig. 5. The survival of *Cognettia sphagnetorum* (Enchytraeidae) in both distilled water and three different dilutions of sulphuric acid. Optimal osmotic potential of the solutions was produced by adding sodium chloride.

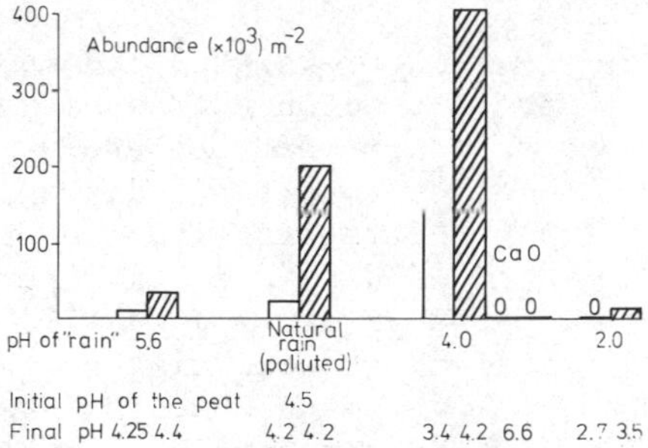

Fig. 6. Population density of *Cognettia sphagnetorum* (Enchytraeidae) in peat from lysimeters watered with "rain" of different acidity. Open columns: groundwater table at 10 cm. Hatched columns: groundwater table at 50 cm.

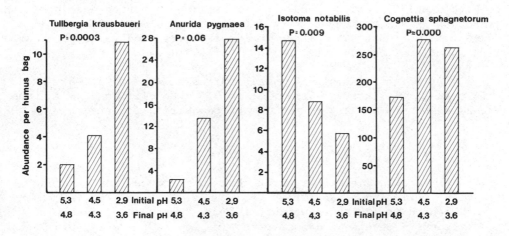

Fig. 7. The immigration of three species of collembols and one enchytraeid species, into raw humus samples with different pH values. The pH of the raw humus was adjusted either by sulphuric acid or lime. The humus was then sterilised with gamma irradiation.

EFFECTS ON DECOMPOSITION OF ORGANIC MATTER

Lodgepole Pine Needles

Six preweighed lodgepole pine needles from experiment A-1, were placed on glasswool, moistened with distilled water, in a petri dish (Ishac and Hovland[6]). The two halves of the dish were taped together to prevent drying. The needles were incubated at 15 or 25°C. Three replications from selected treatments were sampled after 75 days and six after 90 days. The decomposition was estimated as loss of dry weight (105°C). Chemical analyses of needles were carried out according to standard methods (Ogner et $al.$[16]).

The percentage decomposition of the pine needles from different experimental plots are given in Table 3. The decomposition was greater at the pH 3 treatment than at the pH 5.6 treatment. It was less on the needles from limed plots treated with rain at pH 3 than from unlimed plots treated with rain at pH 3. At pH 5.6 there was no significant difference between limed and unlimed plots. Needles from trees watered with 50 mm month^{-1} decomposed faster than needles from plots watered with 25 mm month^{-1} (P = 0.03). Chemical analyses of the needles for calcium and other elements did not explain the differences found between the decomposition rates of the needles.

Needles from the control plots (25 m "rain" month^{-1}, pH 5.6, no lime), were also incubated, in flasks, with dilute sulphuric acid. Three g of needles were mixed with 50 ml of dilute sulphuric acid to pH values 1.0, 1.8, 3.5 and 4.0. To ensure an active decomposition process in all series, the needles were innoculated with spores of $Trichoderma$ $harzianum$. Three replicates were used for each treatment. The needles were incubated at 25°C for 105 days.

The weight loss significantly increased (P < 0.05) when the initial pH was increased from 1.8 to 3.5. No difference was found between pH 3.5 and 4. At pH 1 no decomposition occurred.

Spruce Needles

Parts of a laboratory experiment on decomposition of spruce needles will also be described (Hovland et $al.$, in prep.). Green spruce needles (70 g dry weight, 105°C), were placed in poly-ethylene lysimeters as demonstrated in Figure 8. During two weeks all lysimeters were watered with distilled water. After this moistening period the lysimeters were watered twice weekly with distilled water (pH 5.6) and distilled water added with sulphuric acid to pH 3 and 2. The water did not contain neutral salts. To one series of lysimeters was added 100 mm month^{-1} of water and to another series 200 mm month^{-1}. The number of replications per treatment was eight. The lysimeters were incubated in darkness at

Table 3.   Decomposition (% weight loss), of lodgepole pine needles
            from different experimental plots. The temperature and
            time of incubation are given in the table.
            Residual mean square = 2.7.

| Treatment | | | % weight loss ($\bar{x}$) | | | |
| | | | 15°C | | 25°C | |
Irrigation mm month^{-1}	pH	Liming kg CaO/ha	75 days (n = 3)	90 days (n = 6)	75 days (n = 3)	90 days (n = 6)
25	5.6	–	14.5	17.3	22.0	20.2
25	3.0	–	16.6	18.9	23.0	26.1
25	5.6	3000	14.3	16.8	19.9	23.4
25	3.0	3000	13.7	14.6	16.8	21.4
50	3.0	–	17.4	20.4	30.0	30.3

20°C and approximately 90% relative humidity. The eluate was
analysed regularly, but results from this part of the study will not
be considered in the present paper.

    After 3 months one half of the lysimeters were dismantled and
the needles weighed and analysed for chemical content. The remaining
lysimeters were dismantled after 9 months and the needles analysed.
The inorganic chemical analyses were standard procedures described
by (Ogner et al.[16]). Analysis of monosaccharides was based on
acid hydrolysis (24 h, 1N H_2SO_4, 100°C). Therefore, the monosaccharides originate from the oligo- and polysaccharides in the
needles. The measurements were performed with isopropylidene
derivatives by gas chromatography (Ogner et al.[17]).

    The effect of "rain" pH and "rain" quantity on the decomposition
of the needles was relatively small (Fig. 9). After 3 months the
mean among the treatments ranged from 73% to 78% and after 9 months
the range was from 61% to 66%. With 100 mm month^{-1} of "rain" the
decomposition rate during the first 3 months was greatest in the
pH 3 and pH 2 treatments. During the next 6 months the decomposition
slowed down in these treatments and after 9 months there were no
significant di-ferences between the treatments. With 200 mm "rain",
the weight loss of the needles was smallest at the pH 3 and pH 2
treatments after both 3 and 9 months. The effects mentioned were

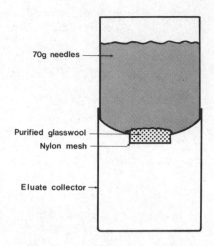

Fig. 8. Design of lysimeters used in the study of decomposition of spruce needles.

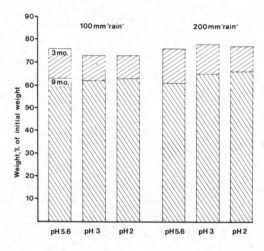

Fig. 9. The effect of acidity and the quantity of "rain" on the decomposition of Norway spruce needles.

small but statistically significant. The difference among the "rain" levels was highly significant after 3 months, but insignificant after 9 months.

During the experiment the CO_2 evolution was measured regularly with a CO_2 analyser. The respiration data support the results from the weight measurements.

The initial content, in g per 100 g needles, of monosaccharides was: 7.38 of glucose, 1.75 of galactose, 5.50 of mannose and 1.52 of xylose. During the first 3 months most of the glucose disappeared and very little seemed to be available for decomposition in the ensuing 6 months (Table 4). The loss of glucose, after 3 months and 100 mm of "rain", was largest at the pH 3 and pH 2 treatments. This

result agrees with the observed weight losses. In the period from
3 to 9 months, the decomposition of galactose, mannose and xylose
diminished with the application of highly acid "rain". This result
might explain the reduced decomposition rate of the acidified
needles during the period from 3 to 9 months.

The original content in mg per 100 g needles (dry weight), of
elements was: K 474, Mg 87, Ca 628, Mn 245, N 954 and P 167. The
loss of these elements through leaching is shown in Table 5.

Table 4. The effect of "rain" quantity and acidity on the
           decomposition of monosaccharides in the needles after
           3 and 9 months treatment.

"Rain" treatment		Monosaccharides, % of initial content							
mm month^{-1}	pH	Glucose		Galactose		Mannose		Xylose	
		3mo	9mo	3mo	9mo	3mo	9mo	3mo	9mo
	5.6	32	28	76	61	69	50	75	59
100	3.0	31	26	74	55	70	49	77	59
	2.0	29	27	70	60	66	61	75	59
	5.6	37	30	85	55	76	47	84	59
200	3.0	34	30	80	65	74	61	83	72
	2.0	33	27	80	59	74	63	85	71

The leaching of the metal cations was, in general, largest for K
followed by Mg, Mn and Ca. The leaching for all these elements was
strongly influenced by the rain acidity. After 9 months very small
quantities of K, Mg and Mn were left in the needles treated with
pH 2 "rain". P was also leached in relatively large quantities.
The leaching was significantly influenced both by the quantity and
acidity of the rain. The N was resistant to leaching. After 9
months only about 10% of the N had disappeared from the needles.
There was no clear effect of the rain acidity or the amount of rain
on the N content. Since no NO_3 and only minute quantities of NH_4
were found in the eluate, it would appear that the incubation
period was too short for examining the effects on nitrogen
mineralisation.

Table 5.   The effect of rain quantity and acidity on the elemental content of the needles after 3 and 9 months treatment.

"Rain" treatment mm month⁻¹	pH	Elements in the needles, % of initial content											
		K		Mg		Ca		Mn		N		P	
		3mo	9mo	3mo	9mo	3mo	9mo	3mo	9mo	3mo	9mo	3mo	9mo
100	5.6	56	28	77	67	92	85	83	71	89	86	59	43
	3.0	50	18	70	56	92	83	72	68	90	86	57	38
	2.0	45	8	59	13	88	42	62	20	94	90	62	34
200	5.6	41	20	72	66	92	90	77	73	94	93	50	41
	3.0	36	11	59	36	91	75	66	52	89	90	48	32
	2.0	30	4	39	2	83	17	44	3	91	87	52	31

Cellulose and Wood Material

Pieces of cellulose (5 x 3 x 0.1 cm), with a mean dry-weight of
1.17 g and sticks of aspen (*Populus tremula*), (4.2 x 0.2 x 0.2 cm),
with a mean dry-weight of 79 mg were used in the field experiments
(Hovland & Abrahamsen 1976). The cellulose pieces and aspen sticks
were placed in the field in September 1974 and collected one year
later. From each treatment-plot 12 cellulose pieces and 36 aspen
sticks were collected. Each type of material from the individual
plots, was pooled. The material was cleaned of soil and plant
remnants, dried at 105°C and weighed. Weight loss was used as a
measure of decomposition.

At the unlimed plots the decomposition rate decreased by the
application of very acid water (pH 2.5), in one (A-2), of the four
experiments (Fig. 10). Liming increased the decomposition rate
very significantly in two of the tree experiments, in which lime
had been applied. There was a significant interaction between liming
and acidification. In experiments A-3 and B-1, increased
acidification decreased the decomposition rate at the two highest
lime levels. The most decomposed pieces (4500 kg CaO ha^{-1}), had a
higher content of K, Ca, Mg, total N and P than the least decomposed
pieces (0 kg Ca ha^{-1}).

The decomposition of the aspen sticks was not significantly
influenced by any of the treatments.

Mixed Raw Humus

The experiment on invasion of soil animals into nylon bags
containing mixed and sterilised raw humus (Chapter 3.2.), was also
used to analyse the decomposition rate of the humus material (Hågvar
& Abrahamsen, in prep.). The dry weights of the bags used in the
arthropod extractions were measured. The results show that the
decomposition rate decreased significantly (P $\sim$ 0.000) with increased
acidification of the humus material (Tab. 6).

DISCUSSION

The soils used in the studies described are all quite acid.
Many of the invertebrates seem to tolerate and even prefer more
acidic conditions than those occurring naturally. In agreement with
this very few animals became more abundant when soil acidity was
reduced by liming. This observation is in accordance with Hill
et al.[18], who recorded reduced abundance of oribatid mites in
raw humus after liming. Lundkvist[19], reported a decreased
abundance of *Cognettia sphagnetorum* in an acidification experiment
in a Scots pine forest in Sweden. However, in this experiment the

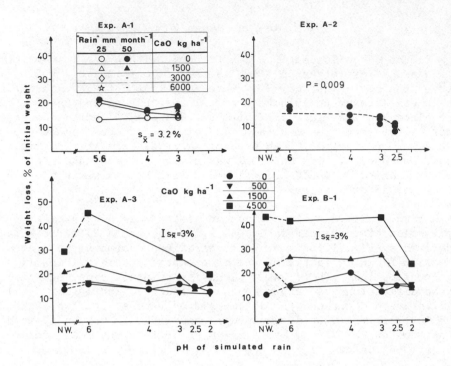

Fig. 10

sulphuric acid applied was very concentrated and according to the
author, the results could be due to shock effects. Since the most
acid "rain" in the present study resulted in similar effects, the
two studies appear to support each other.

The increased abundance of some soil animals, especially
certain collembol species, is difficult to explain. One possible
explanation could be an increase in the food reserves. The soil
meso- and microfauna are supposed to feed, to a large degree, on
fungi. As also demonstrated in this study fungus species are able
to grow in quite acidic conditions. Alexander[20] (p.56),
mentions that many fungi grow readily at pH values as low as 2 to 3.
Because fungi appear to be dominant in acid habitats, they are
responsible for a considerable part of the biochemical transformations. Alexander (*op. cit*) considers that this is not a result of
optimal pH conditions for the fungi, but rather a consequence of
the lack of microbiological competition for the food reserves.

Table 6.   The effect of soil pH_{H_2O} and "rain" acidity on the decom-
           position of mixed raw humus, measured as weight loss.
           The weight losses are expressed as a percentage of the
           initial weight and are given with their 95% confidence
           limits.

pH of rain	5.3	4.3	3.5
Initial pH of humus	5.3	4.5	2.9
Final pH of humus	4.8	4.3	3.6

Weight of humus, %

of initial weight	81.9 $\pm$ 1.9	88.1 $\pm$ 1.1	90.0 $\pm$ 0.6

Another possible explanation for the increased abundance is
reduced numbers of predators. A tendency for such a reduction was
observed for predatory mesostigmate mites in the most acidified
plots. However, these predatory mites live primarily in the
uppermost layer of the raw humus, yet an increased abundance of
collembols and oribatids was observed even in deeper soil layers.

The decomposition studies showed that the initial decomposition
of plant remains was only influenced to a small extent by the degree
of acidification. The result is supported by a study on tulip poplar
leaves carried out by Witkamp[21], who found that variation in
pH values between 4.0 and 8.2 had little influence on the production
of CO_2 during the initial stages of leaf decomposition. He also
observed that the pH of three-day old cultures with *Trichoderma
viride* were almost entirely modified by the fungus. Thus the initial
pH of the culture had an insignificant effect on its final pH.

On the other hand the experiment concerned with incubation of
raw humus material, indicates that decomposition of humus material
is pH dependent. The effect of liming on carbon turnover in acid
soils has long been recognised (Alexander[20]; Russel[22]), but
the effect of acidifying acid soil has, to our knowledge, only been
studied by Tamm, et al.[4]. They found, in agreement with our
results, that incubation of raw humus material with powdered sulphur
or sulphuric acid reduced CO_2 production.

The effects discussed cannot be applied directly to the acid
rain problem. The effects, in general, were produced by much greater
hydrogen ion concentrations than those found in precipitation. If

progress is to be made it would appear necessary to employ long term experiments that cover all the various phases of decomposition. Any artificial rain used should have a composition similar to that of natural precipitation.

REFERENCES

1. Royal Ministry for Foreign Affairs and Royal Ministry of Agriculture, Air pollution across national boundaries. The impact on the environment of sulfur in air and precipitation. Sweden's case study for the United Nations conference on the human environment. Stockholm, 96 pp. (1971).

2. N. Malmer, Om effekterna på vatten, mark och vegetation av ökad svaveltillförsel från atmosfären. En översikt från ekologiska utgångspunkter. (On the effects of water, soil and vegetation from an increasing atmospheric supply of sulfur. A survey on ecological bases.) Statens Naturvårdsverk, Solna. SNV PM 402: 125 pp. (1973).

3. C. O. Tamm, Acid precipitation: Biological effects in soil and on forest vegetation, Ambio 5:235 (1976).

4. C. O. Tamm, G. Wiklander, and B. Popović, Effects of application of acid to poor pine forest, Water, Air, and Soil Pollution 8:75 (1977).

5. G. Abrahamsen, K. Bjor, and O. Teigen, Field experiments with simulated acid rain in forest ecosystems. SNSF-prosjektet FR 4/76, Oslo-Ås, 15 pp. (1976).

6. Y. Ishac, and J. Hovland, Effects of simulated acid precipitation and liming on pine litter decomposition, SNSF-project IR 24/76, Oslo-Ås, 20 pp. (1976).

7. O. Modess, Zur Kenntnis der Mykorrhizabildner von Kiefer und Fichte, Symb. Bot. Upsal. 5(1):1 (1941).

8. G. Lindeberg, and M. Lindeberg, Effect of pyridoxine on the growth of Marasmius perforans, Arch. Microbiol. 49:86 (1964).

9. D. M. Guttormsen, Eddiksyre som veksthemmende faktor hos Boletus variegatus. (Acetic acid as growth-inhibiting factor of Boletus variegatus.) Thesis. Univ. Oslo, Oslo, 66 pp. (1973).

10. S. Hågvar, and G. Abrahamsen, Eksperimentelle forsuringsforsøk i
     skog. 5. Jordbunnszoologiske undersøkelser.
     (Acidification experiments in conifer forest. 5. Studies
     on the soil fauna.) SNSF Project IR 22/77, Oslo-Ås.
     47 pp. (1977).

11. S. Hågvar, Eksperimentelle forsuringsforsøk i skog. 6.
     Virkning av syrebehandling og kalking på to vanlige grupper
     av jordbunnsdyr: Collemboler og midd. (Acidification
     experiments in conifer forests. 6. Effects of
     acidification and liming on Collembola and Acarina.)
     SNSF-project IR in press, Oslo-Ås (1978).

12. F. B. O'Connor, Extraction of enchytraeid worms from a
     coniferous forest soil, Nature (London) 175:815 (1955).

13. A. Macfayden, Soil arthropod sampling, Adv. Ecol. Res. 1:1
     (1962).

14. G. Abrahamsen, Studies on body-volume, body-surface area,
     density and live weight of Enchytraeidae (Oligochaeta).
     Pedobiologia 13:6 (1973).

15. F. H. Brække, Ionetransport og svovelomsetning i torvmark.
     I. Effekten av sur nedbør på torvmonolitter med
     permanente grunnvannsspeil og temperaturregulering.
     (Ion transport and sulphur turnover in peatland. I.
     Effects of acid rain on peat monoliths with controlled
     temperature and ground water relationships.) SNSF-
     project IR, Oslo-Ås (in press).

16. G. Ogner, A. Haugen, M. Opem, G. Sjøtveit, and B. Sørlie,
     Kjemisk analyseprogram ved Norsk institutt for
     skogforskning. (The chemical analysis program at The
     Norwegian Forest Research Institute.) Meddr. Norsk inst.
     skogforsk. 32:207 (1975).

17. G. Ogner, A. Haugen, M. Opem, C. Sjøtveit, and B. Sørlie,
     Kjemisk analyseprogram ved Norsk institutt for
     skogforskning. Supplement I. (The chemical analysis
     program at The Norwegian Forest Research Institute.
     Supplement I.) Meddr. Norsk inst. skogforsk. 33:85 (1977).

18. S. B. Hill, L. J. Metz, and M. H. Farrier, Soil mesofauna and
     silvicultural practices, in: "Forest soils and forest land
     management: 119." Proceedings of the Fourth North
     American Forest Soils Conference. B. Bernier and
     C. H. Winget, eds., Les Presses de l'Université Laval
     (1975).

19. H. Lundkvist, Effects of artificial acidification on the
         abundance of Enchytraeidae in a Scots pine forest in
         northern Sweden, in: "Soil Organisms as Component of
         Ecosystems". Proc. 6th Int. Coll. Soil Zool., Ecol. Bull.
         (Stockholm) 25:570. U. Lohm and T. Persson, eds.
         (1977).

20. M. Alexander, "Introduction to soil microbiology". Second
         edition, John Wiley & Sons, Inc., New York. 467 pp. (1977).

21. M. Witkamp, Environmental effects on microbial turnover of
         some mineral elements. Part I. Abiotic factors,
         Soil Biol. Biochem. 1:167 (1969).

22. E. W. Russel, "Soil conditions and plant growth". 10th
         edition, Longman, London, 849 pp. (1973).

EFFECTS OF ACIDITY ON MICROORGANISMS AND MICROBIAL PROCESSES IN

SOIL

Martin Alexander

Laboratory of Soil Microbiology, Department of Agronomy,

Cornell University, Ithaca, New York, USA 14853

Microbiologists have long been interested in the effects of acidity on the residents and activities of terrestrial ecosystems. For decades, scientists in many countries have explored the changes brought about by the acidification of soils associated with the use of fertilizers and have examined the relative rates of several transformations as influenced by pH of the habitat. An enormous literature was developed in the early phases of soil microbiology. However, such investigations fell into disfavor, and much of the research therefore terminated. Only recently has it become fashionable to work in this area once again, as it became evident that similar stresses were induced owing to acid precipitation, and once again there is renewed interest. It is now evident, moreover, that large gaps in our knowledge exist, and though the information base is large, many information deficiencies remain so that it is not frequently possible to assess the impact of acidity on the residents or the activities they catalyze.

Six major microbial groups reside in the underground ecosystem: bacteria, actinomycetes, fungi, algae, protozoa and viruses. The viruses are obligate parasites, and research on them is limited owing to the lack of knowledge of their interactions with their various hosts, be they higher animals, plants or microbial cells. Catalogs of the genera and species of these several groups exist, with particular attention to the major inhabitants. These dominant species have been found in widely dissimilar soils, and hence the soil ecologist is presented with a large array of both heterotrophic and autotrophic microorganisms, few of which show a distinct biogeography. Although the ecologist is concerned with the inhabitants, their distribution, and factors affecting them, few but the soil microbiologists are interested in community structure.

363

Chief interest is directed to the transformations that are effected
by these organisms, and hence my attention will focus on community
function rather than structure---the chemical reactions brought
about by the organisms rather than their identities.

Microbial processes in soil may be said to have four roles in
the function of the larger ecosystem that encompasses the soil
itself and the overlying vegetation.

(1) First, microorganisms are essential for plant growth.
Nearly all of the nitrogen and most of the phosphorus and sulfur,
as well as other nutrient elements, are bound in organic combination. In this form, the elements are largely or entirely unavailable for utilization by higher plants. It is only by heterotrophic
activity that the vast store of nitrogen and the reserve of phosphorus and sulfur are made available to higher plants. Thus, processes
that lead to the conversion of the organic forms of these elements
to the inorganic state are crucial for maintaining plant life in
natural or agricultural ecosystems. The fact that nitrogen is
limiting to food production in much of the world and governs primary
productivity in many terrestrial habitats is an indication of the
key role of these degradative processes.

(2) Microorganisms are also critical for creating and maintaining many of the characteristics associated with soil structure. A
good structure is necessary for root development, proper aeration,
and water movement. The contribution of the microflora is linked
partly with the formation of humus---a product of their acitvities
---and the humus in turn provides soil with many of the physical
characteristics as well as with the nutrients that are so essential
for the development of economically important as well as natural
plant populations. In addition, microorganisms bind together small
soil particles into units known as aggregates, and these aggregates
owe their origin partially to microbial excretions as well as to the
physical binding from the filaments of fungi and actinomycetes.

(3) The microflora is also critical in preventing the excessive
accumulation of organic matter, with the likely build-up in the concentration of toxic materials arising from the degradation of plant
remains. Were there to be excessive organic matter in soils, moreover, anaerobiosis with the consequent inhibition of growth of higher
plants would ensue. This is evident in poorly drained areas where
organic matter accumulates. Many of the products of partial degradation or anaerobic decomposition of plant remains are toxic, and a
large number of aromatic compounds as well as simple aliphatic acids
are phytotoxic and do considerable damage to the growth of plants.
Were it not for adequate microbial activities, these phytotoxins
would likely accumulate far more frequently than they do at the
present time.

(4) The maintenance of environmental quality also requires an active microflora. Society now introduces a variety of toxic substances into soil. The degradation of some of these is well known, particularly among the pesticides, and there is now growing interest in the application to land of sludge, sewage, and large quantities of animal manure associated with feedlot operations. The destruction of synthetic organic compounds and the natural waste materials relies on microorganisms, because no other agency is able to destroy organic compounds and convert them to the inorganic forms that are environmentally acceptable. It has also become evident in recent years that the levels of carbon monoxide and volatile hydrocarbons in the atmosphere associated with anthropogenic and natural sources do not build up in the atmosphere owing to the ability of microbial populations in soil to destroy them at remarkably rapid rates. A major stress on soil communities, therefore, might ultimately be reflected in a diminution in the capacity of the microbial populations to maintain local and regional environmental quality.

A large number of distinct processes are involved in these four categories of microbial activity. In some instances, the processes result in the conversion of the organic complexes of an element to the inorganic form. In other instances, the transformations entail an oxidation of the reduced form of an element to the more oxidized state, as in the cases of nitrogen, sulfur, iron and other elements. Important reactions are also associated with the reduction of higher oxidation states of elements to the more reduced forms, and denitrification, methane formation, and sulfide accumulation are examples of reduction processes. Of considerable local, regional and global significance are the reactions that lead to the volatilization of different compounds of many elements and the return of volatile forms of elements to soil. The latter processes are known as fixation reactions, and of prime concern at the present is nitrogen fixation, a microbiologically mediated process whose importance is becoming increasingly evident with the rising cost of nitrogen fertilizer. These various transformations have examples in the cycles of carbon, nitrogen, phosphorus, sulfur, iron and other elements.

Microbiologists are notorious for studying their organisms one population at a time and in pure culture. Many ecologists, knowing little better, encourage us to do so in the hope that such information will have environmental relevancy. Admittedly, such investigations yield an enormous amount of basic information on the behavior of microorganisms and help explain what is observed under natural conditions. However, such an approach has many limitations. For example, it is extremely rare that the active population in a soil process is known. In the absence of such knowledge, it is impossible to assess whether the population studied in isolation is indeed the active species in the original habitat. Furthermore, the physical or chemical stresses that are studied in this way are

rarely the ones that have an influence on the microbial population
in nature. This results from the fact that microorganisms **exist**
not in the habitat defined by the size of the physical probe introduced into the soil to measure some particular physical parameter
or by the extraction solution used to measure a chemical property
but rather their habitat is at a more microscopic scale. This is
well illustrated by the apparent localization of acid-sensitive
actinomycetes in the profile of acid soils. The data suggest that
these nontolerant organisms exist in microsites where the pH is
higher than in the ambient environment, a more alkaline region that
might be associated with the local release of ammonia during the
decomposition of nitrogenous compound[1].

The microenvironmental effect of acidity is also illustrated
in studies of soil enzymes, which, if excreted by the organism, are
frequently associated with clay particles. These colloidal particles have a negative charge, and they are thus surrounded by a
cloud of hydrogen ions, the concentration of which is higher than
that in the bathing fluid. The pH adjacent to the particle surface
is thus lower than that in the surrounding solution, and an enzyme
associated with the particle will be functioning at a different pH
than the enzyme at some distance from the clay surface. Hence, the
measured pH for enzymatic activity is frequently different from the
actual pH, a difference associated with the fact that the pH of the
solution rather than at the clay surface is being measured. Similar effects of the microenvironment on the function of microorganisms themselves rather than the individual enzymes have been observed[2].

It is important to bear in mind that many, and probably most,
microbial transformations in soil may be brought about by several
populations. Thus, the elimination of any one population is not
necessarily detrimental inasmuch as a second population not affected by the stress may fill the partially or totally unfilled niche.
For example, the conversion of organic nitrogen compounds to inorganic forms is characteristically catalyzed by a number of species,
often quite dissimilar, and a physical or chemical perturbation affecting one of the species may not seriously alter the rate of the
conversion. On the other hand, a few processes are in fact carried
out, so far as it is now known, by only a single population, and
elimination of that species could have serious consequences. Examples of this are the nitrification process, in which ammonium is
converted to nitrate, and the nodulation of leguminous plants, for
which the bacteria are reasonably specific according to the leguminous host.

Considerable attention has been directed to finding sensitive
processes or indicator microorganisms of environmental stress. The
availability of such indicators would be advantageous in that they
would serve to highlight when there is a detrimental change induced

in a particular environment. Most attempts to find sensitive
organisms or processes have come from investigations of pesticide
effects, but recently there has been a resurgence of interest in
the influence of heavy metals because of the disposal of sludge
containing these elements onto land areas. To serve as an indicator of pH stress, the single population or group of populations
must be acid sensitive, and the sensitivity of these organisms
must be appreciably greater than that of other species.

One of the best indicators is the nitrification process
because the responsible organisms, presumably largely autotrophic
bacteria, are sensitive both in culture and in nature to increasing
acidity[3]. Although nitrification will sometimes occur at pH values
below 5, characteristically the rate falls with increasing acidity
and often is undetectable much below pH 4.5. Limited data suggest
that the process of sulfate reduction to sulfide in soil is markedly
inhibited below pH values of 6[4], and studies of the presumably
responsible organisms in culture attest to the inhibition linked
with the acid conditions. It has also been known for some time
that the blue-green algae are not present in acid soils although
they have both adequate moisture and are exposed to sunlight. These
organisms are not too easy to enumerate by conventional counting
procedures, but their activity can be assessed by measuring the
rate of photosynthesis using $^{14}CO_2$. However, other algae will also
bring about photosynthesis in soil, so that a more specific test for
the activity of these organisms is to measure nitrogen fixation in
the light and in the dark, the light-induced process being affected
by the blue-greens. Some of our recent data attest to the sensitivity of these organisms to acidity, the inhibition being noted in
the rates of both CO_2 fixation and nitrogen fixation[5]. Both of these
reactions are simply measured in soil, the first by the incorporation of the radioisotope into the soil organic fraction, the second
by the reduction of acetylene to ethylene. The latter process
mimics nitrogen fixation and can be evaluated simply by gas
chromatography.

Studies of the effects of acidification from nitrogen fertilizers or sulfur amendments to soil and comparisons of the populations
in soils of dissimilar pH values attest to the sensitivity of
bacteria to increasing hydrogen ion concentrations. Characteristically, the numbers of these organisms decline, and not only is the
total bacterial community reduced in abundance but so too are individual physiological groups. The actinomycetes, which taxonomically
are also considered to be bacteria, are generally less abundant also
as the pH falls. On the other hand, the relative abundance of fungi,
assessed by counting procedures, rises, an increase that may be
associated with the lack of competition from other heterotrophs at
the lower pH[3,6]. The pH of the soil not only influences the microbial
community at large but also those specialized populations that colo-
nize the root surfaces[7].

The chemistry of humus is largely unknown and hence the organisms responsible for its degradation are poorly characterized. It is likely that a variety of dissimilar heterotrophs is responsible owing to the heterogeneity of the substrate. Nevertheless, acidity is generally linked with decreased rates of humus decomposition, and liming will promote the conversion of the humic compounds to carbon dioxide. The extent of the suppression varies with the soil and with the nature of the organic materials, and in some instances there is rapid degradation at moderate acidities and in some instances even rapid decomposition at extreme acidities, such as at pH 1.8[8].

The rapid escalation in cost of nitrogen fertilizer and the growing awareness that food production in the developing countries is often limited by inadequate nitrogen supply have prompted renewed attention to biological nitrogen fixation. From the economic viewpoint, the most attractive nitrogen-fixing system is that involving leguminous plants and members of the genus Rhizobium. For our present purposes, one need only refer to quite early literature showing the effect of pH on both legume growth and nodule production. It was reported as early as 1922 that soybeans would grow at values as low as pH 3.9 on fixed nitrogen compounds, but they only became nodulated at pH values of 4.6 or above[9]. Thus, nodulation, which is the essential prelude to the development of the nitrogen-fixing symbiosis, is especially sensitive to these acidic circumstances. Studies of alfalfa and black medick (Medicago lupulina) show that they need pH values of 6.5 or greater when using atmospheric nitrogen, although they grow at pH 5.5 in soils receiving fertilizer nitrogen[10]. In the case of red clover, nodulation is frequently absent or poor at pH 5.2 or below and is improved at higher pH values[11]. In addition, the numbers of the infective root-nodule bacterium are often reduced and become vanishingly small as the pH values fall below 5.0[12]. Hence, nitrogen inputs from the legume-Rhizobium symbiosis will likely be remarkably lower as acidity makes its effect known. The poor growth of legumes in soils of low pH may be an effect of acidity per se, an influence of the existence of toxic levels of manganese, iron and aluminum (which are more soluble in soils of low than of high pH) or the result of deficiencies of molybdenum or some other plant nutrient which is less available at the acid reaction. The influence might be on the host directly, the processes of nodulation or nitrogen fixation, or the development of the root-nodule bacteria in the root zone.

The same type of harm is found in nonsymbiotic nitrogen fixation. Early trials by total nitrogen analysis and involving samples amended with unrealistically high levels of carbohydrate suggested the sensitivity, and more recent studies with the acetylene-reduction technique confirm these early findings. Not only nitrogen fixation in soil but also in forest litter will be reduced, witness the report that the application of water at pH 5.0 markedly reduced the rate of acetylene reduction by litter samples[13]. Acid forest soils are notably poor in this activity although acid peats sometimes show modest acetylene-reduction rates. Acid environments probably owe their nitrogen fixation to heterotrophs, clearly not particularly active, and they seem to be largely devoid of blue-green algae[14].

The conversion of organic nitrogen in soil to ammonium, by contrast with nitrification, does not show a marked pH sensitivity, and frequently little influence of pH is noted over the moderate ranges commonly encountered in agricultural soils[3]. This probably results from the existence of a broad range of fungi as well as bacteria and actinomycetes that contribute to the process so that nontolerant species are readily replaced. The greater resistance of the organic nitrogen mineralization sequence then the nitrification process is reflected in the finding that frequently ammonium levels rise with time in acid soils, whereas nitrate is commonly the dominant form of inorganic nitrogen that is freely available to plants in nonacid soils. Of course, leaching and plant uptake will alter the relative dominance of the two nitrogen ions.

Denitrification is gaining increased attention because of the concern with nitrate as a pollutant and the possible impact of one of the products on the ozone shield against ultraviolet radiation. Denitrification is the conversion of nitrate to N_2 and nitrogen oxides. It is brought about by bacteria that use nitrate as an electron acceptor. Characteristically these bacteria are sensitive to the hydrogen ion, especially in culture. Moreover, an inverse relationship exists between acidity and the number of denitrifying bacteria, at least at the few sites studied[15]. Nevertheless, nitrification in some soils is still rapid at pH values of 4.7, but it is likely that the dominance of bacteria in this transformation makes it a sequence that would likely be reduced appreciably and eliminated entirely with increasing acidities.

From 15 to 85% of the phosphorus in soil is organic, and this large reserve is thus unavailable to higher plants. Hence, an inhibition of phosphorus mineralization in lands that are phosphorus deficient would have a serious impact on the growth of agricultural crops, tree growth and primary productivity. Therefore, it is worth mentioning that the microbial conversion of organic to inorganic forms of phosphorus in soil appears to be lower in the more acid environments[17], but the data are too limited to warrant generalizations. Similarly, the breakdown of organic sulfur compounds in humus to yield the products that plants can assimilate for growth is influenced by the same variable[18], and soils with pH values of 5.0 degrade humus sulfur only slowly.

Not only the rates of microbial processes but also the products may be altered. These changes may be of more than minor signifi-

cance because the new products may exhibit different toxicities
to plants, animals and the functioning of ecosystems. A classic
instance of a change in products with pH is found in denitrification, and here there occurs a greater accumulation of nitrogen
oxides[19]. The emission of these oxides from soil is now believed
to contribute significantly to atmospheric pollution (as NO) and
possibly to a reduction in the ozone shield that protects the
biosphere from damaging ultraviolet radiation (N_2O). It is widely
known that products of pesticide breakdown may persist or be quite
toxic, and new products associated with acidification similarly
may indeed be more persistent or more toxic than the metabolites
generated under more neutral conditions.

Acidity also has a variety of effects on the incidence and
severity of diseases of plants caused by subterranean microorganisms.
The low pH may favor increased growth of the pathogenic bacterium
or fungus, it can alter the ability of the plant to resist invasion
by the parasite or it can affect the interactions between pathogens
and neighboring microorganisms. In some instances, the low pH
probably has a direct effect on the pathogen because its behavior
in nature and in culture is almost the same. In other cases, the
influence of acidity probably is related to an altered resistance of
the plant or to a change in the capacity of the indigenous microflora to control the parasitic species. The classic case of
disease control associated with acidity is potato scab, and the
sensitivity of the actinomycete in culture is associated with the
scarcity or absence of scab in soils more acid than pH 5.4[20]. On
the other hand, the root rot of poinsettia caused by Thielaviopsis
basicola is greatly reduced in severity when the host is grown in
soils of pH 5.5 or less, but this is not an effect on the pathogen
because the fungus grew well below pH 5.0[21]; hence, one must in
this and similar instances evoke the hypothesis that the resistance
of the host is changed or that the microflora normally suppressing
the pathogen is benefited by acidity. Conversely, certain diseases
are more severe as the pH falls, and in these instances acidification
is linked with increasing harm to plants[22].

Few explanations supported by adequate data exist to account
for the changes in disease incidence or severity linked with pH,
but it is known that the abundance of antagonistic microorganisms
is regulated by the same variable[23]; in addition, some of our
investigations have focussed on the importance of protozoa in modifying the population density of free-living and parasitic bacteria[24]
Predatory protozoa are susceptible to a variety of environmental
perturbations, and acidity is no exception.

A change in an environmental variable such as pH will alter
other properties of soil, and often it is difficult to ascertain
the precise cause of the observed pH effect; that is, whether it
is a direct influence of the hydrogen ion concentration or some

other modification of the habitat. For example, the toxicity of low-molecular-weight fatty acids increases as the pH falls[25], and undoubtedly similar changes apply to other molecules. It is not clear whether organic acids are detrimental to microbial activity in aerated soils, but we have recently noted that the suppression of microbial metabolism in flooded circumstances is related to the interaction between pH and fatty acid accumulation and that the toxicity is correlated with the increasing concentration of the undissociated organic acid. An enhancement of the inhibition by other types of toxicants can also be attributed to acidity; for example, we have recently reported that the marked suppression of blue-green algae by low concentrations of SO_2. or the bisulfite associated with it, is markedly greater as the pH declines[26].

Moreover, plant physiologists and agronomists have known for some time that iron, aluminum and manganese are more toxic in acid environments, a suppression that is linked with the increasing solubility of these cations at higher hydrogen ion concentrations. Microbiologists have not devoted much interest to the reasons for the pH suppression of microorganisms or microbial activities, but it is quite likely that one or more of these cations may be the inhibitor, at least at extreme acidities. For example, aluminum at concentrations of 10 ppm may reduce the numbers of bacteria, and somewhat higher concentrations may suppress fungi[27]. The decomposition of organic matter in soil is depressed with increasing levels of aluminum in areas at pH 4 or below, and it has been postulated that organic matter may sometimes accumulate in soils because of their high content of available aluminum. Soils contain inhibitors of the germination or outgrowth of fungal spores, and aluminum has been ascribed a role in this toxicity[29]. Although the information is scanty, aluminum appears to have a detrimental effect on soil bacteria even in culture[30].

Legumes are particularly susceptible to inhibition by toxic cations and, as pointed out above, part of their poor growth in acid soils may be attributable to cation toxicity rather than to the hydrogen ion per se. For example, the addition of manganese to acid soils reduces nodulation and nitrogen fixation by beans, although nodulation and nitrogen fixation were quite good even in a soil of pH 4.4[31]. Such data suggest that manganese, as well as aluminum and probably iron, must be considered in attempts to explain the detrimental effects of acidity on symbiotic nitrogen fixation.

In addition, the inhibition of nitrogen fixation by nodulated legumes and free-living microorganisms sometimes is attributable to the reduced availability of specific nutrients. Particularly outstanding in this regard is molybdenum, which is required by all nitrogen-fixing organisms. This element rarely is in insufficient supply in areas of near neutral pH, but chemical reactions

lead to its poor availability under acid circumstances. Thus, low
yield and reduced nitrogen fixation in acid soils may occasionally
be alleviated simply by the addition of this element, and molybdenum fertilization is sometimes practiced and is often quite
useful[32].

Given this body of information, is it appropriate to state that
acidification is necessarily bad? In many instances, the answer
appears to be unequivocally yes. Many processes that are important
for plant growth are clearly suppressed as the pH declines. However
particularly disturbing in our attempts to make generalizations are
the many observations that the inhibition noted in one soil at a
given pH is not observed in another soil at the same pH. The harm
in the first instance is incontrovertible, but the reason for the
lack of injury in the second is as yet totally unclear. In addition
the rates of some processes are not significantly diminished by
modest acidification, but one cannot afford to be sanguine because
of the few soils that have been examined and the obvious local
differences among soils even in the processes that have been well
studied. Furthermore, short term trials of any stress factor may
be misleading because microorganisms may become acclimated to
changes in pH. A few instances of such adaptation have been
described, although certain microorganisms are, as far as is now
known, unable to adapt to a much wider pH range[33]. The capacity of
some microorganisms to acclimate to increased acidity suggests that
studies should be conducted with environments that have been maintained at different pH values for some time, an approach that has
not been usually used, the studies typically having been done with
soils maintained only for short periods at the greater acidity.
The existence of acclimated populations would be reflected in the
finding of species whose pH optima differ from those in comparable
but nonacid soils, and evidence is available that populations in
certain acid soils have become acclimated to the peculiarity of
their surroundings, and they show different pH ranges for growth
and activity than their counterparts in more neutral sites[34].

From this brief review, I hope it is apparent that the
information base on the effects of acidification on microorganisms
and microbial processes is reasonably large. However, the areas
our ignorance are even far larger. One of the deficiencies in our
attempts to establish the ecological consequences of acidification
is the few soils examined and the consequent inability to generalize
from one locality to another. What is observed in one soil is
frequently not noted in another. Furthermore, the consequences of
greater acidity even in the subterranean ecosystem are totally
unclear. Hence, despite the progress that soil microbiologists have
made during the past several decades, considerably more information
is required, not only to satisfy the curiosity of microbial ecologists but also to provide the data that are necessary to ensure that
soils continue to function as a nutrient and physical base for

growing agricultural crops and natural plant communities and to guarantee environmental quality and food production for future generations.

REFERENCES

1. S. T. Williams and C. I. Mayfield, Soil Biol. Biochem. 3:197 (1971).

2. A. D. McLaren and J. J. Skujins, Can. J. Microbiol. 9:729 (1963); J. Skujins, A. Pukite and A. D. McLaren, Soil Biol. Biochem. 6:179 (1974).

3. W. S. Dancer, L. A. Peterson and G. Chesters, Soil Sci. Soc. Am. Proc. 37:67 (1973); J. W. White, F. J. Holben and C. D. Jeffries, Soil Sci. 37:1 (1934).

4. W. E. Connel and W. H. Patrick, Jr., Science 159:86 (1968).

5. R. S. Wodzinski, D. P. Labeda and M. Alexander, J. Air Pollut. Control Assoc. 27:891 (1977); J. T. Wilson and M. Alexander, unpublished observations.

6. M. Krol, W. Maliszewska and J. Siuta, Pol. J. Soil Sci. 5:25 (1972); K. Steinbrenner, Zent Bakteriol., Abt. II, 119:448 (1965).

7. E. Welte and G. Trolldenier, Naturwissenschaften 48:509 (1961).

8. G. Abrahamsen, R. Horntvedt and B. Tveite, Water, Air and Soil Pollut. 8:57 (1977); W. F. Hirte, Zent. Bakteriol., Abt. II, 125:639, 647 (1970).

9. O. C. Bryan, Soil Sci. 13:271 (1922).

10. E. G. Mulder, T. A. Lie, K. Dilz and A. Houwers, Symp. 9th. Intl. Cong. Microbiol., Moscow, p.133 (1966).

11. K. Dilz and E. G. Mulder, Neth. J. Agric. Sci. 10:1 (1962).

12. W. A. Rice, D. C. Penney and M. Nybor, Can. J. Soil Sci. 57:197 (1977).

13. R. Denison, B. Caldwell, B. Bormann, L. Eldred, C. Swanberg and S. Anderson, Water, Air and Soil Pollut. 8:21 (1977).

14. F. Dooley and J. A. Houghton, Brit. Phycol. J. 8:289 (1973).

15. C. L. Valera and M. Alexander, Plant Soil 15:268 (1965).

16. D. M. Ekpete and A. H. Cornfield, Nature 208:1200 (1965).

17. R. L. Halstead, J. M. Lapensee and K. C. Ivarson, Can. J. Soil
     Sci. 43:97 (1963).

18. F. E. Nelson, Soil Sci. Soc. Am. Proc. 28:290 (1964).

19. J. Wijler and C. C. Delwiche, Plant Soil 5:155 (1954).

20. F. M. Blodgett and F. B. Howe, New York (Cornell) Agric. Expt.
     Sta Bull. 581 (1934).

21. D. F. Bateman, Phytopathology 52:599 (1962).

22. J. R. Bloom and H. B. Couch, Phytopathology 50:532 (1960).

23. D. D. Kaufman and L. E. Williams, Phytopathology 55:570 (1965).

24. M. Habte and M. Alexander, Appl. Microbiol. 29:159 (1975);
     M. Habte and M. Alexander, Arch. Microbiol. 113:181 (1977).

25. J. M. Goepfert and R. Hicks, J. Bacteriol. 97:956 (1969).

26. R. S. Wodzinski and M. Alexander, J. Environ. Qual. (in press).

27. K. Matsuda and T. Nagata, Nippon Dojo Hiryogaku Zasshi 28:
     405 (1958).

28. V. K. Mutatkar and W. L. Pritchett, Soil Sci. Soc. Am. Proc.
     30:343 (1966).

29. W. H. Ko and F. K. Hora, Soil Sci. 113:42 (1972).

30. A. A. Zwarun, B. J. Bloomfield and G. W. Thomas, Soil Sci.
     Soc. Am. Proc. 35:460 (1971).

31. J. Dobereinger, Plant Soil 24:153 (1966).

32. A. J. Anderson and D. V. Moye, Aust. J. Agric. Res. 3:95
     (1952); M. B. Parker and H. B. Harris, Agron. J. 54:480
     (1962).

33. D. J. Kushner and T. A. Lisson, J. Gen Microbiol., 21:96
     (1959); F. Mendez-Castro and M. Alexander, Rev. Latioamer.
     Microbiol. 18:151 (1976).

34. C. T. Corke and F. E. Chase, Soil Sci. Soc. Am. Proc. 28:68
     (1964); S. T. Williams, F. L. Davies, C. I. Mayfield and
     M. R. Khan, Soil Biol. Biochem. 3:187 (1971).

SOIL ORGANISMS AND LITTER DECOMPOSITION IN A SCOTS PINE FOREST -

EFFECTS OF EXPERIMENTAL ACIDIFICATION

Erland Bååth, Björn Berg, Ulrik Lohm, Björn Lundgren,
Helene Lundkvist, Thomas Rosswall, Bengt Söderström
and Anders Wirén[a]

Swedish Coniferous Forest Project, Department of Ecology and Environmental Research, Swedish University of
Agricultural Sciences, S-750 07 Uppsala, Sweden

INTRODUCTION

The effects of increasing acidification in different ecosystems
is today a topic under debate. Several investigations are being
carried out to evaluate different questions within this relatively
new problem area. In forest ecosystems the research efforts are
focused mainly upon effects on vegetation and soil chemistry. Other
areas, e.g. the soil biota, have been given little attention, perhaps
owing to the fact that soil biological processes are relatively
poorly understood but also because the study of soil organisms is
more laborious than is the study of vegetation or soil chemistry.

The aim of the present study was to examine the extent to which
soil biological properties, such as amounts of different soil organism populations as well as decomposition rates, were affected in a
field experiment with dramatically increased acidification.

MATERIALS AND METHODS

The investigation was performed in an established optimum nutrition experiment with a randomized block design on Scots pine
(*Pinus sylvestris* L.) on morain in Northern Sweden. The textural
soil type is a well developed iron podzol[1, 2]. The present study

a) Authors listed in alphabetical order

covered three blocks, where in some plots the soil acidity had been changed by addition of dilute sulphuric acid (Table 1).

Decomposition rates were determined on Scots pine needle and root litters. Brown needle litter was sampled from about 15-year-old trees. Fresh root litter with a diameter of 2 to 3 mm was collected and cut into 10 cm pieces. The litters were put into litter bags made of nylon net with a mesh size of about 1 mm and placed in the field in October 1976. Weight loss after one year was determined by drying to constant weight at 85°C. Analyses of nitrogen and lignin were carried out.

Fungal mycelium, both total[3] and fluorescein diacetate (FDA) active[4], was measured in the A_{01}/A_{02}, A_2 and B horizons whereas bacterial biomass was determined only in the A_{01}/A_{02} horizon. Number and biomass of bacteria were measured by the acridine orange staining method[5], and in addition a size classification of the cells was made.

Physiological characteristics of the bacterial populations isolated from six soil samples from the A_{01}-A_{02} horizon were determined with a multipoint test procedure and the results were treated with a factor analytical method[6].

The cores for enchytraeid sampling were taken down to 10-15 cm depth and divided into A_{01}/A_{02}, A_2 and B horizons. Extractions were carried out with a modified Baermann technique[7]. Microarthropods were extracted from soil cores taken down to 10-12 cm depth with the Madfadyen high gradient extractor[8].

RESULTS AND DISCUSSION

In the present paper only preliminary results are presented and a full paper will follow. There were notable differences in thickness between the corresponding soil layers with and without acidification. After seven years of treatment with acid the A_{00} horizon had decreased to 40 per cent (from 2 to 1.2 cm in the "acid 3" plots) with a similar tendency for the less acidified plots. For the humus layer (A_{01}/A_{02}) there was a similar tendency whereas a significant increase took place in the bleached soil layer A_1/A_2. The "acid 3" plots showed an increase of 50 per cent (from 6.1 to 9.1 cm) and the "acid 1" plots about 40 per cent. The bleached soil profile had thus moved downwards indicating increased leaching. The level of organic matter (loss on ignition) also decreased considerably in A_1/A_2 - down to 55 and 67 per cent of the control for "acid 1" and "acid 3", respectively. The pH of the soil horizons only decreased slightly in the acidified plots (Table 1).

The decomposition rate of Scots pine needle and root litter
was lower after acidification (Table 2). For needle litter the
weight losses after one year were 95 and 90 per cent of that for
the control and for root litter 79 per cent. There were no significant differences in chemical composition between litter decomposed in acidified plots and in control plots[9]. A possible consequence of the reduced weight loss rate would be larger amounts of
litters stored in the upper soil layers and consequently more nutrients stored in a form not available to plants.

The measurements of fungal mycelium and bacterial biomass
could support the results above. There was a decrease of active
mycelium (FDA) in the whole soil profile of the acidified plots
indicating a reduced decomposing potential (Table 2). The bacterial
biomass showed similar changes with an even more pronounced biomass
decrease than for the fungi (down to 25 per cent). In the acidified
plots the average size of bacteria decreased to about half of
that in the control.

The increase in total mycelium could be an effect of a lowered
decomposition rate of fungal hyphae. The changed numbers of soil
microarthropods might also have had an influence on this process.

The changed acidity caused decreases in the enchytraeid population, which was dominated by *Cognettia sphagnetorum* (Vejd)
(Table 2), a result which was in accordance with that of Hågvar
and Abrahamsen[10] and of Lundkvist[11], the latter in an earlier
experiment finding a 90 per cent decrease of the population in an
acidified plot. Other soil animals also appeared to be affected.
A species of the genus *Trachytes* (Mesostigmata) and adults of
Oppia obsoleta (Pauli) showed a decrease in the acidified plots.
The dominant collembolan species *Tullbergia krausbaueri*, Börner,
showed a significant increase in the acidified plots compared to
the control (Table 2). Hågvar and Abrahamsen[10] found an increase
for Collembola in one of their coniferous forest study sites. The
results from the separate investigations were thus in agreement.
Most of the species showed no significant difference between
control plots and the acidified ones.

Populations of soil bacteria from control plots and "acid 3"
plots were compared with regard to their physiological/biochemical
abilities. Significant differences were detected when the material
was analyzed by a factor analytical method. The bacterial populations from "acid 3" plots consisted to a larger proportion of spore
formers (49%) than did the populations from the control plot (10%),
and they had a larger proportion of chitinolytic, proteolytic and
amylolytic bacteria. The three populations from the "acid 3" plots
differed between themselves much more than did those from the
control plots. The position of the six populations along two of
the factor axes which showed significant differences (5% level)
between the two sets of populations are shown in Figure 1.

Table 1. Doses and years of application of acid in the invest-
          gated parts of the block[a]. Measured pH (H_2O) on soil
          suspensions in 1976.

Year	"Acid 1" (H_2SO_4 kg x ha^{-1})	"Acid 3" (H_2SO_4 kg x ha^{-1})	Control
1971	50	50 + 50 + 50	
1972	50	50 + 50 + 50	
1973	50	50 + 50 + 50	
1974	50	150	
1975	50	150	
1976	50	150	
A_{01}/A_{02}	4.2	4.2	4.6
A_1-A_2	4.6	4.2	4.7
B	5.0	5.1	5.2

Table 2. Some *in situ* measured changes in weight loss, microbial
          biomasses and soil fauna between acidified plots and
          control plot. The relative values are given as percent-
          ages of the control. A significant difference to the
          control at the 95 % level is indicated by x.

	Control	"Acid 1"	"Acid 3"
Weight loss			
needle litter	100	95	90
root litter	100	n.d.	79x
Microorganisms in the A_{01}/A_{02} horizon			
bacterial numbers	100	n.d.	52x
bacterial biomass	100	n.d.	25x
bacterial cell size	100	n.d.	48
Fungal total mycelium[a]	100	114	122
FDA-active mycelium	100	85	59
Soil fauna			
Enchytridae (total)	100	76	11x
Acari (total)	100	95	74
Collembola (total)	100	128	144
Tullbergia krausbaueri	100	310	525x

a) Including both dead and live mycelium

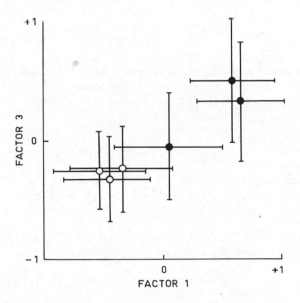

Figure 1. Position of the bacterial populations from control
          soils (o) and "acid 3" soils (●) in relation to
          factors 1 and 3 after varimax rotation (n= 33, N = 451).
          The standard deviations are given by bars.

          Factor 1 is characterized by: presence of spores,
          ability to hydrolyse starch and chitin, production
          of acid from sucrose, ability to grow at pH 9, small
          ability for ammonification, phosphatase and catalase
          production and low resistance to penicillin.

          Factor 3 is characterized by low ability to produce
          acid from a range of sugars (e.g. cellobiose, maltose
          mannose, glucose, galactose and xylose).

The results presented in this paper are only preliminary and a more extensive report will appear shortly where the observed facts of experimental acidification will be discussed further. Although the relevance of the results obtained in this investigation in relation to the normal field situation could be a matter of discussion, it seems clear that acidification can have a marked influence on the soil biota.

ACKNOWLEDGEMENTS

The investigation was performed within the Swedish Coniferous Forest Project supported by the Swedish Natural Science Research Council and Swedish National Environmental Protection Board.

REFERENCES

1. H. Holmen, Å. Nilsson and G. Wiklander, Department of Forest Ecology and Forest Soils, (Stockholm) Research Notes 26, 34 pp. (1976).

2. C.O. Tamm, G. Wiklander and B. Popovic, Water, Air, and Soil Pollution 8:75 (1977).

3. P.C.T. Jones and J.E. Mollison, Journal of General Microbiology 2:54 (1948).

4. B.E. Söderström, Soil Biology and Biochemistry 9:59 (1977).

5. G. Trolldenier, Zentralbl. Bakt. Abt. II, 127:25 (1972).

6. T. Rosswall and E. Kvillner, Advances in Microbial Ecology 2:1 (1978).

7. F.B. O'Connor, in: "Progress in Soil Zoology." P.W. Murphy, ed., Butterworths, London (1962).

8. T. Persson and U. Lohm, Ecological Bulletins (Stockholm) 23 211 pp. (1977).

9. E. Bååth, B. Berg, U. Lohm, B. Lundgren, H. Lundkvist, T. Rosswall, B. Söderström and A. Wirén, In manuscript.

10. S. Hågvar and G. Abrahamsen, Ecological Bulletins (Stockholm) 25:568 (1977).

11. H. Lundkvist, Ecological Bulletins (Stockholm) 25:570 (1977).

EFFECTS OF SULPHUR DEPOSITION ON LITTER DECOMPOSITION AND

NUTRIENT LEACHING IN CONIFEROUS FOREST SOILS

T. M. Roberts[1], T. A. Clarke[1], P. Ineson[1] and T. R. Gray[2]

Department of Botany[1], University of Liverpool, Liverpool, Lancashire, England; Department of Biology[2], University of Essex, Cochester, Essex, England

INTRODUCTION

In most areas of the U.K. dry deposition of SO_2 and $(NH_4)_2SO_4$ are as important as "acid precipitation" in determining the acidifying potential of sulphur and nitrogen inputs to soils[1]. Indeed, the acidification of urban parkland soils in Merseyside has been ascribed primarily to the dry deposition of sulphur dioxide[2]. The SO_2 concentration in urban areas has decreased markedly in the last 20 years but there is some evidence that rural concentrations have marginally increased due to more effective dispersal[3]. The historical trends of rainfall acidity are even less certain in rural areas but there is some evidence for an improvement in urban areas (e.g. the mean annual rainfall pH was 3.8 between 1930 and 1940 in central Merseyside compared to 4.2 in 1976-77). The effects of increasing the deposition of sulphur compounds to vegetation and soils in rural areas has been the subject of considerable debate[4,5]. It is generally agreed, however, that the effects on agricultural soils will be small due to the extensive use of lime and the relatively high base saturation of these soils. Liming is not generally practical in afforested areas and it has been suggested that those soils which are poorly buffered and have a low cation exchange capacity will be most susceptible[6]. It has also been suggested that acid rain may affect forest productivity by reducing litter decomposition and thereby decreasing the rate of nutrient cycling[7].

Consequently, an experimental field site was set up in 1977 at Delamere Forest in central England (Grid Ref. SJ 583689) and studies of the sulphur cycle and the effects of acid applications are now underway. The site is located at a 50-year old _Pinus_

sylvestris/Pinus nigra stand planted on a podsol with low cation
exchange capacity and base saturation. The organic horizon (0–5cm)
can be clearly differentiated into L, F_1, F_2 and H layers. The
A_n horizon (5–10cm) is underlain by a poorly differentiated Eg
horizon and the Bs horizon has red/brown iron deposits to a depth
in excess of one metre but there is no continuous iron–pan.
Chemical analysis of the mineral horizons gave the following
results:

pH (0.01M $CaCl_2$) = 3–3.8: CEC = 2.6 –9.6 m.eq/100g:

Organic Content = 0.5 – 5.0%

METHODS AND RESULTS

Nutrient Fluxes at Delamere Forest

Precipitation, both in the open and beneath the pine canopy
was collected in 20cm diameter funnel rain gauges erected 1.25
metres above ground (8 gauges in the open and 10 beneath the
canopy). A lysimeter trench was built in a clearfelled area
adjacent to the pine stand. Four 20cm diameter funnel–type
lysimeters containing just the L horizon (fresh pine needles)
and four lysimeters containing the L + F + H horizon were located
in the lysimeter trench in the open and also in a smaller trench
established under the canopy. Four undisturbed soil monoliths
were taken from beneath the canopy and sealed in 50 cm. deep by
30 cm. diameter free–draining lysimeters in the open trench 6
months before measurements began. Daily sulphur dioxide
concentrations were measured by bubbling air through hydrogen
peroxide solutions and titrating the acidity against N/250
sodium borate. Precipitation and lysimeter leachates were
collected every two weeks. The pH was determined immediately and
subsamples stored at 4oC. Nitrate/nitrite was measured by the
sulphanilamide/copper hydrazine method; ammonium by the
hypochlorite/nitroprusside method and phosphate by the ammonium
molybdate/ascorbic acid method using a Technicon Autoanalyser.[8]
Sulphate was determined by the manual turbidimetric method.[9]
Calcium and magnesium were measured by atomic absorption
spectrophotometry – sodium and potassium by flame emission
spectrometry. The data presented here is for the period from
December 1977 to April 1978 and may, therefore, give an overestimate of the annual values (Table 1).

The volume of throughfall was 50% of the amount of
precipitation in the open and 44% of the open precipitation
passed through the organic layer lysimeters under the canopy.
Between 77% and 98% of the rainfall passed through the organic
layer lysimeters in the open compared to 70% which percolated
through the soil monolith lysimeters.

Table 1   Ion Fluxes (kg/ha/annum) at a 40 year old Pine Stand
           in Central England

Ion	Precipi-tation	LEACHATE L-Horizon	L+F+H	Soil (0-5cm)	Through fall	LEACHATE L+F+H
Hydrogen	1.0	1.2	2.1	1.6	1.4	2.1
Sulphur	20	31	50	119	70	100
Ammonium-N	9	6	27	31	32	53
Nitrate-N	3	4	6	25	6	5
Calcium	3	32	25	30	12	49
Sodium	47	31	44	48	58	62
Potassium	5	18	27	41	12	36

Sampling period = December 1977 to April 1978 inclusive.
Details of the lysimeters used to provide leachates are given
in the text.

    The hydrogen ion input averaged 1kg/ha/annum and there was
considerable enrichment beneath the pine canopy and further
enrichment beneath the organic horizons. Temporal fluctuations
in the acidity of leachate from the L-layer lysimeters were
significantly correlated with the biweekly rainfall collections.
This was not so for the organic-layer and soil monolith leachates
as these horizons are more strongly-buffered.

    The sulphur input in precipitation averaged 20kg/ha/annum.
There was appreciable enrichment beneath the canopy and further
enrichment occurred beneath the organic horizons due to the dry
deposition of SO_2 or mineralisation of organic sulphur. The
average SO_2 concentration during the study period was $50\mu g/m^3$.
If the enrichment beneath the L-layer in the open was entirely
from the dry deposition of SO_2, this would give a velocity of
deposition of 0.1cm/s. This calculation does not, of course,
take into account the dry deposition of SO_2 to the rainfall gauge
which may be as high as 0.05-0.1cm/s. This Vg to pine litter of
0.15-0.2cm/s compares favourably with experimental values of about
0.2cm/s determined experimentally for acid soils.[10] If the
enrichment in throughfall were entirely from dry deposition this
would give a Vg of 0.5cm/s. This compares with values of 0.2-2.6
determined experimentally for pine canopies[11,12]. This may be an
underestimate as the dry deposition to the funnel beneath the
canopy may be less than in the open. There may also be a small
amount of dry deposition to the forest floor and understorey
vegetation.

The NH_4-N input in precipitation averaged 9kgN/ha/annum.
This increased to 32kg/ha/annum beneath the canopy due to crown
leaching or dry deposition of $(NH_4)_2 SO_4$ aerosol. There was some
immobilisation of ammonium in the L-layer in the open lysimeters
but there was considerable mineralisation of ammonium in the F and
H layers both in the open and beneath the canopy.

The nitrate input averaged 3kgN/ha/annum. There was no
indication of nitrification in the L-layer or organic layer
lysimeters but the nitrate-N from the soil monolith lysimeters
had increased to 25kg/ha/annum. This was presumably due to a
low rate of autotrophic nitrification at alkaline microsites in
the acid mineral horizons or possibly heterotrophic nitrification.[13]

The input of calcium averaged 3kg/ha/annum and the most
marked enrichment occurred in the L-layer leachate. There was no
further mobilisation of calcium from the mineral horizons.

The input of sodium in precipitation averaged 47kg/ha/annum
with a small enrichment beneath the canopy. The sodium input
was not fixed or further sodium released in any horizon and the
sodium output approximately balanced the input. In contrast,
considerably more potassium was leached from the organic horizons
than was deposited in precipitation and some potassium was also
released from the mineral horizons.

Effects of Acid Applications at Delamere Forest

This part of the study was designed to examine the effects
of acid treatments applied at frequent intervals and low
concentrations on nutrient leaching and litter decomposition. The
experiments consist of applications of 25 and 50 kgS/ha/ann as
sulphuric acid solutions, (pH 3.1 and 2.7) at a rate of 5mm every
2 weeks. These treatments approximately double the average input
of sulphur in the open and in the throughfall respectively. In
addition, some areas were treated with elemental sulphur at the
same doses as above in order to determine whether the effects on
litter decomposition found in incubation studies[14] actually occur
in the field. These treatments were applied to L-layer, organic
layer and soil monolith lysimeters in the trench located outside
the canopy. Similarly, 4m^2 plots containing 3 year old Pinus
sylvestris transplants and 2m^2 plots of forest floor beneath the
canopy were treated. All treatments were replicated four fold.

Nutrient Leaching

The rates at which the acid applications leached through the
soil horizons during the first 5 months after the start of the
treatments are shown in Figure 1 and 2. The hydrogen ions applied
were not fixed by the L-layer and appeared in the leachate shortly

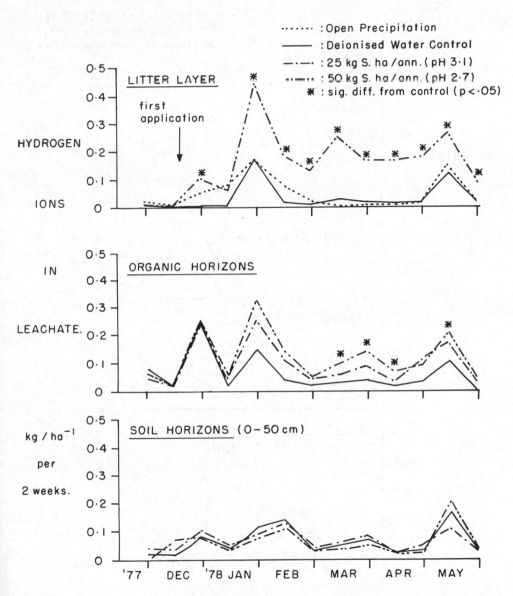

Figure 1. Effect of acid applications on the rate of hydrogen
leaching through the Delamere podsol.

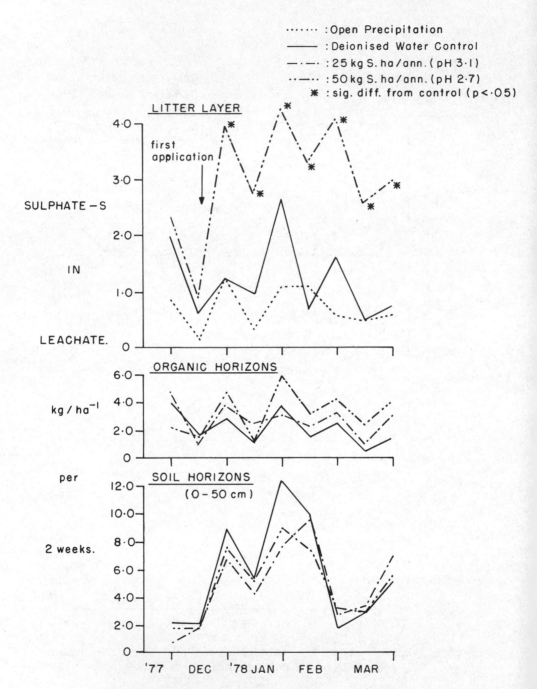

Figure 2. Effect of acid applications on the rate of sulphur
           leaching through the Delamere podsol.

after application. Some buffering occurred in the F + H layers but
the applied hydrogen appeared in the leachate after 6 weeks. The
application of acid had no effect on the acidity of leachate from
the mineral horizons. The applications of ground sulphur have had
no effect on the acidity of any leachates.

The sulphur applied as acid was not fixed by the pine needles
in the L-layer. There was some retention by the F + H layers but
the applied sulphur began to appear in the leachate after 6 weeks.
The applications had no effect on the sulphur content of leachate
from the mineral horizons. The sulphur applied as powder did not
appear as sulphate in either the L-layer or organic layers.

The addition of sulphuric acid have so far had no effect on
the rate of cation removal from any horizon except possibly for
the mobility of ammonium in the L-layer and organic-layer lysimeters.
Some ammonium in precipitation was immobilised as it percolated
through the control L-layer lysimeters but large amounts were
released by mineralisation in the F and H layers. The acid
applications slightly reduced the amount of ammonium immobilised
in the L-layer and increased rate of release in the F and H
layers. Mineralisation and litter-bag studies in progress should
indicate whether these trends are due to changes in mineralisation
or immobilisation. The acid treatments had no effect on the
ammonium leaching through the mineral horizons nor on the rate of
nitrification in the mineral horizons. The elemental sulphur
treatments had no effect on the mobility of ammonium in any of the
lysimeters (Figure 3).

Litter Decomposition

The decomposition of pine needles as affected by the acid
treatments was studied by means of the nylon mesh-bag technique.[15]
In addition, two mesh bag sizes were chosen (1mm and 5μm) to
compare the decomposition process in the presence and absence of
soil animals. A randomized block design with replicates was used
for this experiment, there being six replicates per bag size
sampling date. At intervals of two months, these bags were
collected and returned to the laboratory where the moisture content
and dry weight (105°C to constant weight) of the litter were
determined. Also at two-monthly intervals, samples from the L, F_1,
F_2, and H organic layers were removed for respiration analysis,
(Infra Red Gas Analysis at 20°C, flow rate 0.8 litres/min).
Measurements of bacterial and fungal contributions to L-layer
respiration were also determined using a modification of the
technique of Anderson and Domsch.[16] Soil pH (using a 1:2 paste of
litter:distilled water) and the moisture content of each organic
horizon was also determined.

The application of acid solutions (pH 3.1 and 2.7) had no

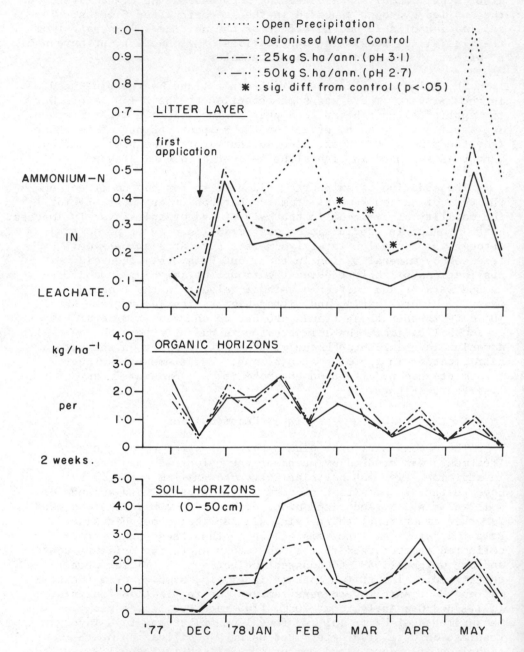

Figure 3.   Effect of acid application on the rate of ammonium
leaching through the Delamere podsol.

effect on the pH of the organic layers in the treated field plots
beneath the canopy (Table 2). Nor did the respiration measurements
on these treated plots show any significant effect of acid
applications during the first six months of the study. In addition,
there was no shift from the 90%:10% balance of fungal and bacterial
respiration found in the control plots. In contrast, the litter
bags taken from these plots did show a significant increase in
weight loss with increasing acidity of applied solution (Figure 4).
This 15% increase in the overall rate of decomposition of fresh
needles was similar for the small mesh as well as the large mesh
bags.

DISCUSSION

The input of sulphur in precipitation at the study site was
considerably in excess of that required for forest growth. There
was considerable enrichment of sulphur in the throughfall and in
the organic-layer lysimeters. Calculations of the velocity of
deposition assuming that most of the sulphur enrichment was from
the dry deposition of SO_2 gave values that were comparable with
values obtained experimentally. There was a net export of
ammonium, calcium, sulphur and potassium from the forested area.
The acidity of precipitation had little effect on the acidity of
leachates through the soil profile.

The acid applications increased the acidity of leachates from
the L-layer lysimeters but there was no effect on the organic-
layer or soil monolith leachates. Similarly, the acid applications
did not change the pH of any organic horizon in the field plots.
These observations agree with the comments of Wiklander[17] that
acid rain is unlikely to lead to further acidification of strongly
acid soils. The sulphate applied fairly rapidly leached through
the organic horizons but was readily fixed by the mineral horizons.
This observation is in agreement with studies which report
extensive adsorption of sulphate by sesquioxide coatings on
mineral particles in weathered soils.[18] The acid applications had
no effect after the first 6 months on the rate of leaching of
calcium, potassium, sodium and magnesium from the organic or
mineral horizons. This is in contrast to the results of Abrahamsen
et al.,[19] who found that acid solutions (pH 6,4,3 and 2) applied
to a Norwegian podsol over a 2-year period resulted in increased
leaching of calcium, magnesium, potassium and aluminium. The
acid application in litter-bags was also accelerated by the acid
treatments although there was no change in respiration rates in
treated field plots. Swedish studies[14] have shown a similar
increased in ammonium availability but the extremely acid
applications (pH <2) used in this study decreased the rate of
respiration. However, Norwegian incubation studies[20] using more
realistic acid treatments (pH 3-4) have shown an increase in
cubations carried out in distilled water.

Table 2. Effect of acid applications on pH, moisture content and respiration rate of the organic horizons beneath a pine canopy.

Sampling Time		2 months				4 months				6 months			
Treatment		0	DW	25	50	0	DW	25	50	0	DW	25	50
Profile	Parameter												
L	pH	3.88	3.96	3.88	3.98	4.05	4.11	4.02	4.05	4.04	4.08	3.98	3.92
	% moisture	47.9	46.5	47.8	45.0	31.2	30.2	29.0	29.7	52.8	53.3	53.2	55.3
	Respiration	25.4	25.8	24.3	22.9	6.44	5.40	5.35	4.43	15.51	15.61	17.12	17.00
F_1	pH	4.00	4.00	3.90	4.00	3.99	4.05	3.94	3.93	3.84	3.83	3.90	3.79
	% moisture	72.9	67.0	74.0	75.6	62.8	64.4	66.6	66.2	65.2	67.6	69.3	68.2
	Respiration	28.8	21.9	27.9	28.3	12.5	17.1	14.1	14.3	16.5	19.3	15.9	14.6
F_2	pH	4.03	4.16	3.82	4.00	4.05	4.16	3.94	3.89	3.89	3.85	3.61	3.64
	% moisture	72.4	72.4	71.5	70.4	71.4	71.7	73.9	71.2	69.0	78.2	74.8	77.1
	Respiration	5.9	5.9	4.6	5.1	3.5	3.7	3.2	3.5	2.1	3.7	1.8	2.7
H	pH	3.78	3.71	3.59	3.67	3.66	3.69	3.70	3.67	3.49	3.59	3.41	3.31
	% moisture	63.1	62.8	64.9	64.7	55.2	55.7	68.9	61.5	55.6	57.9	54.6	62.2
	Respiration	2.1	2.1	1.8	2.2	0.6	0.7	0.8	1.4	0.6	0.7	0.7	0.8

Respiration expressed in cm^3 CO_2 per hour per 100 gm D.W.

0 = No Treatment   25 = 25 kgS/ha/ann (pH 3.1)
DW= Deionised Water Control   50 = 50 kgS/ha/ann (pH 2.7)
Each value is the mean of 4 replicates.

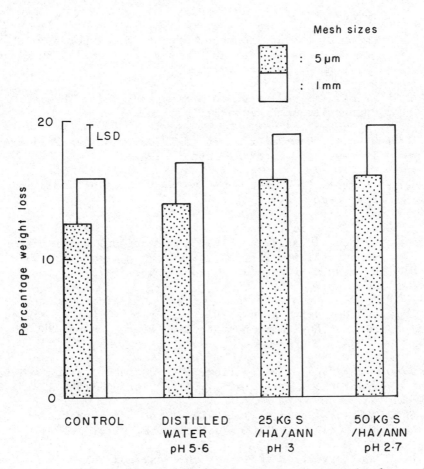

Figure 4. Effect of acid applications on the weight loss of litter bags.

It is clear from these studies that episodes of acid precipitation with a pH of around 3-3.5 may produce small changes in the rate of pine litter decomposition and nitrogen availability. However, the effects of these changes on forest productivity cannot yet be quantified.

REFERENCES

1. Department of the Environment, Effects of airborne sulphur compounds on forests and freshwaters, Pollution Paper No.7, HMSO, London (1976).

2. C. M. Vick, Air pollution on Merseyside, Ph.D. Thesis, University of Liverpool (1975).

3. Warren Spring Laboratory, National Survey of Air Pollution, 1961-7, Vol. 2, HMSO, London (1972).

4. G. E. Likens, F. H. Bormann, and N. M. Johnson, Acid rain, Environment 14:33 (1972).

5. S. Odén, The acidity problem - an outline of concepts, in: "Proceedings of The First International Symposium on acid precipitation and the forest ecosystem". Ohio State University, Columbus, Ohio, 36 pp. (1975).

6. N. Malmer, On the effects on water, soil and vegetation of an increasing atmospheric supply of sulphur. National Swedish Environment Protection Board Report, SNV PM 402E (1974).

7. C. O. Tamm, and E. B. Cowling, Acidic precipitation and forest vegetation, Water, Air, Soil Pollution 7:503 (1977).

8. S. E. Allen, "Chemical Analysis of Ecological Materials", Blackwell Scientific Publications, Oxford (1974).

9. H. L. Golterman, "Methods for chemical analysis of freshwaters", IBP Handbook No. 8, Blackwell Scientific Publications, Oxford (1969).

10. J. A. Garland, Deposition of gaseous sulphur dioxide to the ground, Atmos. Environ. 8:75-79 (1974).

11. A. Martin, and F. R. Barber, CEGB Report No. SSD/MID/N44/75, (1975).

12. J. A. Garland, and J. R. Branson, The deposition of sulphur dioxide to pine forest assessed by a radioactive tracer method, Tellus 29:445 (1977).

13. W. V. Bartholomew, and F. E. Clark, eds., "Soil Nitrogen", American Society of Agronomy, Inc., Wisconsin (1965).

14. C. O. Tamm, G. Wiklander, and B. Popovic, Effects of application of sulphuric acid to poor pine forest, in: "Proceedings of the First International Symposium on acid precipitation and the forest ecosystem", Ohio State University, Columbus, Ohio (1975).

15. K. L. Bocock, O. Gilbert, C. K. Capstick, D. C. Twinn, and J. S. Waid, Changes in leaf litter when placed on the surface of soils with contrasting humus types. I. Losses in dry weight of oak and ash leaf litter, J. Soil Sci. 11:1 (1960).

16. J. P. E. Anderson, and K. H. Domsch, Measurement of bacterial and fungal contributions to respiration of selected agricultural and forest soils, Can. J. Micro. 21:314 (1975).

17. L. Wiklander, The acidification of soil by acid precipitation, Grundförbättring 26:155 (1973).

18. D. W. Johnson, and D. W. Cole, Sulphate mobility in an outwash soil, Water, Air, and Soil Pollution 7:489 (1977).

19. G. Abrahamsen, R. Hornvedt, and B. Tveite, Impacts of acid precipitation on coniferous forest ecosystems, Water, Air, and Soil Pollution 8:57 (1977).

20. F. H. Braekke, ed., "Impact of acid precipitation on forest and freshwater systems", SNSF - project report FR 6/76 (1976).

SMELTER POLLUTION NEAR SUDBURY, ONTARIO, CANADA, AND EFFECTS ON FOREST LITTER DECOMPOSITION

B. Freedman and T.C. Hutchinson

Dept. of Botany and Institute for Environmental Studies

University of Toronto, Toronto, Ontario, Canada, M5S 1A1

INTRODUCTION

Sudbury, Ontario (46° 21' N. 80° 59' W, 259 metres above sea level), was the study area, together with its surrounding forest plant communities. The data were collected along a transect running SSE of Sudbury, centering on the 380 m smokestack of the International Nickel Co. Ltd., at Coppercliff, Ontario.

The study area is entirely within the Canadian Shield, and was glaciated in the most recent of the Pleistocene glacial advances, which ended some 10,000 to 12,000 years ago in this area. The Sudbury district lies within the Great Lakes-St. Lawrence Forest Region (60), and is characterized by mixed broad-leaf-conifer forests. Soils in the area range from strongly to weakly podzolic, depending on forest cover type. They are normally thin (up to 15 cm), but glacial till several metres deep occupies many of the valley floors.

The area is a very important one for mineral extraction, especially for nickel, copper, iron, cobalt, silver and the platinum group. From the early days of the mining-smelting developments in the area, to the present, heavy metal and sulfur pollutants have been emitted in very large quantities. In early times, smelting was done in open pits by creating huge "sandwiches" of sulfide ore and timber, which were ignited and allowed to oxidize pyrolytically, after which the nickel and copper concentrates were gathered for refining. This process generated choking, phytotoxic ground level plumes of sulfur dioxide and heavy metal particulates which devastated surrounding plant communities, and consequently led to severe erosion of soils from slopes and

395

exposure of naked bedrock, which was blackened by the acid plumes.
This environmental degradation was further assisted by intensive
logging to fuel the roast beds, and by an increased incidence of
fire due to high fuel loads resulting from accumulation of slash
and/or dead, dry timber.

These open pit roasting techniques were gradually replaced
by smelter facilities, which vented pollutants into the atmosphere
through relatively tall smokestacks. Most of the present
Coppercliff smelter was constructed by 1930, and pollutants were
vented through a 155 m smokestack. Two other 152 m stacks were
built in 1936. In 1972, the three smokestacks at Coppercliff were
replaced by a single 380 m "superstack" (the world's tallest),
through which most of the present atmospheric pollutants are
dispersed.

Another Sudbury area smelter is located near Coniston which
was constructed in 1913, but was closed in 1972. Prior to 1913
a roast bed was used. The third smelter is at Falconbridge, begun
in 1930 with two stacks of 55 m and one of 40 m.

The 380 m "superstack" at Coppercliff emitted 1.2×10^6 mt
(metric tons) in 1977, this being markedly down from the total INCO
emissions of 2.5×10^6 mt in 1970 (Ferbuson, INCO, pers. comm.,
April 1978). These compare to total Canadian emissions of $6.0 \times$
10^6 mt in 1972 (4). Even at the relatively lower 1977 emission
rates, the Coppercliff smokestack is the world's largest point
source emitter of SO_2 (69), and is a significant contributor to
the total global anthropogenic SO_2 output, estimates of which
vary from $35-40 \times 10^6$ mt/year (62) to 91×10^6 mt/year (40). Thus
current emissions from the 380 m superstack at Coppercliff comprise
almost 20% of total Canadian emissions from all sources, and from
1 to 3% of total world anthropogenic emissions. Present-day
emissions from the Falconbridge smelter are some 0.33×10^6 mt/year,
which is about 25% of the Coppercliff total (Ferguson, pers.
comm., April 1978).

Emissions of particulates in 1977 from the Coppercliff stack
were some 1.0×10^4 mt. down from total INCO emissions of 3.4×10^4
mt in 1970 (Gormley, INCO, pers. comm., Feb. 1977). These
particulates are mainly iron oxides, although copper and nickel
are also significant. On two stack sampling dates in 1977 (June
6 and July 8), emissions of iron were 3.3 mt, copper 0.9 mt, and
nickel 0.7 mt respectively. These particulates (90% on a particle
frequency basis) are mainly less than 7 microns in size, as this
size fraction is not efficiently trapped by the Cottrell dust
collectors, which are the main particulate control devices used
at Coppercliff (Ferguson, pers. comm., April 1978).

It is the purpose of the present study to examine some of the

effects of the large nickel smelter near Sudbury, Ontario on
surrounding forested ecosystems. The impact of the Coppercliff
smelter of the International Nickel Company, Canada Limited will
be examined in terms of 1) present inputs of the major emitted
pollutants to areas at various distances from the smelter, and
burdens of these pollutants in soils and vegetation, and 2)
effects of pollutant burdens in forest soils on litter
decomposition processes, examined in terms of i) litter standing
crop relative to rates of litter input, ii) populations of litter
decomposers, and iii) activity indices related to rates of litter
breakdown.

METHODS

The studies were carried out along a SSE transect centering on
the 380 m smokestack of the Coppercliff smelter, west of the city
of Sudbury. This transect does not impinge on any ore bodies,
and passes through relatively uninhabited areas.

Bulk Deposition Measurements

     This was measured as wet plus dry deposition at sites along a
SSE transect during 30 day intervals in the growing seasons of
1976 and 1977. Bulk collections were made in open-topped
cylindrical polyethylene containers (area of exposure = 200 cm^2).
During each sampling interval, two replicate samplers were
exposed on a 60 cm high wooden platform (after 7 and 59). After
the exposure period, the pH in the solution in the collector was
measured in the field, and the solutions were then filtered through
a 0.45 micron filter. Sulfate in the filtrate was analyzed
turbidimetrically (5,6). Metals in solution were measured by
flameless atomic absorption using a graphite furnace and an atomic
absorption spectrophotometer fitted with deuterium arc background
correction (55,56). Particlates retained by the 0.45 micron filter
papers were measured after a heat-assisted nitric-perchloric acid
digest, followed by flame atomic absorption analysis. Total
metal deposition was calculated as the sum of particulate plus
dissolved metals. Replicated standard reference materials were
analyzed to test for systematic errors in the analytic procedure
(see 22).

Sulfation Measurement

     Sulfation measurements were made at all sites where bulk
deposition was collected, by exposing two replicate sulfation
plates beneath the 60 cm high wooden support platforms. Sulfation
plates contain lead dioxide, which reacts with sulfur dioxide in

the air to form lead sulfate. Following exposure, the sulfate can
be extracted and analyzed by a simple turbidimetric determination
(8,31,72).

Pollutant Burdens in Soils and Vegetation

Soil reaction was measured on field-wet samples by creating
a slurry with distilled water, and measuring the pH with a
combination electrode. Total heavy metals or bases were
determined on dried and homogenized soil or plant samples, digested
by a hot nitric-perchloric acid mixture, and analyzed by flame
atomic absorption spectrophotometry (22). Ammonium acetate and
distilled water-extractable heavy metals in soils were determined
using methods modified from (5). Total organic sulfur, and
ammonium acetate-extractable sulfate were measured using published
methods (11). Total sulfur in plant tissue was analyzed by the
method of (70). Analyses of standard reference materials were
made in conjunction with the total metal and base analyses, and
total plant sulfur analyses, to check for systematic errors
(see 22).

Soil Litter Studies

i) Litter Standing Crop Relative to Litter Inputs. The
standing crop of leaf litter was collected in nine forested stands
at various distances from the Coppercliff smelter in May 1976.
Litter was sampled by randomly placing fifteen 0.25 x 0.25 m
quadrats per site, and handsorting material in the field into
"litter" (01, recognizable to tree species) and "duff" (02,
recognizable as leaf fragments, as opposed to amorphous organic
matter, or humus), after (18). Leaf litter only was collected,
with twigs, cones, etc. being excluded. Collected material was
dried and weighed, and the total dry weight of 01 + 02, and the
ratios of 01:02 calculated.

Measurement was also made of litter input to four forested
stands (2 close to the smelter, 2 control sites) by exposing 6
randomly-placed Im x Im litter traps for a period of one year (May
1976 to May 1977). The litter traps were constructed after (48)
and (50). Litter was collected from the traps at monthly intervals
during the snow-free season, but only a single collection was
made over the period of continuous snow cover. A Litter
Accumulation Index (LAI) was calculated for these four sites as
the ratio of standing crop of litter (i.e. total 01 + 02) to litter
fall.

ii) Litter Bag Decomposition and Base Loss. Leaf decomposition
can be studied by exposing material in inert litter bags in natural

habitats (12,35,44). In this study, 20 cm x 13 cm envelopes were
constructed from fibreglass screening and filled with known
weights of leaves of three Sudbury area tree species (<u>Betula</u>
<u>papyrifera</u>, <u>Populus</u> <u>tremuloides</u>, and <u>Pinus</u> <u>strobus</u>), which were
collected as fresh leaf material at an uncontaminated forest site.

These litter bags were placed on the soil surface at five
forested stands (three contaminated and two control sites) in June
1976, and replicated collections were made after various exposure
intervals. Collected material was hand-sorted to remove insects
or obvious insect fecal material, after which oven-dry weights
were determined, and weight loss calculated. The relative loss of
basic cations (calcium, magnesium, and potassium) from litter at
various sites was determined by comparing initial amounts to that
remaining after exposure in the field. Amounts of basic cations
were determined from analyzed concentrations and total weights of
material.

iii) <u>Soil CO_2 Flux Measurements</u>. CO_2 flux was determined for
soil samples collected from forest sites at various distances from
the smelter. Briefly, soil CO_2 flux (used as an index of soil
respiration) was determined using an infrared gas analyzer by
measuring the rate of increase of CO_2 concentration in a closed,
recirculating chamber system into which a soil sample has been
deposited (22,44,66). This rate of change of concentration, when
multiplied by the total system volume, yields a volumetric rate of
CO_2 evolution which can be standardized to soil weight.

Soil samples for CO_2 flux determinations were collected in May
of 1977 from the 02-A1 (duff-humus) soil horizon of mesic mixed
forest communities along a SSE transect, and these were analyzed
for both CO_2 flux and acid phosphatase activity.

iv) <u>Soil Acid Phosphatase Assays</u>. Soil enzyme activity has
been used an an index of soil metabolism in a number of studies
(reviewed recently by (67)), including several recent studies on
metal-polluted soils (e.g. 20,38,74). The method used in this
study to assay acid phosphatase activity (see 22 for description)
was modified from (17).

Soil samples for acid phosphatase assays were collected in
May of 1977 from the 02-A1 (duff-humus) soil horizons of mesic
mixed forest communities along a SSE transect, and these were
assayed for both acid phosphatase activity and CO_2 flux. The soil
samples were also analyzed for heavy metals (copper, nickel, and
zinc), so that the relationship of enzyme activity and CO_2 flux
to heavy metal content could be determined.

v) <u>Soil Microfaunal Collections</u>. Soil microfauna were
collected using standard aluminum soil corers (internal diameter =

3.5 cm, height = 7.0 cm) from four forest sites (2 close to
Coppercliff, 2 controls) on two dates in July 1976. Sixty
replicate cores were collected per site on each date, and
arthropods were extracted using a modified Berlese–Tullgren high
gradient funnel system, with heat and chemical (naphthalene)
repellant, over two day post-collection periods (33). For
purpose of analysis, the arthropods were broadly grouped as Mites
(Acarina), Springtails (Collembola), and Others (mostly Coleoptera,
Diptera, and Hymenoptera).

    vi) <u>Soil Microfungal Populations</u>. Soil fungal populations
were estimated by the dilution plate count method (11,54). A
Rose–Bengall agar (with streptomycin to inhibit bacteria) was the
nutrient medium upon which fungi were cultured. Plates were incubated at 28°C +1°C for 8 days, after which the number of fungal
colonies were counted. Identification to genus of the commonly-
isolated colonies was done by Dr. J. Morgan-Jones of the Dept. of
Botany, University of Toronto. The tolerance of soil fungal populations to several heavy metals was tested by adding copper, nickel,
or both (as soluble chloride salts) to the nutrient agars, (giving
a concentration range of 0–100 ppm total metal) and noting the
relative numbers and sizes of colonies isolated in these treatments
compared with controls.

    vii) <u>Litter Perturbation Experiment</u>. In this experiment,
litter (01 + 02) was collected from a control mixed-forest site,
air-dried, homogenized in a Waring blender, passed through a 5 mm
sieve, and then thoroughly mixed. Aliquots of litter of about 75
grams airdry weight were then placed in 2 litre polyethylene containers, with perforated lids to allow atmospheric exchange, and
had enough copper chloride and/or nickel chloride added to give
concentration ranges of from 70 to 10,000 ppm d.w. nickel or 210
to 10,000 ppm d.w. copper. The lower concentrations used represent
the initial concentrations in the litter. The metals were added in
solution, with the water being sufficient to bring the litter to
80% water holding capacity. The samples were incubated at 28°C
+2°C for 60 days inside a darkened growth chamber. Water loss was
determined every 2 or 3 days gravimetrically, and enough distilled
water added to return the litter to 80% water holding capacity.
At the end of the experiment (60 days), the dry weight loss and CO_2
flux from the litter were determined and expressed on a relative
(percent of control) basis.

OBSERVATIONS AND DISCUSSION

Present Pollutant Inputs and Burdens

    Measurements made during three 30-day intervals in the
growing season of 1976, and three in 1977, indicate that inputs
of sulfate, iron, nickel, and copper and rates of sulfation were

relatively high at sites closer to the Coppercliff smelter,
compared with sites further away. Data for a representative
sampling period in the growing season of 1977 are summarized in
Figure 1 (see also 22). With respect to pollutant metal
deposition, the relative inputs indicated iron >> nickel≈copper.
Concentrations of iron, copper, nickel, and sulfate were also
elevated in snow samples, indicating that inputs of these
pollutants also occur during the winter months (see 22).
Relatively high rates of smokestack-derived pollutant deposition
at sites close to metal smelters has been well documented in the
recent literature, including a number of studies in the Sudbury
area (32,34,47,57).

From the average patterns of deposition and sulfation observed
over the 6 sampling periods in 1976 and 1977, first order estimates
were made of the percentages of SO_2-S, SO_4-S, nickel, copper and
iron emitted from the Coppercliff superstack that are deposited
within a 60 km radius circle. For the purpose of making these
calculations, several assumptions were made: 1) that the pattern
seen along the SSE transect is typical of other compass directions,
2) that the time period sampled is representative of deposition
rates during the rest of the year, and 3) that the only significant
deposition of smelter-derived pollutants originates with the
Coppercliff smelter, i.e. the Falconbridge smelter and other sources
of emissions do not interfere along the chosen transect. This
latter assumption means that the calculation may overestimate the
fallout from the superstack.

Using the mean pattern of sulfate deposition at sites at
various distances from the smelter, a first-order estimate was
made of the total annual bulk deposition of sulfate within a 60
km radius. Similarly, an estimate was made of the dry deposition
of SO_2 using SO_2 concentration data derived from the mean patterns
of sulfation. In this calculation of dry SO_2 deposition, deposition
velocities of 5.0 mm sec^{-1} and 8.0 mm sec^{-1} were used to estimate
deposition within 12 km, and within 12-60 km of the smelter
respectively. These deposition velocities were suggested by D.
Fowler (Institute of Terrestrial Ecology, Midlothian, U.K., pers.
comm., Sept. 1978) as being typical of dry SO_2 deposition to largely
devegetated, and vegetated surfaces respectively. Using these
methods, total sulfur depositions of 0.47×10^7 kg SO_4-S/year and
11.9×10^7 kg SO_2-S/year respectively are calculated to be deposited
within a 60 km radius of the Coppercliff smelter. These calculated
depositions include "background" SO_4 and SO_2 deposition, or the
sulfate or SO_2 deposition which would occur in the absence of the
smelter emissions. To calculate net smelter-derived sulfate
deposition, background sulfate deposition is approximated by using
the mean deposition data obtained at the 60 km site in the present
study (i.e. 3.2 kg SO_4-S/ha-year). This compares well as a control
with values of 3.0 kg SO_4-S/ha-year obtained at the Experimental

Lakes Area of northwestern Ontario (65). Similarly, background SO_2
concentrations were also estimated from the mean values at the 60 km
site (i.e. 6.6 µg SO_2/m^3, compared with values of 12.8 µg/m^3 for
rural areas of southern Sweden (27). When these background estimates
are subtracted from the total SO_4-S and SO_2-S deposition data for
the 60 km circle, net smelter-derived depositions of 0.12 x 10^7 kg
SO_4-S/year, and 1.23 x 10^7 kg SO_2-S/year are derived. These
depositions, expressed as a percentage of the total 1977 emissions
from the Coppercliff smelter of 6 x 10^8 kg SO_2-S/year, account for
0.2% and 2.1% respectively of the emitted sulfur. Thus, based on
the present data, it appears that only 2.3% of the sulfur emitted
from the Coppercliff smelter is deposited within a 60 km radius
of the smelter.

Similar calculations for nickel, copper, and iron depositions
were made using the mean patterns of bulk-deposition. Again, background deposition was calculated using the data of the 60 km site.
Such calculations provide estimates of the percentage of emitted
metals deposited within a 60 km radius of the Coppercliff smelter
of 42% for nickel, 40% for copper, and 52% for iron. This is a
clear indication that deposition of these emitted particulates is
a more rapid and complete process within a 60 km radius of the
smelter than occurs for SO_2, a gaseous emission. It also indicates
that despite the more efficient gravitational fallout of particulates, up to 60% of those emitted are carried beyond 60 km. More
than 97%of the emitted sulfur is carried beyond this distance.

Thus, virtually all of the sulfur emitted from the Coppercliff
smelter is exported over longer distances, and it contributes to
the acidity of precipitation over a large area. Similar observations with respect to the Coppercliff smokestack were made by
Gormley (1977, pers. comm.), who calculated that 3% of the total
emitted sulfur was deposited as sulfate within a 100 km radius
circle. Similarly, Kramer (41) calculated that 0.63% of the total
sulfur emitted from the Coppercliff smelter was deposited within
an area of 3900 km^2 of elevated sulfate deposition. At a site in
Alberta, Canada, it was found that less than 2% of the sulfur
emitted at relatively low levels from sour gas (i.e. natural gas
desulfurizing) plants was deposited within a 10 km radius circle
(61).

The calculated deposition within a 60 km radius of the smelter
for nickel, copper and iron were 42%, 40%, and 52% respectively
of the total emissions from this source. Kramer (41) calculated
that 69% of the emitted nickel and 42% of the emitted copper were
deposited within their areas of elevated deposition. Kramer's
total iron deposition accounted for more than 100% of the iron
emitted from the Coppercliff smelter. He attributed this apparent
anomaly to interference from a nearby iron ore reduction plant,
which also emits large quantities of iron particulates.

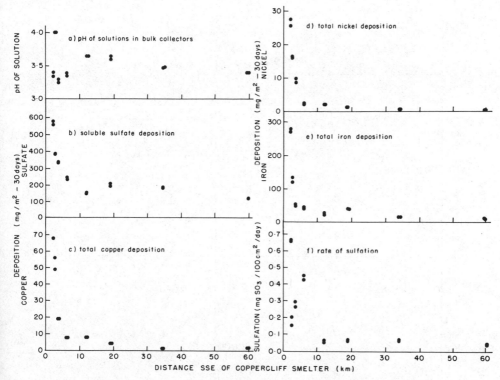

Figure 1. Bulk deposition and sulfation measurements during a
representative sampling interval (July 18 - Aug, 1977)
along a SSE transect centering on the Coppercliff Smelter.
Duplicate samples are shown.

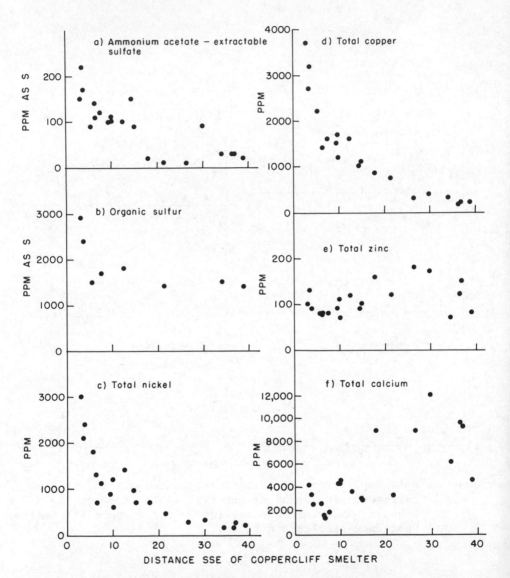

Figure 2.   Analyses of forest litter (0 – 7 cm depth).   Each
             point represents the mean of 6 replicate determinations.

Interference from this source may also have influenced the measurements of iron deposition described in the present study. It is
clear that the emitted metal particulates are deposited more
rapidly than the sulfur, but that about 50% are nevertheless
carried beyond 60 km from the smelter.

The pH of the solution in the bulk collectors did not vary
consistently with distance from the smelter on any sampling date
(i.e. Figure 1a), or with sulfate concentration in the solution.
The pH's were all, however, quite acid (all pH's measured were
less than 4.5, most were less than 4.0), as compared with the pH
of distilled water in eqilibrium with atmospheric carbon dioxide
(i.e. pH 5.65), and are indicative of a severe regional acid precipitation problem. The low pH of precipitation (both event and
bulk-collected) in the Sudbury area has been previously noted in
the literature (e.g. 19,34), as has the acidity of precipitation
on Precambrian Shield areas throughout a much larger area of
south-central Ontario (19). Bulk-collected pH measurements are
known to compare favourably with the pH of event-sampled rainfall
collected during the same period, as evidenced by the data of
Dillon et al. (19), who compared these two sampling methods for
areas of Precambrian Shield in south-central Ontario. The
apparent lack of a correlation between the sulfate concentration
and the pH of event or bulk-collected rainfall (even though
sulfate is the dominant anion) in precipitation at sites not
proximate to oceans (24,43,58) has been noted in other recent
studies (e.g. 10,19,39,42,43,58). This is probably due to a
combination of 1) the presence of other acid anions, and 2)
additional buffering substances in the solution.

The highest sulfation values observed approximate the Ontario
Ministry of the Environment criterion for maximum allowable levels
of sulfation of 0.7 mg $SO_3/100$ cm^2-day (57).

Forest soils close to the Coppercliff smelter contain elevated
total and ammonium acetate-extractable levels of copper, nickel
and iron, and of ammonium acetate-extractable sulfate and total
organic sulfur, as compared to sites further away (Figure 2, Table
1). These elevated soil concentrations are undoubtedly due to
the fact that copper, nickel, iron and sulfur are pollutants that
have been emitted by the Coppercliff smelter over long periods of
time. Elevated soil burdens of copper, nickel, iron and sulfate
have been noted in other studies done around Sudbury area smelters
(e.g. 25,32,34,46,47).

Whitby and Hutchinson (32) have shown that the low pH and
high concentrations of total and water-soluble copper, nickel and
aluminum in the devegetated mineral soils surrounding the then-
operating Coniston smelter were highly toxic to bioassay vegetation. This may also be the case for the high levels of total and
ammonium acetate-extractable nickel and copper in the organic soil

horizons of forested sites close to the Coppercliff smelter, although no plant bioassays have yet been done with these soils.

Zinc, considered here as a control (non-emitted) metal, does not show any trend with distance from Coppercliff (Figure 2e). Calcium, however, shows a trend to higher concentrations in forest soils at greater distances from the smelter (Figure 2f). The reasons for this pattern are not clear, but could be due to 1) higher rates of leaching of calcium at sites closer to the smelter under the influence of acidic precipitation and/or competitive replacement by copper, nickel, or iron at cation exchange sites in the litter (e.g. 63), 2) differences in the calcium concentration of the soil litter derived from the different plant species at the polluted relative to the control sites, or 3) differences in the geochemistry of the bedrock and glacial till which contribute to the chemistry of the forest litter. Magnesium, another base, does not show this trend with distance from the smelter (22). Costescu (16) found no distance-related trends in either calcium or magnesium in mineral soils collected along transects centering on the Coniston smelter.

No trend with distance was observed in either the pH of forest litter or of exposed hilltop mineral soil (Figure 3). This has occurred in spite of the high inputs of SO_2 and sulfate to the sites closer to the smelter over a long period of time, and in spite of intense fumigations by sulfuric acid mists and SO_2 in the early years of the smelting developments in the area. The lack of present effect on soil pH is probably due to four main factors: 1) the natural buffering capacities of the soils, 2) the added buffering capacity due to inputs of pollutant cations (i.e. copper, nickel, iron), 3) the relatively high leachability of sulfate (29,36) from most soils where precipitation exceeds evaporation, leading to a loss of much of the deposited sulfate to drainage systems, and 4) the fact that the acid inputs are only incremental to the amounts of acidity produced naturally in soils via the nitrogen and sulfur cycles, when reduced nitrogen or sulfur compounds are oxidized, releasing hydrogen ions (23, 58).

Pollutants emitted by the Coppercliff smelter (i.e. copper, nickel, and sulfur) are found in high concentrations in unwashed vegetation samples of an array of species at sites closer to the smelter, relative to sites further away (Table 2). Copper and nickel are elevated in vegetation to a greater extent than is sulfur. This is partly due to the fact that sulfur, an essential plant nutrient, is found in relatively high concentrations even in uncontaminated foliage, while copper and nickel are found in only trace quantities. Bowen (13) cites representative level of sulfur in angiosperms of 3400 ppm, while copper and nickel mean levels were reported as concentrations of 14 ppm and 2.7 ppm. The high tissue concentrations of copper, nickel and sulfur could have derived from 1) particulate impaction onto plant

TABLE 1 - Total and extractable metals in 02-A1 (duff-humus) soil samples collected at sites at various distances SSE of the Coppercliff Smelter. Mean ± S.D., n = 6 replicate determinations per site.

SITE	ANALYSIS	NICKEL (ppm)			COPPER (ppm)			IRON (ppm)			ZINC (ppm)		
		MEAN	S.D.	%	MEAN	S.D.	%	MEAN	S.D.	%	MEAN	S.D.	%
1-0 (3.5 km) Denuded	Total	220	200		210	110		20100	8100		73	34	
	Ammon. Acet.	31	31	14.3	39	37	18.6	110	40	0.5	4	4	5.5
	Water	30	30	13.8	7	9	3.3	8	4	0.04	3	3	3.6
1 (3.5 km) Forested	Total	1400	400		2400	100		18100	5500		100	20	
	Ammon. Acet.	550	290	39.4	1400	1000	57.6	590	470	3.3	44	18	43.1
	Water	9	6	0.6	7	4	0.3	6	2	0.04	1	1	1.1
2 (5.8 km) Forested	Total	1900	1400		2600	800		22900	14300		100	30	
	Ammon. Acet.	280	100	14.9	700	190	27.0	560	190	2.4	25	7	26.0
	Water	4	1	0.2	3	1	0.1	9	3	0.04	1	1	1.4
7 (29.7 km) Forested	Total	590	190		470	170		11100	4800		170	50	
	Ammon. Acet.	60	14	10.2	25	9	5.3	160	50	1.5	69	6	40.4
	Water	4	1	0.6	3	1	0.6	8	2	0.07	2	1	1.1
8 (34.3 km) Forested	Total	460	130		450	100		10800	2700		120	30	
	Ammon. Acet.	28	11	6.1	13	5	2.9	70	20	0.6	38	21	33.0
	Water	1	0.3	0.3	0.7	0.3	0.2	5	2	0.05	1	0.3	0.5

Total = nitric-perchloric acid digest

Ammon. Acet. = 1.0 M Ammonium Acetate, pH 4.8

Water = distilled water extract

% = % extractable of total metals

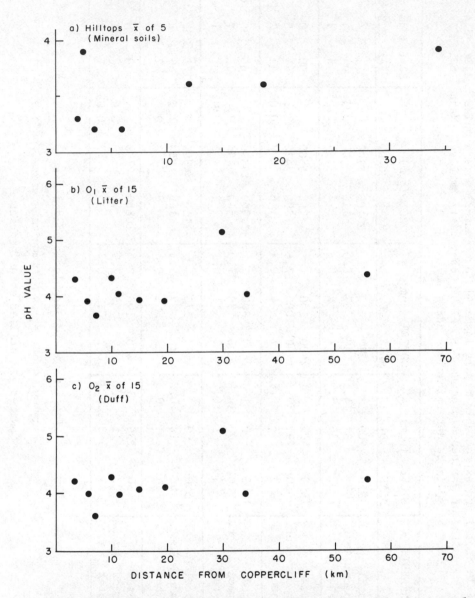

Figure 3.   pH litter and duff of forested sites, and of mineral
            soils, from exposed hilltops and slopes at various
            distances from Coppercliff smelters.

surfaces,   2) absorption from contaminated soils followed by translocation to leaves, or   3) direct absorption from the atmosphere via stomata.   Routes 1 and 2 would have been major contributing factors to the copper and nickel contents, and both routes have been shown to occur in the Sudbury context (16).   All three routes would have operated to produce the elevated sulfur content of plant tissue close to the smelter (15,29,36,64,81). Elevated levels of copper, nickel and sulfur in vegetation have been noted in other studies done in the Sudbury area (e.g. 32,34, 46,47,51,73).

The bases calcium, magnesium and potassium were also analyzed in plant tissues (17 species) at various distances from the smelter (see 22).   No consistent trends with distance were apparent. Some species (i.e. Acer rubrum, Agrostis stolonifera, Diervilla lonicera and Polygonum cilinode) exhibited decreasing foliar concentrations of calcium and magnesium with increasing distance from the smelter.   None of the species examined exhibited decreased calcium, magnesium, or potassium foliar concentrations at sites closer to the smelter.   Such a pattern might have been expected due to the higher inputs of acid anions (i.e. sulfate) at sites closer to the smelter, (see Figure 1).   High inputs of acid anions (usually sulfate, and usually accompanied by low solution pH) have been shown to lead to accelerated rates of base leaching from leaves in both field and laboratory experiments involving perter-bations with artificial "acid rain" or "acid mist" (e.g. 1,2,3,14, 21,83,84).   Similar effects on base leaching from soils following acid rain perturbations in the field have also been reported (e.g. 2,3,52,53,71).

Soil Litter Decomposition Studies

i) Litter Standing Crop and Accumulation Ratio.   Collections of litter standing crop (i.e. O1 + O2) at nine forested stands located at various distances from the Coppercliff smelter indicate a trend towards higher standing crops at sties closer to the smelter (Table 3).   This litter is not accumulating as O1 (litter) relative to O2 (duff) (Table 3), as was found by Watson et al (79), who considered this observation to be evidence of decreased O1 mineralization due to high levels of lead and zinc in soils near a lead smelter.   The increased litter standing crop at sites closer to the smelter soes not appear to be due to differences in the amount of leaf litter input at the various sites (Table 4), i.e. the differences in the calculated Litter Accumulation Index between the two "contaminated" forest sites (i.e. Sites 1 and 2 in Table 4) and the two "control" sites (i.e. Sites 8 and 9) are largely due to differences in litter standing crop, and not to differences in litter fall.   Accumulation of litter in forests near smelters has previously been noted in the literature (e.g. 37,45,68,78,79).

TABLE 2 – Analysis of homogenized plant tissues collected SSE of
Coppercliff.

a) SULFUR   TOTAL SULFUR CONCENTRATION (ppm x 10^3, dry weight)

SPECIES	1.6 km	3.0 km	3.5 km	5.8 km	11.9 km	21.3 km	34.3 km	60.0 km
Acer rubrum	1.8	1.4	1.0	1.6	1.2	0.8	0.8	1.0
Agrostis stolonifera	4.0	5.3	3.4	2.6	3.3	3.4	3.2	2.2
Epilobium angustifolium	2.1	1.9	4.1	4.0	2.9	1.3	1.7	1.0
Populus tremuloides	4.9	3.5	4.3	5.0	3.5	4.5	4.6	2.7
Solidago canadensis	3.0	2.7	3.6	2.1	2.5	1.6	1.3	1.3

b) NICKEL   TOTAL NICKEL CONCENTRATION (ppm x 10^2, dry weight)

SPECIES	1.6 km	3.0 km	3.5 km	5.8 km	11.9 km	21.3 km	34.3 km	60.0 km
Acer rubrum	1.0	0.3	0.7	0.4	0.3	0.1	0.07	0.02
Agrostis stolonifera	0.7	0.8	1.3	0.3	0.4	0.3	0.1	0.06
Epilobium angustifolium	0.8	1.1	0.9	0.6	0.3	0.2	0.1	0.07
Populus tremuloides	0.7	1.0	1.0	3.7	0.6	0.6	0.5	0.04
Solidago canadensis	1.7	2.0	1.7	2.0	0.7	0.3	0.3	0.04

c) COPPER   TOTAL COPPER CONCENTRATION (ppm x 10^2, dry weight)

SPECIES	1.6 km	3.0 km	3.5 km	5.8 km	11.9 km	21.3 km	34.3 km	60.0 km
Acer rubrum	1.3	0.8	0.4	0.4	0.2	0.2	0.02	0.09
Agrostis stolonifera	0.7	0.7	-0.5	0.2	0.2	0.08	0.08	0.06
Epilobium angustifolium	1.1	1.6	0.8	0.4	0.4	0.1	0.1	0.1
Populus tremuloides	0.9	0.4	0.3	0.3	0.1	0.1	0.1	0.1
Solidago canadensis	1.8	1.1	0.8	0.3	0.4	0.1	0.1	0.2

TABLE 3 - Standing crop of 01 (litter) and 02 (duff) at forested
sites SSE of the Coppercliff smelter. n = 15 replicate
0.25 x 0.25m quadrats per site.

STANDING CROP (Mean $\pm$ S.D.)

SITE	DISTANCE	01	02	01 + 02 *	01 / 02 **
		(grams x 10^2 per square metre)			
1	3.5 km	4.4 $\pm$ 3.6	8.7 $\pm$ 4.9	13.1 $\pm$ 6.6	0.57 $\pm$ 0.43
2	5.8	7.3 $\pm$ 4.5	12.5 $\pm$ 7.9	19.8 $\pm$ 10.3	0.83 $\pm$ 0.76
3	7.2	4.4 $\pm$ 3.3	9.3 $\pm$ 2.8	13.7 $\pm$ 4.9	0.52 $\pm$ 0.38
4	10.2	3.5 $\pm$ 2.6	6.0 $\pm$ 3.2	9.5 $\pm$ 4.3	0.64 $\pm$ 0.39
5	11.3	2.0 $\pm$ 1.3	7.5 $\pm$ 2.5	9.4 $\pm$ 2.5	0.33 $\pm$ 0.29
6	15.1	2.1 $\pm$ 0.9	6.2 $\pm$ 2.4	8.4 $\pm$ 2.6	0.41 $\pm$ 0.24
7	19.6	3.5 $\pm$ 1.8	3.7 $\pm$ 2.0	7.2 $\pm$ 2.5	0.95 $\pm$ 0.40
8	29.7	2.3 $\pm$ 1.0	3.9 $\pm$ 1.1	6.2 $\pm$ 1.3	0.64 $\pm$ 0.38
9	34.3	2.6 $\pm$ 1.1	3.0 $\pm$ 1.1	5.5 $\pm$ 1.8	0.92 $\pm$ 0.38

*regression of (01 + 02) vs distance: $Y = -1.82X + 15.3$ (r= 0.79, p<.01)

**regression of 01/02 vs distance: $Y = 0.0086X + 0.52$ (r= 0.42, n.s.)

All of these studies documented increases of litter standing crop
at forest sites close to the investigated smelters. However, only
one study (68) related these differences to relative rates of
litter input. Strojan found only a slight reduction in litter
input at his contaminated sites, and considered that the
observed differences in litter standing crop were due to decreased
litter mineralization. This agrees with our own observations.

ii) <u>Litter Bag Decomposition Studies</u>. Direct measurements of
leaf litter decomposition were made by exposing leaf tissue of
three tree species (<u>Betula papyrifera</u>, <u>Populus tremuloides</u>, and
<u>Pinus strobus</u>) in litter bags at contaminated and control forest
sites at various distances from the Coppercliff smelter.
Decomposition rates (i.e. net weight loss) were measured after
various exposure intervals in the field.

After the maximum exposure interval (488 days), the leaf dry
weight loss of all three species was greater at the two control
sites than at the three contaminated sites (Table 5). However,
the differences between all three contaminated sites (i.e. sites
1, 2, and 3) and the two control sites (sites 8, 9) were not all
significant in the analysis of least significant difference. Only
one other study (68) has reported on the relative weight loss of
leaf tissue exposed in litter bags in the field in a smelter
context. This author (68) found weight losses at a contaminated
site (up to 26,000 ppm zinc, 2300 ppm lead, 900 ppm cadmium in
02 litter) to be only about one half that observed at a control
site.

TABLE 4 - Litter Accumulation Index for four forested sites SSE of
the Coppercliff smelter.

SITE	DISTANCE	TOTAL LEAF LITTER FALL	TOTAL 01 + 02 LITTER	LITTER ACCUMULATION INDEX
		(gm/m^2)	$(gm \times 10^2/m^2)$	$(\times 10^2)$
1	3.5 km	206 ± 67	13.1 ± 6.6	6.4 *
2	5.8	201 ± 30	19.8 ± 10.3	9.9 *
8	29.7	259 ± 27	6.2 ± 1.3	2.4
9	34.3	223 ± 37	5.5 ± 1.8	2.5

* significantly larger than sites 7, 8 at $P < 0.05$ t-test.
a) Total Leaf Litter Fall estimated by 6 quadrats per site (1m x 1m).
b) Total 01 + 02 Litter estimated by 15 quadrats per sits (.25m x .25m).

The relative net losses of calcium, magnesium and potassium from leaves of the three tree species after the various exposure intervals have also been examined (see 22). This was done to examine whether any relation might be found between these parameters and the relatively high rate of anion input close to the smelter (i.e. high sulfate loading, but no pH difference in bulk-collected precipitation at sites closer to the smelter relative to the sites further away). None of the species showed consistent, statistically significant differences in these parameters between the two control sites and the three contaminated sites closer to the smelter. The relative rates of base loss were potassium >> magnesium > calcium at the various sampling intervals. This trend agrees with the observations of (9), (27), and (68), who found these bases to be lost from tree leaves confined in litter bags in the same series order.

iii) <u>Soil CO_2 Flux Measurements</u>. Soil CO_2 flux, considered as an index of overall soil respiration processes, was determined for soil samples collected from forest sites at various distances from the Coppercliff smelter. Consistently lower rates of CO_2 flux were measured from soil samples collected at sites closer to the smelter, relative to sites further away (Table 6). However, within-site variation is high in these analyses, with a coefficient of variation of up to 66%. This high variation could be due to such factors as spatial heterogeneity in microbial populations or activity, or water content in the soils, which was not standardized prior to the assay.

Other **studies** have looked at CO_2 flux from soils as an index of soil respiration. In a series of papers investigating a site in Sweden contaminated by copper, zinc and other heavy metals, a decreased CO_2 flux from contaminated spruce needle mor soils relative to uncontaminated soils was noted (20,63,74,75,76). These effects were correlated with soil heavy metal burdens.

iv) <u>Soil Acid Phosphatase Assays</u>. Acid phosphatase assays were made on contaminated and control duff-humus soil samples collected from forest sites at various distances from the Coppercliff smelter. The assay of soil acid phosphatase activity is considered to be an index of overall soil respiratory activity. The technique used gives an approximation of extracellular activity, plus an unknown fraction of the total intracellular acid phosphatase activity (67). Acid phosphatases are important soil enzymes, since assimilation by plants or microorganisms of phosphate from organic compounds (30 to 70% of the total soil phosphate is organically-bound) is preceded by the hydrolysis of the bound phosphate by extracellular phosphatase enzymes (67).

The data indicate a large decrease in mean acid phosphatase activity in contaminated sites close to the smelter, relative to

TABLE 5 – Relative dry weight loss of tree leaves in
15 mesh litter bags placed at forest sites at
various distances from the Coppercliff smelter.
Mean percent loss ± S.D.

a) Betula papyrifera

SITE	DISTANCE	51 DAYS	117 DAYS	314 DAYS	488 DAYS
1	3.5 km	30 ± 1	44 ± 2	52 ± 1	63 ± 3
2	5.8	28 ± 2	43 ± 2	51 ± 5	57 ± 3
3	7.2	26 ± 1	34 ± 3	46 ± 3	58 ± 12
8	29.7	40 ± 4	51 ± 1	61 ± 6	67 ± 10
9	34.3	36 ± 3	55 ± 5	60 ± 8	72 ± 10
n =		5	4	4	5
ANOVA significance		n.s.	0.01	0.01	0.01
L.S.D. .05		–	5	5	7

b) Populus tremuloides

SITE	DISTANCE	51 DAYS	117 DAYS	314 DAYS	488 DAYS
1	3.5 km	15 ± 0.4	26 ± 0.7	35 ± 2	39 ± 3
2	5.8	18 ± 0.2	25 ± 1	31 ± 0.4	36 ± 1
3	7.2	20 ± 1	25 ± 1	32 ± 1	36 ± 2
8	29.7	21 ± 1	30 ± 0.9	36 ± 0.6	46 ± 5
9	34.3	22 ± 0.8	28 ± 2	39 ± 2	47 ± 2
n =		5	4	4	5
ANOVA significance		0.01	n.s.	0.01	0.01
L.S.D. .05		4	–	4	5

c) Pinus strobus

SITE	DISTANCE	51 DAYS	117 DAYS	314 DAYS	488 DAYS
1	3.5 km	11 ± 0.2	17 ± 0.4	32 ± 0.5	40 ± 2
2	5.8	14 ± 0.5	17 ± 0.4	31 ± 2	38 ± 4
3	7.2	20 ± 0.5	24 ± 2	35 ± 2	40 ± 0.7
8	29.7	16 ± 0.4	28 ± 0.8	37 ± 4	45 ± 6
9	34.3	13 ± 0.1	21 ± 0.5	32 ± 0.5	50 ± 4
n =		5	4	4	5
ANOVA significance		0.01	0.01	n.s.	0.01
L.S.D. .05		3	4	–	7

control sites further away (Table 6, Figure 4). However, within-
site variation is large, so that the relative activities of the
three contaminated sites (3.0, 3.7 and 5.2 km) and the three
control sites (34.3, 47.8 and 50.2 km) are not all significantly
different in the analysis of least significant difference (i.e.
the rate of phosphatase activity at the contaminated site at 5.2
km was not significantly lower than that of the control site at
47.7 km (Figure 4).

The CO_2 flux from aliquots of the same soil samples assayed
for acid phosphatase activity was also measured. Differences in
CO_2 flux exist between the sites, although it is notable that the
contaminated site at 5.2 km exhibits rates comparable to those of
the control site at 50 km (Table 6). Soils from a devegetated
mineral-soiled hilltop close to the smelter (Site A-0 at 3.0 km)
were found to have a much lower rate of both CO_2 flux and acid
phosphatase activity than soils from a comparable, adjacent
forested site (A, 3.0 km). This is due to the low microbial
activity in the mineral soils from the devegetated hilltop.

Concentrations of copper, nickel and zinc were also analyzed
in aliquots of the soil samples assayed for CO_2 flux and acid
phosphatase activity (Table 6). Table 7 summarizes linear
regressions and correlation coefficients calculated from the data
summarized in Table 6. All correlations presented are significant,
most at $p < .001$. Total soil copper is more highly negatively
correlated with both CO_2 flux and enzyme activity than is nickel,
while copper plus nickel gives an intermediate correlation. This
may be related to the observations of Horsfall (30), who found
(in a series of laboratory perterbation experiments) that copper
had a more potent fungicidal effect than nickel.

Other authors have investigated the effects of heavy metal
pollution in forest soils near smelters on soil acid phosphatase
activity. Tyler (74,75,76) found lower soil acid phosphatase
acitvity at sites contaminated by copper and zinc, relative to
uncontaminated sites, at a smelter at Gusum, Sweden. He noted
(75) that it was unclear whether the decreased enzyme activity
in the metal-polluted soils was due to 1) lower enzyme concentrations in soil, or 2) inactivation of soil enzymes by heavy
metals, either of which would contribute to lower activity in
assays.

Tyler (77) also found decreased acid phosphatase activity in
soils perturbed by various concentrations of vanadium (up to 100
ppm) in a controlled, laboratory experiment. In another laboratory
perturbation experiment (38), the effects of various concentrations
of twenty separate trace metals on soil acid phosphatase activity
were examined. These authors noted decreases in activity following
addition of copper and nickel, and noted that copper had approximately twice the inhibitory effect of nickel on a concentration
basis.

TABLE 6 – Summary of acid phosphatase activity, CO_2 flux, and total metal analyses of O2-A1 (duff-humus) soil samples collected from sites at various distances SSE of the Coppercliff smelter on June 2, 1977. Mean ± S.D., n = 10 replicate collections per site.

SITE	DISTANCE		CO_2 FLUX RATE/F.W.	RATE/D.W.	ENZYME ACTIVITY RATE/F.W.	RATE/D.W.	TOTAL COPPER	TOTAL NICKEL (ppm dry weight)	TOTAL ZINC
A-0 *	3.0 km	MEAN	0.06	0.07	8.5	9.2	210	220	70
		S.D.	± 0.02	± 0.02	± 10.0	± 10.8	± 110	± 200	± 30
A	3.0	MEAN	0.18	0.26	41.6	63.9	2400	1400	100
		S.D.	± 0.09	± 0.14	± 16.2	± 19.3	± 700	± 400	± 20
B	3.7	MEAN	0.31	0.58	46.2	78.5	2600	1900	100
		S.D.	± 0.16	± 0.38	± 19.3	± 44.7	± 800	± 1400	± 30
C	5.2	MEAN	0.38	0.72	73.9	144.8	1400	1200	90
		S.D.	± 0.17	± 0.40	± 58.5	± 114.0	± 400	± 400	± 20
D	34.3	MEAN	0.59	0.98	191.7	316.5	470	590	170
		S.D.	± 0.19	± 0.38	± 47.7	± 82.4	± 170	± 190	± 50
E	47.8	MEAN	0.51	0.94	105.5	205.6	450	460	120
		S.D.	± 0.15	± 0.32	± 49.3	± 114.7	± 100	± 130	± 30
F	50.2	MEAN	0.39	0.62	180.2	248.7	230	270	300
		S.D.	± 0.15	± 0.28	± 75.5	± 122.4	± 70	± 100	± 70

CO_2 flux: μl CO_2/minute - gm f.w. or d.w.
Enzyme Activity: μg p-nitrophenol produced / 20 min.-gm f.w. or d.w.
* devegetated hilltop. All other sites are forested.

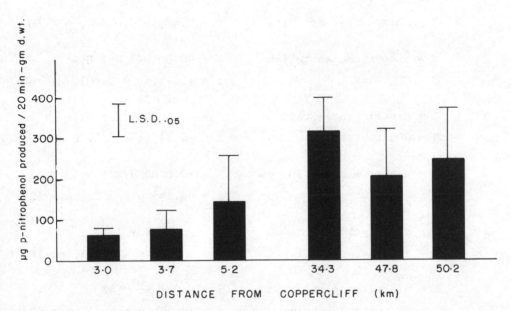

Figure 4 – Acid phosphatase activity of duff–humus soil samples from
forest communities near Coppercliff, Ont.
$\overline{X} \pm$ S.D., n = 10 replicate determinations per site.

TABLE 7 – Linear regressions and correlation coefficients
calculated using data on soil acid phosphatase
activity, CO_2 flux, and total soil nickel and
copper. Data base was 10 replicate determinations
for each of six forested sites at various
distances from the Coppercliff smelter.

1   X = CO_2 flux   Y = 216 X + 18
         Y = phosphatase activity   r = 0.48, p < 0.001, n = 60

2   X = copper concentration   Y = -0.000077 X + 0.49
         Y = CO_2 flux   r = -0.41, p < 0.001, n = 60

3   X = nickel concentration   Y = -0.000048 X + 0.44
         Y = CO_2 flux   r = -0.21, p < 0.1, n = 60

4   X = copper + nickel conc.   Y = -0.000038 X + 0.48
         Y = CO_2 flux   r = -0.34, p < 0.01, n = 60

5   X = copper concentration   Y = -.044 X + 162
         Y = phosphatase activity   r = -0.63, p < 0.001, n = 60

6   X = nickel concentration   Y = -.037 X + 142
         Y = phosphatase activity   r = -0.40, p < 0.01, n = 60

7   X = copper + nickel conc.   Y = -.023 X + 159
         Y = phosphatase activity   r = -0.55, p < 0.001, n = 60

phosphatase activity: ug p-nitrophenol produced / 20
                      min - gm f.w.

CO_2 flux : ul CO_2 evolved / minute - gm f.w.

nickel, copper concentrations: ppm dry weight

Other soil enzymes have also been assayed with respect to the
effect of heavy-metal polluted soils near smelters. At the Gusum,
Sweden site already mentioned, a decrease in dehydrogenase activity
in the copper and zinc-contaminated soils close to the smelter
was noted (63). At the same site, decreased urease activity in
the contaminated soils was found, but no effect on b-glucosidase
(74,75). Also at Gusum, decreased starch amylase activity was
documented in the contaminated soils (20). Finally, at a mine
waste site contaminated by lead and zinc, decreased urease activity
in contaminated soils was found, relative to controls (82).

v) Soil Microfaunal Collections. Soil samples for microfaunal
analysis were collected from two contaminated and two control
forest sites at various distances from the Coppercliff smelter,
and extracted using a modified Berlese-Tullgren funnel system.
The data are highly variable within sites, with most standard
deviations being greater than their associated means (Table 8).
The July 17, 1976 data indicate significantly lower (p <.05) mite
and springtail populations at the two contaminated sites (3.5 and
5.5 km) relative to the control sites (29.9 and 34.3 km). This
pattern was not, however, found in the July 14, 1976 data.

Watson (78,79) investigated the impact of a metal-contaminated
(mainly lead and zinc) site near a lead smelter in Missouri on
the microarthropod fauna. She found that reduced values of microarthropod density, diversity, richness, and biomass occurred
simultaneously with elevated concentrations of heavy metals in
litter at sites closer to the smelter. Lead, zinc, copper and
cadmium were found in elevated concentrations in microarthropods
from contaminated sites, and predatory microarthropods were found
to concentrate lead, zinc and cadmium relative to detritivore or
fungivore microarthropod prey species. These authors also found
large spatial variation in their microarthropod collections, with
a maximum coefficient of variation of 330% being observed for
density data collected from the O1 litter horizon, and 67% from
the O2 duff horizon. They felt this high variation to be due to
physical soil factors, such as temperature or relative humidity.
In another study, lower numbers and diversity were found of
microarthropods inhabiting leaf litter exposed in litter bags at
metal-contaminated sites close to a smelter near Palmerton,
Pennsylvania, compared to control forest sites (68).

vi) Soil Microfungal Populations. Populations of soil fungi
were estimated for two contaminated and two control forest sites
at various distances from the Coppercliff smelter (Table 9).
Although within-site variation is large (maximum coefficient of
variation = 109%), there do appear to be lower mean population
of fungi at the contaminated site closest to the smelter (3.5 km
site). However, none of these differences were significant in the
analysis of variance. Populations at the other contaminated site

TABLE 8 - Numbers of soil microarthropods extracted per
          core from forest soils collected at sites at
          various distances from the Coppercliff smelter.
          Mean ± S.D., n = 60 replicate soil cores per site.

a) Collection Date: July 14, 1976

SITE	Collembola	Mites	Others
3.5 km	1.9 ± 2.3	1.1 ± 1.3	0.3 ± 0.6
5.5	1.1 ± 1.3	2.4 ± 2.8	0.4 ± 0.7
29.9	2.1 ± 2.3	3.3 ± 3.4	0.8 ± 1.1
34.3	1.3 ± 1.9	2.2 ± 2.5	0.5 ± 0.7
ANOVA significance	0.05	0.01	n.s.
LSD $_{.05}$	0.7	0.6	-

b) Collection Date: July 17, 1976

SITE	Collembola	Mites	Others
3.5 km	1.5 ± 1.9	2.5 ± 2.1	0.4 ± 0.7
5.5	1.3 ± 1.5	3.3 ± 3.1	0.5 ± 0.7
29.9	2.7 ± 2.5	5.1 ± 3.8	0.6 ± 0.7
34.3	4.4 ± 4.1	5.0 ± 3.5	0.8 ± 1.5
ANOVA significance	0.01	0.01	n.s.
LSD $_{.05}$	1.0	1.2	-

are not consistently different from those of the two control sites
(Table 10). For all sites, the fungal colonies isolated were
predominantly of the genus Penicillium (73% of total colonies),
followed by Trichoderma (13%), Mortierella (10%, and Mucor (22%).

Carter (unpublished data, pers. comm., 1978) also looked at
fungal populations isolated from similar soil samples collected
from the same forest sites investigated in this study. He noted
high spatial variation in his population estimates, and also did
not find statistically significant differences in fungal populations
between contaminated and control sites when his data were expressed
on a soil fresh weight basis.

TABLE 9 – Total microfungal populations (soil dilution
method, Rose-Bengall agar with streptomycin)
in forest soils from two contaminated and two
control sites near Sudbury, Ontario.

(Colonies x 10^5 / gm f.w.)

SITE	June 7, 1976	July 1, 1976	July 8, 1976
3.5 km	1.4 + 0.5	1.1 + 0.3	1.8 + 0.5
5.8	4.3 + 4.7	2.6 + 1.3	3.3 + 0.7
29.7	1.9 + 1.1	2.7 + 2.9	4.1 + 2.6
34.3	4.0 + 1.9	2.5 + 1.4	3.7 + 1.8
n (soil samples)	5	4	4
ANOVA significance	n.s.	n.s.	n.s.

Other studies have looked at the effects of metal-polluted
soils near smelters on microbial populations. Jordan and
Lechavalier (37) investigated a zinc smelter in Pennsylvania and
found decreased numbers of fungi, actinomycetes, and bacteria in
contaminated soils close to the smelter. Hartman (28) looked at
fungi in soils contaminated with cadmium, copper, lead and zinc
near a smelter in Montana, and found decreases in fungal populations and species diversity. Williams et al. (82) looked at a
lead and zinc-contaminated smelter site in England, and found
reduced fungal populations in both soils and litter. Martin (45)

looked at sites near a smelter in England contaminated by zinc, lead, and cadmium and found decreased numbers of phylloplane fungi at sites closer to the smelter.

The effects of copper, nickel, or both in various concentrations in the nutrient agar on the numbers and growth rate of colonies isolated from contaminated and control forest sites were also investigated (see 22). For all sites, the numbers of colonies isolated decreased with increasing concentrations of copper and nickel. In addition, the colony diameters were relatively small at the highest metal concentrations (10% of control at 100 ppm nickel or copper after 8 days incubation), indicating slower mycelial growth rates. "Tolerant" fungi were isolated from both contaminated and control forest soils, and no differences appeared between the sites in the relative numbers of tolerant fungi isolated. Similar observations have also been made by Carter (unpublished data, pers. comm., 1978), who isolated fungi on a nickel-perturbed agar from soils collected from the same forest sites near Sudbury.

Jordan and Lechevalier (37) investigated microbial populations in contaminated soils near a zinc smelter in Pennsylvania. They found a decrease in the numbers and growth rates of fungal colonies isolated from soils when the nutrient agar was amended by adding various concentrations of zinc. Notably, they also isolated tolerant fungi from control sites, and little difference was seen in the relative levels of zinc tolerance of fungi between the contaminated and control forest soils. They made similar observations for actinomycetes and bacteria. Hartman (28) examined fungal populations at sites contaminated with cadmium, copper, lead and zinc at a smelter site in Montana. He compared colonies of Fusarium oxysporum isolated from contaminated and control soils in his area with respect to tolerance to these metals in the growth medium, and found the smelter populations to be relatively tolerant, although both "ecotypes" were capable of growth in the presence of the metals. He noted that concentrations of metals in the mycelium reached very high levels, with up to 18,150 ppm copper, 14,300 ppm cadmium, 29,000 ppm zinc, and 132,000 ppm lead being reached.

vii) Litter Perturbation Experiment. In this laboratory experiment, a homogenate of litter collected from an uncontaminated mixed forest site was amended with various concentrations of copper and/or nickel to yield total concentrations of up 10,000 ppm of each or both. Weight loss and CO_2 flux were then measured after a 60 day incubation period to determine whether these parameters were affected by these metals in the litter.

The mineralization of the litter was retarded by the presence of both copper and nickel in the litter (Figure 5). Both metals began to exert a negative effect when present in the litter to concentrations of 1000 ppm. Such total nickel concentrations in litter are frequently observed within 15 km of the smelter (Figure 2). When copper and nickel are each present in the litter in concentrations of 500 ppm, they together exert a negative effect on mineralization, even though separately they had no effect at 500 ppm. This indicates a possible synergistic mechanism of effect for the two metals. At higher concentrations of the two metals together, the negative effect is only slightly less than when the metals are present separately.

The data for CO_2 flux, measured at the end of the 60 day incubation period, indicate that copper and/or nickel will depress the rate of CO_2 evolution from the litter (Figure 6). Note that these data are much more variable than the data for litter mineralization (coefficient of variation of control = 37%, versus 3% for weight loss). At most metal concentrations, a depression of CO_2 flux was measured relative to the control litter. When both metals are present together, the effect was greater than that observed when the metals are present separately.

In both the mineralization and CO_2 flux determinations, copper exerted a somewhat more negative effect than did nickel at the same concentration. This agrees with the observations of (30), that copper has a greater fungicidal effect than nickel and of (38) who noted that copper had twice the inhibitory effect of nickel on soil acid phosphatase activity.

SUMMARY

Inputs of smelter-derived pollutants are relatively high at sampling sites closer to the Coppercliff smelter, compared with control sites up to 60 km SSE. These present-day inputs, plus major historical inputs since the initiation of smelting activity in the late 19th century, have resulted in significant increases in certain metal pollutants in soils and vegetation at forest sites close to the smelter as well as increased sulphur levels. An accumulation of litter standing crop has been shown to occur at forested sites close to the smelter. This litter accumulation may be due to relatively low rates of mineralization of litter at the contaminated sites. Lower rates of decomposition of leaves in litter bags occurs at contaminated sites, together with relatively low rates of soil metabolic activity. These include soil CO_2 flux and acid phosphatase activity. Lower populations of soil micofungi and micoarthropods also occur. However, not all of the above differences between contaminated and control sites were statistically significant, due to large within-site spatial variation in the various parameters measured. In a laboratory experiment involving the addition of copper and/or nickel to a litter homogenate, a depression of litter

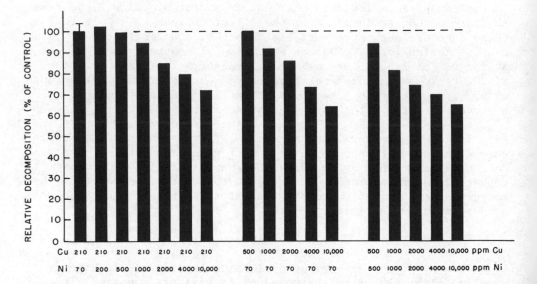

Figure 5.   Relative decomposition after a 60 day incubation of control
           litter, and litter amended with various concentrations of
           Copper and/or Nickel.
           Control, M ± S.D., n=6 replicates.
           All other determinations x̄, n=2 replicates.

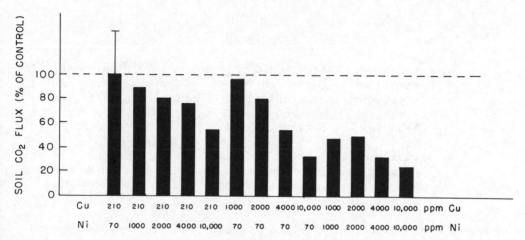

Figure 6. Relative CO_2 flux from litter after 60 day incubation of control litter and litter amended with various concentrations of Copper and/or Nickel.
Control $\bar{x} \pm$ S.D., n=6 replicates.
All other determinations $\bar{x}$, n=2 replicates.

mineralization and CO_2 flux was shown to occur at metal concentrations similar to those observed at contaminated sites in the field.

Sulfur emissions from the Coppercliff smelter result in elevated dry deposition of SO_2, sulfate concentration in precipitation, sulfate and organic sulfur contents of forest litter, and elevated sulfur concentrations in plant foliage. However, there were no discernible distance to smelter-related changes in bulk-collected rainfall pH, soil litter pH, mineral soil pH, base leaching from litter in litter bags, or concentrations of bases in foliage. Overall, the negative effects attributable to acidic precipitation on terrestrial communities surrounding the smelter appear to be small, relative to the effects of sulfur dioxide and heavy metal residues in soils.

REFERENCES

1.   G. Abrahamsen, K. Bjor and O. Tiegen, Field experiments with simulated acid rain in forest ecosystems. I. Soil and vegetation characteristics, experimental design, and equipment. Research Report No. 4, SNSF-Project. 1432 Ås-NLH, Norway, 15 pp. (1975).

2.   G. Abrahamsen, K. Bjor, R. Horntvedt and B. Tveite, Effects of acid precipitation on coniferous forest. Research Report FR6/76. SNSF-Project. 1432 Ås-NLH, Norway, 63 pp. (1976).

3.   G. Abrahamsen, R. Horntvedt and B. Tveite, Impacts of acid precipitation on coniferous forest ecosystems. in: "Proceedings of the First International Symposium on Acid Precipitation and the Forest Ecosystem" U.S. Dept. of Agriculture. Forest Service. General Technical Report NE-23, p. 991 (1976).

4.   Air Pollution Control Directorate, A national inventory of air pollutant emissions. Summary of emissions for 1972. Air Pollution Control Directorate. Report EPS 3-AP-75-5. Environment Canada Ottawa, (1976).

5.   S.E. Allen, H.M. Grimshaw, J.A. Parkinson and C. Quarmby, "Chemical Analysis of Ecological Materials" Blackwell Sci. Publ. London, (1974).

6.   American National Standards Institute, Standard methods of test for sulfate ion in water and waste water. in: "1976 Annual Book of ASTM Standards" Washington, D.C. p. 425 (1976).

7. American National Standards Institute, Standard method for
        collection and analysis of dustfall (settleable
        particulates). in: "1976 Annual Book of ASTM Standards"
        Washington, D.C., p. 506 (1976).

8. American National Standards Institute, Evaluation of total
        sulfation in atmosphere by the lead peroxide candle.
        in: "1976 Annual Book of ASTM Standards" Washington,
        D.C., p. 533 (1976).

9. P.M. Attiwill, The loss of elements from decomposing litter.
        Ecology 49:142 (1968).

10. M. Benarie and P. Detrie, Assessment of an OECD study on
        long range transport of air pollutants (LRTAP) involving
        some aspects of air chemistry. in: "Studies in Environ-
        mental Science" Vol 1, M. Benaire, ed., (1978).

11. C.A. Black, D.D. Evans, J.L. White, L.E. Ensminger and F.E.
        Clark (eds.), "Methods of Soil Analysis" Monograph
        No. 9. American Society of Agronomy, Inc. Madison,
        Wisconsin. (1965).

12. K.L. Bocock, O. Golbert, C.K. Capstick, D.C. Twinn, J.S.
        Waid and M.J. Woodman, Changes in leaf litter when
        placed on the surface of soils with contrasting humus
        types. I. Losses in dry weight of oak and ash leaf
        litter. J. Soil Sci. 11:1 (1960).

13. H.J.M. Bowen, "Trace Elements in Biochemistry" Academic
        Press. New York (1966).

14. D.W. Cole and D.W. Johnson, Atmospheric sulfate additions
        and cation leaching in a Douglas Fir ecosystem. Water
        Resources Research 13(2):313 (1977).

15. R. Coleman, The importance of sulfur as a plant nutrient in
        world crop production. Soil Science 101(4):230 (1966).

16. L.W. Costescu, The ecological consequences of airborne
        metallic contaminants from the Sudbury smelters. Ph.D.
        Thesis. Dept. of Botany, University of Toronto. (1974).

17. R.M. Cox, Properties of some enzymes of zinc tolerant and
        non-tolerant clones of Anthoxanthum odoratum. Ph.D.
        Thesis, University of Liverpool. Liverpool, U.K.
        (1976).

18.   R.F. Daubenmire. 1974, "Plants and Environment. A Textbook
       of Plant Autecology" J. Wiley and Sons, Toronto.
       (1974).

19.   P.J. Dillon, D.S. Jefferies, W. Snyder, R. Reid, N.D. Yan,
       D. Evans, S. Moss and W.A. Scheider, Acidic precipi-
       tation in south-central Ontario: recent observations.
       Ontario Ministry of the Environment. Rexdale, Ontario.
       (1977).

20.   A. Ebregt and J.M.A.M. Boldewijn, Influence of heavy metals
       in spruce forest soil on amylase activity, CO_2
       evolution from starch, and soil respiration. $\underline{Plant\ and}$
       $\underline{Soil}$. 47:137 (1977).

21.   J.A.W. Fairfax and N.W. Lepp. The effects of a simulated acid
       precipitation on cation losses from a range of tree
       litters. $\underline{in}$: "Proceedings of the Kuopio Meeting on
       Plant Damages Caused by Air Pollution" L. Karenlampi
       ed., Kuopio, Finland, p. 123 (1976).

22.   B. Freedman, Effects of smelter pollution near Sudbury,
       Ontario, Canada on surrounding forested ecosystems.
       Ph.D. Thesis. Dept. of Botany, University of Toronto.
       (1978).

23.   C.R. Frink and G.K. Voigt, Potential effects of acid
       precipitation on soils in the north temperate region.
       $\underline{in}$: "Proceedings of the First International Symposium
       on Acid Precipitation and the Forest Ecosystem"
       U.S. Dept. of Agriculture. Forest Service. General
       Technical Report NE-23 p. 685 (1976).

24.   J.N. Galloway, G.E. Likens and E.S. Edgerton, Hydrogen ion
       speciation in the acid precipitation of the northwestern
       United States. $\underline{in}$: "Proceedings of the First Inter-
       national Symposium on Acid Precipitation and the Forest
       Ecosystem. U.S. Dept. of Agriculture. Forest Service.
       General Technical Report NE-23, p. 383 (1976).

25.   E. Gorham and A.G. Gordon, Some effects of smelter pollution
       northeast of Falconbridge, Ontario. $\underline{Can.\ J.\ Bot}$.
       38:307 (1960).

26.   J.R. Gosz, G.E. Likens and F.H. Bormann, Nutrient release
       from decomposing leaf and branch litter in the Hubbard
       Brook Forest, New Hampshire. $\underline{Ecological\ Monographs}$
       43:173 (1973).

27.   P. Grennfelt, C. Bengstar and L. Skarby,   An estimation of
         the atmospheric input of acidifying substances to a
         forest ecosystem.   This symposium.

28.   L. Hartman, Fungal flora of the soil as contaminated by
         varying concentrations of heavy metals.   Ph.D. Thesis
         University of Montana. (1976).

29.   M.E. Harward and H.M. Reisenauer, Reactions and movement
         of inorganic soil sulfur. Soil Science 101(4):326
         (1966).

30.   J.G. Horsfall, "Principles of Fungicidal Action" Chronica
         Bot. Co. Waltham, Mass. (1956).

31.   N.A. Huey, The lead dioxide estimation of sulfur dioxide
         pollution. J. Air Pollution Control Assoc. 18(9):610
         (1968).

32.   T.C. Hutchinson and L.M. Whitby, Heavy metal pollution in
         the Sudbury mining and smelting region of Canada. 1.
         Soil and vegetation contamination by nickel, copper and
         other metals. Environmental Conservation 1(2):123
         (1974).

33.   T.C. Hutchinson, J. Hellebust and M. Telford. Oil spill
         effects on vegetation and soil microfauna and Norman
         Wells and Tuktoyaktuk, N.W.T. Environmental-Social
         Committee, Northern Pipelines. Task Force on Northern
         Oil Development, Report No. 74-14. Information Canada.
         Ottawa. (1974).

34.   T.C. Hutchinson and L.M. Whitby, The effects of acid
         precipitation and heavy metal particulates on a boreal
         forest ecosystem near the Sudbury smelting region of
         Canada. in: "Proceedings of the First International
         Symposium on Acid Precipitation and the Forest Ecosystem"
         U.S. Dept. of Agriculture. Forest Service. General
         Technical Report NE-23, p. 745 (1976).

35.   V. Jensen, Decomposition of angiosperm tree leaf litter.
         in: "Biology of Plant Litter Decomposition" Vol. 1.
         C.H. Dickinson and G.J.F. Pugh, eds., Academic Press,
         New York, p. 69 (1974).

36.   H.V. Jordan and L.E. Ensminger, The role of sulfur in soil
         fertility. Advances in Agronomy 10:407 (1958).

37.   M.J. Jordan and M.P. Lechavalier, Effects of zinc smelter
        emissions on forest soil microflora. Can. J. Microbiol.
        21:1855 (1975).

38.   N.G. Juma and M.A. Tabatabai, Effects of trace elements on
        phosphatase activity in soils. Soil Sci. Soc. of Amer.
        Jour. 41(2):343 (1977).

39.   S. Kasina, Methods of wet deposition research of sulfur
        compounds in the region of Cracow. Contribution to
        Workshop Meeting to Consider Methods Involved in Studies
        of Acid Precipitation to Forest Ecosystems. Edinburgh,
        U.K. (1977).

40.   W.W. Kellog, R.D. Cadle, E.R. Allen, A.L. Lazrus and E.A.
        Mortell, The sulfur cycle. Science 175(4022):587
        1972.

41.   J.R. Kramer, Fate of atmospheric sulfur dioxide and related
        substances as indicated by chemistry of precipitation.
        Unpublished manuscript. Dept. of Geology, McMaster
        University. Hamilton. (1975).

42.   S.Y. Krupa, M.R. Coscio and F.A. Wood. Evidence for
        hydrogen ion donor systems in rain. in: "Proceedings
        of the First International Symposium on Acid Precipi-
        tation and the Forest Ecosystem" U.S. Dept. of Agri-
        culture, Forest Service, General Technical Report
        NE-23, p. 371 (1976).

43.   G.E. Likens, F.H. Bormann, R.S. Pierce, J.S. Eaton and N.M.
        Johnson. "Biogeochemistry of a Forested Ecosystem"
        Springer-Verlag, N.Y., (1977).

44.   A. Macfayden, The soil and its total metabolism.
        in: "Quantitative Soil Ecology: Population, Production,
        and Energy Flow" I.B.P. Handbook No. 18. Blackwell
        Sci. Publ., Oxford, p. 1 (1971)

45.   M.H. Martin, Effects of heavy metals on the environment
        around smelters. Presented at N.E.R.C.: Air Pollution
        Research Liaison Committee, Lancaster. U.K., Sept.
        (1977).

46.   P.C. McGovern and D. Balsillie, SO_2 levels and environmental
        studies in the Sudbury area during 1971. Air quality
        Branch, Ontario Ministry of the Environment, Toronto
        (1972).

47. P.C. McGovern and D. Balsillie, Effects of SO_2 and heavy
    metals on vegetation in the Sudbury area (1974). Air
    Quality Branch, Ontario Ministry of the Environment,
    Toronto (1975).

48. A. Medwecka-Kornas, Plant litter. in: "Methods of Study in
    Quantitative Soil Ecology: Population, Production, and
    Energy Flow" I.B.P. Handbook No. 18. Blackwell Sci.
    Publ., Oxford, p. 24 (1971).

49. C.S. Miller, Decomposition of coniferous leaf litter.
    in: "Biology of Plant Litter Decomposition" Vol. 1
    C.H. Dickinson and G.J.F. Pugh, eds., Academic Press,
    N.Y., p. 105 (1974).

50. P.J. Newbould, "Methods for Measuring the Primary
    Productivity of Forests" I.B.P. Handbook No. 2.
    Blackwell Sci. Publ., Oxford, (1967).

51. E. Nieboer, H.M. Ahmed, K.J. Puckett and D.H.S. Richardson,
    Heavy metal content of lichens in relation to distance
    from a nickel smelter in Sudbury, Ontario. Lichenologist.
    5:292 (1972).

52. S. Odén and R. Anderson, The longterm changes in the chemistry
    of soils in Scandinavia due to acid precipitation.
    Section 5.1 of Sweden's Case Study for the United Nations
    Conference on the Human Environment. Royal Ministry
    for Foreign Affairs, Royal Ministry for Agriculture,
    Stockholm, (1971).

53. L.N. Overrein, Sulfur pollution patterns observed; leaching
    of calcium in forest soil determined. Ambio. 1(4):145
    (1977).

54. D. Parkinson, T.R.G. Gray, J. Holding and H.M. Nagel-de-Boois,
    Heterotrophic microflora. in: "Methods of Study in
    Quantitative Soil Ecology: Population, Production,
    and Energy Flow" I.B.P. Handbook No. 18. Blackwell
    Sci. Publ., Oxford, p. 34 (1971).

55. Perkin-Elmer, "Analytical methods for atomic absorption
    spectrophotometry using the HGA graphite furnace"
    Perkin-Elmer, Norwalk, Conn., (1975).

56. Perkin-Elmer, "Analytical methods for atomic absorption
    spectrophotometry" Perkin-Elmer. Norwalk, Conn.
    (1976).

57.   R.R. Potvin and D. Balsillie, Air quality monitoring report
       for the Sudbury area (1975). Air Quality Branch,
       Ontario Ministry of the Environment, Toronto, (1976).

58.   I.R. Reuss, Chemical/biological relationships relevant to
       ecological effects of acid rainfall. U.S. Environmental
       Protection Agency. Pub. EPA-600/3-75-032. Corvallis,
       Oregon. (1975).

59.   T.M. Roberts, W. Gizyn and T.C. Hutchinson, Lead contamin-
       ation of air, soil, vegetation and people in the
       vicinity of secondary lead smelters. in: "Trace
       Substances in Environmental Health - VIII" D.D.
       Hemphill, ed., University of Missouri, Columbia, p. 155
       (1974).

60.   J.S. Rowe, Forest Regions of Canada. Dept. of the Environment.
       Canadian Forestry Service, Pub. No. 1300. Information
       Canada, Ottawa (1972).

61.   R.D. Rowe, A sulfur budget for sour gas plants. in:
       "Proceedings of the First International Symposium on
       Acid Precipitation and the Forest Ecosystem" U.S. Dept.
       of Agriculture. Forest Service. General Technical
       Report NE-23, p. 241 (1976).

62.   Royal Ministry for Foreign Affairs, Royal Ministry of
       Agriculture, Air Pollution across national boundaries.
       The impact on the environment of sulfur in air and
       precipitation. Kungl. Boktryckeviet P.A. Norstedt et
       Soner, Stockholm, (1971).

63.   A. Ruhling and G. Tyler, Heavy metal pollution and decomp-
       osition of spruce needle litter. Oikos 24:402 (1973).

64.   J.A. Schiff and R.C. Hodson, The metabolism of sulfate.
       Ann. Rev. Plant Physiol. 24:381 (1973).

65.   D.W. Schindler, R.W. Newbury, K.G. Beaty and P. Campbell,
       Natural water and chemical budgets for a small
       Precambrian lake in central Canada. J. Fish. Res. Bd.
       Can. 33:3526 (1976).

66.   Z. Sestak, J. Catsky and P.G. Jarvis, "Plant Photosynthetic
       Production. Manual of Methods" W. Junk Publ. The
       Hague. (1971).

67.   J. Skujins, Extracellular enzymes in soil. CRC Critical
       Reviews in Microbiology. 4(4):383 (1976).

68.   C.L. Strojan, Forest litter decomposition in the vicinity
        of a zinc smelter. Oecologia 32:203 (1978).

69.   P.W. Summers and D.M. Whelpdale. Acid precipitation in
        Canada. in: "Proceedings of the First International
        Symposium on Acid Precipitation and the Forest
        Ecosystem" U.S. Dept. of Agriculture. Forest Service.
        General Technical Report NE-21, p. 411 (1976).

70.   M.A. Tabatabai and J.M. Bremmer, A simple turbidimetric
        method of determining total sulfur in plant materials.
        Agronomy Journal 62:805 (1970).

71.   C.O. Tamm, G. Wiklander and B. Popović. Effects of
        application of sulfuric acid to poor pine forests.
        in: "Proceedings of the First International Symposium
        on Acid Precipitation and the Forest Ecosystem" U.S.
        Dept. of Agriculture, Forest Service, General Technical
        Report NE-23, p. 1011 (1976).

72.   F.W. Thomas and C.M. Davidson, Monitoring sulfur dioxide
        with lead peroxide cylinders. J. Air Poll. Cont. Assoc.
        11(1):24 (1971).

73.   F.D. Tomassini, K.J. Puckett, E. Nieboer, D.H.S. Richardson
        and B. Grace, Determination of copper, iron, nickel,
        and sulfur by X-ray fluorescence in lichens from the
        Mackenzie Valley, Northwest Territories and the Sudbury
        District, Ontario. Can. J. Bot. 54(14):1591 (1976).

74.   G. Tyler, Heavy metal pollution and soil enzymatic activity.
        Plant and Soil. 41:303 (1974).

75.   G. Tyler, Effects of heavy metal pollution on decomposition
        in forest soils. National Swedish Environment
        Protection Board, Solna, Sweden (1975).

76.   G. Tyler, Heavy metal pollution, phosphatase activity and
        mineralization of organic phosphorus in forest soils.
        Soil Biol. Biochem. 8:327 (1976).

77.   G. Tyler, Influence of vanadium on soil phosphatase activity.
        Jour. Environ. Qual. 5(2):216 (1976).

78.   A.P. Watson, Trace element impact on forest floor litter in
        the New Lead Belt region of southeastern Missouri.
        in: "Trace Substances in Environmental Health - IX"
        D.D. Hemphill, ed., University of Missouri. Columbia,
        p. 227 (1975).

79.   A.P. Watson, R.I. Van Hook, D.R. Jackson and D.E. Reichle.
        Impact of a lead mining-smelting complex on the forest
        floor litter arthropod fauna in the New Lead Belt
        Region of Missouri. Oak Ridge National Laboratory,
        Environmental Sciences Division Pub. No. 881. (1976).

80.   L.M. Whitby and T.C. Hutchinson, Heavy metal pollution in
        the Sudbury mining and smelting region of Canada.   II.
        Soil toxicity tests. Environ. Conserv. 1(3):191
        (1974).

81.   D.C. Whitehead, Soil and plant-nutrition aspects of the
        sulfur cycle. Soils and Fertilizers. 27(1):1 (1964).

82.   S.T. Williams, T. McNeilly and E.M.H. Wellington, The
        decomposition of vegetation growing on metal mine waste.
        Soil Biol. Biochem. 9:271 (1977).

83.   T. Wood and F.H. Bormann, Increases in foliar leaching caused
        by acidification of an artificial mist. Ambio
        4(4):169 (1976).

84.   T. Wood and F.H. Bormann, Short-term effects of a simulated
        acid rain upon the growth and nutrient relations of
        Pinus strobus L. in: "Proceedings of the First Inter-
        national Symposium on Acid Precipitation and the Forest
        Ecosystem" U.S. Dept. of Agriculture, Forest Service,
        General Technical Report NE-23, p. 815 (1976).

EFFECTS OF RAINFALL ACIDIFICATION ON PLANT PATHOGENS

D.S. Shriner and E.B. Cowling[2]

Environmental Sciences Division, Oak Ridge National

Laboratory[1], Oak Ridge, Tennessee 37830

INTRODUCTION

Wind-blown rain, rain splash, and films of free moisture play important roles in the epidemiology of many plant diseases. The chemical nature of the aqueous microenvironment at the infection court is a potentially significant factor in the successful dissemination, establishment, and survival of plant pathogenic microorganisms. Acidic rainfall has a potential for influencing not only the pathogen, but also the host organism, and the host-pathogen complex. Although host-pathogen interactions add a degree of complexity to the study of abiotic environmental stress of plants, it is our hope, through the use of a combination of general concepts, theoretical postulations, and experimental data, to describe the potential role that rainfall acidity may play in the often subtle balance between populations of plants and populations of plant pathogens.

The direct effects of acidic precipitation on vegetation are becoming increasingly better understood[1,2]. The indirect consequences of both acute and chronic exposure of vegetation to

---

[1]Operated by Union Carbide Corporation under contract W-7405-eng-26 with the U.S. Department of Energy. Publication No. _____, Environmental Sciences Division, ORNL.

[2]Department of Plant Pathology, North Carolina State University, Raleigh, North Carolina 27650.

acidic precipitation are very complex, however. Their effect is
variable in time, and involves a variety of potential
interactions which are only partially understood[3].

MECHANISMS BY WHICH ACIDIFICATION MIGHT
INFLUENCE THE DISEASE PROCESS

Based on a generalized disease cycle typical of many plant
diseases, we can hypothesize particular stages in that disease
cycle where the stress of acid rain might be of greatest
significance to disease development. These include:

1. The primary inoculum over winters in plant debris,
    reproduces asexually in crop residues moistened by
    spring rains. Spores are disseminated by wind-blown
    rain and water splash.

2. The spore lands on susceptible host tissue. Free
    moisture on the leaf surface is essential to spore
    germination, and external growth prior to penetration.

3. Penetration of the host occurs directly through a
    primary host barrier to penetration, the cuticle, or
    through stomatal openings, or through wounded tissues.

4. Colonization occurs in host tissues. The pathogen is
    "buffered" by host tissues, and less directly influenced
    by external environmental factors, and thus more
    dependent on host metabolism.

5. Secondary inoculum is produced in infected tissues,
    disseminated by wind-blown rain and water splash.

6. High humidity, and free moisture on leaf surfaces favors
    development of an epidemic.

7. Inoculum persists over winter in crop residues.

Certainly, this is a generalized and simplified example of a
disease cycle, yet in each of the stages involved, we see clear
potential for the chemical composition of rainwater to play a
significant role in the development of tne disease interaction.
In stages one, two, five, and six, the pathogen is in intimate
contact with a solution which may, because of its acidity, or
other chemical constituents, be inhibitory to the pathogen. In
stages three and four, the possible direct effects of rain
acidity on the host plant include: (a) advanced weathering of the
cuticular defense barrier[4,5]; (b) interference with normal

function of guard cells[6], and hence, potentially altered stomatal penetration; (c) direct injury, with potential invasion of wounded tissues[7]; and (d) potential leaching of water-soluble spore germination inhibitors from host tissues[8].

In all stages of the disease cycle, other direct and indirect effects on the host plant may dramatically influence the fitness, or disease-proneness of that host. Among these possible additional effects are included:

1. Non-symptomatic disturbance of normal carbon and nutrient metabolism, including alteration of leaf and root exudation processes, and accelerated leaching of nutrient elements from foliage.

2. Synergistic interactions with other environmental stress factors, such as gaseous air pollutants (SO_2, O_3, etc.), light and temperature.

We will focus on many of the above potential phenomena, and review the evidence to date to support or reject the hypothesis that acidic rains may significantly affect plant-parasite interactions.

CASE EXAMPLES: EVIDENCE FOR AN EFFECT

If we pursue the disease-cycle analogy, we can examine step-by-step evidence for the potential interaction of acidic rain with plant disease processes:

Inoculum Dissemination

Virtually hundreds of species of fungi and bacteria are disseminated by rain splash and wind-blown rain[9]. The inability of many bacteria to survive, or remain infective in acidic media is also well established[10]. The relative resistance of bacteria, yeast, and mold fungi to acid conditions is reported to be in the approximate ratio of 1:4:5 respectively.

Leben[12] has investigated the use of acidic buffer sprays for their potential usefulness as disease control agents. In that work, Leben demonstrated that when acidic buffer sprays were applied to tomato foliage prior to inoculation with Alternaria solani (Ell. and Mart.) Jones and Grout, reductions in incidence of early blight were 53-79 per cent. Yarwood[13] has also described therapy of certain powdery mildews due to acidic conditions on the leaf resulting from applications of sulfuric, nitric, and acetic acids. In separate experiments with acidified

media, Leben[12] reported significant growth reductions on nearly
all of 15 species of plant pathogenic fungi at pH 4.1 and severe
growth reductions on all 15 species at pH 3.2.

In experiments with simulated acidic rain, Shriner[14] found
the acidity of the water in which the inoculum was disseminated
significantly influenced the subsequent success of the bacterial
bean pathogen, Pseudomonas phaseolicola (Burkh.) Dows. Bacteria
which remained in an infection droplet of pH 3.2 for as little as
2 hours were found incapable of infecting either healthy or
acidic rain-injured bean foliage.

Host Defenses

Substantial volumes exist[15,16,17] which review the role of
the plant cuticle and epicuticular waxes as barriers to
penetration by plant pathogens. Martin and Juniper[15] and Purnell
and Preece[8] discuss the weathering of epicuticular waxes on leaf
surfaces by rain. Purnell and Preece concluded that little of
the wax removed by rain from mature, fully expanded leaves is
regenerated. Other data[5] suggest that at least for one species,
willow oak, the rate at which the weathering of the leaf surface
progresses may also be a function of the acidity of the rainfall
impacting that surface.

The significance of the weathering of the leaf surface in
terms of plant disease development is several-fold:

(1)   Weathered surfaces may actually pose a less formidable
      mechanical barrier to direct penetration by pathogens.

(2)   Weathered surfaces increase in wettability[15]. This
      increase in wettability may subsequently result in
      significantly greater numbers of water-borne propagules
      being retained on the leaf surface in a position to
      effect successful penetration[15].

(3)   Weathered leaf surfaces have been shown to have
      increased retention of certain fungicides[15], and other
      particulates[18]. (Weathering also makes a wide range of
      plant cuticles more permeable to weed killers[15].)

A different type of effect on host defense barriers
potentially induced by acid rains is the creation of infection
courts on the leaf surface as the result of direct injury to leaf
tissues. Such direct, visible injury may occur from exposure to
rainfall acidity at pH 3.4 or below[2]. Shriner[14] found disease
severity increased when beans inoculated with P. phaseolicola had

been predisposed by exposure to simulated rains of pH 3.2.
Numerous small necrotic lesions were thought to have provided the
pathogen with increased opportunity for penetration of the host
tissues.

For those organisms which are facultative parasites, the
added stress of acid precipitation would appear to provide the
necessary advantage to the pathogen to result in successful
penetration and colonization. From other data[14], however,
obligate parasites (whose fortunes are more directly dependent
upon those of the host) appear to be inhibited in their disease
development by the stress of acidic rain on the host plant. Such
an effect may have been the indirect result of imbalanced host
metabolism.

The Disease Interaction

Several lines of evidence have been developed in recent
years for the indirect effect of gaseous air pollutants on plant
disease development as a result of effects on host plants[19].
Changes in the chemical composition of foliage, e.g., sulfur content,
pollutant-induced production of phytoalexins, and altered supplies
of metabolites are suggested mechanisms by which the growth and
development of the pathogen within the host may be affected[20,21,22].

No direct evidence of similar effects has been reported for
acidic rain/host-parasite interactions. However, observations of
Shriner[14,23] of effects on root-knot nematode infection of
Phaseolus vulgaris 'Red Kidney' and Rhizobium nodulation of the
same species by simulated acid rain could be accounted for by
alterations in host physiology induced by simulated rain of pH
3.2.

Other Factors Influencing Interaction
of Acid Rain with Plant Disease Development

Several researchers have suggested the importance of dry
particulate deposition on foliar surfaces in relation to disease
incidence. Cement kiln dust deposits have been reported to
increase the incidence of a fungus leaf-spot on sugar-beet
leaves[24], and limestone processing dust appeared to stimulate
leaf spot infections on wild grape and sassafras leaves.

Smith and Dochinger[25] have reported that leaf surfaces of
trees lining urban streets contained greater than normal burdens
of certain particle-borne metal contaminants and that these
particles were effectively retained by the leaf surfaces. Other

works by Lindberg et al.[26] have shown similar retention of fly ash particles by leaf surfaces in a more remote forest canopy.

The potential would appear to be great for particles lodged on leaf surfaces to interact chemically with either ambient gaseous pollutants, or precipitation. Such particles could effectively increase the surface area of leaves available for reaction with gaseous pollutants, and, depending upon their source, either neutralize, or further acidify rain water contacting the leaf surface. While the ultimate effects of such particle-leaf interaction on plant disease processes are currently unknown, such interactions certainly occur in nature, and may be important in the final response of a plant to complex mixtures of air pollutants and plant pathogens.

SUMMARY

During the process of dispersion, deposition, and initial growth on the leaf surface, plant pathogenic fungal spores and bacteria are quite vulnerable, and often at the most tenuous phase of their life history. The data reviewed suggest that acid rain in the leaf surface microenvironment can be a critical factor in disease establishment. Understanding the positive and negative potentials of these interaction phenomena in field situations remains a challenge to our ultimate understanding of the effects of all forms of biotic and abiotic stress on plant growth and yield.

REFERENCES

1. J. N. Galloway, and E. B. Cowling. J. Air Poll. Cont. Assoc. 28:229 (1978).

2. J. S. Jacobson, and P. Van Leuken. Proc. Fourth Inter. Clean Air Congress, Tokyo, May 16-20 p.124 (1977).

3. C. O. Tamm, and E. B. Cowling, Water, Air and Soil Poll. 7:503 (1977).

4. D. S. Shriner, Ph.D. Thesis, N.C. State Univ., Raleigh, N.C., 79 p. (1974).

5. D. S. Lang, D. S. Shriner, and S. V. Krupa, Proc. 71st Ann. Mtg. APCA, Houston, Paper #78-7.3 (1978).

6. L. S. Evans, N. F. Gmur, and F. DaCosta, Amer. J. Bot. 64:903 (1977).

7. D. S. Shriner, M. E. DeCot, and E. B. Cowling, Proc. Am.
    Phytopath Soc. 1:112 (1974).

8. T. J. Purnell, and T. F. Preece, Physiological Plant Pathology
    1:123 (1971).

9. J. C. Walker, "Plant Pathology", 2nd Ed., McGraw-Hill,
    New York (1957).

10. W. W. Umbriet, "Modern Microbiology", W. H. Freeman and Co.,
    San Francisco (1962).

11. P. J. W. Saunders, in: "Ecology of Leaf Surface Micro-
    organisms", Academic Press, New York (1971).

12. C. Leben, Phytopathology 44:101 (1954).

13. C. E. Yarwood, Proc. 2nd Internat. Congr. Crop Protection,
    22 p. (1951).

14. D. S. Shriner, Phytopathology 68:213 (1978).

15. J. T. Martin, and B. E. Juniper, "The Cuticles of Plants",
    St. Martin's Press, New York, (1970).

16. T. F. Preece, and C. H. Dickson, "Ecology of Leaf Surface
    Micro-organisms", Academic Press, New York (1971).

17. C. H. Dickinson, and T. F. Preece, "Microbiology of Aerial
    Plant Surfaces", Academic Press, New York (1976).

18. W. H. Smith, in: "Microbiology of Aerial Plant Surfaces",
    Academic Press, New York (1976).

19. A. S. Heagle, Ann. Rev. Phytopath. 11:365 (1973).

20. L. H. Weinstein, D. C. McCune, A. L. Alvisio, and P. Van Leuken,
    Environ. Pollut. 9:145 (1975).

21. W. J. Manning, W. A. Feder, P. M. Papia, and I. Perkins,
    Environ. Pollut. 1:305 (1971).

22. D. T. Tingey, and U. Blum, J. Env. Qual. 2:341 (1973).

23. D. S. Shriner, Env. Exp. Bot. (in press).

24. W. J. Manning, Environ. Pollut. 9:87 (1975).

25. W. H. Smith, and L. S. Dochinger, in: "Better Trees for
    Metropolitan Landscapes", Washington, D.C. (1975).

26.   S. E. Lindberg, D. S. Shriner, R. R. Turner, and L. K. Mann,
        <u>in</u>: ORNL-5257 (1976).

VARIATIONS IN PRECIPITATION AND STREAMWATER CHEMISTRY

AT THE HUBBARD BROOK EXPERIMENTAL FOREST DURING 1964 TO 1977

Gene E. Likens[1], F. Herbert Bormann[2], John S. Eaton[1]

Section of Ecology and Systematics[1],
Division of Biological Sciences,
Cornell University, Ithaca, New York 14853

School of Forestry and Environmental Studies[2],
Yale University
New Haven, Connecticut 06511

INTRODUCTION

In 1975 at the First International Symposium on Acid
Precipitation and the Forest Ecosystem in Ohio, we summarized the
changes relative to sulfur, nitrogen and hydrogen ion input in
precipitation during 1964 to 1974 for the Hubbard Brook Experimental
Forest in New Hampshire (Likens et al.[1]). In some ways this paper
is an update of those data.

The Hubbard Brook Ecosystem Study began late in 1962 and data
on chemistry of precipitation have been collected since 1963. This
continuous record of precipitation chemistry is the longest in the
United States. However, even with this seemingly extensive record
it is very difficult to analyze trends in precipitation chemistry.
Indeed, data collected over a 2 or 3, or even 5-yr period may be
very misleading as to conclusions about the overall trend. As new
points are added to the record we find that the original conclusions
may or may not hold. A major question is — Do these changes
represent variations on a long-term theme, or have new trends
developed?

The Hubbard Brook Experimental Forest is located in the White
Mountain National Forest in northern New Hampshire. This area is
particularly suited for monitoring changes in regional precipitation
chemistry as it is remote from urban and industrial centers, yet
is generally downwind from them.

HYDROLOGIC FLUX

Since the inputs and outputs of water directly affect the chemical inputs and outputs for an ecosystem, the flux of the water must be known quantitatively. Precipitation averages about 130 cm/unit area-yr at Hubbard Brook. Of this about 62% is lost as streamflow and the remainder is lost as vapor via evapotranspiration (Likens et al.[2]). Average monthly precipitation is fairly constant throughout the year at Hubbard Brook, however, the monthly streamflow pattern is quite different (Figure 1). Streamflow during the summer months is very low, primarily because of evapotranspirational water losses from the forest, whereas during the spring streamflow is very high as the snowpack melts. About 30% of the total annual streamflow occurs during the month of April alone; some 54% of the annual streamflow occurs during the months of March, April and May (Likens et al.[2]). In contrast only about 13% of the annual streamflow occurs from June through September.

TRENDS IN PRECIPITATION CHEMISTRY

Long-term precipitation chemistry at Hubbard Brook is dominated by hydrogen ion and sulfate on an equivalency basis (Table 1). Ammonium and nitrate are the second-most abundant cation and anion, respectively. In contrast, calcium and sulfate dominate the stream water draining from the ecosystem and a number of other chemical changes occur as the precipitation flows through the ecosystem. For example, nitrogen and phosphorus are accumulated within the system, and concentrations in stream water are generally less than those in precipitation. Concentrations of calcium, sodium, potassium, magnesium and aluminum are increased in stream water as these chemicals are generated from weathering of minerals within the system (Likens et al.[2]).

The mean annual pH of precipitation from 1964 to the present has been variable but shows no statistically consistent trend (Figure 2). These data also illustrate that short-term records can be very misleading. Data from 1968-69 to 1970-71 would have given the opposite conclusion about trends than would data from 1970-71 to 1973-74. Neither reflect the long-term trend accurately.

In contrast to the trend in annual pH, annual sulfate concentrations in precipitation have tended to decrease from 1964 to 1977 (Figure 3). The negative slope (0.07 mg/liter-yr) of a regression line fitted to these data is very highly significant with a correlation coefficient of 0.71. The slope of the regression line, however, was not statistically significant after the first 10 years (Likens et al.[1]). There have been a number of changes in the control of sulfur emissions to the atmosphere since about 1969-70

Table 1.   Annual mean concentrations (weighted for volume) of
           dissolved substances in bulk precipitation and stream
           water for watersheds 1 - 6 of the Hubbard Brook Experi-
           mental Forest during 1963 to 1977.

SUBSTANCE	PRECIPITATION		STREAM WATER	
	mg/l	µeq/l	mg/l	µeq/l
H^+	$0.070^{a,b}$	69.8	$0.011^{a,c}$	11.3
NH_4^+	0.19	11.0	0.03^b	1.83
Ca^{++}	0.14	7.18	1.61	80.3
Na^+	0.11	4.74	0.84	36.4
Mg^{++}	0.04	3.12	0.36	29.9
K^+	0.06	1.51	0.22	5.62
Al^{+++}	—^{d}	—	0.21^e	23.3
$SO_4^=$	2.74^b	57.0	6.25^b	130
NO_3^-	1.46^b	23.6	2.14^b	34.5
Cl^-	0.42^c	11.9	0.54^c	15.2
$PO_4^\equiv$	0.008^f	0.268	0.0022^g	0.10
HCO_3^-	~0.006^h	0.098	0.92^i	15.1
Dissolved silica	0.02^f	—	4.50^b	—
Dissolved organic carbon	2.4^j	—	1.0^k	—
pH	4.15		4.96	
TOTAL	7.66	(+) 97.3 (−) 92.9	18.6	(+) 189 (−) 195

[a] Calculated from weekly measurements of pH
[b] 1964 - 1977
[c] 1965 - 1977
[d] Not determined, trace quantities
[e] 1964 - 1970, and 1976-77 for W6 only
[f] 1972 - 1977
[g] 1967 - 1968, 1972 - 1977
[h] Calculated from H^+ − HCO_3^- Equilibrium
[i] Watershed 4 only, 1965 - 1970
[j] 1973 - 1974 (Jordan and Likens[3])
[k] Watershed 6 only, 1967 - 1969

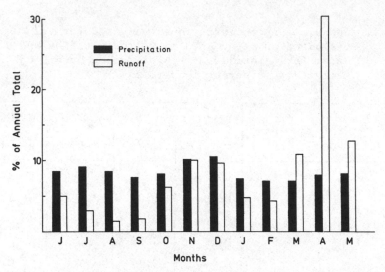

Figure 1.   Monthly precipitation and streamflow as percent of annual total for the Hubbard Brook Experimental Forest during 1956 to 1974.

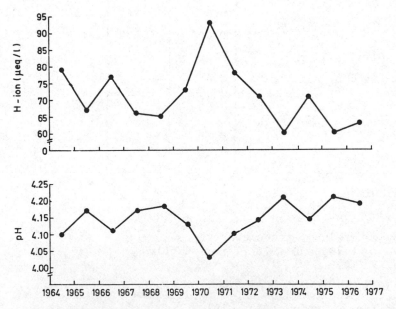

Figure 2.   Annual hydrogen ion concentration and pH of precipitation (weighted for volume) at the Hubbard Brook Experimental Forest from 1964 to 1977.

in the eastern United States, but our data are not adequate to
indicate how these changes are related to the sulfate concentrations
at the remote Hubbard Brook site.

The annual nitrate concentrations in precipitation increased
rather markedly after 1964 [a value of about 0.8 mg NO_3/liter can
be extrapolated from isopleth maps given by Junge[4], for 1955-56],
but have not increased much since about 1970-71 (Figure 3).   The
next several years of record will be very informative in determining
the trend for concentration of nitrate in precipitation at Hubbard
Brook.

Data from other locations in the eastern United States show
clearly that nitrate concentrations increased after about 1945
(Figure 4).   Data from the Cornell University Agriculture Experiment
Stations at Ithaca and Geneva, New York show that the ammonium
concentrations were neither equal to or greater than the nitrate
concentrations from about 1915 to 1945, but after 1945, although
the ammonium concentrations were relatively unchanged, the nitrate
concentration increased significantly (Likens[5]).   The ammonium
concentration has changed very little during the period of our study
at Hubbard Brook (Figure 3).

AVERAGE MONTHLY CONCENTRATIONS IN PRECIPITATION

Average monthly concentrations of hydrogen ion, sulfate,
nitrate and ammonium show variable patterns on a long-term basis
(Figure 5).   Typically, summer rains are much more acid than winter
snows.   Also concentrations of sulfate and ammonium are lowest in
winter and highest in summer; the seasonal pattern for hydrogen
ion, sulfate and ammonium are strongly correlated.   Concentrations
of nitrate tend to be relatively constant on a monthly basis
(somewhat higher in spring), and are not correlated with
concentrations of hydrogen ion, sulfate or ammonium.

ANNUAL INPUTS IN PRECIPITATION

Input (concentration times volume of water) of hydrogen ion to
the Hubbard Brook Experimental Forest has changed appreciably during
the past 13 years (1964 to 1977).   In Figure 6 the solid line in
the upper graph is a regression for all of the data for 13 years
and its slope is not statistically different from zero.   However,
it appears that there may have been two trends in terms of hydrogen
ion input.   That is, during the first 10 years there was an increase
in the input of hydrogen ion, but during the last few years there
may have been a leveling off or decrease in input (Figure 6).   The
positive slope of the regression line for the first 10 years (1964
to 1974) is significant (Likens et al.[1]).   The slope of the regression

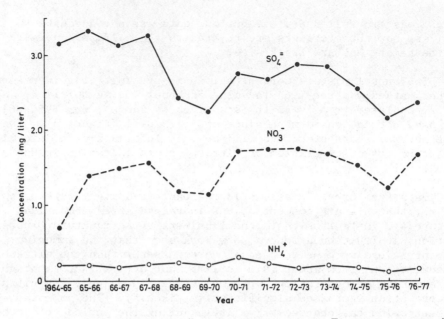

Figure 3.   Annual weighted concentrations of $SO_4^=$, NO_3^- and NH_4^+
in precipitation at the Hubbard Brook Experimental Forest from 1964
to 1977.

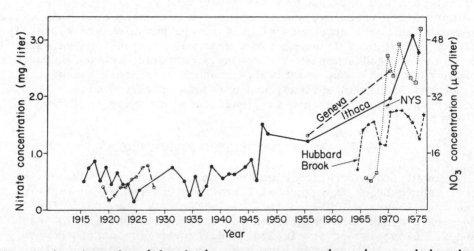

Figure 4.   Annual weighted nitrate concentrations in precipitation
at various locations in the eastern United States.   Data from Ithaca
and Geneva, New York from Likens (1972).   Data from New York State
(NYS) is the average of 9 stations operated by the U.S. Geological
Survey in New York and Pennsylvania.

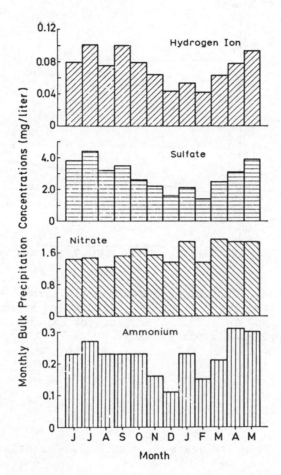

Figure 5. Average monthly concentrations of hydrogen ion, sulfate, nitrate and ammonium in precipitation at the Hubbard Brook Experimental Forest during 1964 to 1977.

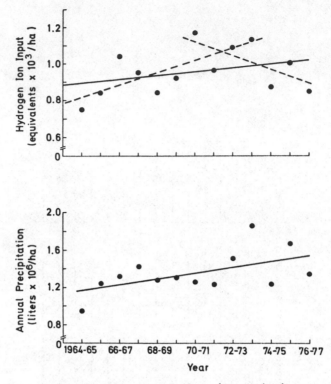

Figure 6.   Annual input of hydrogen ion (in kg/ha) and annual precipitation (in liters/unit area) for the Hubbard Brook Experimental
Forest during 1964 to 1977. (See text for explanation of regression
lines on hydrogen ion input graph.)

line for the last seven years is negative and statistically
significant.

The annual input of hydrogen ion obviously is related to the
amount of precipitation (Figures 6 and 7) because the input is
calculated by multiplying the concentration times amount of water.
However, there is appreciable scatter of points around the
regression line; correlation coefficient of 0.63 (Figure 7). It is
noteworthy that the annual hydrogen ion concentration was
relatively constant even though the amount of precipitation varied
by two-fold during 1964 to 1977 (Figure 8).

Indeed it would appear that there has been an increase in the

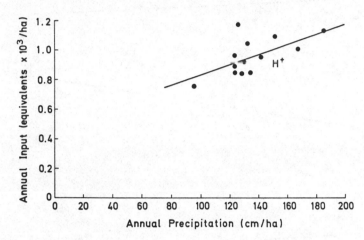

Figure 7. Relationship between the annual hydrogen ion input, and
the annual precipitation input for the Hubbard Brook Experimental
Forest during 1964 to 1977.

amount of precipitation during our period of study (Figure 6). In
fact 1964-65, at the beginning of our study, was the driest year
on record and 1973-74 was the wettest (Likens et al.[2]). However,
if a longer record of precipitation is considered (Likens et al.[2]),
no case can be made for a trend of increasing amount of
precipitation. The apparent trend (Figure 6) is an artifact of the
improbable occurrence of an extremely dry year at the beginning and
of a very wet year at the end of the period of observation. Again
this points up the problem of generating misleading conclusions
from short-term records.

The relationship between annual sulfate input and hydrogen
ion input (Figure 9) is largely dependent upon the data for the
driest year and the wettest year. Without those two years the
relationship is not very convincing; in fact, it was not
statistically significant after 10 years (Likens et al.[1]).

There is a much stronger correlation (r = 0.75) between annual
nitrate input and hydrogen ion input (Figure 10). The nearly 1:1

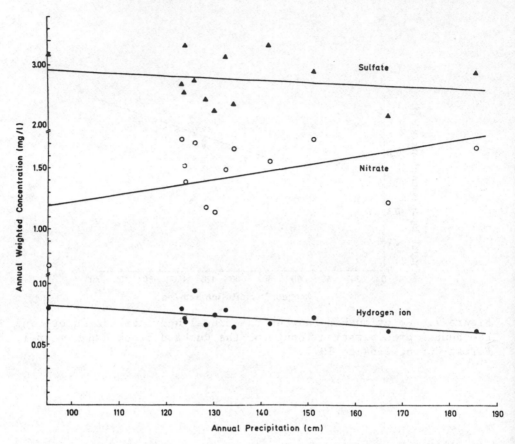

Figure 8. Relationship between the annual weighted concentration of hydrogen ion, nitrate and sulfate and amount of annual precipitation during 1964 to 1977.

relationship between inputs of hydrogen ion and nitrate strongly suggests that nitric acid is a crucial component in recent changes in acidity of precipitation at Hubbard Brook.

The annual inputs of sulfate and nitrate are directly correlated with the amount of precipitation (Figure 11). However, the relationship for nitrate is much less variable than the relationship for sulfate. This relationship adds support to the hypothesis that recent changes in the inputs of hydrogen ion are largely dependent on changes in the inputs of nitrate, even though sulfuric acid is the dominant acid in precipitation at Hubbard Brook (Table 1, Figure 12). The apparent direct relationship between the annual amount of precipitation and the annual volume weighted concentration

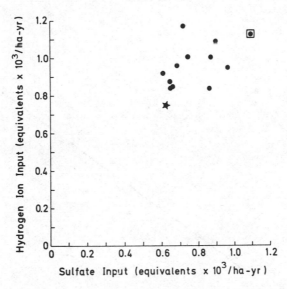

Figure 9.   Relationship between the annual hydrogen ion input and the annual sulfate input during 1964 to 1977 for the Hubbard Brook Experimental Forest.   ★ = driest year; ◉ = wettest year.

of nitrate (Figure 8) is of much interest because of the multiplying effect on inputs, but more data are required to validate this relationship.

From 1964 to 1977 there has been a general downward trend in the sulfate contribution to the total anion equivalents (Figure 12). Since about 1970 the decline has been somewhat less or reversed. The proportior of nitrate to the total has more or less increased throughout the period.   Two conclusions can be drawn: (1) There has been an increasing importance of nitric acid in precipitation at Hubbard Brook as we reported earlier (Likens et al.[1]), and (2) the change has been somewhat less after about 1970 and is probably related to the slower increase in nitrate concentration in precipitation after 1970 (Figure 3).   The proportion of hydrogen ion to the total cations has increased throughout the period

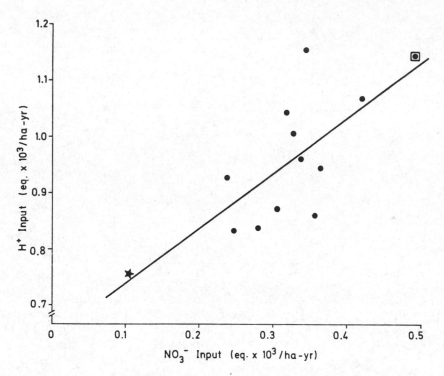

Figure 10. Relationship between the annual hydrogen ion input and
the annual nitrate input during 1964 to 1977 for the Hubbard Brook
Experimental Forest. ★ = driest year; ◉ = wettest year.

(Figure 12), but this is primarily a function of the decrease in
total cation equivalents of which hydrogen ion is the major
component.

INPUT—OUTPUT FLUX

The forested ecosystem at Hubbard Brook is very effective at
neutralizing the inputs of acid precipitation (Figure 13). Annual
inputs of hydrogen ion in precipitation exceed outputs in drainage
waters by about an order of magnitude (Likens et al.[2]). The annual

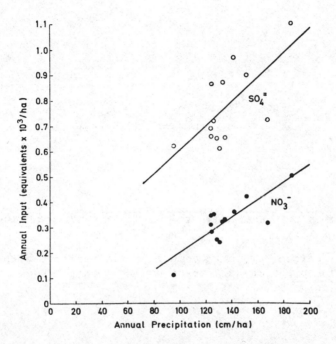

Figure 11.   Relationship between the annual input of sulfate or
nitrate and the annual amount of precipitation during 1964 to 1977.

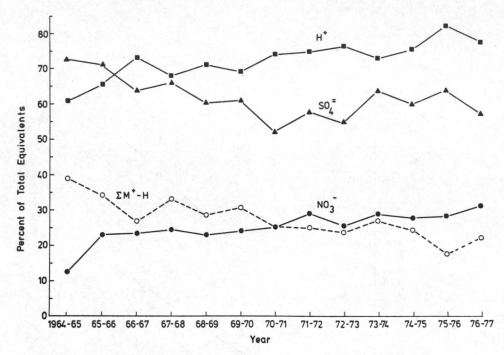

Figure 12. Percent of ionic composition of precipitation for the Hubbard Brook Experimental Forest during 1964 to 1977. ΣM^+ is sum of all cations.

input of sulfur in bulk precipitation was 37.1 kg/ha and the annual output in drainage water was 54.6 kg/ha during the period 1964 to 1977. Total inputs were somewhat greater (cf. Eaton et al.[6]).

The monthly flux of sulfate at Hubbard Brook shows an interesting pattern (Figure 14). The monthly precipitation input exceeds the streamwater output during the summer months and then during all the rest of the months the output exceeds the input. The two "cross-overs" coincide with the period of activity/inactivity of the deciduous vegetation (about the first of October as the leaves fall; and about the middle of May as the buds are breaking in the spring). Thus there appears to be a very strong biological

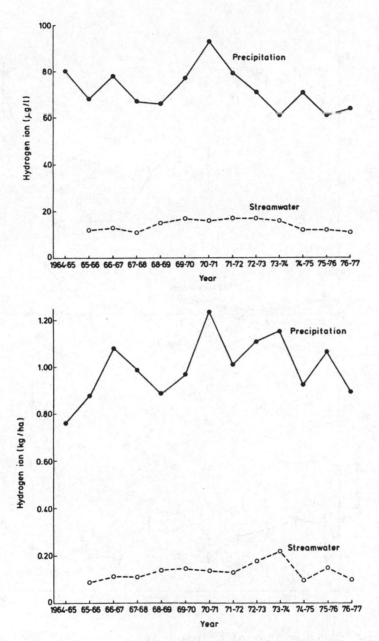

Figure 13. Annual concentrations (μg/litter) and input and output (kg/ha) of hydrogen ion for Watershed 6 of the Hubbard Brook Experimental Forest during 1964 to 1977.

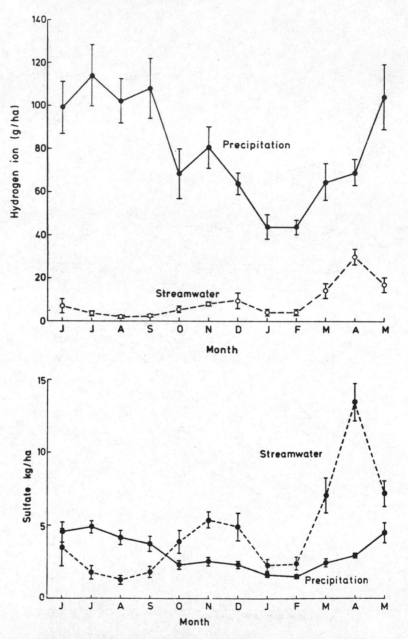

Figure 14. Monthly flux of hydrogen ion and sulfate for forested watershed ecosystems of the Hubbard Brook Experimental Forest. Mean values + one standard deviation of the mean for the period 1964 to 1977. Solid line is input in precipitation and dashed line is output in stream water.

component to the sulfur flux for forested ecosystems at Hubbard Brook.

Hydrogen ion has a very different pattern, i.e. precipitation inputs exceed streamwater outputs by a large amount during each month of the year (Figure 14). The output of hydrogen ion during March, April and May is increased as the volume of runoff is increased from melting of the snowpack. Long-term, average monthly concentrations are only slightly increased during this period (Figure 15), but initial runoff from the melting snowpack may be appreciably more acid for one to a few days. Nevertheless, the forested ecosystem at Hubbard Brook is very effective at neutralizing the precipitation input of strong acids during each month (Figure 15). The hydrogen ion concentration in rainfall during the summer averages about 86 μEq/liter; summer rainfall is much more acidic than winter-time precipitation (47 μEq/liter). Nevertheless, the stream water consistently has a hydrogen ion concentration of about 9 to 14 μEq/liter throughout the year (Figure 15).

The forested ecosystem at Hubbard Brook thus moderates the impact of acid precipitation as judged by streamwater outputs. The system also is very effective at filtering out a variety of heavy metals such as lead (Siccama and Smith[7]). The question is — How long can the ecosystem continue to be an effective filter for toxic substances before filtration capabilities are weakened and/or destroyed? When the natural system no longer serves as an efficient filter, maintenance of water quality may depend on substitution of energy-rich technological filters (Bormann and Likens[8]).

For example, the ecosystem functions quite differently when it is disturbed. In response to deforestation at Hubbard Brook, concentrations of most of the dissolved chemicals in stream water increased dramatically, except for sulfate which decreased (Likens et al.[9]). This has been the pattern in all the clear cutting operations that we have observed in northern New England, commercial clear cutting, strip cutting and deforestation (Martin et al.[10]). We believe that part of the explanation for this decrease in stream-water concentrations of sulfur is the removal of impaction surfaces provided by the forest canopy. Thus, as the biotic portion of the ecosystem is stressed (or destroyed) regulation of quality and quantity of runoff may be greatly impaired. However, this temporary loss of biotic regulation of ecosystem biogeochemistry sometimes may be viewed as a feedback mechanism promoting rapid biotic recovery of the system (Likens et al.[11]; Bormann and Likens[12]).

ACID PRECIPITATION

Acid precipitation (i.e. pH < 5.6) currently occurs over a wide area of the eastern United States (Figure 16). In 1955 and 1956

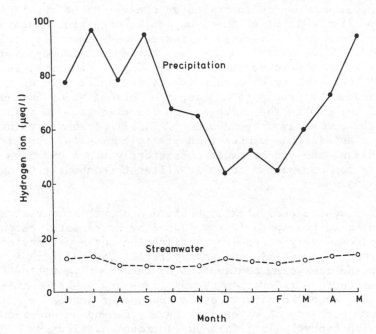

Figure 15. Weighted monthly concentration of hydrogen ion for precipitation and streamwater for Watershed 6 of the Hubbard Brook Experimental Forest during 1965 to 1977.

large areas, particularly the northeastern United States, were subjected to acid precipitation. The lowest annual pH values were about 4.4. By 1972-73 rather marked changes had occurred. For example, the 5.6 isoline no longer appeared on the map for the United States east of the Mississippi, and based on these data (Figure 16) the concentrations of hydrogen ion in the northeastern United States had increased appreciably. Annual weighted averages of about 4.1 appeared at several localities. Not many of the locations are exactly the same in 1972-73 as in 1955-56, and there

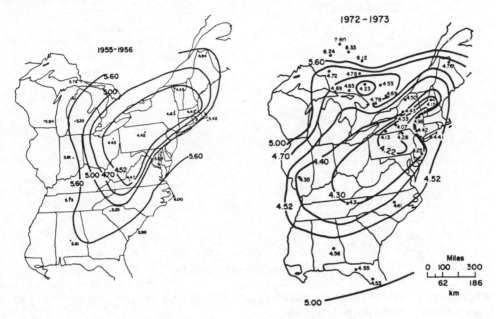

Figure 16. Average pH of annual precipitation in the eastern United States during 1955-56 and 1972-73 (modified from Likens 1976).

are large areas with no data, but there are very few points that do not fit the general isoline distributions.

One final point: The acidity of precipitation may have increased sharply in the northeastern United States during the mid 1950's (Likens[13]); whereas more recent changes, if any, have been much smaller (Figure 16). However, these changes are usually reported in pH units. Since pH is expressed on a logarithmic scale, it should not be forgotten that relatively few hydrogen ions are required to change the pH 5.6 to 4.6 in a poorly buffered solution like precipitation (Figure 17), whereas relatively large quantities of hydrogen ions are necessary to reduce the pH from 4.6 to 3.6 (values

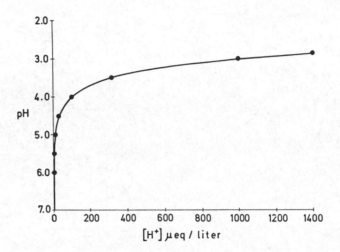

Figure 17.   Relationship between pH and hydrogen ion concentration
for dilute aqueous solutions.

frequently recorded in present-day precipitation). Thus if atmospheric sources of acidity were increasing slowly, continued declines
in average annual pH values of precipitation would be progressively
much less pronounced as the pH value decreased. Small decreases in
the pH of precipitation from current levels might lull us into
complacency about the acid precipitation problem. However, underlying these fractional decreases in pH are relatively large
increases in hydrogen ion concentration and inputs to natural
ecosystems. Last August we observed the most acidic rainstorm on
record at Hubbard Brook. A 0.1-cm storm on 24 August 1977 had a
pH of 2.85, or more than 1400 μEq/liter of hydrogen ions!

REFERENCES

1. G. E. Likens, F. H. Bormann, J. S. Eaton, R. S. Pierce, and
    N. M. Johnson, Hydrogen ion input to the Hubbard Brook
    Experimental Forest, New Hampshire during the last
    decade, in: "Proc. The First Internat. Symp. on Acid
    Precipitation and the Forest Ecosystem", L. S. Dochinger
    and T. A. Seliga, eds. USDA Forest Service General
    Tech. Report NE-23, p.397 (1976); Also, Water, Air, and
    Soil Pollution 6:435.

2. G. E. Likens, F. H. Bormann, R. S. Pierce, J. S. Eaton and
    N. M. Johnson, "Biogeochemistry of a Forested Ecosystem"
    Springer-Verlag New York Inc., 146 pp. (1977).

3. M. J. Jordan, and G. E. Likens, An organic carbon budget for
    an oligotrophic lake in New Hampshire, U.S.A.,
    Verh. Internat. Verein. Limnol. 19(2):994 (1975).

4. C. E. Junge, The distribution of ammonia and nitrate in rain
    water over the United States, Trans. Amer. Geophys. Union
    39:241 (1958).

5. G. E. Likens, The chemistry of precipitation in the central
    Finger Lakes Region. Water Resources and Marine Sciences
    Center Tech. Report 50, Cornell University, Ithaca,
    New York, 61 pp. (1972).

6. J. S. Eaton, G. E. Likens, and F. H. Bormann, Wet and dry
    deposition of sulfur at Hubbard Brook. Proceedings from
    the NATO Advanced Research Institute on Effects of Acid
    Precipitation on Terrestrial Ecosystems, Toronto, Ontario
    (1978).

7. T. G. Siccama, and W. H. Smith, Lead accumulation in a northern
    hardwood forest, Environ. Sci. & Tech. (1978).

8. F. H. Bormann, and G. E. Likens, The fresh air-clean water
    exchange, Natural History 86(9):62 (1977).

9. G. E. Likens, F. H. Bormann, N. M. Johnson, D. W. Fisher, and
    R. S. Pierce, Effects of forest cutting and herbicide
    treatment on nutrient budgets in the Hubbard Brook
    watershed-ecosystem, Ecol. Monogr. 40(1):23 (1970).

10. C. W. Martin, R. S. Pierce, G. E. Likens, and F. H. Bormann,
    Commercial clearcutting affects nutrient cycles and
    stream chemistry in the White Mountains of New Hampshire.
    Submitted for publication (1978).

11. G. E. Likens, F. H. Bormann, R. S. Pierce, and W. A. Reiners,
     Recovery of a deforested ecosystem, Science 199(4328):492
     (1978).

12. F. H. Bormann, and G. E. Likens, "Pattern and Process of a
     Forested Ecosystem", Springer-Verlag, New York Inc. (1979).

13. G. E. Likens, Acid precipitation, Chemical and Engineering News
     54:29 (1976).

EFFECTS OF ACID DEPOSITION UPON OUTPUTS FROM TERRESTRIAL TO AQUATIC ECOSYSTEMS

Eville Gorham[1] and William W. McFee[2]

Department of Ecology & **Behavioral** Biology[1], University
of Minnesota, Minneapolis, Minn. 55455
Department of Agronomy[2], Purdue University, West
Lafayette, **Ind.** 47907

INTRODUCTION

Linkages among ecosystems are receiving much attention
currently (Likens and Bormann[1]; Hasler,[2]) and are especially
important for aquatic ecosystems, which serve as receptors for the
diverse outputs of terrestrial ecosystems and are strongly affected
by them. The influence of acid precipitation upon terrestrial
outputs may be direct, where elements deposited from the atmosphere
to the land pass over or through the soil to streams and lakes;
or it may be indirect, where elements added in precipitation
accelerate or retard the processes of soil weathering, leaching
and organic decomposition.

COMPONENTS OF ACID PRECIPITATION

The increasing acidification of streams and lakes in northwestern Europe and northeastern North America is a combined effect
of acid precipitation falling upon the lakes and upon their
drainage basins. In southern Scandinavia only a few lakes
measured in the period 1920-1940 had pH levels below 5.5, whereas
in 1974, about 40% of 155 lakes in southern Norway were below
that level (Wright and Henriksen,[3]). In both Norway and
Sweden, declines in pH of 0.02-0.05 units/year have been recorded
over recent decades (Gjessing et al.,[4]; Wright and Gjessing,[5].
In the Adirondack mountains of the northwestern United
States, 4% of the mountain lakes about 600 m elevation had pH
values below 5.0 in the period before 1940, whereas in the mid-
1970's 51% were below that value (Schofield,[6]). The average
pH decline was 0.05 units/year (Wright and Gjessing,[5]). The

balance between acid deposition directly on the lake surface and
acid deposition transferred from the land has not been assessed.
It will be determined not only by the ratio of drainage area to
lake area, but also by the absorption of sulfur and nitrogen
oxides and the impaction of acid droplets and particles upon
terrestrial vegetation, and by the degree of neutralization
occurring in the canopy and in the soil (Nihlgard,[7]; Eaton,
Likens and Bormann,[8]).

Acid precipitation contains a variety of heavy metals,
particularly in the vicinity of metal smelters (Gorham,[9])
and these too may be deposited directly in lakes or be transferred to them from the land. Concentrations of Pb and Zn in
Norwegian lakes are distinctly higher in the south than in the
central and northern parts of the country (Henriksen and Wright,[10]
presumably owing to air pollution from industrial Europe.

In Lake Constance, Cd, Hg, Pb and Zn are enriched 3- to 4-
fold in surface sediments, again most likely because of air
pollution from coal combustion (Müller, Grimmer and Böhnke,[11]).
In lake sediments in the northeastern U.S.A., Ag, Au, Cr, Ni,
Pb, Sb and V exhibit increased rates of deposition in recent
times, presumably for the same reason (Galloway,[12]; Galloway
and Likens,[13]). In many instances the balance between direct
fallout on the lakes and transfer from the land is uncertain,
because of unmeasured impaction and the possibility of enhanced acid
leaching of heavy metals from terrestrial soils. However, in the
case investigated by Galloway,[12] and Galloway and Likens,[13]
atmospheric deposition appears to be the predominant cause of
recent enrichment in heavy metals, because other metals (Al, Fe,
Ca, K, Mn) which might be expected to leach from acidified soils
are not enriched in the same way.

Acid precipitation also contains a variety of plant nutrients
such as Ca, K, S, N (Gorham,[14]) and possibly P (Gorham,[9]).
These likewise may be transferred from land to water, but in
addition most are likely to be enriched by leaching from
terrestrial soils upon which acids have been deposited.

Finally, acid precipitation contains a wide range of organic
molecules, including alkanes, polycyclic aromatic hydrocarbons,
phthalic acid esters, and fatty acid ethylesters, as well as a
variety of commonly used industrial chemicals such as polychlor-
inated biphenyls. These are reported from Norway in concentrations
of one to a few hundred ng/l (Lunde et al.,[15]). How much
passes from land to water is unknown, but Müller, Grimmer and
Böhnke," record enrichments of up to 50-fold for several
polycyclic aromatic hydrocarbons in the recent sediments of Lake
Constance.

In all the above-mentioned instances, there is an urgent need for mass-balance studies to determine the degree to which the acids, heavy metals, nutrients and organic molecules reaching aquatic ecosystems have originated from direct precipitation upon the lake surface, transfer from atmospheric deposition upon the land, or -- in some cases -- enhanced soil leaching due to acid precipitation. The most difficult requirement in determining chemical budgets is the accurate measurement of absorption and impaction of elements and molecules upon the vegetation canopy, and their subsequent transfer. In determining mass balances of some elements (N, S), gaseous emissions from soil may also be very difficult to assess. The problems are well illustrated by studies of the Hubbard Brook Experimental Forest (Likens, et al.,[16]).

An attempt to compare deposition with absorption plus impaction of hydrogen ions in a German beech forest (Mayer and Ulrich,[17] suggests that they are approximately equal (about) 1 kequiv/ha/yr) and that about 40% of the total is buffered by the canopy. Litter decomposition produces a further substantial input of hydrogen ions (0.7 kequiv/ha/yr) to the mineral soil but only a small amount (0.30 kequiv/ha/yr) escapes to seepage below the rooting zone. Assessments of this kind are, however, subject to many difficulties, particularly in measuring gaseous absorption and particulate impaction (Galloway and Parker, this meeting), so that the estimates can only be regarded as very approximate.

MATERIALS LEACHED FROM TERRESTRIAL SOILS BY ACID PRECIPITATION

Acids deposited in precipitation may undergo three possible reactions. (1) In arid regions, or where soils are young (as in recently glaciated areas), they will be neutralized by free bases such as calcium or magnesium carbonates in the soil. (2) Where soils have no free bases, but an appreciable cation-exchange capacity, some hydrogen ions will replace metal cations on organic or inorganic exchange complexes. (3) Some hydrogen ions will also react to varying degrees with silicates or other minerals in the soil (cf. Burger,[18,19]; Kramer,[20,21]), releasing soluble metal cations and silica and neutralizing some proportion of the acid. Hydrogen ions remaining after ion exchange and mineral weathering may be transferred in runoff or ground-water to aquatic ecosystems along with the metal cations and silica released by action of the acid. In small drainage basins on granitic bedrock, the output/input ratio of hydrogen ions ranges between 0.02 and 0.50 (Gjessing et al.,[4]).

On agricultural lands where liming and fertilization are routine, acid precipitation is readily counteracted by liming and its acid input is small compared to the potential acidity of N

fertilizers and natural N and S transformations (Reuss,[22] ;
Frink and Voigt,[23]; McFee and Kelly,[24]). Acid precipitation
at pH 4 could lower the base saturation of a typical midwestern
U.S. soil appreciably over 100 years and increase soil acidity by
a few tenths of a pH unit (-0.3 pH unit indicates a doubling of
acidity) if left unamended (McFee, Kelly and Beck,[25]). In this
way acid deposition accelerates the natural tendency of soils in
humid climates, where precipitation exceeds evaporation, to lose
metal cations by leaching and become more acid. Because natural
soil acidification, though very slow, is generally accompanied by
deleterious side effects (such as nutrient impoverishment and
reduction in biological activity) any acceleration is generally
viewed as undesirable.

That acid precipitation is neutralized to some degree by even
poorly buffered soils is shown by recent Norwegian data (Wright
and Henriksen,[3]), which indicate that no lakes are currently
quite as acid as the precipitation falling upon them, and more
than 50% are less than half as acid as the precipitation. A
comparison of the most dilute mountain lakes in the English Lake
District with rain there of approximately the some ionic
concentration, 150 μequiv/l (Gorham,[26]), reveals that the lake
waters are 18 μequiv/l lower in hydrogen ion and 0.21 μequiv/l
higher in Ca + Mg than the rain, presumably owing to ion exchange
and weathering in the soils of the drainage basins.

How much of the recent increase in total ions present in
south Swedish lakes and discharged by Swedish rivers (Malmer,[27,28])
can be ascribed to atmospheric deposition, and how
much to the increased leaching consequent upon such deposition,
has yet to be calculated. However, it has been shown that in
small drainage basins on granitic bedrock the net losses of Ca +
Mg + Al (between 10 and 90 k equiv/km^2/yr) are directly related
to the amounts of hydrogen ion input which have been neutralized
or retained in the watershed (Gjessing et al.,[4]). Aluminium
is very scarce in precipitation, and the high concentrations in
strongly acid lakes of south Norway (> 200 μg/l as compared to < 20
μg/l in circumneutral lakes) must be due to acid weathering of
aluminosilicates in the soils (Wright and Gjessing,[5]) and
displacement of Al ions into solution from cation-exchange
complexes in the soil. A similar effect is evident in the
Adirondack Mountains of the northeastern U.S.A. (Galloway,[12]

Experimental leaching of a forest podzol in field lysimeters,
by simulated rain acidified with sulfuric acid over the pH range
6 to 2, revealed a distinct increase in output of Ca, Mg, K and
Al at each unit decrease in pH, most notably between pH 3 and 2
(Abrahamsen et al.,[29]). Effects on a podzol-brown earth were
less marked, particularly for K and Al, but other studies cited
by the above authors have yielded opposite results. There are

theoretical reasons for expecting non-calcareous, poorly buffered brown earths to be more affected by acid precipitation than already rather acid podzols (Malmer,[27]; Wiklander,[30]). The soils that will release the most basic cations in response to acid precipitation are the slightly acid soils with considerable exchange capacity (Wiklander,[31]). These cations Ca, Mg, K, etc. are seldom considered detrimental to the receiving stream. However aluminum, which may be toxic to fish, may be leached from some soils and that becomes most likely in very acid soils that are rich in alumino-silicate minerals. Within the normal range of acidity in soils of humid regions (pH 8.2 - 4.0), the solubility of aluminum compounds such as Gibbsite, amorphous $Al(OH)_3$ and Kaolinite increases rapidly below pH 5.0 (Norton,[32]). Some acid podzol soils are very low in Al-bearing minerals, but most contain Al that can be removed by acid leachate. Old, highly weathered soils are usually rich in Fe and Al oxides and are probably the most likely to release additional Al when treated with acid rain, but this has not been demonstrated.

Extrapolation of marked leaching effects at pH 3 and 2 to those of "normal" acid rain at or near pH 4 is difficult, but as Malmer,[28] has pointed out, over several decades precipitation at pH 4 may have an effect equivalent to less prolonged experimental leaching at pH 3. On the other hand, a problem with such research is our inability to accelerate all of the other processes that would occur over several decades. For example, during that period, the vegetation would cycle significant quantities of cations from deep in the soil to the surface. There would be inputs from the atmosphere of other materials, and weathering of rocks and minerals by other than acid processes would be proceeding. Moreover, as Bache (this meeting) has remarked, some effects of leaching at pH 3 might never occur at all at pH 4.

Near the metal smelter at Wawa, Ontario, acid precipitation has caused very severe leaching of Ca from the devastated forest soils (Gordon and Gorham,[33]) with consequent enrichment of lakes in the area. Beyond 30 km downwind of the main smelter plume, lake concentrations of Ca are about 0.5 m equiv/l. At 12 km, where the forest is severely damaged by SO_2 fumigation, Ca is elevated to about 1.1 m equiv/l. It declines to about 0.75 m equiv/l close to the smelter, where the even heavier acid loading has presumably, already leached away a good deal of the easily weatherable Ca from soil minerals.

Lake waters near the metal smelters of Sudbury, Ontario, are greatly enriched in Cu and Ni, which are smelted there (Stokes, Hutchinson and Krauter,[34]; ct. Gorham,[9]). Although much of the lake input appears to come by atmospheric deposition (Hutchinson and Whitby,[35]), the lake water mean for Ni (2,400 µg/l) greatly exceeds the maximum concentration recorded in

precipitation (830 μg/l), so that leaching of mine spoil seems a
likely additional source.

Where acid rain is a result of SO_2 fumigations sufficient to
devastate the vegetation, as at Sudbury (Gorham and Gordon,[36])
severe soil erosion occurs and transfers large amounts of particulate matter to rivers and lakes. At Sudbury erosion has
affected more than 1300 km^2, and river sediments approximately 80
km to the SW are high in Cu and Ni (Hutchinson et al.,[37])
How far this is owing to transport of enriched sediment particles
from the Sudbury area, and how far to sorption by local sediments
of dissolved metal ions from Sudbury, is at present unknown, but
sediment transport seems likely to be much more important.

The effects of acid precipitation upon mobilization of the
plant nutrients, N and P, from the soil are little known at
present. Nitrogen especially may be enriched in acid precipitation
and can exert not only a fertilizing effect but also an acidifying effect (Reuss,[38]). The Norwegian experiments
(Abrahamsen et al.,[29]) have shown no consistent effects of acid
leaching with sulfuric acid upon N mobilization, and P was not
investigated. In southern Norwegian lakes subject to enhanced
acid deposition, P is seldom much enriched (Hendrey and Wright,[39])
and may come from local point sources. Phosphorus chemistry
in soils is complicated, but the P available for plant uptake is
usually maximized in slightly acid to neutral soils. Apatite,
one of the phosphatic minerals in soil, is weatherable by dilute
sulfuric acid (Burger,[18,19]), but in very acid conditions
secondary minerals of iron and aluminum phosphates form, reducing
P availability to plants and presumably reducing leaching to
aquatic systems. In alkaline soils with high Ca activity, P is
immobilized as relatively insoluble calcium phosphates. Because
of the importance of P as the chief limiting nutrient in fresh
waters (Schindler,[40]) its transfer from land to water under
the influence of acid deposition deserves more thorough investigation.

Transfer of organic matter from terrestrial to aquatic ecosystems may also be influenced by acid deposition. According to
Schnitzer (this meeting), fulvic acids of low molecular weight
change their structure and become more soluble as soils acidify,
and can therefore be leached. Because fulvic acids form stable
complexes with metal ions and interact with clay minerals, the
transfer of these materials may also be affected (see also L.
Peterson, this meeting).

The amount of nutrient elements leached from a soil may be
greatly influenced by the rate and extent of microbial decay of
organic materials and the conversion of inorganic forms of such
elements as N and S to more or less soluble forms. Most soil

bacteria that have been studied show sensitivity to acidity, fungi to a lesser degree (Alexander, this meeting). Acid precipitation, through its influence on soil microorganisms, especially those inhabiting surface layers, is likely to influence the output of important nutrients.

MATERIALS RETAINED IN TERRESTRIAL ECOSYSTEMS
UNDER AN ACID PRECIPITATION REGIME

Prominent among the materials prevented from reaching aquatic ecosystems by prior interaction with terrestrial ecosystems are the acids deposited from the atmosphere, which may be substantially neutralized -- or exchanged for metal cations -- in passing through plant canopies (Eaton, Likens, and Bormann,[8] and through thin non-calcareous soils (Gjessing et al.,[4]). In short-term leaching experiments with sulfuric acid (Abrahamsen et al.,[29]) which did not alter greatly the soil pH, not all the added SO_4 appeared in the leachate, possibly because of adsorption. Over the long-term, acidification will itself lead to an increased capacity for adsorption of anions such as SO_4 and PO_4 (Reuss,[22,38]).

Heavy metals supplied from atmospheric sources or by acid leaching are likely to be trapped to some degree by vegetation, as shown in the elegant studies of accumulation in mosses by Rühling and Tyler,[41,42]. They may also become bound to organic or clay exchange sites in the soil, or be complexed by organic molecules. They may also be retained in the soil if they are deposited as relatively insoluble particles. According to Parker, McFee and Kelly,[43]. Zn has accumulated to 2,500 ppm in the top 2.5 cm of a sandy urban soil in NW Indiana, U.S.A., and is being deposited at about 1 kg/ha/yr; but only 0.15 kg/ha/yr is leaching through the soil profile. In subalpine forests of New Hampshire, unexpectedly high concentrations of Pb have been observed in the soil humus layers, apparently accumulating from distant air pollution because of heavy precipitation combined with high winds and abundant plant surfaces for aerosol impaction (Reiners, Marks and Vitousek,[44]). The power of soils to sequester heavy metals under various atmospheric loadings and at various pH levels deserves investigation by lysimeter experiments such as those of Abrahamsen et al.,[29]).

The supply of nutrient elements from terrestrial to aquatic eco-systems may be lessened as well as enhanced by acid deposition, with the balance between increase and decrease dependent upon the complex interplay of diverse biogeochemical processes and hence not well understood. If acidification retards litter decomposition on land, as it does to a modest degree in experiments in aquatic habitats (Leivestad, et al.,[45]), then recycling and leaching of

several nutrient elements will be inhibited. Acidification is
also likely to increase sorption of PO_4-P by hydroxy-Al and
hydroxy-Fe complexes in the soil (Reuss,[22,38]), and so
inhibit its transfer to aquatic ecosystems. Acid deposition is
likely to influence the N cycle in forest soils in diverse and
opposing ways, and Tamm,[46] has proposed an interesting model
with both negative and positive feedback loops as a basis for
further investigation. In general, the expectation is that
increased acidity will inhibit N cycling (and release to the
forest biota, or to aquatic ecosystems) by increasing immobilization
of stable forms of N in the B horizon.

Retarded litter decomposition in a highly acid environment
is likely to affect the export of organic molecules to aquatic
ecosystems in ways which are difficult to predict. If decay is
inhibited overall, export will decline; but if the production of
coloured "humic" substances is retarded less than their further
decomposition, then acidification might well result in increased
colour of the waters percolating through soils to streams and lakes
As already noted, fulvic acids are especially likely to be
affected.

If soil acidification is exceptionally severe, as for
instance near the Sudbury metal smelters, the very nature of the
organic matter in the soil may be changed owing to sulfonation of
the previously carboxylated benzene rings of the "humic acids"
(Hutchinson and Whitby,[47]). Such altered molecules, which
have an unusually low N content, will have greatly altered (and
strengthened) capacities to bind metals, and so will retard
their transport from terrestrial to aquatic ecosystems. Their
influence upon soil weathering remains to be investigated, but
could be pronounced.

The transfer of the diverse organic molecules found in
atmospheric precipitation (Lunde et al.,[15]) from land to water
will presumably also be influenced by changes in soil structure
and adsorptive properties consequent on increasing acidification.
However, the subject does not appear to have been investigated.

FACTORS INFLUENCING THE RESPONSE OF AQUATIC
ECOSYSTEMS TO ATMOSPHERIC DEPOSITION

The vulnerability of aquatic ecosystems to atmospheric pollutants, either deposited directly or washed from the land, will depend
upon a variety of climatic, geologic, topographic, morphometric,
biotic and anthropogenic factors, most of which are equally
important in determining the vulnerability of terrestrial
ecosystems.

Chief among the climatic factors is the amount of
precipitation available for rainout and washout of pollutants. Wind
direction and wind speed are very significant as well (Gorham,[14]
and wind speed and humidity may both influence gaseous
absorption and particulate impaction. Temperature is important
insofar as it determines the balance between rain and snow. Snow
appears to have a lower efficiency than rain for scrubbing various
pollutants from the atmosphere (Herman and Gorham,[48]; Gorham,[9]
Fowler, this meeting). However, where it accumulates over
the winter as snowpack, it may provide a sudden pulse of severe
pollution to streams and lakes during spring thaw (Hultberg,[49]
because the soluble materials are frozen out as intercrystalline
brines and are washed out early in the melting process (Gorham,[50]).
The ratio of precipitation to evaporation is yet another
factor, determining the degree to which terrestrial soils are
leached, influencing the rate at which precipitation passes
through streams and lakes, and affecting the entrainment of dust
particles to the atmosphere.

Geologic factors (including patterns of glaciation) are
important in several ways. The areas most sensitive to acid
precipitation are those with hard, crystalline bedrock and very
little surficial soil, a large part of Norway being a prime
example. The water chemistry of streams and lakes draining such
areas strongly reflects the chemistry of atmospheric precipitation
(Gorham,[14,51]). **Where the soil mantle is thicker, its**
texture, mineralogy, and organic content become very significant,
and water chemistry is strongly influenced by soil chemistry
(Gorham,[14]; Kramer,[21]). **Soils with free calcium carbonate**
(or other carbonates) are not susceptible to acidification, and
non-calcareous soils are also relatively resistant if they are
fine-textured, with clay and organic matter providing a high
cation-exchange capacity and buffering the soil against the
effects of acid deposition. Soil minerals differ greatly in their
response to acid weathering, and also in their content of
carbonate impurities (Burger,[18,19]). They differ greatly as
well in their content of heavy metals which can be mobilized by
acid depositon.

Soil texture and structure will affect the pathways of water
through the soil. Coarse soils with plentiful stones and boulders,
or with numerous old root channels, may allow rapid percolation
and a very incomplete equilibration between water and soil
(Troedsson,[52]; Tamm and Troedsson,[53]; Gorham,[14]). In
this way streams and lakes may be acidified even where soil bases
are theoretically sufficient to neutralize acid deposition, and
effects upon streams and lakes may be more rapid than those upon
terrestrial ecosystems.

Topography has considerable influence. Upland headwater

lakes and streams, where soils are thin and precipitation is heavy, are especially susceptible to acidification. Lower in the hydrologic network, where soils are thicker, there is more likelihood of alkaline groundwater input. Topography further determines lake depth and the ratio of watershed area to lake area, both of which (along with precipitation and evaporation) determine the residence time of water in lakes.

Biotic factors must also be considered. The height, type and duration of leaf canopy have a significant effect upon gaseous absorption and particulate impaction from the atmosphere. The magnitude of transpiration, which varies with the plant cover, influences the ratio of precipitation to evaporation. Soil organisms also vary greatly in their sensitivity to acid deposition, and will thus influence ecosystem response. For example, microorganisms that control the conversion of N and S forms in the soil are sensitive to changes in acidity and Abrahamsen et al.,[29] have shown changes in soil fauna numbers under acid treatments.

Anthropogenic factors must be taken into account. Spatial and temporal patterns of urban/industrial development play a major role in determining the impact of acid deposition upon both terrestrial and aquatic ecosystems. The kinds of energy resources in use and the controls on atmospheric emissions greatly influence the nature and extent of atmospheric deposition. The degree of agricultural activity, such as liming and fertilization, will also have a substantial role in ameliorating the impact of acid deposition (Reuss,[22]; Frink and Voight,[23]), as may dustfall from windblown soil (Gorham,[9]). Whether aquatic impacts can be ameliorated more effectively by liming drainage basins, or the lakes and streams themselves, has yet to be investigated and deserves experimental study.

Finally, natural episodic phenomena may be locally or temporarily significant. Forest fires, tornadoes, etc., may entrain into the atmosphere alkaline particulate material capable of neutralizing acid deposition. On the other hand, natural (and sometimes long continued) fires in deposits of coal, lignite, or peat may cause acid deposition capable of severe degradation of adjacent ecosystems (e.g., the Smoking Hills of Arctic Canada see Hutchinson, this meeting). Vulcanism may also be an important contributor to local acid rain (Bottini,[54]).

Investigations of the complex effects of acid deposition upon aquatic ecosystems are best pursued by means of observations and experiments upon whole watersheds, because the vulnerability of lakes and streams to acid precipitation and its associated pollutants rests so largely upon the nature of the adjacent land and its biota. Integration of aquatic with terrestrial research programs should therefore be a matter of the highest priority for the future.

Such integration would also allow better comparison of differences in the rate at which acid deposition affects terrestrial and aquatic ecosystems. Present evidence indicates considerable effects upon sensitive aquatic ecosystems over a time scale of decades, whereas for comparable terrestrial ecosystems the time scale could well be an order of magnitude greater.

CONCLUSION

The following expression summarizes the measurements required for a thorough assessment of the effect of atmospheric deposition upon any elemental or molecular transfer from terrestrial to aquatic ecosystems:

$$P + I\ (\pm\ G)\ \pm\ B\ \pm\ S\ +\ E\ =\ 0$$

where P = precipitation, I = particle impaction, G = gaseous absorption or emission, B = biomass retention or release, S = soil retention or release, E = erosion, and 0 = output from the terrestrial ecosystem. Biomass retention or release may be either short- or long-term, and involve the vegetation canopy, plant roots, and the organic portions of the soil, both living and non-living. Output may involve only stream outflow in sealed basins, but will include groundwater seepage in others. Where watersheds are not sealed by impervious bedrock, groundwater seepage may also have to be measured as an input factor in the above expression if an adequate mass balance is to be obtained.

Given the number of terms in the above expression (some encompassing diverse gains and losses), and the fact that only P and 0 are commonly measured, it is not at all remarkable that our understanding of the influence of acid deposition upon transfers from terrestrial to aquatic ecosystems is so incomplete.

REFERENCES

1. G.E. Likens and F.H. Bormann, Linkages between terrestrial and aquatic ecosystems. _BioScience_ 24:447 (1974).

2. A.D. Hasler, "Coupling of land and water systems." Springer Verlag, New York. (1975).

3. R.F. Wright and A. Henriksen, Chemistry of small Norwegian lakes, with special reference to acid precipitation. Internal Report, SNSF Project, IR 33/77, Part 1, 40 p., Oslo, (1977).

4. E.T. Gjessing, A. Henriksen, M. Johannessen and R.F.
    Wright, Effects of acid precipitation on freshwater
    chemistry. Research Report, SNSF Project FR6/76,
    Part 3, p. 64, Oslo, (1976).

5. R.F. Wright and E.T. Gjessing, Acid precipitation: changes
    in the chemical composition of lakes. Ambio
    5:219 (1976).

6. C.L. Schofield, Effects of acid precipitation on fish,
    Ambio 5:228 (1976).

7. B. Nihlgård, Precipitation, its chemical composition and
    effect on soil water in a beech and spruce forest in
    South Sweden. Oikos 21:208 (1970).

8. J.S. Eaton, G.E. Likens and F.H. Bormann, Throughfall and
    stem-flow chemistry in a northern hardwood forest.
    J. Ecol. 61:495 (1973).

9. E. Gorham, Acid precipitation and its influence upon aquatic
    ecosystems -- an overview. Water, Air & Soil Pollut.
    6:457 (1976).

10. A. Henriksen and R.F. Wright, Concentrations of heavy metals
    in small Norwegian lakes. Internal Report, SNSF
    Project, IR 33/77, Part 3, 29 p., Oslo, (1977).

11. G. Müller, G. Grimmer and H. Böhnke, Sedimentary record of
    heavy metals and polycyclic aromatic hydrocarbons in
    Lake Constance. Naturwissenschaft 64:427 (1977).

12. J.N. Galloway, Air pollution: impact on aquatic systems,
    in: New Directions in Century Three: Strategies for
    Land Use. Proc. Soil Conser. Soc. Amer., p. 211. S.C.S.A.
    Ankeny, Iowa, (1977).

13. J.N. Galloway and G.E. Likens, Atmospheric enhancement of
    metal deposition in Adirondack Lake sediments. Research
    Technical Completion Report, Project A-067-NY,
    Office of Water Research and Technology, U.S. Dept. of
    Interior, 40 p. (1977).

14. E. Gorham, The influence and importance of daily weather
    conditions in the supply of chloride, sulphate, and
    other ions to fresh waters from atmospheric precipitation.
    Phil. Trans. Royal Soc. London, B, 241:147 (1958).

15. G. Lunde, J. Gether, N. Gjøs, and M.-B.S. Lande, Organic
    micropollutants in precipitation in Norway. Research
    Report, SNSF Project, FR9/76, 17 p., Oslo, (1976).

16.   G.E. Likens, F.H. Bormann, R.S. Pierce, J.S. Eaton and
         N.M. Johnson, "Biogeochemistry of a forested ecosystem."
         Springer Verlag, New York, (1977).

17.   R. Mayer and B. Ulrich, Acidity of precipitation as in-
         fluenced by the filtering of atmospheric sulphur and
         nitrogen compounds -- its role in the element balance
         and effect on soil. U.S.D.A. Forest Service General
         Technical Report NE-23, p. 737 (1976).

18.   D. Burger, Calcium release from 11 minerals of fine-sand
         size by dilute sulfuric acid. Can. J. Soil Sci.
         49:11 (1969).

19.   D. Burger, Relative weatherability of calcium-containing
         minerals. Can. J. Soil Soc. 49:21 (1969).

20.   J.R. Kramer, Geochemical and lithological factors in acid
         precipitation. U.S.D.A. Forest Service General Technical
         Report NE-23, p. 611 (1976).

21.   J.R. Kramer, Acid precipitation. in: "Sulfur" J.O. Nriagu,
         ed., J. Wiley and Sons, (1978).

22.   J.O. Reuss, Chemical/biological relationships relevant to
         ecological effects of acid rainfall. EPA - 660/3-75-032,
         (1975).

23.   C.R. Frink and G.K. Voigt, Potential effects of acid precipi-
         tation on soils in the humid temperate zone. U.S.D.A.
         Forest Service General Technical Report NE-23, p. 685
         (1976).

24.   W.W. McFee and J.M. Kelly, Air Pollution: Impact on Soils
         in New Directions in Century Three: Strategies for
         Land Use. Proc. of Soil Conser. Soc. Amer. 1977 mtg.,
         p. 203 S.C.S.A. Ankeny, Iowa, (1977).

25.   W.W. McFee, J.M. Kelly and R.H. Beck, Acid precipitation
         effects on soil pH and base saturation of exchange
         sites. Water, Air & Soil Pollut. 7:401 (1977).

26.   E. Gorham, Ecological aspects of the chemistry of atmospheric
         precipitation. Special Publication, National Center
         for Atmospheric Research, (in press) (1978).

27.   N. Malmer, On the effects on water, soil and vegetation
         of an increasing atmospheric supply of sulphur. National
         Swedish Environment Protection Board, SNV-PM-402E, 98 p.,
         Stockholm, (1974).

28.   N. Malmer, Acid precipitation: chemical changes in the soil.
           Ambio 5:231 (1976).

29.   G. Abrahamsen, K. Bjor, R. Horntvedt and B. Tveite, Effects
           of acid precipitation on coniferous forest. Research
           Report, SNSF Project, FR 6/76, Part 2, p. 36, Oslo.
           (1976).

30.   L. Wiklander, Leaching of plant nutrients in soils 1.
           General principles. Acta Agric. Scand. 24:349 (1974).

31.   L. Wiklander, The acidification of soil by acid precipitation.
           Grundförbättring 26(4):155 (1973/74).

32.   S.A. Norton, Changes in chemical processes in soils caused
           by acid precipitation. U.S.D.A. Forest Service General
           Technical Report NE-23. p. 711 (1976).

33.   A.G. Gordon and E. Gorham, Ecological aspects of air pollution
           from an iron-sintering plant at Wawa, Ontario.
           Can. J. Bot. 41:1063 (1963).

34.   P.M. Stokes, T.C. Hutchinson and K. Krauter, Heavy metal
           tolerance in algae isolated from polluted lakes near
           Sudbury, Ontario, smelters. Water Pollut. Res. in Can.
           8:178 (1973).

35.   T.C. Hutchinson and L.M. Whitby, Heavy metal pollution in
           the Sudbury mining and smelting region of Canada.
           1. Soil and vegetation contamination by nickel, copper,
           and other metals. Envir. Conservation 1:123 (1974).

36.   E. Gorham and A.G. Gordon, The influence of smelter fumes
           upon the chemical composition of lake waters near
           Sudbury, Ontario, and upon the surrounding vegetation.
           Can. J. Bot. 38:477 (1960).

37.   T.C. Hutchinson, A. Fedorenko, J. Fitchko, A. Kuja, J. van
           Loon and J. Lichwa, Movement and compartmentation of
           nickel and copper in an aquatic ecosystem. in:
           "Environmental Biogeochemistry", vol. 2, J.O. Nriagu,
           ed., Ann Arbor Science Publishers, p. 565 (1976).

38.   J.O. Reuss, Chemical and biological relationships relevant
           to the effect of acid rainfall on the soil-plant system.
           U.S.D.A. Forest Service General Technical Report NE-23,
           p. 791 (1976).

39. G.R. Hendrey and R.F. Wright, Acid precipitation in Norway: effects on aquatic fauna. J. Great Lakes Res. 2 (Suppl. 1):192 (1976).

40. D.W. Schindler, Biogeochemical evolution of phosphorus limitation in nutrient-enriched lakes of the Precambrian Shield. in: "Environmental Geochemistry" vol. 2, J.O. Nriagu, ed., Ann Arbor Science Publishers. p. 647. (1976).

41. A. Rühling and G. Tyler, An ecological approach to the lead problem. Botaniska Notiser, Lund 121:321. (1969).

42. A. Rühling and G. Tyler, Regional differences in the deposition of heavy metals over Scandinavia. J. Appl. Ecol. 8:497 (1971).

43. G.R. Parker, W.W. McFee and J.M. Kelly, Metal distribution in a forested ecosystem in urban northwestern Indiana. J. Envir. Qual. 7 (in press) (1978).

44. W.A. Reiners, R.H. Marks and P.M. Vitousek, Heavy metals in subalpine and alpine soils of New Hampshire. Oikos 26:264 (1975).

45. H. Leivestad, G. Hendrey, I.P. Muniz and E. Snekvik, Effects of acid precipitation on freshwater organisms. Research Report, SNSF Project, FR6/76, Part 4, p. 87, Oslo, (1976).

46. J.O. Reuss, Sulfur in the soil system. in: "Sulfur in the environment," Missouri Botanic Garden, St. Louis, p. 51-61. (1975b).

47. T.C. Hutchinson and L.M. Whitby, The effects of acid rainfall and heavy metal particulates on a boreal forest ecosystem near the Sudbury smelting region of Canada. Water, Air & Soil Pollut. 7:421 (1977).

48. F.A. Herman and E. Gorham, Total mineral material, acidity, sulfur and nitrogen in rain and snow at Kentville, Nova Scotia. Tellus 9:180 (1957).

49. H. Hultberg, Thermally stratified acid water in late winter -- a key factor inducing self-accelerating processes which increase acidification. U.S.D.A. Forest Service General Technical Report NE-23, p. 503. (1976).

50.   E. Gorham, Soluble salts in a temperate glacier. Tellus
          10:496 (1958).

51.   E. Gorham, Factors influencing supply of major ions to inland
          waters, with special reference to the atmosphere. Geol.
          Soc. Amer. Bull. 72:795 (1961).

52.   T. Troedsson, Vattnet i skogsmarken. Kungliga Skogshögskolans
          Skrifter, No. 20, 215 p. (1955).

53.   C.O. Tamm and T. Troedsson, A new method for the study of
          water movement in soil. Geologiska Föreningens i
          Stockholm Förhandlingar 79:581 (1957).

54.   O. Bottini, Le pioggie caustiche nella regione vesuviana.
          Ann. Chim. Applic. 29:425 (1939).

EFFECTS OF ACID LEACHING ON CATION LOSS FROM SOILS

T. C. Hutchinson

Department of Botany and Institute for Environmental

Studies, University of Toronto, Ontario, Canada, M5S 1A1

INTRODUCTION

As attention has become increasingly focused on the widespread occurrence of acidic precipitation, the question of the impact of this precipitation on plant growth and on soil microbial activity has been addressed. Since soils of a podsolic nature cover a large proportion of the areas identified as having aquatic lake systems sensitive to acidic rains, it is also of concern to understand the impact of acid precipitation on soil chemistry and, especially, on the leachability of constituent elements.

The ability of soils to exchange both cations and anions is generally very considerable. This exchange capacity is clearly much greater in terrestrial than in the aquatic systems. Podsolic soils are naturally acidic (pH 3.8–5.2), and have vertical profiles that have developed as a result of natural, selective leaching processes. The removal of iron compounds from the upper A horizon, and their re-deposition in the B horizon are typical of these podsols. The acid leaching is, at least partially, a function of the percolation of acidic organic acids resulting from microbial decomposition of conifer needles and litter. The soils themselves are well buffered by humic and fulvic acids, together with clay minerals, these having numerous cation and anion exchange sites.

A metal-loading effect often occurs in smelter situations, or following the addition of contaminated sewage sludges. This often leads to a decrease in concentration at deeper horizons within the profile (e.g. Hutchinson and Whitby[1], Hutchinson et al.[2]). The high cation exchange capacities of mosses and lichens are illustrated by numerous studies, including those of Rühling and Tyler[3], which show

481

strong retention of such metals as lead and zinc by these plants.
Other studies have shown that high concentrations of metals occurred
in the organic horizons of soils, especially in areas of airborne
deposition (Rühling and Tyler[4], Strojan[5]). The residence times in
soils of such metallic pollutants is of considerable interest. In
addition, the effects of acidic leachates on cation removal which
might result from increasingly acidic precipitation needs to be
known, since the solubility of many essential plant nutrients is
pH-dependent. This is also true of the solubility of certain
constituent elements of the clay minerals, e.g. Al, Fe^{+3}, and Mn.

     A small number of recent studies address the latter question.
The Norwegian SNSF programme is examining it in detail. Tyler[6]
has very recently reported on the leachability of Mn, Zn, Cd, Ni,
V, Cu, Cr and Pb in two organic forest soils. Some of his data are
discussed in the present paper. Wallace et al.[7] considered effects
of soil acidification on leaching and plant uptake, and found
increased manganese accumulation in tissues. Norton[8] predicted
that at pH's as low as 3.0, Mn, Al, and Fe^{+3} could be expected to
dissociate from clay minerals, and to go into solution.

     In the present study, the leaching of a number of elements,
(including Al, Mn, Fe, Zn, Ca and S) from soils in the area of the
Smoking Hills in the western Canadian Arctic, and from some smelter
polluted soils near Sudbury, Ontario, are compared by analyses of
stream outflow chemistry. In addition, the effects of controlled
acid leaching on soils from these sites are compared, with
particular reference to the relation of the elutant pH to mineral
losses. Finally, Tyler's[6] significant study of metal retention in
the organic horizons of two mor soils when subjected to acid
leaching is considered. The relative retention of selected elements
in these acidic soils can be related to soil composition.

MATERIALS AND METHODS

Field Sites - The Smoking Hills, N.W.T.

     The area is on Cape Bathurst, at 70°14'N and 127°10'W.
Spontaneous burning of bituminous shales occurs along 40 km of sea-
cliffs producing intense sulphur fumigations over adjacent tundra.
The burns have been ongoing for at least several hundred years, and
have produced extreme acidities in both tundra soils and ponds
(Hutchinson et al.[9]). The low rainfall in the area is extremely
acidic near to the source of the burns. A stream drains a watershed close to the major burns, and after heavy rains, temporary
pools also form in the fumigation area. In the summers of both
1976 and 1977, samples were collected for analyses of pH, conductivity, and sulphate content. The samples were taken along the

stream, from its point of origin about 600 m from the burning cliff, to a distance of 4 km. Samples were also taken from the temporary pools 3 days after rain. Samples were transported to Toronto for analyses of a number of elements by atomic absorption spectro-photometry or neutron activation. Chloride was determined using a specific ion electrode.

Sudbury Area Samples

The Sudbury area has been subjected to fumigations by sulphur emitted by nickel-copper smelters over about 70 years. Airborne metallic contaminants from the smelter have also accumulated in soils over a wide area. Especially significant are nickel and copper, but cobalt and arsenic also occur (Hutchinson and Whitby[10]). A stream drains a slag heap located at Coniston which, as it descends, joins another small stream draining a less contaminated area. Samples were taken along these streams in September 1977, and were chemically analysed as described above. All samples were filtered through a 0.45 μm filter.

Leaching Experiments

Soil samples were obtained from an area affected by the main fumigations at the Smoking Hills, as well as from an infrequently fumigated area about 3.5 km distant. As well, samples from the Sudbury area were collected; from a relatively uncontaminated forested site 12 km west of the Copper Cliff smelter, and from a depauperate site adjacent to the Coniston slag heaps. The soils were dried, passed through a 2 mm sieve, and then loosely packed into glass columns (2.5 cm i.d., 25 cm long). A leaching solution was continuously added from an overhead reservoir, and samples of percolate were collected incrementally for chemical analysis. A range of acidities was used for the leaching, i.e. pH 2.5, 4.5 and 5.7. These acidities were produced using Analar sulphuric acid.

RESULTS

The chemical analyses of filtered pond water from the Smoking Hills area (arranged according to their pH ranges) are given in Table 1. The water chemistry of the ponds for most elements clearly relates to their H^+ ion status. In the most acidified ponds, high concentrations of some of the heavy metals (Zn, Ni, Cd) and clay-derived elements such as Al, Fe and Mn occur. Extreme concentrations of sulphate also occur. These ponds are all strongly influenced by the local SO_2 fumigations, either directly or through their water-sheds. For Al, Fe, Zn, Ni and Cd, concentrations only fall below 1 ppm in those ponds with pH values of 4.5 and above. Manganese

Table 1.   Chemical analysis of filtered pond water from the Smoking Hills area, arranged according to their H^+ status. A total of 46 ponds are included. Geometric means of the data are given in mg/l. From Hutchinson et al.[9].

pH Range	Al	Fe	Mn	Zn	Ni	Cd	As	Ca	SO_4	Cl	n
1.5 to 2.5	270	500	61	14	6.3	0.520	0.130	301	8200	59	4
2.5 to 3.5	5.5	18	15	0.45	0.21	0.022	0.005	157	890	88	14
3.5 to 4.5	1.1	1.2	3.6	0.12	0.04	0.011	0.005	44	156	27	9
4.5 to 5.5	<0.6	0.5	23	0.03	0.04	0.001	0.004	182	813	83	1
5.5 to 6.5	<0.2	0.2	1.8	0.05	0.06	0.012	0.006	249	713	13	1
6.5 to 7.5	<0.8	<0.04	0.7	0.08	0.02	0.001	0.004	90	360	46	4
7.5 to 8.5	<0.7	0.1	<0.5	0.04	0.004	0.003	0.005	49	106	55	8
8.5 to 9.5	<0.2	<0.04	<0.2	<0.03	0.007	<0.001	0.006	31	34	21	3
9.5 to 10.5	<0.2	0.2	1.2	5.3	0.01	<0.001	0.005	16	29	17	2

shows an extended range of solubility, with >1 ppm occurring up to
pH 6.5. These high metal concentrations contrast with control ponds
in watersheds distant from the fumigations, where pH's range from
6.5 to 10.5. Obviously, the elevated concentrations of such
potentially-toxic elements as Al, Zn, Ni and Cd are likely to exert
a powerful selective effect on biota. This has been illustrated in
studies of the biota of these ponds (Hutchinson et al.[9]) which are
severely depleted overall, but are themselves tolerant to these
extreme acid and heavy metal stresses.

The chemical data for the stream samples and for the temporary
pools are given in Table 2. The high solubility of many of the
elements is paralled by the very low pH of many of the temporary
pools. The losses of such elements as Al, Mn, Fe and Zn from the
acidified soils are reflected by the upper stream analyses. The
stream also carries high concentrations of Ca and sulphate. At a
distance of 1200 m downstream, the streamwater pH rose from a low
of 3.4 to 6.9, due primarily to diluting additions of snowmelt and
drainage waters from adjacent calcareous soils. At about 1500 m
downstream, a shale outcrop was drained by a small stream. The
acidic drainage resulting from this outcrop added further quantities
of sulphate, aluminium, manganese, zinc, etc. to the stream. There
is a striking parallel between the chemical composition of the
acidified tundra ponds and the temporary pools formed in the same
area following rain events. The high solubility of many elements
at the low pH's is emphasized.

Sudbury Stream Analyses

The data for the streams at Coniston are shown in Table 3.
The high concentrations of Fe, Mn and Zn in the acidic streams are
shown, as are the locally-emitted metals Ni and Cu. Extremely high
sulphate levels are also apparent, even in the non-slag stream.
Drainage from the entire sulphur-fumigated local area appears to be
involved in this. Ca does not appear to be pH dependent and occurred
at <10 ppm at all sites. The concentrations of Al are also elevated,
occurring in the ppm range. These elevated concentrations of
phytotoxic metals have caused depletion of the biota in Sudbury area
lakes (Whitby et al.[11]). In some polluted lakes algal strains occur
that are adapted to elevated levels of certain metals such as copper,
nickel and cobalt (Stokes et al.[12]).

Soil Leaching Experiments

Representative data for the four soils are shown in Table 4 and
Figs. 1 and 2. The initial pH's of the soils used were 3.5 and 7.4
for the fumigated and control soils from Smoking Hills; 3.3 for the
polluted Coniston soil and 3.6 for the relatively uncontaminated
forest soil.

Table 2.   Chemical analyses of stream and temporary pools. Collected
           from the Smoking Hills area in June 1977. Data are given
           in mg/l.

Site	pH	Al	Fe	Mn	Zn	Ni	Ca	SO$_4$	Cl
a) Stream									
Pool of origin	4.0	2.3	0.84	5.83	0.26	0.04	79.1	376	61
1st flow	3.9	2.3	1.17	2.90	0.22	0.08	44.0	226	21
Pool in stream	3.55	2.4	1.57	3.63	0.26	0.08	46.3	248	28
100 m downstream	3.40	6.3	9.61	2.26	0.56	0.12	24.9	250	19
250 m "	4.05	1.7	0.84	2.11	0.22	0.04	27.6	144	21
1200 m "	6.90	0.2	0.24	0.26	0.05	0.02	17.6	52	8
Stream above shale outcrop at 1450 m	6.90	0.1	0.07	0.33	0.03	0.02	46.6	105	14
Small stream from shale	4.55	6.7	0.89	0.60	0.34	0.19	48.5	226	8
Main stream 1500 m after inflow at shale	6.15	0.2	2.72	0.35	0.17	0.04	24.4	160	4
b) Temporary pools after rain									
40 m from cliff burn	2.25	70.0	20.8	66.8	2.51	1.37	420.6	3800	350
Pool or stream 640 m from cliff	3.00	9.0	6.8	8.9	0.69	0.26	132.2	722	64
Pool 160 m from cliff	2.9	8.1	9.0	8.0	0.45	0.18	322.5	1285	110
Pool 540 m from cliff	2.9	5.3	17.1	6.0	0.35	0.10	29.0	413	32
Pool 500 m from cliff	2.9	4.8	22.4	5.5	0.33	0.09	70.9	410	33
Pool 400 m from cliff	2.1	18.0	22.6	3.1	0.79	<0.04	27.7	590	83
Control Pools									
1	8.7	< 0.1	1.3	0.3	<0.01	<0.02	105.2	307	150
2	8.8	0.3	0.2	0.0	<0.02	<0.02	35	60	23
3	8.8	0.1	0.3	0.1	nd	nd	35	61	43

Table 3.   Chemical analyses of a stream draining the Coniston slag
           heap at Sudbury, and one with which it merges. Data are
           given in mg/l.

Slag stream	pH	Al	Fe	Mn	Zn	Co	Cu	Ni	Ca	SO$_4$
1. Stream above confluence	3.8	1.6	11.6	0.7	0.34	0.91	0.06	29.3	6.7	538
2. " " "	3.6	<0.04	39.4	1.1	0.26	1.5	0.02	53.6	16.0	616
3. " " "	3.8	1.04	29.1	1.0	0.22	1.3	0.06	40.5	15.9	558
4. " below "	5.9	0.4	13.6	0.8	0.16	1.0	0.05	30.1	18.0	433
Non-slag stream	7.0	0.4	0.3	0.3	0.02	<0.03	0.04	1.3	22.0	222
" "	7.6	1.0	0.7	0.03	<0.01	<0.03	0.04	0.3	15.3	350

     The soils all initially yielded leachate having elevated
elemental composition compared with later washings, irrespective of
the pH of the elutant. Following this, a constant pattern was
established. In the Smoking Hills control soil, the leachate had a
pH >7.0, even when leached with a solution of H$_2$SO$_4$ at pH 2.5, until
15 litres had been passed through the column (representing 120
flushings of the column). The pH then dropped rapidly as cation
exchange sites became saturated with H$^+$ ions. The amounts of Al,
Mn, and Fe leached from this soil remained very low while the
leachate had a high pH. This limited elemental loss was in marked
contrast to the soil from the fumigation-zone at the Smoking Hills,
in which all leachates removed large quantities of Al. A
concentration of >1 ppm was maintained in leachate from the pH 2.5
extractions, even though the actual extracted solution had a pH
of >4.0. (Fig. 2.)

     The relatively uncontaminated forest soil collected west of
Copper Cliff showed a relationship between the pH of the extractant
and the quantity of Al leached. An elutant of pH 2.5 was very
effective in removing Al from the soil column.

     Mn was also readily eluted by the acidic extractant (pH 2.5)
from soils from both the fumigated Smoking Hills site, and from
the uncontaminated Sudbury forest site. Much lower concentrations
of Mn were removed at pH 2.5 from the Coniston or Smoking Hills
control sites, as well as by the distilled water or pH 4.5
extractant. Fe was leached in high concentrations from the three

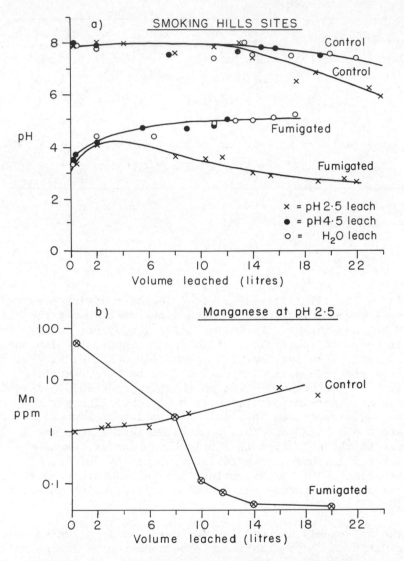

Fig. 1. a) pH and b) manganese concentrations
of the percolate of a control and a fumigated
Smoking Hills soil when leached with a solution
of H_2SO_4 at pH 2.5. (In 1a solutions of pH
4.5 and 5.7 were also used.)

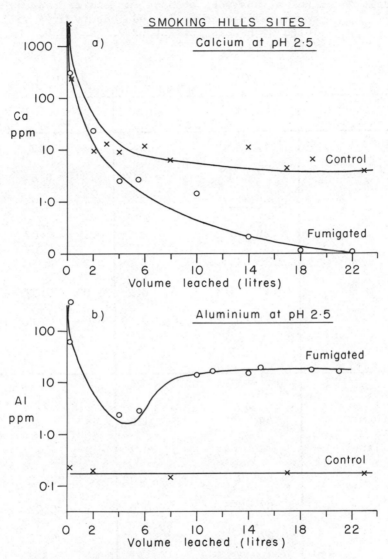

Fig. 2. a) Calcium and b) aluminium concentrations of the percolate from a control and a fumigated Smoking Hills soil when leached with a solution of H_2SO_4 at pH 2.5.

Table 4.   Chemical analyses of some cations extracted from a
polluted and a non-polluted Sudbury area soil when
eluted with solutions of pH 2.5 and 4.5.   Data are
given in mg/ℓ.

SUDBURY SOILS

Amount Leached At pH 2.5	Control				Polluted			
	Ca	Mn	Al	Cu	Ca	Mn	Al	Cu
0–100 ml	62	50	57	0.38	94	12	164	–
1 litre	8	10	14	0.26	12	0.6	14	28
2	0.22	1.9	15.5	0.29	0.18	0.02	15	17.3
4	0.26	1.7	13	0.28	0.19	0.07	6	4.6
8	0.15	1.4	5	0.06	0.26	0.16	8	1.3
14	<0.11	0.04	10.5	0.25	0.09	0.02	6.5	1.1
18	<0.11	0.16	10	0.20	0.05	0.02	3.5	0.8
22	0.2	–	9	–	0.10	–	4	–
At pH 4.5								
0–100 ml	54	30	50	1.5	38	4	40	–
1 litre	3	4	4	0.3	2	0.1	2	2.5
2	<0.07	0.7	<3	<0.02	5.8	<0.02	<3	1.7
4	<0.07	0.4	<3	<0.02	12.5	<0.02	<3	0.8
8	<0.07	0.3	<3	<0.02	–	<0.02	<3	0.8
14	<0.07	0.08	<3	<0.02	0.07	<0.02	<3	0.5
18	<0.07	0.1	<3	<0.02	–	<0.02	<3	–
22	–	0.1	<3	–	0.07	<0.02	<3	–

acidic soils, especially from the Coniston soil where initial
concentrations of over 100 ppm occurred in the leachate. These data
clearly relate to the pond and stream chemistries illustrated for
the two areas in Tables 1-3. The red solution colour, together
with the low pH employed as extractant, suggest that the ionic form
leached was ferric iron.

The data for Ca in the leachate (Table 4) showed a high availability in the Smoking Hills control soil, when exposed to all three
leachates. The fumigated soil from the Smoking Hills also released
Ca and, despite its low pH, clearly had available Ca for exchange.
The two Sudbury soils had low Ca concentrations in the leachates.
These soils also released very small quantities of Zn, while the
fumigated Smoking Hills soil released quantities in the ppm's.
This can be correlated with the high concentrations noted in the
temporary pools and in the acidic tundra ponds (Tables 1 and 2).

Experiments by G. Tyler

In the experiments described in Tyler[6], samples of organic
needle mor were collected from a spruce forest surrounding a brass
foundry, homogenized, and then loosely packed into a plastic cylinder
(100 mm, basal area 464 cm^2). A control set of soil samples were
taken from spruce forest sites located 20-30 km from the foundry
mill. Solutions acidified with H_2SO_4 were used to leach the columns
at pH's of 2.8, 3.2, 3.4 and 4.2 (the higher pH of 4.2 was selected
as typical of rainwater in southern Sweden). The columns were
percolated continuously, and samples were taken for analyses at
regular intervals. A total of 125 ℓ was passed through each column.
The polluted site had a higher soil pH than the control due to the
high concentrations of copper and zinc.

It is not appropriate to present Tyler's data in detail.
However, attention is drawn to a) the experimental approach, which
allows estimation of theoretical residence times for various elements
at various "rainwater" acidities, and b) a comparison of soils of
different type, and with different metal loadings. Figs. 2 and 3
illustrate some of his data. The elements differ markedly in their
leachability. The acidity of the percolate is a key factor in
determining retention times.

In the control podsol, Mn, Zn, Cd and Ni were found to be
readily leached by a pH 2.8 solution (Fig. 2), while V, Cu, Cr and
Pb were retained much more strongly. A similar pattern was seen
with the polluted soil, except that the copper (which had accumulated
in the soil as a result of emissions from the foundry) was much more
readily extractable. The striking effect of the acidity of the
leaching solutions is shown in Fig. 3, which compares the efficacy
of solutions from pH 2.8 to 4.2 to leach copper from the polluted

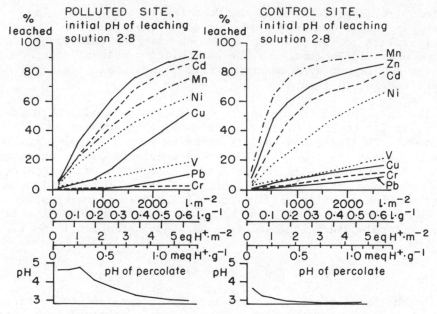

Fig. 3.   Amount of metal elements released from two mor soils
           (% of the total content) during course of leaching
           experiment and (lower) pH of the percolate. pH of
           the leaching solution was 2.8 (after Tyler[6]).

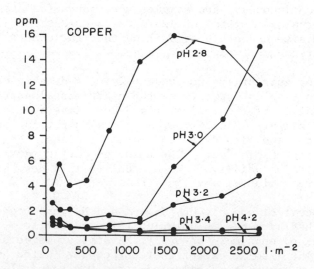

Fig. 4.   Copper concentrations of the percolate from the polluted
           soil during the course of the experiment at different pH
           of the leaching solution (after Tyler[6]).

soils. One would expect the polluted Sudbury soil to behave in a similar manner, with contamination of local water bodies by copper in acid drainage (See Table 4). The 10% residence times estimated by Tyler for various elements ranged from 3 years for Mn, to 70-90 years for Pb in the control soil, and from 2 years for V to >200 years for Pb in the polluted soil, at a rainwater acidity of pH 4.2. The calculated residence times for most elements (except V and Cr) decreased with increasing rainwater acidity.

DISCUSSION

The present study examines the effects of acid leaching on metal-contaminated, and non-contaminated soils. The solubilities of many of the elements examined are clearly high in acid soils, and in soils leached by acid drainage. In the leaching experiment using the control Smoking Hills soil, limited loss of the "clay elements" Al, Fe and Mn caused the acidic input to be converted to a base-rich alkaline discharge. The water analyses of the Smoking Hills area ponds of higher pH (>7.0) correlates with the acid soil leaching experiments, in that only low concentrations of Al, Fe, Mn and Zn occurred in water samples (Table 1). The fumigated soil of pH 3.3 readily released high concentrations of these elements, as well as elevated levels of Ni. This also correlates well with the local pond water chemistry (Table 1).

Similarly, in the Sudbury area, where extensive SO_2 fumigations have occurred together with metal particulate emissions, acidic streams carry high loads of the clay metal elements, as well as high concentrations of the metals emitted by the smelter, e.g. Cu and Ni.

One probable effect of acid rains is the mobilization of previously deposited heavy metal accumulations from smelters, sludges, automobiles, etc. The acidity of the rainwater, together with the type of soil involved and the specific element being considered, will be modifying factors in this process. Many examples of such elements appearing in increased concentrations in acidified waters already occur. Schofield's[13] report on the possible role of aluminium in fish mortality in acidified Adirondacks Lakes is one example. Reports of the generally toxic nature of Sudbury area lakes due to elevated heavy metals concentrations are another (Stokes et al.[12]). The levels noted for Al in Schofield's studies, and in Wright et al.[14], are low compared with those reported for the Smoking Hills ponds or Sudbury streams. In all cases, the probabilities of watershed sources of Al are high. A second possibility is Al mobilisation from sediments as a lake acidifies. The pH-solubility curve for Al shows rapid increases in solubility below pH 4.5, so that the problem of metal toxicity will accelerate as water bodies acidify below pH 4.5.

The retention time of heavy metals in soils under various acidic precipitation regimes was considered in the artificial leaching experiments of Tyler[6]. The metals differed markedly in their residence times, but these differences decreased with increasing percolate acidity. Pb showed a 10% residence time of 70-90 years when leached with a solution of pH 4.2, while a stronger acid leach of pH 2.8 decreased the 10% retention time to 20 years. Zn and Mn showed even greater reduced retention times, compared with Pb.

During the course of this leaching the metals were lost in solution. While in the soils, as well as once into the water bodies, the solubilised metals may be toxic to plant growth. The highest concentrations will occur in soil solutions, so that it is within soils that one might look for potentially phytotoxic effects. The major limiting effect of Al on plant growth in acid soils is well known (e.g. Magistad[15], Rorison[16]). The increased acidities of Sudbury area soils, as a result of major smelter inputs of sulfur have not only caused concentrations 'available' of Cu, Ni and Co in soils within 10 km of the smelters to be directly phytotoxic to plant growth, but have released from the clay minerals concentrations of Al which are themselves toxic to plants (Whitby and Hutchinson[17]). A few species of plants have been able to invade the inner contaminated areas since SO_2 levels decreased in 1972 following the building of the 380 m Superstack at Copper Cliff. Populations of one of these, the grass Deschampsia cespitosa, have been assayed for their tolerances to the pollutant metals Cu and Ni, and to the acid-release metal Al. When compared with populations of the same grass from control areas, it was found that the Sudbury populations were significantly more tolerant to all three elements in solution (Cox and Hutchinson[18]). This suggests that in areas of acid precipitation in which soil pH's drop below 4.0, the development of Al-tolerance may occur in plants. However, only a limited number of plants are likely to adapt in this way.

Another major effect of acid leaching of soils is one which has been dealt with extensively at this symposium and elsewhere. It is the phenomenon of the loss of bases (e.g. Ca, Mg, K) from soils as a result of cation exchange processes in the soil. Studies by Rorison, Nilsson, Likens et al., Wiklander, Gorham and McFee, Bache, and Mayer and Ulrich in the present volume all refer to this process. The effect is well documented. Irrespective of the pH of the soil, acid inputs have caused losses through leaching of bases. Acid rains may cause the availability of these bases to increase, but a long-term depletion of part of the nutrient reserves of the soil may also inevitably occur. Short-term cation nutrient availability may ultimately be followed by base insufficiency, with attendant effects on forest growth. Well documented cases of such effects do not exist at present. Base leaching from soils is a naturally-occurring process. However, increased acidity of rainfall

can be expected to accelerate the rate of this process, although
certain potentially compensating processes such as mineralization
will also be affected. It is also possible that overall effects of
increased mobility of essential and non-essential elements will be
buffered by soil microbial populations. As well, all of the
elements involved have deficiency, adequacy and toxicity levels
which are species-specific. The 'soil as a black box' is not an
adequate approach. The modifying influence of canopy and of soil
on ultimate quality of watershed discharges needs much greater
attention.

ACKNOWLEDGEMENTS

The financial assistance of the National Research Council of
Canada and the Federal Department of Indian and Northern Affairs is
gratefully acknowledged in these studies, as is the skilled technical
assistance of S. Aufreiter, H. Schabelski and B. Como.

REFERENCES

1. T. C. Hutchinson, and L. M. Whitby, Heavy-metal pollution in the
Sudbury mining and smelting region of Canada, I, Soil and
Vegetation contamination by nickel, copper and other metals,
Environmental Conservation 1:123 (1974).

2. T. C. Hutchinson, M. Czuba, and L. M. Cunningham, Lead, cadmium,
zinc, copper and nickel distributions in vegetables and
soils of an intensely cultivated area and levels of copper,
lead and zinc in the growers, in: "Trace Substances in
Environmental Health", D. D. Hemphill, ed., Univ.
Missouri, Columbia 8:81 (1974).

3. A. Rühling, and G. Tyler, Sorption and retention of heavy metals
in the woodland moss Hylocomnium splendens, Oikos 21:92
(1970).

4. A. Rühling and G. Tyler, Heavy metal pollution and decomposition
of spruce needle litter, Oikos 24:402 (1973).

5. C. L. Strojan, Forest leaf litter decomposition in the vicinity
of a zinc smelter, Oecologia 32:203 (1978).

6. G. Tyler, Leaching rates of heavy metal ions in forest soil,
Water, Air and Soil Pollution 9:137 (1978).

7. A. Wallace, E. M. Romney, and J. Procopiou, Mobilization of
nutrients in soil by acids of sulfur and chelating agents,
in: "Mineral Cycling in South Eastern Ecosystems",
F. G. Howell, J. B. Gentry, M. H. Smith, eds., ERDA Symp.
Series, 740513: p.687 (1975).

8. S. A. Norton, Changes in chemical processes in soils caused by acid precipitation, in: "Proc. 1st Internat. Symp. on Acid Precipitation and the Forest Ecosystem", L. S. Dochinger and T. A. Seliga, eds., USDA Forest Service General Tech. Report NE-23:711 (1976).

9. T. C. Hutchinson, W. Gizyn, M. Havas, and V. Zobens, The effect of long-term lignite burns on arctic ecosystems at the Smoking Hills, N.W.T., in: "Trace Substances in Environmental Health", D. D. Hemphill, ed., Univ. Missouri, Columbia 12: In Press (1978).

10. T. C. Hutchinson, and L. M. Whitby, The effects of acid rainfall and heavy metal particulates on a boreal forest ecosystem near the Sudbury smelting region of Canada, Water, Air and Soil Pollution 7:421 (1977).

11. L. M. Whitby, P. M. Stokes, T. C. Hutchinson, and G. Myclik, Ecological consequences of acidic and heavy-metal discharges from the Sudbury smelters, Canadian Mineralogist 14:47 (1976).

12. P. M. Stokes, T. C. Hutchinson, and K. Krauter, Heavy metal tolerance in algae isolated from contaminated lakes near Sudbury, Ontario, Can. J. Bot. 51:2155 (1973).

13. C. L. Schofield, Acid precipitation: effects on fish, Ambio 5:228 (1976).

14. R. F. Wright, T. Dale, E. T. Gjessing, G. R. Hendrey, A. Henriksen, M. Johannesen and I. P. Muniz, Impact of acid precipitation on freshwater ecosystems in Norway, in: "Proc. 1st Internat. Symp. Acid Precipitation and the Forest Ecosystem", L. S. Dochinger and T. A. Seliga, ed., USDA Forest Service General Tech. Report NE-23:459 (1976).

15. O. C. Magistad, The aluminium content of soil solution and its relation to soil reaction and plant growth, Soil Sci. 20:181 (1925).

16. I. H. Rorison, The effect of extreme soil acidity on the nutrient uptake and physiology of plants, "Symp. on Acid Sulphate Soils", Pub. 18, Vol. 1, H. Dost, ed., Internat. Institute for Land Reclamation and Improvement, Wageningen (1972).

17. L. M. Whitby and T. C. Hutchinson, Heavy-metal pollution in the Sudbury mining and smelting region of Canada, II, Soil toxicity tests, Environmental Conservation 1:191 (1974).

18. R. M. Cox, and T. C. Hutchinson, Metal co-tolerances in the
        grass _Deschampsia_ _cespitosa_, _Nature Lond_. 279:231 (1979).

STUDIES OF ACID RAIN ON SOILS AND CATCHMENTS

J.E. Rippon

Central Electricity Research Laboratories

Kelvin Avenue, Leatherhead, Surrey KT22 7SE, U.K.

INTRODUCTION

The quality of surface waters depends not only on the composition of the precipitation and dry deposition falling in the catchment but also on the time sequence and reactions that occur as the rain passes through the soil. These processes are physical, chemical or biological, and will depend on the structure and topography of the soil, the chemical composition as determined by the parent material, and on the vegetation and organic layers at the soil surface.

The object of the work being undertaken at CERL is to understand how soil processes are affected by acid rain and how these, in turn, affect surface water quality. The programme falls into two related projects:

1) Field Work
   Two small catchment studies, at Snake Pass, Derbyshire, and on the Tillingbourne, Surrey.

2) Experimental Work
   (a) Lysimeter studies of leaching processes
   (b) Chemical/microbiological/biochemical studies as soils are leached in pot-scale experiments.

None of these projects is far enough advanced for the presentation of detailed results or conclusions. The projects are described here and preliminary results given, with emphasis on H^+ and $SO_4^=$ movements in the systems. Brief mention will also

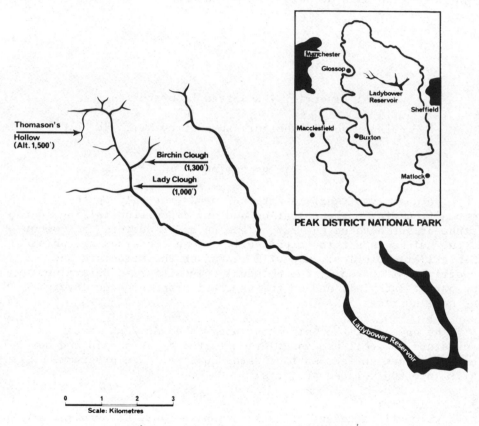

Fig. 1   The Snake Pass Catchment

be made of the results of the first six months of acid watering
on microbiological and biochemical processes.

CATCHMENT STUDIES

Snake Pass

The Catchment   The original objective of the study was to
identify acid streams in the UK and follow the climatic and
seasonal fluctuations of water quality and biota.   There is
little information in the UK on the chemistry of acid streams, and
changes in water quality that occur as the stream passes through
the catchment.   The streams chosen for study rise above Snake Pass
in the Peak District of Derbyshire, a region of carboniferous
limestone overlain by millstone grit.   The area lies between the
industrial conurbations of Manchester to the south west and
Sheffield to the east; the ambient sulphur dioxide concentrations
in the catchment are below 40 $\mu g\ m^{-3}$.

The streams sampled rise in peat at about 500 m, but then
flow through afforested areas.   The head stream, Thomason's
Hollow, is in open moorland with a catchment of 0.49 km^2;   Lady
Clough lies in the flood plain below a region of mainly coniferous
forest, with a catchment of 6.44 km^2 (Fig. 1).

Stream water samples have been collected every other week for
chemical analysis from September 1976 to December 1977.   In 1976
rainfall data were obtained from the Severn Trent Water Authority,
but in 1977 a sampler was placed near Thomason's Hollow, and
weekly samples taken for analysis.   Invertebrate samples were
collected monthly.

Results   The results have shown that the first order stream
(Thomason's Hollow) has a pH consistently below pH 4.5 and often
below pH 4.0 (Fig. 2), whereas Lady Clough, 150 m lower, is much
more variable, being near neutrality at times of low flow, but
acid ($\sim$ pH 4.0) at times of high flow (Fig. 3).

The summer of 1976 was exceptionally hot and dry and ended in
September with very heavy rain; the rain pattern in 1977 showed no
prolonged dry spells, with a total rainfall of 1,453 mm.   The rain
was always acid $\sim$ pH 4.2, and contained between 68 and 102 $\mu eq\ l^{-1}$
sulphate.   Bimonthly weighted mean values for H^+ and $SO_4^=$ are
given in Table 1.

The large output of sulphate from Thomason's Hollow and
Lady Clough in January/February 1977 was largely due to the very
heavy rainfall in these months, but could also have included the
residue of the dry deposits accumulated during the long dry
summer of 1976.

TABLE 1 HYDROGEN AND SULPHATE BALANCES – SNAKE PASS (Bimonthly weighted means)

Date	Stream	Hydrogen				Sulphate				Balance	
		Rain In μeq l⁻¹	Rain In Eq ha⁻¹	Output μeq l⁻¹	Output Eq ha⁻¹	Rain In μeq l⁻¹	Rain In Eq ha⁻¹	Output μeq l⁻¹	Output Eq ha⁻¹	H^+ Eq ha⁻¹	$SO_4^=$ Eq ha⁻¹
Sept/Oct 76	TH*		230**	410	530		350**	680	890		
	LC		230	24	60		350	474	1090		
Nov/Dec 76	TH			259	580			392	880		
	LC			20	30			446	710		
Jan/Feb 77	TH	64	220	237	1020	85	290	276	1190	TH −1310	−1550
	LC			76	310			310	1260	LC −230	−1910
Mar/April 77	TH	37	70	204	580	96	190	295	840		
	LC			64	210			347	1130		
May/June 77	TH	51	100	224	170	103	200	363	280		
	LC			33	20			411	210		
July/Aug 77	TH	51	100	92	90	103	200	221	210	TH −110	−90
	LC			2	0.6			346	120	LC +199	−60
Sept/Oct 77	TH	13	30	142	80	69	150	279	150		
	LC			13	10			410	280		
Nov/Dec 77	TH	45	140			97	290				
	LC										

* TH – Thomason's Hollow　　　LC – Lady Clough

** Estimated from Goyt Valley data[1]

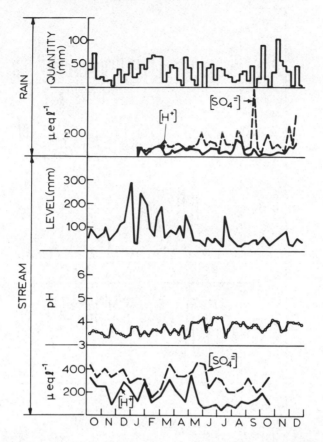

Fig. 2 Seasonal changes in rain and stream composition Thomason's Hollow, Snake Pass.

In the winter there is a net loss of hydrogen and sulphate from both catchments. In the summer months (May-October) the input of hydrogen and sulphate in rain virtually balanced the loss in the stream in Thomason's Hollow; in Lady Clough however, there was a net retention of hydrogen within the catchment, and a small loss of sulphate. The lower hydrogen ion concentration in Lady Clough compared with Thomason's Hollow is balanced by higher cation concentrations.

In the winter months the volume of water falling on the Lady Clough catchment as rain is equalled by the volume leaving in the stream. The reduced flow in the summer may be accounted for by evapotranspiration.

It is generally assumed that on balance the sulphate ion is not retained in the system[2]. The excess sulphate leaving these catchments must, in the absence of a known geological source, have originated as dry deposit. The mean concentrations of hydrogen and sulphate in the streams are:

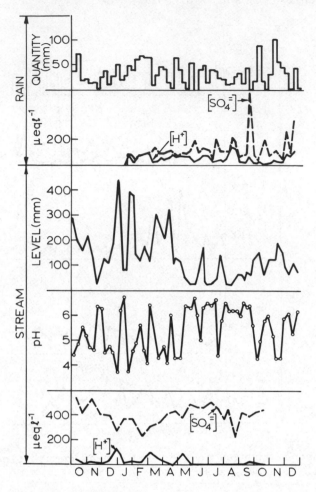

Fig. 3. Seasonal changes in rain and stream composition Lady Clough, Snake Pass

$$H^+ \quad \begin{matrix} \mu eq\ 1^{-1} \\ SO_4^= \end{matrix}$$

	H^+	$SO_4^=$	
Rain	45	91	Jan – Dec 1977
Thomason's Hollow	225	308	} Nov 1976 – Oct 1977
Lady Clough	60	346	

Assuming that rain volume and stream output volume balance in winter, and there is no evaporation these figures suggest that about two thirds of the sulphate leaving the catchment has been derived from dry deposit.

A very rough estimate of the annual dry deposition, using an ambient SO_2 concentration of 40 μg SO_2 m^{-3}(3) and a deposition velocity of 8 mm s^{-1}(1), gives a value of 50.4 kg S ha^{-1} yr^{-1} which

TILLINGBOURNE

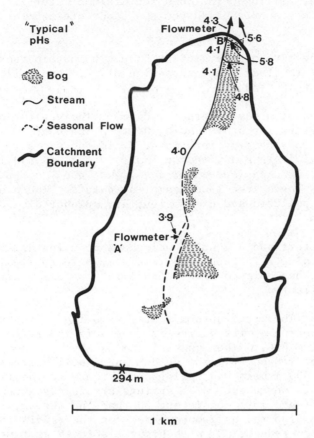

Fig. 4.   The Tillingbourne Catchment with typical pH values.

may be compared with a figure of 21.2 kg S ha^{-1} for the wet deposition.

Benthic macroinvertebrate samples taken monthly show that seasonal changes in numbers followed a similar pattern in Thomason's Hollow and Lady Clough and that the total numbers of organisms were also very similar.   The variety of species identified was, however, lower in the consistently acid stream, and in particular, Ephemeroptera were entirely absent.

Tillingbourne

The Catchment   The Tillingbourne commences as a small seasonal stream on the North slopes of Leith Hill in Surrey. Leith Hill (294 m) consists of a layer of Lower Greensand (Hythe Beds) overlying impermeable clay.   Springs occur at the junction

of the two formations and bogs where the clay approaches the
surface. The Tillingbourne does not drain a water-tight catchment,
and some water is extracted from the sandstone aquifer for potable
use.

The objective of the project is not to obtain mass balances
of ions, but to study the interaction of rain and soil, especially
with regard to H^+ and $SO_4^=$

The area of study covers $\sim$ 1.0 km^2. The vegetation includes
mixed coniferous forest (*Pinus sylvestris* and *Larix decidua*) on
the upper slopes merging into mixed deciduous trees (largely
Betula pendula, Alnus glutinosa, Fagus sylvatica, and *Ilex
aquifolium*); bracken (*Pteridium aquilinum*) and *Vaccinium myrtillus*;
and *Sphagnum* bog at the lower sampling points. The area is not
regarded as polluted and has ambient air sulphur dioxide levels
of $\sim$ 20 $\mu g \ m^{-3} (4)$

Flow meters and a V-notch weir have been installed together
with a recording rain gauge and rainfall collectors, although
storm damage has delayed flow measurements. Soil water samples
are also collected.

Results The first autumn/winter sampling period shows that
the Tillingbourne at its source is more acid (pH 3.7 - 4.0) than
rain (pH 4.1 - 4.6). The upper stream is seasonal and some
acidity is picked up from the acid A and B soil horizons (pH 3.6).
After about 500 m bogs and springs contribute to the flow and the
stream is more permanent. The springs are fed by ground water and
have a higher and more consistent pH than the stream, $\sim$ pH 5.6 - 5.9.
The bogs are also fed by ground water but the acidifying action of
the vegetation reduces pH, with highest acidity at times of highest
flow. The overall effect of these inputs, however, is to raise
the pH as it flows down the catchment, Fig. 4 shows "typical" pH
values obtained in September 1977.

Sulphate in rain is 50 - 105 $\mu eq \ 1^{-1}$. In the stream sulphate
concentration falls progressively down the catchment. In the
seasonal upper part of the stream where input is largely inter-flow
values of 780 $\mu eq \ 1^{-1}$ have been recorded, but ground water contains
only about 370 $\mu eq \ 1^{-1}$ and the water draining from the bogs has
even lower concentrations (Fig. 5). The decreasing sulphate
concentrations down the catchment are probably partly due to
dilution, but bacterial sulphate reduction in the bogs may also
be important. This reduction in sulphate concentrations was not
seen in the Snake Pass streams, where the bogs do not appear to
be anaerobic and gradients are steeper. An estimate of dry
deposition to a catchment of such varied vegetation can only be
very approximate. If a mean ambient SO_2 concentration of 20 $\mu g \ m^{-3}$
is taken, and a deposition velocity of 8 $mm \ s^{-1}$ a value of

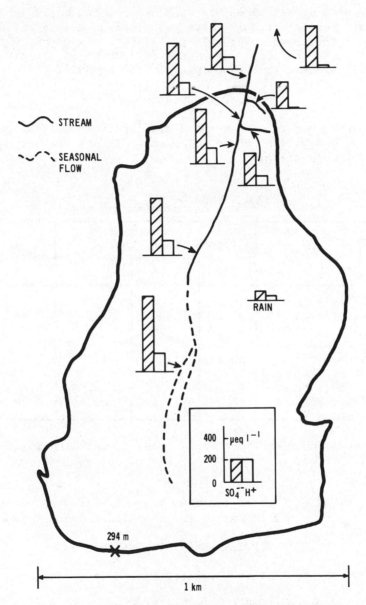

Fig. 5. Mean sulphate and hydrogen ions in the Tillingbourne 28.9.77 – 21.12.77.

25.2 kg S ha^{-1} yr^{-1} is obtained.

The area of catchment above the upper flow meter is 0.254 km^2. A sulphate budget for July–January indicates an apparent net retention of sulphate over this period with a measured wet input of 516 kg SO$_4$, a calculated dry input of 1330 kg SO$_4$, but an estimated stream output of only about 150 kg SO$_4$.

Dry Deposition

Table 2 compares the wet and dry inputs of sulphur to the Snake Pass and Tillingbourne catchments with values for Southern Norway for comparison.

TABLE 2 SULPHUR DEPOSITION

	S Deposition kg ha^{-1} yr^{-1}			Dry/Total %
	Dry	Wet	Total	
Tillingbourne* (July 77–Jan 78)	25.2	13.5	38.7	65.2
Snake Pass (Jan – Dec 77)	50.4	21.2	71.6	70.4
Birkenes, Norway[2] (1974–1975)	11.7	18.4	30.1	38.9

*Values doubled to estimate annual figure; wet deposition will be overestimated.

In the UK dry deposition thus accounts for $\sim$ 70% of the total sulphur inputs, in contrast to the much lower value for Norway.

LYSIMETER LEACHING EXPERIMENTS

The Lysimeters – Collection and Installation

The lysimeters used in the leaching studies contain undisturbed cores of a podzolic Lower Greensand soil (Folkestone beds). It would have been advantageous for comparative purposes to use soils of similar parent material to the soils of Southern Norway, but in the UK these only occur in Scotland. Furthermore the terrain is too hilly and the soils too thin and stony for the collection of the Agricultural Research Council "monolith" lysimeters that we have used. Greensand is acid and has the

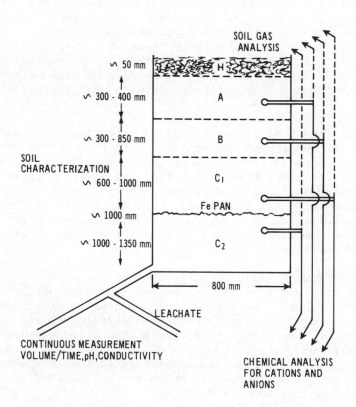

Fig. 6.   Lysimeter experiments.   Lysimeters watered with 10 mm, 3 times
          a week H_2SO_4 pH3 on nos. 2 and 4.   Distilled water pH 5.6 on
          nos. 1 and 3.

advantage of being relatively homogeneous through the soil depth.
The A and B horizons are each of approximately 300 mm depth, which
facilitated insertion of soil water collectors in each horizon.

The cores were collected from a Forestry Commission site at
Headley, west of Farnham in Surrey. The vegetation included
bracken (*Pteridium aquilinum*) and *Deschampsia flexuosa*, with mixed
deciduous and coniferous trees, mainly *Betula pendulosa*, *Pinus
sylvestris*, and *Tsuga hetorophylla*. Four monoliths were collected
in 1976, along a N-S line with 1500 mm between centres. Fibreglass
cylinders, 1350 mm deep and 800 mm diameter fitted with a steel
cutting ring on the bottom and a pressure plate on top were slowly
forced into the soil using a 10-ton jack and a jig. The lysimeters
have been installed at CERL in a trench with the surface at ground-
level and back-filled with sand. They were watered with distilled
water in addition to natural rain for $\sim$ 15 months prior to the start
of "acid rain" watering.

At the same time as the monoliths were taken soil samples were
collected for subsequent chemical analysis from each of three faces
adjacent to the monolith. Although there were differences in the
depth of the horizons in each monolith the horizons could be clearly
differentiated as humus, a dark gray A horizon with leached sand
grains, and a brownish B horizon. What has been called the C
horizon, a yellow/brown sand, was divided into two layers by an
'iron-pan' or concretion about 10 mm thick at a depth of between
850 - 1120 mm (Fig. 6).

In addition to leachate collection at the base of the
lysimeters soil water is also extracted from each horizon using
ceramic cup probes. The cups 54 x 21 mm diameter (Soil Moisture
Corp.) were bonded to a plastic tube, and stainless steel tubing
used to connect the probes to a vacuum line after passing through
a collecting tube. A vacuum of $\sim$ 500 mm is used to extract the
soil water; 60 ml samples are usually extracted in a few hours
from the A and C_2 probes but 15 hours may be required to collect
sufficient samples from the B and C_1 layers.

Fresh litter from an equivalent area (0.5 m^2) at the original
site has been placed on the lysimeter surfaces in March 1977 and
January 1978.

The watering regime used is equivalent to 10 mm of rain to
each lysimeter on Mondays, Wednesdays and Fridays, although with
occasional gaps, giving a total volume of 750 l rainfall
equivalent to $\sim$ 1500 mm per year. The distilled water used as
control on lysimeters 1 and 3 has a pH $\sim$ 5.6, and the acid rain
is sulphuric acid at pH 3.0. The acid input to "acid" lysimeters
is then about 10 times the natural wet deposit in S. Norway
(1500 mm, pH 4) and about 20 times that at Headley (800 mm, pH 4).

Soil Properties

The soil samples collected as the monoliths were taken give the base-line soil properties for all future tests. The results of the analyses are given in Table 3. For each horizon 12 samples were available, one each on the east, north and west sides of each of the four cores. These samples show some differences between them - No.2, in particular, has the most uneven soil horizons. The soils have very low cation exchange capacities and in this respect are poorer than the soils used in the SNSF lysimeter studies.

The total volume of the lysimeters is 679 l with an estimated total pore volume of $\sim$ 360 l. Plug flow predicts a period of approximately six months for an applied substance to appear in the leachate. Initial high volumes of Ca in the leachate can be attributed to cement used to seal the lysimeters but this has now been leached. The cations, Na, K, and Mg have remained fairly constant in concentration, but the Fe, Al and Mn values tend to fluctuate. The pH of the leachate was initially affected by the cement, but is now stable at about 4.5.

The Effects of Acid Watering

The hydrogen and sulphate inputs and losses from the lysimeters during the period of acid watering are given in Table 4. With only three quarters of the pore volume through, it is not expected that the sulphate "wave" would be detected in the leachate.

Since April 1976 L2 has lost 8.8 g of sulphate compared with 8.4 g from L4, and the recent increased loss from L4 may represent leakage down the wall on a few waterings.

The pore volumes above each probe have been calculated and the time at which increased levels of sulphate should be collected at the probes in L2 and L4 determined. These predicted times agree very well with the actual times shown in Fig 7A and B, confirming that sulphate moves through the soil column as a plug flow. Examination of the concentrations however, indicate that some interactions occur. In the A layer the $SO_4^=$ concentrations in the acid treated lysimeters (2 and 4) vary $\pm$ 20% above and below the level predicted assuming no interactions. The sulphate concentration in the soil water in the B layer is below that predicted - indicating some adsorbtion. Adsorbtion also occurs in the C_1 layer, for although the sulphate arrived at the predicted time, the beginning of February for L2 and the end of February for L4, the concentration was lower than expected. The sulphate has not yet penetrated past the iron-pan.

The soil water and leachate samples have always shown fluctuating pH, but as yet there is no obvious effect following

TABLE 3 LYSIMETER SOIL CHEMISTRY (Lysimeters 1-4)

Average for 12 samples ± 95% confidence interval

Soil Horizons	H	A	B	C_1	C_2
pH (1:2.5 distilled water)	4.0 ± .2	3.9 ± .1	4.2 ± .1	4.4 ± .1	4.4 ± .1
Percent loss on ignition at 450C	54.0 ± 15.0	5.7 ± 1.1	2.0 ± .2	1.7 ± .3	1.4 ± .3
Cation Exchange Capacity [1]	41.5 ± 18.0	5.7 ± 1.2	2.4 ± .8	1.7 ± .8	1.3 ± .7
Total Exchangeable Bases [1]	13.1 ± 3.6	0.2 ± .2	0.3 ± .3	0.3 ± .2	0.2 ± .1
Exchangeable Acidity - Hydrogen[1]	1.4 ± .3[2]	0.8 ± .1	0.2 ± .1	0.1 ± .1	0.1 ± .2
Exchangeable Acidity - Aluminium[1]	0.2 ± .2[2]	2.0 ± .2	1.8 ± .3	1.4 ± .3	1.5 ± .4
Total Nitrogen [3]	2.0	0.1	0.04	0.04	

[1] meq 100 g^{-1} oven dry (105C) soil
[2] Average of 10 samples only
[3] Percent oven dry (105C) soil. Estimated values from pot experiment samples.

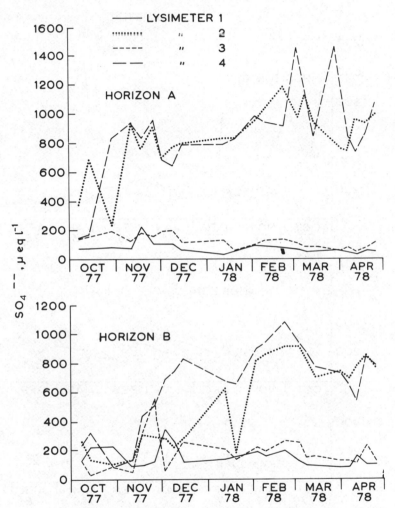

Fig. 7A Sulphate levels in lysimeters 1-4 at horizons A,B,C_1,C_2 and leachate.

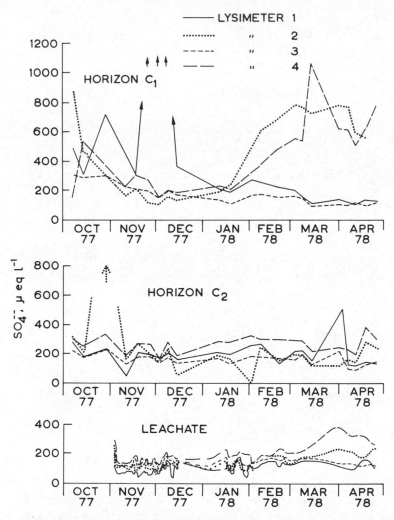

Fig. 7B Sulphate levels in lysimeters 1-4 at horizons A, B, C_1, C_2 and leachate.

TABLE 4   H and SO₄ BUDGETS ON LYSIMETERS
10.10.77 → 10.4.78

Volume of rain, Lysimeters 1 and 3 (distilled water pH 5.65)
= 333.75 1 (≡ 667 mm)
Lysimeters 2 and 4 (H₂SO₄ pH 3)
= 323.67 1 (≡ 667 mm)

	Vol. of leachate collected 1	Hydrogen mg	Sulphate g
Distilled water Input		0.8	0
Output L1	293	12.7	1.8
L3	296	13.6	2.0
H₂SO₄ Input		324	15.5
Output L2	284	7.2	1.9
L4	276	9.9	2.5

TABLE 5 SOIL WATER AND LEACHATE pH MAY 1978

	Lysimeter			
	Control		Acid Rain	
	1	3	2	4
Soil water A horizon	3.98	4.38	3.53	3.88
Soil water B horizon	4.23	4.50	4.08	3.79
Soil water C_1 horizon	4.40	4.18	4.38	4.33
Soil water C_2 horizon	4.25	4.35	4.40	4.23
Leachate	4.40	4.65	4.50	4.65

the application of acid. The mean pH values for the three most
recent samples are given in Table 5.

The sulphate in the soil water must be accompanied by an
equivalent quantity of cations, as there is no corresponding
increase in acidity in soil water of the A, B and C_1 horizons of
L2 and L4. The analysis for cations is still incomplete, but the
Ca concentration in the soil water of the A horizon in lysimeters
2 and 4 is already slightly higher than in the corresponding
horizon of L1 and L3.

Future Studies

The results given here are of a preliminary nature, and it
will be some time yet before the changes caused by acid rain reach
equilibrium. The continued addition of sulphuric acid at pH 3 may
reduce the soil water pH, but how the C.E.C. and exchangeable
acidity of the soil are affected will only be ascertained when
the lysimeters can be opened up.

It is proposed to examine the effects of different watering
regimes, and of different 'rain' compositions. The planned
automatic recording of volume of leachate, and pH should enable a
more complete picture to be obtained of the changes in composition
with time. The use of selective ion probes for the leachate and
soil water lines will also be examined and may provide detailed
automatic monitoring of changes following watering.

POT EXPERIMENT

Rationale

It has been postulated that soil micro-organisms may be
adversely affected by acid rain. If the activity of the
micro-organisms is reduced, nutrient cycling through litter
decomposition will be slowed. The organisms of the N-cycle, in
particular, are thought to be at risk. The lysimeters cannot be
used for destructive microbiological studies, and a pot-scale
experiment has been set up so that the microbial processes can be
investigated.

When the first series of lysimeters was taken sufficient soil
was collected from each horizon to fill ∿ 50 pots 170 mm diameter
and 150 mm deep. Each horizon was well mixed before filling out.
The horizons are being examined individually and in stacks of
H, A, B, and C; H, A and B; and H and A. The pots received
natural rain and distilled water before the acid watering regime
started in July 1977. Half the pots receive distilled water and
the remainder sulphuric acid pH 3. It has not been possible to
water as frequently as for the lysimeters as the pots became
water-logged.

At six-monthly intervals two pots are taken from each
horizon and treatment and the soil subjected to a range of
chemical, microbiological and biochemical tests as indicated in
Table 6.   The leachates are also collected for analysis for
comparison with the lysimeter leachate and probe samples.

TABLE 6 POT EXPERIMENT

Substrates - humus and soils from A, B and C horizons singly and
            stacked.
Treatments - distilled water or H_2SO_4 pH3

Tests at 6 month intervals

1) Chemistry
    Soils analysed for pH, C.E.C., T.E.B. Exchangeable acidity,
    total nitrogen, loss on ignition.

2) Microbiology
    Enumeration of bacteria, fungi and Actinomycetes (6)*
    Growth on media of different pH (8)
                                    { N-cycle (7)
    Enumeration of bacteria of
                                      S-cycle (3)
    Specific substrate - decomposition (8)
                       - utilization (5)
    Enzyme production (6)
    Respiration of undisturbed cores
    n-mineralization
    Loss of tensile strength of buried cotton strips

3) Biochemistry
    Enzyme activities
    Mineralization of ^{14}C-labelled substrates

( )* indicates number of tests in group

Microbiology

The microbial numbers are estimated by the Most Probable
Number method on plates and tubes, and include media for a wide
range of micro-organisms and activities. It is hoped the results
will be more informative than those that would be obtained from a
few labour-intensive, but more precise plate counts.

The results will be examined by the principal components
technique to identify any effect or trend caused by the acid
watering. Principal component analysis on the base-line results

TABLE 7 RESULTS AFTER 6 MONTHS ACID WATERING: POT EXPERIMENT

SINGLE POTS

	H		A		B		C	
	Control	Acid	Control	Acid	Control	Acid	Control	Acid
Soil pH	3.88	3.68	4.03	3.58	5.10*	4.05	4.65	4.20
Loss on Ignition % (450C)	80.9	78.6	4.76	4.75	1.68	1.79	1.58	1.59
Total Nitrogen %	1.9	1.9	0.1	0.1	0.04	0.04	0.04	0.04
Fungi at pH 6 (x 10^7)	53.2	74.6	6.9	5.5	~0.5	~0.6	~0.8	~0.6
Bacteria at pH 6 (x 10^8)	62.4	23.0	11.0	9.3	18.0	3.4	11.6	4.4
Actinomycetes at pH 4 (x 10^6)	85.4	56.7	7.2	3.7	1.7	0.9	0.4	1.7
Azobacter	Not detected		6500	4700	8520	14,040	6700	2900
Carbon respired, 23 days (µg)	3966	3618	117	119	34.9	44.5	46.2	39.3

STACKED POTS

	H		A		B		C	
	Control	Acid	Control	Acid	Control	Acid	Control	Acid
Soil pH	3.88	3.70	3.68	3.70	3.88	3.90	4.18	4.05
Loss on Ignition % (450C)	81.4	80.7	4.8	4.6	1.6	1.8	1.8	1.7
Total Nitrogen %	1.9	1.9	0.1	0.1	0.04	0.04	0.04	0.04
Fungi at pH 6 x 10^7	118	59.6	4.8	3.3	~0.6	~1.0	~0.6	~0.8
Bacteria at pH 6 x 10^8	10.1	12.6	3.0	1.2	1.5	1.2	2.0	4.0
Actinomycetes at pH 4 x 10^6	~16.0	15.9	0.5	0.4	0.6	0.2	0.7	0.7
Azobacter	Not detected		~2000	4300	18,300	12,500	4020	5000
Carbon respired, 23 days (µg)	3443	3183	122	135	34.0	30.7	15.7	16.8

All results based on oven dry (105C) soil
All means of duplicate samples except respiration (4 replicates)
*Abnormally high result due to one sample only (pH 5.75)

obtained before acid watering started has shown, as expected,
significant differences between horizons with no overlap between
horizon populations when the positions of the sample points on the
first three principal components are plotted. The results also
showed greater variability within the litter population. The pots
have just been sampled following six months acid watering but a
full analysis of the data is not yet available. Some results are
presented in Table 7. The soil and leachates from the acid treated
pots had a lower pH than controls in the single layer pots but the
differences were not so clear in the stacked pots.

     The fungal numbers showed a greater decrease through the
horizons than the bacterial numbers. The decrease is presumably
more a reflection of nutrient supply than pH. The bacteria and
fungi were counted on media of pH 4, 5, 6, and 7 and although the
fungi showed no consistent pattern, the bacteria grew better at
the higher pH's with usually an optimum at pH 6. This pH optimum
was also noted in the tests carried out before acid watering
started. Although the average microbial numbers were often lower
in the acid treated pots than the controls there were sometimes
greater differences between duplicate samples than between the
treatments.

     Azotobacter were unusual because they were not isolated from
the humus layers, but occurred in about equal numbers in the
A, B and C horizons.

Respiration Measurements

     Before the pots are emptied and mixed for microbiological
and chemical sampling two cores are removed from each pot using
plastic tubes 150 mm x 20 mm i.d. with holes drilled in them to
prevent the cores becoming anaerobic. The cores are weighed and
placed in glass tubes each fitted with a rubber bung, glass tap
and gas sampling system. The tubes are incubated at $20^{\circ}C$ and
the CO_2 production measured at intervals on a Carle 311H
Chromatograph. The tubes are flushed through with air 24 hours
before measuring respiration. Some results are illustrated in
Fig. 8 and Table 7. Respiration rates were some 30 times greater
in the humus layer than in the A horizon and 100 times greater
than in the B and C horizons.

     Variability between cores was greatest in the H and A horizons,
but in no horizon in single or stacked pots was there a significant
difference between acid watered pots and control. A depression
in respiration was observed in the humus pots treated with acid
but the leachate from the H horizon in the acid treated stacked
pots appears to have stimulated growth in the A horizon below.
Whether these trends will continue as the watering proceeds is
not known.

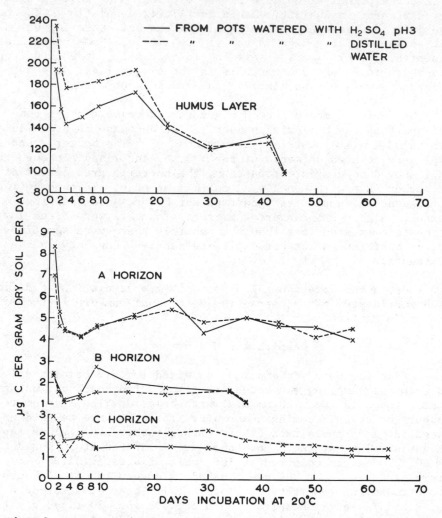

Fig. 8. Carbon dioxide produced in 24h from pot soil cores.

Biochemistry

The soils from the pot experiments are also being used to determine soil enzyme activities and the mineralization rates of ^{14}C-labelled substrates.

The soil enzymes are determined at 37°C under standard conditions in the presence of toluene as a bacteriostatic agent. Although the conditions are not necessarily optimal for each enzyme they are comparable between the different horizons and treatments. The enzymes tested include phosphatase, sulphatase, β-glucosidase, chitinase and xylanase. ^{14}C-labelled substrates are incubated with moist soil (0.3 μ mole g^{-1}) at 20°C. The CO_2 evolved is collected and the radioactivity determined by liquid scintillation spectrometry. The substrates being examined are D-glucose, L-glutamate, urea, benzoic acid, uric acid and acetate.

Preliminary results from the stacked pots, tested by an analysis of variance show no significant changes in enzyme activity or mineralization rates following acid watering, but a markedly higher activity in the humus than in the A horizon, which was in turn higher than in B and C horizons. Some values are given in Table 8.

Leachates

In contrast to the lysimeters the pot leachates have shown an effect of acid watering. The pH of the leachates of all pots has fallen since they were first measured in 1976, and all have shown considerable variation. Table 9 gives the mean hydrogen ion concentration in the pot leachates after 6 months acid watering.

Although there is only one possibly significant difference between the mean hydrogen ion concentrations in the leachates from the control and acid watered pots, the controls are always less acid than the acid treated pots for each horizon. The mean pH of the leachate from the stacked acid treated pots was higher than from any individual horizon. The results from the control pot leachates (Table 9) show that much of the acidity could be attributed to the natural soil acidity. Even with rain at pH 3 much of the hydrogen added as rain is retained in the soils.

The sulphate concentrations in the leachates are lower than expected, except in the leachate from the humus layer. Some interactions of sulphate with the soil must be occurring in the A, B and C layers, as postulated from the lysimeter probe results.

TABLE 8. MEAN ACTIVITIES OF SOME SOIL ENZYMES IN STACKED POTS.

HORIZON	ACTIVITY	PHOSPHATASE		SULPHATASE		β—GLUCOSIDASE		CHITINASE		XYLANASE	
		A	C	A	C	A	C	A	C	A	C
H		89.81	81.23	0.52	0.71	8.22	6.11	2.47	3.60	1.08	1.19
A		6.08	5.96	0.13	0.17	0.14	0.14	0.14	0.14	0.05	0.05
B		0.86	0.92	0.04	0.03	0.04	0.04	0.04	0.04	0.02	0.02
C		0.80	0.62	0.02	0.02	0.04	0.03	0.02	0.02	0.01	0.01

Units of activity: μmoles substrate hydrolysed hr^{-1} g dry $soil^{-1}$ at 37°
* A = acid treatment (pH 3)
  C = control (distilled water)

TABLE 9. RESULTS FROM POT LEACHATE ANALYSES

Horizon	Date sampled	H^+ μeq 1^{-1}		$SO_4^=$ μeq 1^{-1}	
		Acid	Water	Acid	Water
H	10.3.78	507 ± 130	334 ± 74	1436 ± 79	358 ± 91
A	3.3.78	442 ± 237	219 ± 73	877 ± 246	282 ± 97
B	17.3.78	133 ± 52	94 ± 136	912 ± 254	193 ± 109
C	24.2.78	118 ± 15	37 ± 38	1121 ± 233	228 ± 49
H,A,B,C Stacked	23.2.78	67 ± 11	32 ± 21	708 ± 162	244 ± 61

Results are means of between 7 and 11 replicate samples with 95% confidence intervals

DISCUSSION AND CONCLUSIONS

The data given here are preliminary and it is too soon to draw conclusions about the effects of acid rain on soil systems, but valid observations have been made on soil processes, and on the behaviour of H^+ and $SO_4^=$.

The field studies have shown that the acidity of streams decreased with distance from the source whereas sulphate concentrations increased downstream at Snake Pass and decreased in the Tillingbourne. This difference may be due to greater dry deposition in the former, or to sulphate reduction and possibly dilution by ground water in the latter, or to different pathways through soil horizons in the two catchments.

Experimental acid watering of monolith and pot soils suggests some retention of sulphate on soils which is unlikely to occur in the field sites where the rainfall input is less and where a long term equilibrium is established.

The vertical movement of sulphate through the greensand monoliths requires $\sim$ 60 mm of "rain" for each 100 mm depth of soil, the sulphate moving with the water front; it is not known if this soil is representative, and whether vertical flows help to understand interactions in hilly areas where lateral flows are more important.

The interactions of hydrogen ion are more complex. In the monoliths, H^+ is neutralised by other cations even in the acid greensand, but in acid watered pots the higher acidity of leachates suggests that a large part of the H^+ is derived from the soil itself. These differences may result from different times and volumes of "rain" and further more quantitative experiments are planned to elucidate the effects of differing water regimes and quantities.

The microbiological and biochemical tests show no significant effects after 6 months of acid "rain." The pot experiments will continue for a further 2 years and the same tests will be used at field sites.

ACKNOWLEDGEMENTS

My thanks are due to my colleagues for their help with the preparation of this paper, Dr D.J.A. Brown (Snake Pass), Dr K.A. Brown (biochemistry), Mr D.J. Foster (chemistry), Dr R.A. Skeffington (Tillingbourne), Mr J.F. Spencer (soil characterisation), Miss M.J. Wood (microbiology).

The work was carried out at the Central Electricity Research Laboratories and is published by permission of the Central Electricity Generating Board.

REFERENCES

1. P. K. Holland, Initial observations of sulphur transfer
    through the Goyt catchment where the major input is by dry
    deposition, C.E.G.B. North Western Region NW/SSD/RN/7/77
    (1977).

2. E. T. Gjessing, A. Henriksen, M. Johannessen and R. F. Wright,
    Effect of acid precipitation on freshwater chemistry, SNSF
    Research Report FR6/76 (1976).

3. A. Martin, (personal communication).

4. A. R. W. Marsh, (personal communication).

SITE SUSCEPTIBILITY TO LEACHING BY H_2SO_4 IN ACID RAINFALL

D.W. Johnson

Environmental Sciences Division, Oak Ridge National

Laboratory[1], Oak Ridge, Tennessee 37830

INTRODUCTION

One of the major concerns about the terrestrial effects of acid rainfall has been the leaching of nutrient cations from forest soils and the consequences of this to forest growth. A concept that has proven useful in assessing soil leaching due to acid rainfall and many other manipulations is that of anion mobility in the soil[2,3,4]. This concept does not consider cation exchange reactions directly, but the sum of cations must balance the sum of anions, in meq/l, in the soil solution. Total ionic leaching can, therefore, be described by anion flux alone. Although much more basic research has been devoted to cation adsorption and exchange than to anion adsorption and exchange, the anion mobility concept has proven very valuable in explaining soil leaching processes in both undisturbed and disturbed conditions[2,3,4].

The effects of atmospheric sulfuric acid inputs on forest soil leaching have been evaluated from the perspective of the mobility of the sulfate anion in several ecosystems. In this paper, the factors affecting sulfate mobility in soils will be discussed in relation to soil properties and site susceptibility, drawing some examples from specific studies.

---

[1] Operated by Union Carbide Corporation under contract W-7405-eng-26 with the U.S. Department of Energy.

MECHANISMS OF SULFATE ADSORPTION IN SOILS

In general, anion adsorption is associated with hydrous oxides of iron and aluminum in soils, although kaolinite is an adsorber to a lesser extent[5]. There are thought to be two basic mechanisms of anion adsorption: (1) non-specific adsorption, in which anions are held as counter-ions in the diffuse double layer next to positively charged surfaces, and (2) specific adsorption, in which anions penetrate the sesquioxide and enter into co-ordination with two or more ions in the crystal structure [5,6].

Sites for non-specific adsorption lie on the surfaces of sesquioxides, and these surfaces can have a plus, minus, or zero charge depending upon pH[7]. Thus, these surfaces become more negatively charged and adsorb more anions at lower pH values.

Several mechanisms for specific adsorption have been hypothesized. Hingston, et al[6]. coined the term "specific adsorption" to describe the displacement of H_2O or OH^- from sesquioxide surfaces by another anion. In the case of sulfate, this must result in a negatively-charged surface, but Rajan[8] noted only very small increases in negative charge on hydrous alumina after sulfate adsorption. He proposed that after sulfate adsorption an additional reaction occurs in which OH^- is released and a ring structure is formed between sulfate and 2 aluminum ions (Figure 2).

Specific adsorption is a means by which incoming sulfate ions can be immobilized within the soil, but if other anions are displaced in the adsorption process, acid rainfall will still cause a net loss of cations from the soil. The neutralization of the positive charge on sesquioxide surfaces by sulfate adsorption must cause the release of non-specifically adsorbed anions, such as Cl^- and NO_3^-, or the displacement of hydroxyl[8]. According to the anion mobility model, the rate of leaching is then determined by the fate of such displaced anions.

Little anion displacement was found in a sulfuric acid irrigation study on a gravelly outwash soil in Washington state[9]. In that case, sulfuric acid application had virtually no effect on soil leaching past 50 cm, because both H^+ and SO_4^{2-} were totally absorbed within the soil profile (Table 1). There was some mobilization of exchangeable cations from the A horizon, associated with SO_4^{2-} flux through that level. However, with SO_4^{2-} adsorption being virtually complete in the B horizon, these incoming cations were totally adsorbed and little leaching occurred past 50 cm. These data show that leaching was almost completely regulated by soil sulfate adsorption. Some Cl^- mobilization was noted[10], but it was quite insignificant relative to sulfate adsorbed. These data also indicate that sulfate adsorption has enhanced cation

LOW pH:

Sesquioxide surface
adsorbs protons,
becomes − charged,
and provides sites
for non-specific
anion adsorption.

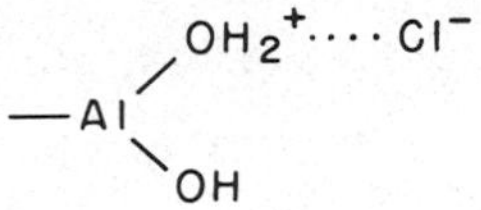

$$-Al \begin{cases} OH_2^+ \cdots Cl^- \\ OH \end{cases}$$

ZERO POINT OF CHARGE:

Sesquioxide surface
has no net charge.

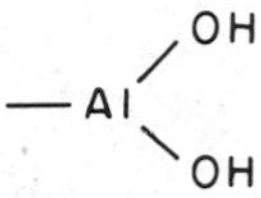

$$-Al \begin{cases} OH \\ OH \end{cases}$$

HIGH pH:

Sesquioxide surface
releases protens,
becomes negatively
charged, and can
adsorb cations.

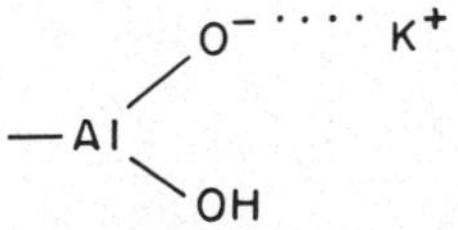

$$-Al \begin{cases} O^- \cdots K^+ \\ OH \end{cases}$$

Figure 1. Schematic diagram of sesquioxide surface showing changes
in surface charge with pH.

LOW pH:

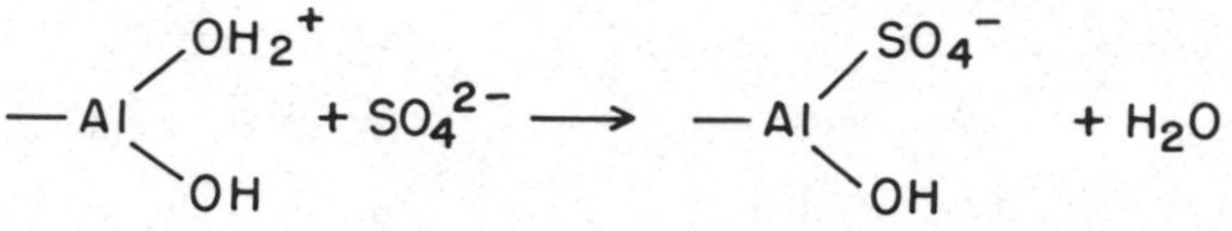

$$-Al\underset{OH}{\overset{OH_2{}^+}{<}} + SO_4{}^{2-} \longrightarrow -Al\underset{OH}{\overset{SO_4{}^-}{<}} + H_2O$$

ZERO POINT OF CHARGE:

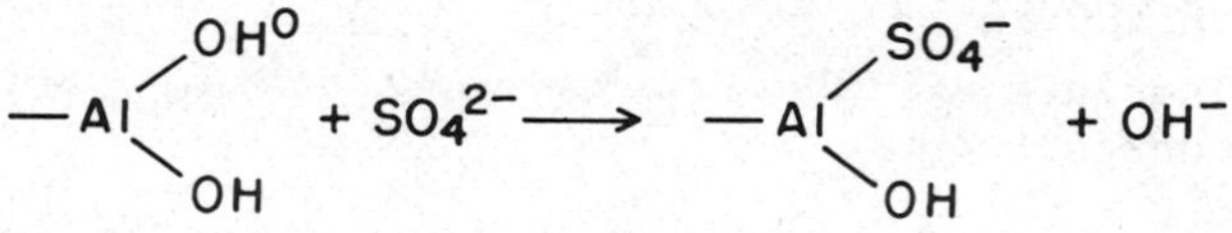

$$-Al\underset{OH}{\overset{OH^0}{<}} + SO_4{}^{2-} \longrightarrow -Al\underset{OH}{\overset{SO_4{}^-}{<}} + OH^-$$

RAJAN'S MODEL:

$$\left[\begin{array}{c} -Al\underset{SO_4}{\overset{OH_2}{<}} \\ -Al\diagdown_{OH_2} \end{array} \right]^0$$

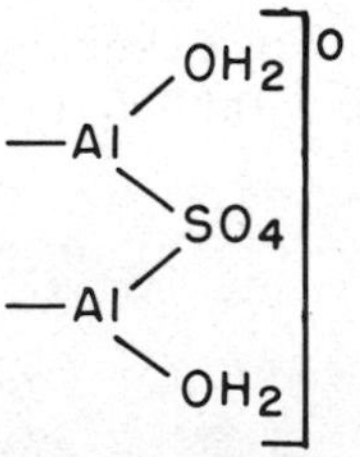

Figure 2. Proposed mechanisms of sulphate adsorption on sesquioxide
           surfaces.

Table 1   Leaching of H^+, SO_4^{2-}, and total cations following
           H_2SO_4 irrigation on a gravelly outwash soil (Johnson
           and Cole[9]). (Values are in equivalents per hectare.)

	A horizon (0-8 cm)	B horizon (8-50 cm)	Total Soil (0-50 cm)
H^+			
Input	35,520	0.2	35,520
Output	0.2	0.0	0
Storage	+35,520	+0.2	+35,520
SO_4^{2-}			
Input	35,520	15,000	35,520
Output	15,000	200	200
Storage	+26,520	+14,800	+35,320
Total Cations (except H^+)			
Input	0	11,200	0
Output	11,200	200	200
Storage	-11,200	+11,000	-200

adsorption as well, since metal cations were immobilized with
SO4 in the B horizon. This is undoubtedly due to the displacement
of the zero point of charge to lower pH values by specific
adsorption of sulfate. This has the net effect of increasing
cation exchange capacity, a phenomenon often noted in connection
with the specific adsorption of phosphate[8].

SOIL PROPERTIES IN RELATION TO SULFATE
ADSORPTION AND SUSCEPTIBILITY TO
LEACHING BY H_2SO_4

Early work by Chao, et al.[11] demonstrated the close relationship between soil sesquioxide content and sulfate adsorption
capacity, and many workers have confirmed this relationship in a
wide variety of soils [12,13,14,15]. Soil sesquioxide content is in
turn related to parent material and the degree of weathering.
Barrow, et al.[15] noted that soils derived from silicaceous rocks
have much lower sulfate adsorption capacity than soils derived
from basaltic materials, presumably due to differences in Fe and
Al contents. Many studies have shown that tropical latosols have
much greater sulfate adsorption capacities than less highly-
weathered temperate soils[12,13,14]. The influence of soil
sesquioxide content on the susceptibility to leaching by acid

Table 2 Sulfur budgets for some research sites subject to acid rain inputs.

Site	Soil type	Sulfur Precipitation	Input Gaseous + Dry $kg\ ha^{-1}\ yr^{-1}$	Sulfur Output	Balance
Hubbard Brook[17] New Hampshire, USA	Podzol, sandy loam, glaciated 10–12,000 yrs.	12.7	6.1	17.6	−1.2
Thompson Site[16] Washington, USA	Podzolic, gravelly outwash glaciated, 10–12,000 yrs.	4.1	?	6.1	−2.0
Solling Site[18] W. Germany	Brown earth, loess loam overlying sandstone.	24.1	17.9(B)* 42.5(S)	25.6 32.4	+16.4 +32.4
Walker Branch[19] Watershed, Tennessee, USA	Red-yellow podzolic, silt loam, derived from dolomitic limestone.	18.1	?	11.5	+7.4
La Selva[14], Costa Rica	Lateric clay, derived from river alluvium.	12.5	?	0.8	+11.7

*B = Beech Forest, S = Spruce Forest

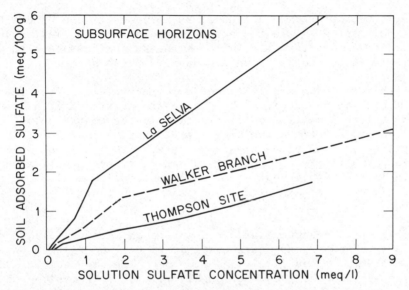

Figure 3. Sulfate adsorption isotherms for soils at the La Selva, Walker Branch and Thompson sites.

rain is quite striking. At temperate A.E. Thompson site[16] in
Washington and northern temperate Hubbard Brook Watershed in New
Hampshire[17], the annual output of sulfate from the ecosystem
exceeds the input (Table 2). In the Hubbard Brook Site, for
which a larger data base is available, the difference is
attributed to gaseous inputs. The soils at these sites differ,
the Hubbard Brook soil being a podzol and the Thompson site soil
a podzolic outwash, but both are glacial and relatively young
(10,000 - 12,000 years old).

In contrast to these sites, there appears to be a sulfur
accumulation in the Solling[18], Walker Branch[19], and La Selva
sites[14], all of which lie on more highly-weathered soils. Dry
deposition is a very important sulfur input pathway to the
Solling site, but the magnitude of such inputs to the Walker Branch
and La Selva sites are not adequately known. It is clear that dry
deposition components would increase the sulfur balance figures of
these as well as the Thompson and Hubbard Brook sites.

Much more information is needed to accurately define the sulfur budgets for these sites, but the patterns are nevertheless
clear: sulfur accumulation is greatest in the most highly-
weathered soils. Given the previous discussions on sulfate
adsorption and soil properties associated with it, it seems both
plausible and likely that sulfur is accumulating as soil adsorbed
sulfate in the highly-weathered, sesquioxide-rich soils. This
hypothesis has been explored in depth and substantially confirmed
for the Solling[20] and La Selva[14] sites.

Soil sulfate accumulation can not go on indefinitely, and
presumably a new steady-state equilibrium will be reached as
soil adsorption sites fill. The response time of a soil to
sulfuric acid leaching can be roughly related to the slope of
its sulfate adsorption isotherm. For example, the sulfate
adsorption per unit change in sulfate concentration is much higher
in the La Selva than in the Thompson site or Walker Branch soil
(Figure 3), thus the effects of a given rate of H_2SO_4 input will
be seen much more quickly in the Thompson soil than in the La
Selva soil. Using sulfate adsorption isotherms such as those in
Figure 3, one can explore responses to H_2SO_4 inputs by construct-
ing simulation models. One such model was constructed using the
finite plate method of Biggar, et al.[21] to simulate soil leaching
following the sulfuric acid irrigation study described earlier[22].
While the simulation produced changes that qualitatively re-
sembled the field situation, it predicted far greater leaching in
upper soil horizons than that which actually occurred (Table 3).
The discrepancy could lie in the assumption of adsorption-
desorption reversibility. Laboratory studies did show that sulfate adsorption was reversible in the study soil, but field

results indicated that over half of the adsorbed sulfate was in a form which could not be extracted with water[24]. Recent investigations have shown that adsorped sulfate, like phosphate, becomes increasingly fixed and unavailable in the soil with time through slow chemical reactions with sesquioxides[25,26]. The magnitude of this slow reaction, or "aging" of soil adsorbed sulfate is not well known as yet, but the existence of this phenomenon serves as a warning to use caution in applying laboratory studies to field situations.

The possibility of channeling as solutions percolate through the soil has been raised in the paper by Bache in this volume, and it deserves consideration here also. If channeling occurs, solutions may not come into equilibrium with the soil and sulfate leaching may be deeper than laboratory studies would predict.

SUMMARY AND CONCLUSIONS

Cation leaching from soils due to atmospheric sulfuric acid inputs cannot occur unless sulfate is mobile in the soil or it displaces another anion that is mobile in the soil. The mobility of sulfate is very much affected by adsorption to sesquioxide surfaces, therefore soils rich in sesquioxides are likely resistant to leaching by sulfuric acid.

Table 3   Simulated and actual sulfate leaching following H_2SO_4 irrigation on a gravelly outwash soil (Johnson and Cole[23]).

Horizon	Sulfate Leaching (eq/ha)	
	Simulated	Actual
A (8 cm)	32,000	15,000
B (50 cm)	150	200

Available data are scanty, but a clear pattern is evident in which forested sites on highly-weathered, sesquioxide rich soils are accumulating sulfur from atmospheric inputs. It is hypothesized that these accumulations are occurring by soil sulfate adsorption.

Laboratory studies can give minimal indices of soil sulfate adsorption capacity, but there are long-term changes in the chemical nature of adsorbed sulfate which result in greater and more permanent retention. Long-term field studies are probably the most suitable way to evaluate a soil's resistance to leaching by sulfuric acid.

REFERENCES

1. G. Backstrand, and H. Stenram, Air pollution across national boundaries: The impact on the environment of sulfur in air and precipitation. Sweden's case study for the United Nations conference on the human environment, p.57 (1971).

2. D. W. Cole, W. J. B. Crane, and C. C. Grier, in: "Forest Soils and Land Management - Proceedings of the Fourth North American Forest Soils Conference", Quebec, p.195 (1975).

3. C. S. Cronan, W. A. Reiners, R. C. Reynolds, and G. E. Lang, Science 200:309 (1978).

4. D. W. Johnson, D. W. Cole, S. P. Geasel, M. J. Singer and R. V. Minden, Arctic and Alpine Research 9:329 (1977).

5. T. Mekaru, and G. Uehara, Soil Sci. Soc. Amer. Proc. 36:296 (1972).

6. F. J. Hingston, R. J. Atkinson, A. M. Posner, and J. P. Quirk, Nature 215:1459 (1967).

7. J. A. Yopps, and D. W. Fuerstenau, J. Colloid Sci. 19:61 (1964).

8. S. S. S. Rajan, Soil Sci. Soc. Amer. J. 42:39 (1978).

9. D. W. Johnson, and D. W. Cole, Water, Air, and Soil Pollut. 7:489 (1977).

10. D. W. Johnson, Unpubl. data (1975).

11. T. T. Chao, M. E. Harward, and S. C. Fang, Soil Sci. Soc. Amer. Proc. 28:632 (1964).

12. E. Bornemisza, and R. Llanos, Soil Sci. Soc. Amer. Proc. 31:356 (1967).

13.   H. Gebhardt, and N. T. Coleman, <u>Soil Sci. Soc. Amer. Proc.</u> 38:259 (1974).

14.   D. W. Johnson, D. W. Cole, and S. P. Gessel, <u>Biotropica</u> (in press).

15.   N. J. Barrow, K. Spencer, and W. M. McArthur, <u>Soil Sci.</u> 108:120 (1969).

16.   D. W. Cole, and D. W. Johnson, <u>Water Resour. Res.</u> 13:313 (1977).

17.   G. E. Likens, F. H. Bormann, R. S. Pierce, J. S. Eaton, and N. M. Johnson, Biogeochemistry of a Forested Ecosystem, Springer-Verlag (1977).

18.   H. Heinrichs, and R. Mayer, <u>J. Environ. Qual.</u> 6:402 (1977).

19.   D. S. Shriner, and G. S. Henderson, <u>J. Environ. Qual.</u> (in press).

20.   P. K. Khanna, and F. Beese, <u>Soil Sci.</u> 125:16 (1978).

21.   J. W. Biggar, D. R. Nielsen, and K. K. Tanji, Trans, ASAE, p.784 (1966).

22.   D. W. Johnson, Processes of elemental transfer in some tropical temperate, alpine and northern forest soils. Ph.D. Thesis, Univ. of Washington, Seattle (1975).

23.   D. W. Johnson, and D. W. Cole, Anion mobility in soils: Relevance to transport from terrestrial to aquatic ecosystems. Ecological Research Series, E.P.A. Corvallis (1977).

24.   D. W. Johnson, Unpubl. data (1978).

25.   N. J. Barrow, and T. C. Shaw, <u>Soil Sci.</u> 124:347 (1977).

26.   F. E. Sanders, and P. B. H. Tinker, <u>Geoderma</u> 13:317 (1975).

ION ADSORPTION ISOTHERMS IN PREDICTING LEACHING LOSSES FROM SOILS

DUE TO INCREASED INPUTS OF "HYDROGEN" IONS - A CASE STUDY

S. Ingvar Nilsson

Department of Plant Ecology, University of Lund

Helgonavagen 5, S-223 62 Lund, Sweden

INTRODUCTION

The question of how the increased amounts of acid deposition over the Scandinavian countries will affect forest primary production is, to say the least, far from being settled.

If the regional acid deposition has any adverse effects on tree growth, they are supposed to be indirect, through alterations of the availability of soil mineral nutrients. From a general knowledge of soil science one can formulate hypotheses concerning the impact on decomposition and mineralization processes, dissolution of precipitated compounds, soil mineral weathering, ion exchange and leaching.

From the forest production point of view, the main interest should be focused on mineral nutrients which are known to be minimum factors. In Scandinavian forests, nitrogen is known to be one of these elements[1,2].

Nevertheless, in order to fully understand the problem the following research strategy is probably the most optimal:

1) Given a certain amount of acid deposition and defined vegetation and soil properties, the first step should be to make quantitative predictions concerning changes of "acid-base" variables in the soil, (pH, base saturation, calcium saturation, acid buffer capacity etc.).

2) The second step would be to look for any significant change in the nitrogen mineralization rate and how it is casually

537

connected with variable changes according to 1). One
crucial question, which has hitherto been practically
neglected, is whether there is any interaction between
increased acid deposition and heavy metals deposited or
already present in the soil. Such an interaction could be
predicted from the solution chemistry of some of these metals
(cf. Tyler[3], who has recently presented a laboratory study
on the subject).

If there is any regional increase in soil acidity, it is most
probably very slight. Most Scandinavian forest soils are podsolic
and have accumulated large stores of "hydrogen" ions during their
pedogenesis. Besides, processes like root uptake and litter
mineralization can account for hydrogen transfers equal to or
larger than the annual acid deposition, (cf. reference 4 and the
literature cited therein). The same is valid for the indirect
"hydrogen" ion input at tree harvest, caused by the export of
mineral nutrients bound in the log biomass. Evaluations of the
atmospheric contribution to the soil acidity, will therefore pose
rather severe problems.

This does not mean however, that the problem can be neglected
at present. There should, on the one hand, be continuous regional
studies concerning soil chemistry and tree growth, and on the
other, field experiments, where the acidification process can be
accelerated at will.

The present study is based on soil samples from such an
experiment and is meant to form a step in the elucidation of
point 1 given above. Acid treatment will of course enhance the
leaching losses of cations adsorbed at the soil colloid surfaces,
as the "hydrogen" ion will interfere with the ion equilibria
between the colloids and the soil solution. These equilibria can
be described with ion adsorption isotherms, where ΔM (the amount
of a certain cation which is desorbed or adsorbed) is a function
of the relative chemical potential (μ) of M in the soil solution
($kcal \cdot mole^{-1}$).

The hypothesis to be tested at the outset of this
investigation was: Does the acidification cause a significant
change in the equilibria between soil solution and soil colloids,
revealed as slope alterations of ion adsorption isotherms?
Calcium adsorption isotherms will be specifically dealth with. A
slope decrease, for instance, means that the calcium ions still
adsorbed to the colloid surfaces, will be more tightly bound than
before the acid treatment and thereby be less available for the
immediate buffering against "hydrogen" ions in the soil solution.

The hypothesis is based on results obtained in previous
laboratory studies, where such alterations were found[5]. A further

description of the adsorption isotherm concept is given under
CALCULATIONS, together with appropriate literature references.

SOIL MATERIAL

Samples were taken from a fertilization experiment in a young
stand of Scots pine (Pinus silvestris L.). A comprehensive account
of the field experiment design is given by Tamm et al[6]. The site,
Lisselbo (160 km N Stockholm), slopes towards the west and is
situated on a glacifluvial eskar.

Three different experiments are being conducted at the site,
two of them solely dealing with the effects of fertilizer
treatment on tree nutrient status and tree growth, whereas the
third includes irrigation, liming and two levels of acid treatment. These four treatments are either combined with, or lack, an
extra addition of NPK-fertilizer. Each treatment is represented
by two 30 x 30 m plots randomly distributed within a block design.

The present study comprises only those plots which have not
received NPK-fertilizer. Effects of liming are under study and
will not be included in this paper.

The site has been irrigated at irregular intervals, during
periods with below average precipitation. Water has been taken
from a nearby oligotrophic lake. Lime corresponding to
1960 kg $Ca \cdot ha^{-1}$ was applied at the outset of the field study (1969).

The two acid treatments correspond to 50 kg $H_2SO_4 \cdot ha^{-1} \cdot yr^{-1}$
and 100 kg $H_2SO_4 \cdot ha^{-1} \cdot yr^{-1}$. Applications were performed from 1970
to 1976 and from 1969 to 1976 respectively.

Soil sampling was performed in the middle of August 1977.

a)   15 soil cores were taken systematically along two diagonals
     in each net plot (20 x 20 m), and combined by soil horizon
     and plot.

b)   In each plot a soil pit was dug within the outer five-metre-
     wide buffer zone. Volumetric samples were taken with steel
     cylinders. In both samplings the A_0-horizon was treated
     separately. The mineral soil was divided into appropriate
     five- or ten-centimetre layers.

LABORATORY METHODS

After sifting through a brass sieve (mesh width two mm) the
samples were analysed for:

a)   Water content (% fresh weight)

b)   Organic matter content as loss on ignition, LI (% dry weight)

c)   Exchangeable amounts of H^+, Na^+, K^+, Ca^{2+} and Mg^{2+} after equilibration with 1M CH_3COONH_4 (pH 7.00), given as $\mu eq.g^{-1}$ (dry weight).

d)   Actual concentrations of H^+, Na^+, K^+, Ca^{2+} and Mg^{2+} in "equilibrium soil solution" (ESS) given in μM.

e)   Ion adsorption isotherms for calcium: "Soil solutions" were prepared from p.a. chloride salts. The cation concentrations were the same as those of ESS, except for Ca^{2+}, the concentration of which was varied in a five-step series from 0 to 200 μM. Soil samples were equilibrated with the appropriate "soil solutions", and from the analysis of H^+, Na^+, K^+, Ca^+ and Mg^{2+} in filtrates and original solutions adsorption isotherms for Ca^{2+} were obtained with ΔM_{Ca} (the amount of calcium adsorbed or desorbed given as μmoles per gram of dry soil) as a function of the relative chemical potential μ of Ca^{2+} in the filtrates (kcal $\cdot$ mole^{-1}). a-b were performed on all samples, c on composite samples only, and d, e, finally, on samples obtained from the soil pits. Analytical details are found in Balsberg[8] and references 5 and 9.

CALCULATIONS

The concept of relative chemical potential was originally given by Woodruff[10]. It has recently been further developed to apply to acid soils[9,11] and is used for interpreting acidification processes[5].

$$\mu = RTln \frac{(\sqrt{a}_{Ca})^4}{a_H \cdot a_{Na} \cdot a_K \cdot \sqrt{a}_{Mg}}$$

a_{Ca} etc. - the cation activity (μM) calculated according to Debye and Hückel, (as given in Hägg[12]; original reference not seen).

R   - universal gas constant in kcal$\cdot$mole$^{-1}\cdot$deg^{-1}

T   - temperature in ^oK

For obtaining adsorption isotherms μCa is plotted agains ΔM_{Ca}:

$$\Delta M = \frac{(C_i - C_e)\cdot v}{1000 \cdot a}$$

C_i - concentrations of Ca^{2+} in the synthetic soil solutions (μM)

C_e - concentration of Ca^{2+} in the filtrates (μM)

a - calculated dry weight of soil samples (g)

v - total volume of equilibrating "soil solution" including the water originally contained in the soil sample. (ml)

Linear regressions between ΔM and μ were found to give the best fit, which is in accordance with Bringmark[9] who worked with a similar kind of soil. In more clayey soils a second degree polynomial has been found to be more appropriate[5,11].

Various regression models have been tested for their summarizing and predictive power. All statistical tests are according to Snedecor and Cochran[13].

The data under c) above were used for the calculation of cation exchange capacity, CEC ($\mu eq \cdot g^{-1}$), the sum of "base" cations, S, ($\mu eq \cdot g^{-1}$), base saturation percentage - $\frac{S}{CEC} \cdot 100$, and calcium saturation percentage, Ca-sat - $\frac{C_{Ca}}{CEC} \cdot 100$.

RESULTS AND DISCUSSION

Test of the Initial Hypothesis

Adsorption isotherms, potential intervals and regression equations are displayed in Figures 1, 2 and 3 for A_0, mineral soil 0-5 cm and mineral soil 5-10 cm respectively. The cationic composition of ESS and the adsorption complex calculated on an equivalent percentage basis is likewise shown in Figures 4 through 6.

It is immediately obvious that there is a large spatial heterogeneity in ion exchange properties, judging from comparisons within horizons and treatments.

From a strictly theoretical viewpoint, an adsorption isotherm which runs from equivalent percentage zero to equivalent percentage 100, calculated on a cation exchange basis should be S-shaped[11], (Figure 7). In acidification studies it should be of most interest to run only through a minor part of the curve, as more drastic changes are difficult to "translate" into a field situation. These minor parts can be approximated to a second, or as in this study, to a first degree polynomial.

One would expect that a diminished calcium saturation percentage should cause a shift from one part of the total isotherm

to another. The saturation percentage interval runs from 14.6
(control plot, A_0-horizon) to 0.2 (irrigation 5-10 cm), all according to Table 1. As there is a pronounced slope increase when
going from mineral soil 5-10 cm to the A_0-horizon, and as the
calcium saturation figures increase in the same direction, but are
rather low taken as a whole, one could expect that a potential
interval shift caused by acid treatment, should reveal itself as
a slope decrease.

A comparison between acidified and non-acidified plots,
(control and irrigation), made for a specified soil horizon, does
not meet the criteria stated above.

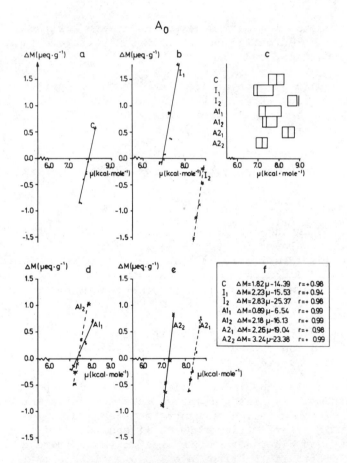

Figure 1. Adsorption isotherms, potential intervals and regression
equations for the A_0-horizon. C; I, A1 and A2 stand for
control, irrigation, 50 kg $H_2SO_4 \cdot ha^{-1} \cdot yr^{-1}$ and 100 kg
$H_2SO_4 \cdot ha^{-1} \cdot yr^{-1}$ respectively. Due to a large amount of
stones, the control plots have been treated as one.

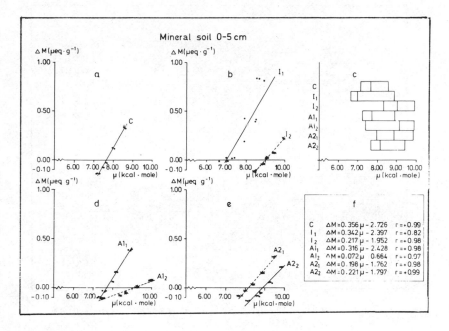

Figure 2. As in Figure 1 for mineral soil 0-5 cm.

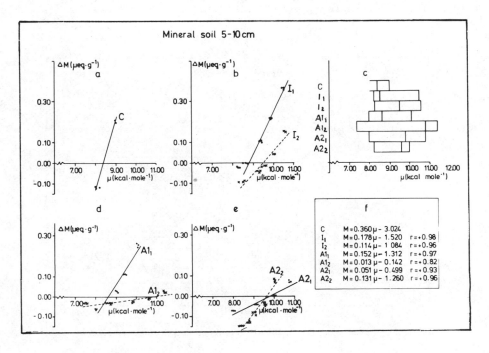

Figure 3. As in Figure 1 for mineral soil 5-10 cm.

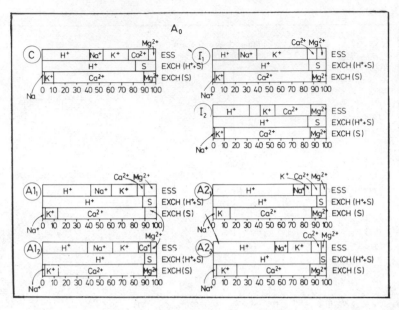

Figure 4. The catonic composition calculated on an equivalent
         percentage basis for "equilibrium soil solution" (ESS),
         exchangeable "hydrogen ions" + exchangeable "base cations",
         (exch. H^+ + S), and exchangeable "base cations" taken
         separately (exch. S) - A_0-horizon.

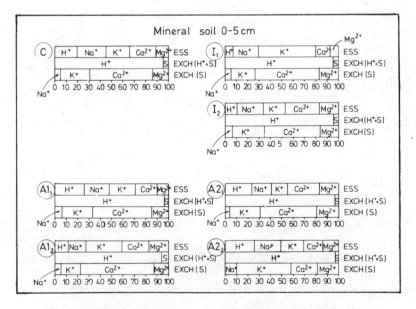

Figure 5. As in Figure 4. - Mineral soil 0-5 cm.

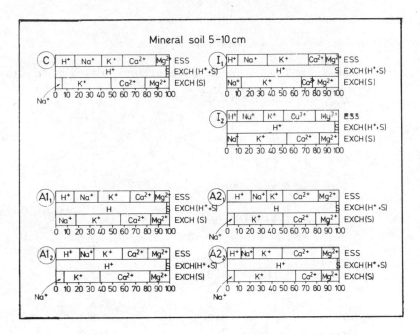

Figure 6. As in Figure 4. - Mineral soil 5-10 cm.

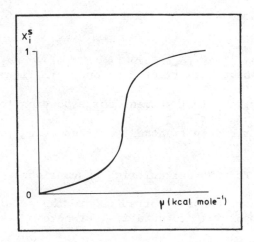

Figure 7. Theoretical ion adsorption isotherm running from
equivalent percentage 0 to equivalent percentage 100.

Table 1.

A_0-horizon

Treatment	pH in ESS	Base saturation %	Calcium saturation %
C	4.2	18.7	14.6
I_1	4.2	15.4	12.0
I_2	4.2	16.3	12.3
Al_1	4.2	11.7	8.9
Al_2	4.2	10.5	7.8
$A2_1$	4.0	7.9	5.7
$A2_2$	4.0	5.8	3.7

Mineral soil 0–5 cm

C	4.6	4.4	2.4
I_1	4.4	4.4	2.5
I_2	4.6	4.3	2.1
Al_1	4.4	3.8	2.0
Al_2	4.2	6.1	3.9
$A2_1$	4.2	3.2	1.6
$A2_2$	4.2	2.5	0.6

Mineral soil 5–10 cm

C	4.8	2.4	0.7
I_1	4.8	1.9	0.2
I_2	4.9	3.0	0.9
Al_1	4.6	2.4	0.7
Al_2	4.6	3.5	1.5
$A2_1$	4.4	2.4	0.7
$A2_2$	4.5	2.1	0.5

C — control, (due to a large amount of stones the control plots have been treated as one, cf. Figures 1–6)

I_1, I_2 — irrigation

Al_1, Al_2 — sulphuric acid treatment corresponding to 50 kg $H_2SO_4 \cdot ha^{-1} \cdot yr^{-1}$

$A2_1, A2_2$ — sulphuric acid treatment corresponding to 100 kg $H_2SO_4 \cdot ha^{-1} \cdot yr^{-1}$

This leads to two contradicting assumptions:

1) A slope decrease has occurred due to the acid treatments, but which is impossible to quantify for heterogeneity reasons.

2) The decrease in calcium saturation has been too small to alter the potential intervals and thereby cause a slope decrease.

An Alternative Approach

An attempt will now be made to use the second assumption stated above. The adsorption isotherm slope shall be regarded as a specific property of the soil exchanger, and its power for predicting changes in calcium saturation percentage at specified acid treatments shall be tested. First of all it will be of interest to relate the isotherm properties to exchanger characteristics, commonly used in soil chemistry as contents of clay and humus or cation exchange capacity. The clay content is most probably of minor interest in the sandy soil under study, and will be neglected. The cation exchange capacity is mostly confined to humus colloids in such soils (cf. references 5 and 14). Linear regressions with adsorption isotherm slope against organic matter content (LI) and CEC yield:

$$\frac{d\ (\Delta M)}{d\mu} = 0.0680 \cdot LI + 0.0639 \qquad\qquad n = 21;\ r^2 = 0.79;\ p < 0.001$$

$$\frac{d\ (\Delta M)}{d\mu} = 0.0087 \cdot CEC - 0.1433 \qquad\qquad n = 21;\ r^2 = 0.83;\ p < 0.001$$

The omission of the clay content seems to be justified as both regressions account for almost the same slope variance.

A closer inspection of the regressions reveals no systematic deviation of data points which could be attributed to acid treatment. Although there is a high level of statistical significance the scatter of the data points is fairly large, indicating that at least one more factor is involved. One such factor, namely the ionic strength, I (μM), will be treated later in this paper.

A visual inspection of Table 1 and Figures 1-3 suggests an inverse relationship between calcium saturation and isotherm slope within a given soil horizon and treatment. Two kinds of linear models are tried to further test this assumption, which also can be stated in energetical terms: A high slope means that less energy is required to withdraw a certain amount of calcium from the adsorption complex, hence the inverse relationship.

For each horizon and treatment, the following calculations are made:

1) $Ca\text{-}sat_{highest\ value}$ - $Ca\text{-}sat_{lowest\ value}$ = diff. Ca-sat

2) $\frac{d(\Delta M)}{d\mu}$ highest Ca-sat value - $\frac{d(\Delta M)}{d\mu}$ lowest Ca-sat value
    = diff $\frac{d(\Delta M)}{d\mu}$

1) and 2) are taken as dependent and independent variables respectively in least square linear regression. The following equations are obtained:

A_0: diff Ca-sat = -0.6 diff $\frac{d(\Delta M)}{d\mu}$ + 0.8 r^2 = 0.58   p > 0.10

0-5 cm: diff Ca-sat = -4 diff $\frac{d(\Delta M)}{d\mu}$ + 0.9 r^2 = 0.98   p > 0.05

5-10 cm: diff Ca-sat = -5 diff $\frac{d(\Delta M)}{d\mu}$ + 0.1 r^2 = 0.35   p > 0.10

A further refinement is obtained by exchanging diff $\frac{d(\Delta M)}{d\mu}$ for diff $\frac{d(\Delta M)}{d\mu} \cdot \frac{I}{LI}$. The new equations are displayed in Figure 8 (see next page).

This approach offers a less ambiguous interpretation than the former one. According to the latter model, isotherm slope and organic matter content could work as predictive tools in the two uppermost horizons, where the effects of acid treatment are most pronounced (cf. Table 1), whereas diff Ca-sat cannot be explained in this manner in the lowermost horizon, where the effects are less obvious. The fact that LI improves the model, at least in the A_0-horizon, clearly indicates that this variable is only one of those determining $\frac{d(\Delta M)}{d\mu}$, which is specifically the case, when the data analysis is split by soil horizons. Nevertheless, the tentative conclusion is that the soil system reacts to the acid treatment in a manner that can be predicted from physico-chemical data of the kind used in this study. The equation parameters and correlation coefficients given above and in Figure 8, should be interpreted in a qualitative manner, (larger than... less than ...), as the data basis in this study is still too scarce to permit a refined quantitative treatment.

Comparisons with earlier studies[5] reveal conflicting evidence as to the constancy of the isotherm slope. Data presented by Bringmark[9] indicate that the ionic strength of the equilibrating solutions could be of great importance as a slope-determining factor. The validity of this conclusion is tested in a multiple linear regression with LI and I as dependent variables and $\frac{d(\Delta M)}{d\mu}$ as the dependent one. I is the median value of the ionic strength interval obtained for each adsorption isotherm.

$\frac{d(\Delta M)}{d\mu}$ = -0.2632 + 0.0590·LI + 0.000652·I

r^2 = 0.89 p < 0.001   n = 21

This could be compared with the regression between $\frac{d(\Delta M)}{d\mu}$ and LI, where 79% of the variance could be accounted for, I accounting for another 10%.

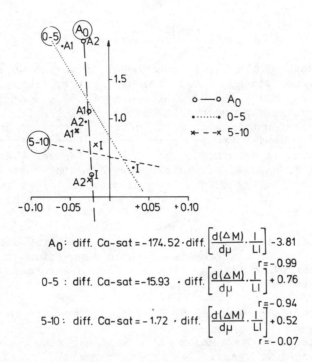

$$A_0: \text{ diff. Ca-sat} = -174.52 \cdot \text{diff.} \left[\frac{d(\Delta M)}{d\mu} \cdot \frac{1}{LI}\right] - 3.81$$

$$r = -0.99$$

$$0\text{-}5: \text{ diff. Ca-sat} = -15.93 \cdot \text{diff.} \left[\frac{d(\Delta M)}{d\mu} \cdot \frac{1}{LI}\right] + 0.76$$

$$r = -0.94$$

$$5\text{-}10: \text{ diff. Ca-sat} = -1.72 \cdot \text{diff.} \left[\frac{d(\Delta M)}{d\mu} \cdot \frac{1}{LI}\right] + 0.52$$

$$r = -0.07$$

Figure 8. Tentative graphical model describing the difference in calcium saturation percentage within treatments as a linear function of the difference between adsorption isotherm slopes equilibrated for soil organic matter (LI).

$A_0: r^2 = 0.98$ $p > 0.05$

$0\text{-}5 \text{ cm}: r^2 = 0.88$ $p > 0.10$

$5\text{-}10 \text{ cm}: r^2 = 0.0049$ $p \gg 0.10$

I probably played a more crucial role in the earlier study referred to[5], as the soil was acidified immediately before the adsorption isotherm determinations. The metal cations desorbed due to this treatment could not "escape" from the closed equilibrating system and probably made a significant contribution to the total ionic strength. The soil material in the present study has been subject to leaching, and thereby the total ion concentration of the soil solution has diminished.

SUGGESTIONS FOR FURTHER TESTING

Refinements of the "difference model" given above should include varying oxidation states of soil organic matter, clay content and clay mineralogy, all these variables are of the utmost importance for the ion exchange properties. Constancy of the isotherm slope is hitherto accepted until evidence pointing in other directions is available. The ionic strength seems to be of minor importance as long as leaching is permitted. It could be assumed that the slope is significantly changed only after very drastic treatments, such as high dosages of lime or sulphuric acid dosages larger than 100 kg $H_2SO_4 \cdot ha^{-1} \cdot yr^{-1}$. As far as liming is concerned, unpublished data suggests such an hypothesis.

In further studies the questions to be answered are: Which is the lowest level of acid treatment, (acid deposition) that can be traced in a certain soil type with reasonable accuracy. What time scales are involved?

Field studies could then proceed along two different lines:

1)   Investigations along an emission gradient with comparable forest and soil types.
      If soil types are too different some kind of standard soil taken from a locality outside the gradient could be put into appropriate vessels placed along the gradient.

2)   The second line involves field experiments of the type presented in this paper, extended to a larger number of soil types and acid treatments, dosages preferably comprising a series from 100 kg $H_2SO_4 \cdot ha^{-1} \cdot yr^{-1}$ and downwards.

From the studies outlined above, it might be possible to classify forest soils according to their "acidification sensitivity" in a more precise manner than has hitherto been made by stating threshold values for acid deposition, below which no acidification effects are discernible. For deposition above the threshold value ion exchange models which include adsorption isotherms[15,16] could be used for a more exact quantification of, for instance, calcium losses.

ACKNOWLEDGEMENTS

This project is sponsored by the National Swedish Environment Protection Board. My sincere thanks to the following persons: Hans Burgtorf, Maj-Lis Gernersson, Ragnhild Ohlin and Mimmi Varga for their skilful technical assistance; Folke Andersson, Lage Bringmark, Ewa Kvillner and Ingrid Stiernquist for reading and commenting upon the manuscript; Carl Olof Tamm for permitting access to the field studies.

REFERENCES

1. C.O. Tamm, Ambio 5: 235 (1976).
2. C.O. Tamm, G. Wiklander and B. Popovic, in: Proceedings of the First International Symposium on Acid Precipitation and the Forest Ecosystem, USDA Forest Service General Technical Report NE-23, 1011 (1976).
3. G. Tyler, Water, Air, and Soil Pollution 9: 137 (1978).
4. F. Andersson and I. Nilsson in manuscript (1978).
5. I. Nilsson, Meddn. Avd. Ekol. Bot., Lunds Univ. 32, (1977).
6. C.O. Tamm, I. Nilsson and G. Wiklander, Research Notes, Dept. of Forest Ecology and Forest Soils, Royal College of Forestry, Stockholm, Sweden, 18 (1974).
7. B. Ulrich and P.K. Khanna, Gottinger Bodenkundliche Berichte 19:121 (1971).
8. A.M. Balsberg (ed.), Meddn. Avd. Ekol. Bot., Lunds Univ. 3:2, (1975).
9. L. Bringmark, Internal Report, Swedish Coniferous Forest Project 52, (1977).
10. C.M. Woodruff, Soil Sci. Soc. Am. Proc. 19: 36 (1955).
11. P.K. Khanna and B. Ulrich, Gottinger Bodenkundliche Berichte 29: 211 (1973).
12. G. Hagg, "Kemisk Reaktionslara", Almqvist-Wiksell, Stockholm (1965).
13. G.W. Snedecor and W.G. Cochran, "Statistical Methods", The Iowa State University Press, Iowa, USA (1967).
14. C.S. Helling, G. Chesters and R.B. Corey, Soil Sci. Soc. Am. Proc. 28: 517 (1964).
15. B. Ulrich, R. Mayer, P.K. Khanna and J. Prenzel, Gottinger Bodenkundliche Berichte, 29: 1 (1973).
16. E. Bosatta and L. Bringmark, Internal Report, Swedish Coniferous Forest Project 43 (1976).

THE SENSITIVITY OF SOILS TO ACID PRECIPITATION

Lambert Wiklander

Department of Soil Science, The Swedish University of

Agricultural Sciences, Uppsala, Sweden

INTRODUCTION

Acid precipitation may lead to adverse effects on the soil and
on the percolating water. The relative magnitude of these effects
varies with the soil properties. In soils with strong buffering
against acid compounds no acidification of the discharge water will
occur. On the other hand, in soils with low buffering capacity, especially if the soil is shallow, the discharge water will become
acidified proportionally to the infiltrating precipitation and to
its acidity.

Other factors of significance for the acidification of soil and
water are climate and topography determining the surface runoff of
snow melt on the frozen soil. The snow may contain considerable
amounts of accumulated H_2SO_4 and in needle forests acidic compounds
from the trees.

Acidification of soils means decrease of pH and of base saturation caused by internal production and external supply of protons
(H^+). A general treatment of acidification of soils and waters must
consider all sources of the acidifying proton and not just the atmospheric supply.

THE SOURCES OF SOIL ACIDIFICATION

H_3O^+ (or H^+) is the acidifying factor in soils. Acid mineral cations, as $Al(OH)_n^{(3-n)+}$, $Fe(OH)_n^{(3-n)+}$, are formed mainly by secondary reactions. The acidifying proton originates from internal and external sources that may be categorized as follows:

1. Uptake of cation nutrients by plants involving an ion exchange between the root acidoids and the soil particles via the soil solution. Exchangeable cation nutrients are replaced by H^+ as a consecutive reaction of the nutrient adsorption by the roots (Figure 1). This biological acidification is an important reaction in fertile soils but a quantitative estimation is difficult to make.

2. Production of CO_2 by plant roots and microorganisms increases the solubility of $CaCO_3$ and the desorption of base cations in soils with pH > 5.5. At pH > 6, where the bonding energy of H^+ to soil particles is very high, H_2CO_3 acts as an energetic and quantitatively very important acidifying factor of soils (Table 1).

$$CO_2 + H_2O = H_2CO_3$$
$$H_2CO_3 = H^+ + HCO_3^- (pK = 6.37)$$
$$Soil-Ca + 2 H^+ = Soil-H_2 + Ca^{2+}$$

The concentration of CO_2 in the soil air may rise to 5-6 volume-% in arable soils. On the other hand, CO_2 does not acidify the drainage water, which in Sweden usually has a pH around 7.

3. Oxidation of NH_4^+ and NH_3, natural and anthropogenic, to HNO_3 and of S, FeS, FeS_2, and H_2S to H_2SO_4 (Figures 2, 3). Strong acidification of the drainage water and recipients often occurs when the soil is rich in reduced S compounds as in gyttja soils or acid sulphate soils. However, the opposite process implying reduction of SO_4 to SO_2 and H_2S and of NO_3 to NO_x and N_2 should not be neglected as it counteracts acidification.

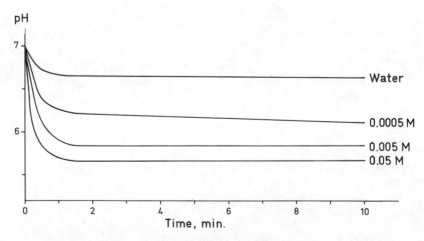

Figure 1. Exchange acidity produced by ion exchange between wheat roots and $CaCl_2$.

$$Root-H_2 + Ca^{2+} = Root-Ca + 2H^+$$

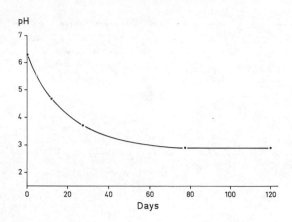

Figure 2. Acidification of a soil by air oxidation of reduced sulphur compounds to H_2SO_4.

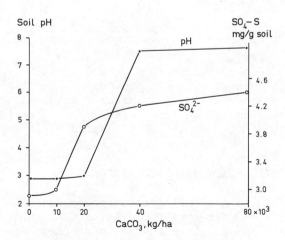

Figure 3. Influence of liming on pH and on rate of sulphur oxidation in the same soil as in Figure 2.

Table 1. The relationship between the pH of drainage water from
cultivated soils and the percentage of HCO_3^- based on $\Sigma HCO_3^-, Cl^-$,
SO_4^{2-}, NO_3^- in the water. Field experiments.

Soils	Non-calcareous			Calcareous	
Soils	Röbäcks-dalen	Gammals-torp	Heagård	Hoby	Marsta
pH	4.6	6.6	7.2	7.3	7.9
% HCO_3^-	7	18	55	60	90

4. Coniferous forests produce very acid litter (Table 2) and
an acid humus layer. Soluble organic acids and fulvoacids are
leached and cause strong acidification and weathering leading to
formation of podsols with low base saturations, sometimes < 5 %.
The main acidifying factors in coniferous forest soils consist of
these acids together with the creation of acid groups, mainly COOH,
by oxidation of the humus material. Acid precipitation plays only
a minor role for the soil pH.

Soluble humus acids have only a slight acidifying effect on
the water recipients.

5. Atmospheric deposition of H_2SO_4 and in minor amounts HNO_3,
NO_x and HCl. After nitrification deposited NH_4 increases the acidification. The atmospheric supply of mineral acids affects the soil
and water to a varying extent, essentially depending on the soil
properties and the salt contents of the atmospheric deposition.

ACIDIFICATION BY PEDOGENIC PROCESSES

In humid climates the soil forming processes sooner or later
lead to acidification of permeable soils unless counteracting measures are taken by man or air-borne salts are deposited. The commencement and rate of acidification depend on the buffering capacity
(mineral composition and texture) and on the internal production
and external supply of H ions. Decrease of pH and base saturation enhances the chemical weathering and formation of more stable compounds
Most primary minerals in soils are unstable even in the neutral pH
range. Weathering, leaching and acidification are stages of the geochemical circulation of elements.

Table 2. Contents of Ca, Mg, K, Na in different litters and peats.

	me/100 g d.m.
Elm, leaf	250
Ash	162
Aspen	164
Maple	144
Oak, eutric soil	145
Oak, dystric soil	89
Birch	97
Beech	60
Pine needles	32
Rawhumus, A_0	9
Sphagnum peat	13
Bog peat	78

These pedogenic processes have been in operation for millions of years, long before the advent of man and the emission of anthropogenic acid compounds. The occurrence of lateritic soils - characterized by low pH, nutrient poverty and high degree of weathering and of fossil podsols offers visible evidence of this natural acidification and soil formation.

ACIDIFICATION AND ION EXCHANGE

Increase of the proton concentration in soil, apart from internal or external sources, leads to adsorption of protons and desorption of base cations or to dissolution of $CaCO_3$ in calcareous soils and enhanced weathering.

The efficiency of H^+ to replace base cations, as Ca^{2+}, Mg^{2+}, K^+, Na^+, can be expressed by the ratio $\Delta Me/\Delta H$, where ΔMe stands for the amount of base cations replaced by a minute amount of H^+ added, ΔH. $\Delta Me/\Delta H$ may vary within the range 1-0.

At pH > 5.5 H^+ is very firmly adsorbed, that is, $\Delta Me/\Delta H \sim 1$, due to pH-dependent negative charges from weak acid groups. At pH < 5.5 the bonding energy of H^+ decreases with the pH due to permanent negative charges from strong acid groups. The behaviour of clay minerals and an illitic soil is shown in Figure 4.

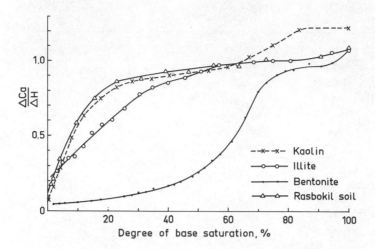

Figure 4. The replacing efficiency of H^+ ($\Delta Ca/\Delta H$) in relation to the degree of base saturation in successive replacement of Ca^{2+} by HCl.

If the soil is denoted R, this ion exchange can be described by the simple formula:

$$R-Me + H^+ = R-H + Me^+$$

and the equilibrium approached by the Donnan equation:

$$\frac{[R-Me]}{[R-H]} = \frac{[Me^+]s}{[H^+]s},$$

[] denoting activity of ions adsorbed or present in the soil solution (s). There is a mutual competition for the negative sites of the soil particles between the base cations (Me^{n+}) and H^+ according to the law of mass action. The concentration of salts in the soil solution and the ratio of Me^{n+} to H^+ in the precipitation is therefore of importance for the replacement of base cations by H^+ and for the consecutive decrease of the base saturation and, thus, for the soil acidification.

The following conclusions may be drawn for noncalcareous soils. If the ratio $[Me^+]/[H^+]$ is the same in the precipitation as in the soil solution no ion exchange will occur. If the ratio is higher in precipitation, the soil pH will increase and if it is lower, Me^+ will be replaced and acidification take place. As a consequence of the latter, even slightly acid rain may have an acidifying effect on soils if the salt content is low. For an estimation of the aci-

difying effect of H$^+$ on a soil, the ratio [Me$^+$]/[H$^+$] in the precipitation must be considered as well as the pH of the soil itself.

The reduction of the soil acidification due to salts in the
'rain' appears in Figures 5 and 6. These results agree with the conclusions that can be drawn from the above ion exchange formula that
atmospheric salts diminish the adverse effects of acid precipitation on soils with low buffering capacity. On the other hand, the
salts have an opposite effect on the discharge water by increasing
the percolation of acids (cf. Table 4). In calcareous soils and in
soils with pH > 6 the atmospheric salts have only a slight influence or none at all on the acidifying effect of mineral acids as
ΔMe/ΔH = 1.

LOSS OF NUTRIENTS IN CULTIVATED SOILS

Acidification of soils is closely connected to leaching of
plant nutrients. Although most cultivated soils are relatively rich
in cation nutrients and have pH > 5, there also occur soils that
are very acid, especially in the subsoil. The rain infiltrating the
soil is normally rather poor in salts but becomes more concentrated
during the passage down to the drainage systems. The rate of salt
increase in the percolating water depends on the soil properties

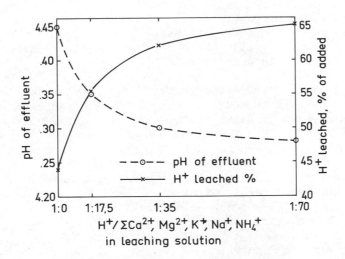

Figure 5. The influence of the ratio H/ΣCa,Mg,K,Na,NH$_4$ in the infiltrating solution (pH 4.07) on the pH of the effluent and on the
leached H$^+$ as % of added H$_2$SO$_4$. Soil: B$_2$ of humo-orthic podsol,
pH 4.4, depth 5.5 cm.

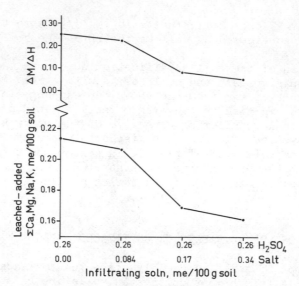

Figure 6. The influence of salts on the net losses of ΣCa,Mg,K,Na in leaching A$_2$ of humo-orthic podsol with salt solutions containing H$_2$SO$_4$ corresponding to 30 kg S/ha. ΔM/ΔH shows the replacing efficiency of H$^+$ after subtraction of the leaching by distilled water.

and is due to dissolution of salts, mineral weathering, humification and replacement of base cations by protons from internal and external sources. In fertile soils there is a strong enrichment of salts in the percolation water.

Figure 7 shows a schematic relationship between the leaching of salts and the soil pH and in Figure 8 there is an experimental verification of part of this relationship. Losses of nutrients in six cultivated soils of different types and with pH of the subsoil ranging 4.3-8.3 are given in Table 3.

Röbäcksdalen is an acid sulphate soil, rich in sulphur compounds that have been partly oxidized to H$_2$SO$_4$ after drainage, pH in subsoil 4.3-4.8. The strong acidification of this rather poor soil has led to an intense weathering and very high losses of S and Mn as well as of base cations. This soil exemplifies the left part of the curve in Figure 7.

The soils Marsta, pH of subsoil 7.0-8.3, and Hoby are calcareous and the main part of the base ion leaching is due to H$_2$CO$_3$ and not to mineral acids (cf. Table 1). Atmospheric deposition of mineral acids on well managed cultivated soils has significance only as nutrient sources of S and N.

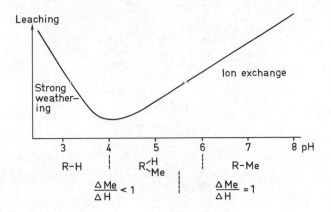

Figure 7. Schematic relationship between the soil pH and the leach-ing of salts indicating the influence of the degree of base satura-tion and chemical weathering.

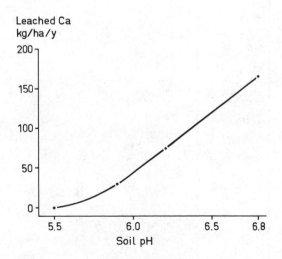

Figure 8. Leaching of Ca, exceeding unlimed treatment, in a liming experiment, Lanna clay soil, related to the soil pH. Mean values of 4 replicates from 28 years after liming.

Table 3. Leaching in cultivated soils, mean values for given period, kg/ha/year. For comparison at-mospheric deposition of nutrients and contents in drainage water from all Sweden, 912 samples.

Locality	Texture	Period years	Precipitation, mm	Ca	Mg	K	Na	Mn	Cl	SO4 -S	NO3 -N	PO4 -P
Röbäcksdalen	Silt	3	653	78	24	19	47	1.9	47	99	5.9	0.039
Marsta	Clay	6	486	77	13	1.6	7	tr.	4.3	3	–	0.033
Lanna	Clay	5	560	51	34	3.3	43		26	18	2.3	0.025
Gammalstorp	Loam	5	710	45	14	3.7	14		13	24	–	0.039
Heagård	Clay-loam	2	630	70	33	16	92	0.054	61	45	8.1	0.240
Hoby	Loam	1	665	257	25	37	49	tr.	59	64	20.8	0.290
From air				10-20	1-10	3-7	5-55	–	7-75	5-22	2-20	–
Drainage water, mg/l				24	11.3	5.2	21	–	27	27	6.2	0.024

SENSITIVITY OF SOILS TO ACID PRECIPITATION

Acidification of a soil does not occur as long as the buffering keeps pH and base saturation constant. The rate of soil acidification is governed by the soil properties, as contents of $CaCO_3$ and ferro-magnesian minerals, texture, kind of litter materials, and water permeability. Other significant factors are the rainfall, the surface runoff, and the ratio $H,NH_4/\Sigma Ca,Mg,K,Na$ of the infiltrating precipitation and of the leachate from forest vegetation.

If the buffering capacity of the soil is too small to neutralize the mineral acids deposited or produced in the soil, acidification of the ground water, watercourses and lakes will take place. This consequence will be especially accentuated in coarse-textured, very acid soils as podsols and acid cambisols, where the base saturation may be less than 5 %. The separate effects on soil and percolating water is summarized in Table 4 and the sensitivity of the soil alone in Table 5.

A few data from leaching experiments will supplement Table 5. Soil columns leached with melted snow adjusted to a pH of 3.3 by H_2SO_4, gave results as shown in Table 6.

The capacity of acid soils to buffer the percolating solution to a certain pH is shown by Figure 9.

Table 4. Acidifying effect of precipitation on soil and water.

Soil pH	$\dfrac{\Delta Me}{\Delta H}$	Soil	Water
8	1	None – slight	None
7	1	Buffering important	None
6	1	Buffering important	None
5	<1	Strong	Slight
4	<1	Medium – slight	Increasing
3	<<1	Very slight	Increasing

Table 5. The sensitivity of different soils to acid precipitation based on: Buffering capacity against pH-change, retention of H^+, replacing power of H^+ for exchangeable base cations ($\Delta Me/\Delta H$), and adverse effects on soils.

| | Calcareous soils | Noncalcareous | | Cultivated soils pH > 5 | Acid soils pH < 5 |
		clays pH > 6	sandy soils pH > 6		
Buffering	Very high	High	Low	High	Moderate
H^+retention	Maximal	Great	Great	Great	Slight
$\Delta Me/\Delta H$	< 1	1	1	≤ 1	≤ 1
Adverse effects	None	Moderate	Considerable	None - slight	Slight

Analysis of the leachates (Table 7) showed decreasing amounts of Ca, Mg, K, Na lost during the leaching despite the pH drop from 6.0 to 4.0 of the infiltrating solution. The volume of percolating solution corresponded to more than 3 years rainfall. Similar results were obtained from another podsol. A slight but noticeable solvation of the organic matter in A_0 in the first leaching phase (pH 6.0) and a higher stability at lower pH might be the explanation of the decreasing base cation losses. Anyhow, in these soils the strong increase of the acidity of the infiltrating solution had no drastic effect on the effluent composition.

Table 6. Leaching soil columns (depth 10 cm) with melted snow, pH 3.3

| Soil | Podsol A_0 | Cultivated | |
		Loam	Loam
Soil pH	3.7	5.2	5.7
Acid recovery, %	78	32	0
$\Delta Me/\Delta H$	0.2	0.7	1

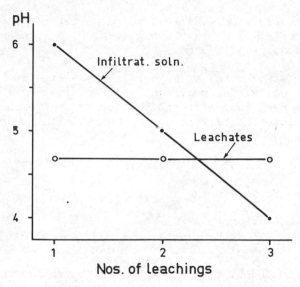

Figure 9. St. Ryd ferri-orthic podsol (A_0, A_1, A_2, B_1) leached with melted snow, pH adjusted to 6.0 (NH_3), 5.0 and 4.0 (H_2SO_4), successively in that order. pH of leachates given. Cf. Table 7.

Table 7. Leaching of St. Ryd podsol and Kongalund acid cambisol with melted snow, pH 5.0 and adjusted to pH 6.0 and 4.0 by NH_3 and H_2SO_4. Soil columns with undisturbed structure. depth 35 cm.

Infiltrating snow water		Leachates					
			Cambisol		Podsol		
pH	C_{Me^+}/C_{H^+}	Nos.	pH	ΣCa,Mg,K,Na me/l	pH	ΣCa,Mg,K,Na me/l	
6	57	Equil. solu.	4.61	1.23	4.62	1.03	
6	57	1	4.71	0.58	4.68	0.46	
5	5.7	2	4.81	0.33	4.67	0.29	
4	0.57	3	4.70	0.24	4.67	0.21	
			Na K Mg Ca		Na K Mg Ca		
Loss in percentage			28 18 20 35		36 23 14 27		

The amount of Ca,Mg,K,Na lost at pH 4 in the podsol was small, 0.21 me/l leachate, or corresponding to about 0.4 ke/ha or 8 kg Ca/ha at a discharge of 200 mm.

The percolation rate was higher in these experiments than normally found in soils. The mobilization of cations by weathering is likely to have been insignificant but for the ion exchange and leaching of solutes there were no physical limitations.

SUMMARY

The active factor in the acidification is H^+. Its sources can be categorized as:

1. Nutrient uptake by plants. The root acidoids adsorb cation nutrients and desorb H^+.

2. CO_2 produced by plant roots and microorganisms.

3. Oxidation of NH_4 and S, FeS, FeS_2, and H_2S to HNO_3 and H_2SO_4.

4. Very acid litter in coniferous forests.

5. Atmospheric deposition of H_2SO_4 and some HNO_3, NO_x, HCl and NH_4 (after nitrification HNO_3).

Based on the ion exchange theory, ion exchange experiments, and leaching of soil samples the following conclusions can be drawn about the acidifying effect on soils by atmospheric deposition of mineral acids.

1. At pH > 5.5 the acids are fully neutralized by decomposition of $CaCO_3$ and other unstable minerals and by cation exchange.

2. At pH < 5.5 the efficiency of the proton to decompose minerals and to replace exchangeable Ca, Mg, K, and Na decreases with the soil pH. Consequently, the acidifying effect of mineral acids on soils decreases but the effect on the discharge water increases in the very acid soils.

3. Salts of Ca, Mg, K, and Na in the precipitation counteract the adsorption of protons and, by that, the decrease of the base saturation. A greater part of the acids percolate through the soil.

The sensitivity of various soils to acid precipitation depends on the buffering capacity and on the soil pH. Noncalcareous sandy soils with pH > 5 are the most sensitive, irrespective of soil type.

Very acid soils are far less sensitive as they are already adjusted to this condition by soil formation and therefore more stable. Easily weatherable minerals have disappeared and the soil pH may be less than that of the precipitation. Many podsols and latosols and very acid cambisols belong to this group. In coniferous forests the acid litter is the main acidifying source.

Clayey soils buffer well and are moderately sensitive. Calcareous soils retain their pH until the carbonate is dissolved.

In cultivated soils, well managed, the mineral acid fallout only means a slight increase of the lime requirement with a cost compensated by the supply of S, N, Mg, K, Na and Ca considered as nutrients.

In all soils with pH > 6 the natural acidification dominates caused by internal production of CO_2 and by the root cation uptake.

THE SENSITIVITY OF SOILS TO ACIDIFICATION

Byron W. Bache

Department of Soil Fertility,
Macaulay Institute for Soil Research

Aberdeen AB9 2QJ, Scotland, U.K.

INTRODUCTION

The sensitivity of a site can be considered as the change produced in some property of the site as a result of an applied stimulus. The applied stimulus here is acid precipitation, but there are many chemical and biological properties of the site which may change as a result of increasing acidification. The biological changes are mainly a response to the chemical changes, so that the chemical changes are the more fundamental. They are also easier to measure, but there are a number of different chemical components relating to acidity, the change in any one of which could be measured. Here, only the soil is being considered, and in particular its "acid-base status". This is best described in terms of the balance between the acid exchangeable cations (Al^{3+} and H_3O^+) and the base exchangeable cations (principally Ca^{2+} and Mg^{2+}). As explained earlier[1], much of the hydrogen is undissociated and therefore not truly exchangeable, but it can be estimated as total acidity by reaction to pH 8.2. The acid-base status is then represented by the degree of base saturation, the soil being considered as fully base saturated at pH 8.2[2,3].

The degree of base saturation of the soil, which refers to the adsorbed and exchangeable cations, is also reflected in the composition of the soil solution, and in particular by the lime potential that can be calculated from it[2] :

$$LP = pH - \tfrac{1}{2}p(Ca,Mg)$$

For strongly acid soils where aluminium is an important cation, there is a case for including the solution Al concentration in such a relationship. This would then give rise to a function :

$$pH + \frac{1}{3}pAl - \tfrac{1}{2}p(Ca,Mg)$$

Or, more simply, the aluminium-base cation balance could be considered alone[4] , to give

$$\frac{1}{3}pAl - \tfrac{1}{2}p(Ca,Mg)$$

But in order to give a direct comparison between all soils, it is preferable to use the simple lime potential, at least until further research has clarified the usefulness of these alternative formulations.

FACTORS DETERMINING SENSITIVITY

The sensitivity of a soil to acidification, i.e. the change produced in its degree of base saturation by a body of water flowing through it, can be estimated as a product of three factors: the difference in lime potential between water and soil, the buffer capacity of the soil, and the fraction of the water body that reacts with the soil.

The difference in lime potential: $\Delta LP = LP_{soil} - LP_{water}$
Where this difference is large, the change will be large, but it seems likely to be related exponentially to ΔLP rather than linearly. Greatest change will therefore occur on neutral soils and with the most acid precipitation. Acid forest soils of the northern boreal zone will be the least affected, because they already have low pH values.

The buffer capacity of the soil profile. Change in soil acidity is inversely related to buffer capacity[1], being greatest in poorly buffered soils. The surface horizons are the most important for immediate effects on vegetation, but in the present context the whole soil profile should be considered. Because the possible effects of acid precipitation on ground water, streams and lakes are also of concern, the buffering effect of all the material between the land surface and ground water should be taken into account, which may include shattered rock and glacial till as well as the pedogenic horizons[5].

Soil buffer capacity is lowest in soils of sandy texture and those with low organic matter contents, and these will change most rapidly. Depth of soil is also important; a shallow soil overlying hard rock has little buffering effect compared to a deep soil.

The fraction of the water body that reacts with the soil, f_w. This is the most difficult factor to quantify, and to relate to practical situations. The important distinction is that between miscible displacement, where $f_w \rightarrow 1.0$, and channelled flow where $f_w \ll 1.0$[1]. Some reaction will occur even with channelled flow, and it may be considerable when combined with dispersion into pores adjacent to the flow paths.

High intensity rainfall is likely to lead to runoff or channelled flow, and hence to low f_w values. The structure of soils, and especially of subsoils, is important here : soils with relatively high macroporosity, such as coarse sandy soils and well-aggregated topsoils with a stable structure, allow water to flow uniformly through them by miscible displacement. Soils with poorly-developed structure, and fine-textured soils with relatively uniform particle size, such as silty clays, have low macroporosity and drainage water travels down relatively few structural cracks or root channels. Compacted layers within the profile impede drainage, allowing a longer time for water to react with soil to equilibrium, but acidification can only proceed if reaction products are leached out.

Perhaps the best way to consider this factor, is simply to recognise that there are situations where incoming water will react less than completely with soil. An obvious corollary should be mentioned:- if acid precipitation does not react to equilibrium with the soil, it may alter the composition of the water bodies into which it subsequently flows.

CONCLUSIONS

These principles should allow a quantitative estimate of the effects of acid precipitation, but this has not yet been done. However, they make it possible to identify sensitive soils in a qualitative manner. The conclusions are similar to those given by Wiklander[6], and they can be summarized as follows. Considerable calcium loss occurs from calcareous soils, and acidification may occur on non-calcareous soils of higher pH values, but under these conditions it is not a serious matter. Where further acidification is likely to have serious consequences, such as in highly un-saturated acidic soils, it is most unlikely to occur. There is, however, an intermediate range of slightly-acid, poorly-buffered shallow sandy soils where relatively little acidification may lead to a striking loss of productivity. Such soils would present no problems in an agricultural situation, because they can be readily limed[7]. They may be more difficult to deal with where

land is in forest, because of the costs involved in liming large
tracts of inaccessible land. Such sites should be carefully
monitored for the effects of acid precipitation, and they present
the most suitable material for experimental study.

REFERENCES

1. B. W. Bache, this symposium.

2. B. W. Bache, Articles on Base Saturation and Soil Reaction in:
     "Encyclopedia of Soil Science", Dowden, Hutchinson and
     Ross, Stroudsburg, Pa. (In Press).

3. R. W. Pearson, and F. Adams, eds. "Soil Acidity and Liming",
     Amer. Soc. Agron., Madison, Wis. (1967).

4. R. C. Turner, W. E. Nichol and J. E. Brydon, Soil Sci., 95:186
     (1963).

5. T. Rosenqvist, Report to the Norwegian Council for Scientific
     Research (1976).

6. L. Wiklander, Grundförbättring, 26:155 (1973/74).

7. C. R. Frink and G. K. Voigt, Water, Air and Soil Pollution,
     7:371 (1977).

SENSITIVITY OF DIFFERENT SOILS TO ACID PRECIPITATION

Leif Petersen

Royal Veterinary and Agricultural University

Chemistry Department, 1971 Copenhagen V, Denmark

INTRODUCTION

The effects on the soil of precipitation containing strong acids will depend on a number of fundamental soil properties, and the effects may be detrimental as well as beneficial. Beneficial effects arise for example because nitrate and sulphate are important plant nutrients and because the strong acids may enhance the liberation of plant nutrients by increased weathering of soil minerals. In alkali soils the addition of acid to the soil may in itself be beneficial, since the pH of these soils is often above the optimum range for plant growth. However, in humid regions the soil pH tends to decrease with time due to production of carbonic and other acids in the soil, and the pH often decreases to values which are below the optimum range for many plants. This natural acidification is enhanced if strong acids are added to the soil with the precipitation. In this paper only this aspect will be dealt with, although the beneficial effects mentioned above should not be overlooked.

In order to predict the ability of a given soil to withstand acidification, information about its base saturation status, its buffer capacity and its possible content of alkaline substances such as carbonates is required. Such information may to a large extent be derived from soil maps which are available at various levels. For a detailed interpretation of the sensitivity of the soils in a given area it will be necessary to rely on large scaled soil maps based on detailed soil surveys. In this paper a very general evaluation of the relative sensitivity to acid precipitation of some of the more important soils of the world will be attempted. A major obstacle to this is the fact that highly different soil

573

classification systems are used in different parts of the world.
In many cases it is extremely difficult or even impossible to
identify equivalent soil entities in the different systems. To
overcome this the discussion will be limited to two soil
classification systems, namely that used in the U.S.A. and a number
of other countries as well, and that used by FAO/UNESCO for
producing a world soil map.[2] In the former system all soils may be
placed in one of the ten soil orders listed in Table 1. The latter
system includes about 25 groups of soils some of which are also
listed in Table 1. In the U.S.A. system each order is subdivided
into several lower categories, and in the FAO/UNESCO system each
group comprises a number of so-called soil units. However, here
the discussion will be limited to the large entities listed in
Table 1.

THE SOILS AND THEIR SENSITIVITY TO ACID PRECIPITATION

Entisols are slightly developed soils with a very weak profile
differentiation. Their properties will to a large extent be
determined by the parent material and the conditions under which
they are formed. They include soils formed on recent alluvial
deposits (Fluvisols), slightly developed soils on loose parent
materials (Regosols), shallow soils on solid rock (Lithosols) and
others. Due to this high degree of variation their sensitivity to
acid precipitation cannot be specified in general terms. Some of
the recent alluvial deposits are heavy clays, frequently with a
content of calcium carbonate, and such soils will not be sensitive
to acid precipitation. On the other hand, other entisols are found
on coarse textured parent materials and these may be rather sensitive
to acid precipitation.

Inceptisols are also slightly developed soils but they show
some profile development, mainly due to weathering (Cambisols).
Like the entisols they vary with respect to the properties which
are important for their sensitivity to acid precipitation. The
clayey inceptisols must be considered rather insensitive to acid
precipitation while more sandy species will be more sensitive.

Aridisols are soils of deserts and other dry regions. In the
FAO/UNESCO system these soils are classified as Yermosols, Xerosols,
Soloncheks, Solonetz etc. Some are saline (soloncheks) or alkaline
(solonetz). As mentioned above the latter may benefit from an
addition of acid since their pH values under natural conditions
usually will be above optimum for plant growth. Due to the low
precipitation in the regions where the aridisols occur they will in
any case receive only small amounts of acid. Hence, these soils
are unlikely to be adversely affected by acid precipitation.

TABLE 1. Soil Orders of the U.S.A. Classification System
and Important Soils of the FAO/UNESCO System.

U.S.A.	FAO/UNESCO
Entisols	Fluvisols, Regosols, Lithosols
Inceptisols	Cambisols
Aridisols	Xerosols, Yermosols, Solonchaks, Solonetz
Mollisols	Chernozems, Phaeozems, Kastanozems
Alfisols	Luvisols
Ultisols	Acrisols
Spodosols	Podzols
Oxisols	Ferralsols
Vertisols	Vertisols
Histosols	Histosols
	Arenosols

Mollisols are soils with a thick dark surface horizon having
a considerable humus content. They occur in continental climates
and are according to the FAO/UNESCO system classified as
Chernozems, Phacozems, Kastanozems etc. These soils have been
subject to only a relatively small degree of leaching and often
they contain calcium carbonate, at least at some depth. Most often
they are medium textured. Their base saturation percentage and
buffer capacity are usually high and they are, therefore, not
likely to be sensitive to acid precipitation.

Alfisols and Ultisols are soils with an argillic horizon, and
they correspond rather closely to the soils which according to the
FAO/UNESCO system are classified as Luvisols and Acrisols
respectively. They have a significant clay content and the Alfisols
(Luvisols) also have a high base saturation percentage. Hence, the
latter are not very sensitive to acid precipitation. Due to a

lower base saturation percentage Ultisols (Acrisols) may include
soils which are more sensitive.

Spodosols are equivalent to Podzols. The possible effects of
acid precipitation on these soils are discussed elsewhere.[3]

Oxisols, which are approximately equivalent to Ferralsols,
are the red soils typically occuring in humid tropical climates.
These soils normally have a considerable clay content but the clay
consists to a large extent of iron and aluminium oxides and of
kaolinite. These compounds all have low cation exchange capacities.
Furthermore, the humus content of these soils is usually low, and
they are highly weathered and leached. They have, therefore,
rather low pH values and a low buffer capacity. Acid precipitation
could, therefore, be expected to have a significant effect on these
soils. They would probably to some extent behave like podzols with
respect to acid precipitation. At present they are probably not
subject to any large input of acid precipitation since only a few
locations in the tropics have a high concentration of industrial
plants causing emission of acids.

Vertisols are black heavy clay soils occuring in tropical
regions as well as in warm regions outside the tropics. They often
contain calcium carbonate or at least have a high degree of base
saturation. The clay consists almost exclusively of smectite which
has a high cation exchange capacity. Hence, these soils have high
buffer capacity and base saturation percentage, and they can hardly
be significantly affected by the amounts of acid usually contained
in the precipitation.

Histosols are soils which consist completely or predominantly
of organic matter (peat). The organic matter may be humified to
a varying extent and the degree of humification will affect the
cation exchange capacity and hence the buffer capacity of these
soils. However, compared with mineral soils the cation exchange
capacity is normally high. With respect to their base saturation
status large variations occur. Some organic soils contain calcium
carbonate and these will not be sensitive to acid precipitation.
Others, e.g. sphagnum peats, may be extremely acid with pH values
about 3 and have a very high content of exchangeable acidity. The
amounts of acidity that can be added with the precipitation is
probably negligible compared with the amounts already present in
the soil, and such soils are, therefore, not likely to be
significantly affected by acid precipitation. In general acid
precipitation must be expected to have a very small effect if any
on histosols due to their high buffer capacity.

Arenosols is a group of soils defined in the FAO/UNESCO system.
The common characteristic of these soils is a sandy texture. This
group will include some more or less acid soils with a low buffer

capacity and these may be relatively sensitive to acid precipitation.

CONCLUDING REMARKS

From the discussion above it appears that the following soil orders of the U.S.A. soil classification system may include soils which are relatively sensitive to acid precipitation: Entisols, Inceptisols, Ultisols, Spodosols and Oxisols. In the FAO/UNESCO system relatively sensitive soils may occur within several of the soil units, some of the more important ones belonging to Regosols, Lithosols, Cambisols, Acrisols, Podzols, Ferralsols and Arenosols.

It should be stressed that each of these soil entities comprise a very large number of soils whose properties vary highly. This is also the case for the properties which are important for the sensitivity to acid precipitation. Any estimation of the sensitivity of a particular soil must necessarily be based on a detailed description of the important properties of this soil. It should also be underlined that the above interpretation is based solely on the ability of the soils to withstand acidification and that it is purely comparative. No conclusions should be drawn from the considerations above as to the absolute effect of acid precipitation on the soils.

REFERENCES

1. Soil Taxonomy. A Basic System of Soil Classification for Making and Interpreting Soil Surveys, Agriculture Handbook No.436, Soil Conservation Service, U.S. Dept. Agric., Washington D.C., (1975).

2. Soil Map of the World 1:5000000, Vol. I, Legend, FAO-UNESCO, Paris, (1974).

3. L. Petersen, Podzolization: Mechanism and Possible Effects of Acid Precipitation,

Rapporteurs Summary

A. WET AND DRY DEPOSITION, INCLUDING MELTING PHENOMENA AND

SNOWPACK CHEMISTRY

M. Benarie and R. E. Munn

The session began with a review by D. Fowler of the mechanisms of wet and dry deposition of sulphur and nitrogen compounds. Wet deposition, which includes rainout and washout, is often parameterized by a washout ratio (concentration in cloud droplets or in precipitation divided by the concentration in air). Dry deposition, which includes the fallout and impaction of particles and the absorption of gases at the surface of the earth, is often parameterized by a deposition velocity, which is a function of both meteorological and biological factors. Dr. Fowler emphasized the uncertainties that surround the practical application of these concepts to determining deposition rates, but nevertheless provided a useful table of typical deposition rates. The table shows the dominance of dry deposition close to source areas, and the increasing contribution made by wet deposition with increasing downwind distance.

The discussion of this paper centred around three main points. Firstly, why are cloud droplets more acid than the ground-level precipitation coming from the clouds? Here the role of ammonia scavenging and of droplet size were mentioned. Secondly, what are the effects of temperature and of other forms of precipitation (snowflakes, supercooled water, ice) on wet deposition rates? The view was expressed that because low temperatures slow down the atmospheric chemical reaction rates, isopleths of wet deposition would be moved out farther from sources. There was also a suggestion that large snowflakes would be effective scavengers. However, this point was debated; a large fraction (50-60%) of the hydrogen ions in a snowflake are incorporated into the supercooled droplet that is the central core of the flake. Finally, there was a discussion of the applicability of the parameterization techniques for estimating deposition. It was generally agreed that over open countryside, and/or above a forest canopy, climatologically useful

estimates of deposition could be made from a knowledge of meteorological conditions and of the nature of the underlying surface,
using the figures in Fowler's Table 3. On any one hour, or even
on any one day, however, the deposition estimates could contain
large discrepancies.

An alternative to parameterization is direct measurement of
deposition rates. This latter approach was surveyed in a paper by
J. N. Galloway, who stressed that the field investigator must be
aware of the ecological effects of acid rains in order to design an
appropriate monitoring system. For episode studies, for example, an
estimate of the monthly deposition rate would be of little value.

Wet deposition over soil and short vegetation can be measured
with a gauge that opens only during rainfall. If the gauge is
exposed also in dry weather conditions, it measures the total wet
and dry deposit into the gauge, but this is an irrelevant quantity
because the gauge is not representative of any natural surface.
There is, in fact, no easy way of measuring dry deposition rates
operationally. (The eddy-correlation approach, which has been
used by B. Hicks and others, is valuable in verifying dry deposition
parameterization techniques but is not sufficiently robust for
routine use.)

For wet deposition, there is the additional problem of
estimating conditions below a forest canopy. Rain droplets falling
from leaf to leaf may pick up considerable contamination before
reaching the soil; other droplets may run down the tree trunks.
This spatial variability in wet deposition needs additional study
in a forest, as indeed it does also to a certain extent over open
countryside.

The subsequent discussion centred around the main points
mentioned above, and Dr. Galloway had an opportunity to amplify his
ideas. It was again stressed that dry deposition could not be
obtained as a difference between total and wet deposition as
measured in a gauge. In fact, the meteorologist could not yet
provide the ecologist with operational estimates of hourly or daily
dry deposition.

The Session continued with two useful case studies; one by
P. Grennfelt of estimates of deposition in a coniferous forest in
southern Sweden; the other by Odén of estimates of the sulphur
budget of Sweden. The aim of P. Grennfelt's paper was to describe
the essential pathways for deposition of nitrogen and sulphur
compounds, and to give a rough estimate of the total input of these
compounds to a coniferous forest ecosystem in southern Sweden.
P. Grennfelt expressed the view that the pathways were complex, and
could not be simplified without loss of essential details, even in
the first approximation. For example, it is not possible to consider

only the SO_2 and sulphate contributions to acidity. The nitrogen compounds probably give an input that on a molar basis, is of at least the same order of magnitude as that of sulphur.

In his Table 2, P. Grennfelt had estimated the frequencies of various vegetation conditions in Southern Sweden, e.g., snow cover 20% of the time; no snow, stomata closed 45% of the time; no snow, stomata open 35% of the time. In the subsequent discussion, the suggestion was made that the case of snow cover should be further subdivided, because there would be periods with snow on the ground but not on the needles. Also the deposition velocities for NO_2 were thought to be too high by Dr. Gravenhorst, who reported values of 0.01 to 0.04 cm s^{-1} over soil and over the ocean. Finally, the general view was expressed that there would be great value in undertaking comparative studies in a deciduous forest.

In the other case study, S. Odén produced an estimate of the sulphur budget for Sweden. Using a box model, he estimated the inputs and outputs of sulphur, giving special emphasis to the flow of sulphur from the rivers to the sea. Assuming no storage in soil and using current estimates of sulphur emissions to the atmosphere and of inputs from fertilizers and weathering, Dr. Odén calculated that the difference between inputs and outputs amounted to 140 kilotonnes of sulphur per year. Odén then used his box model to infer the long-term changes in anthropogenic sulphur in various parts of Sweden during this century (Fig. 4 in Dr. Odén's paper).

In the subsequent discussion, the assumption of zero soil storage of sulphur was questioned, particularly because the mechanisms of sulphur retention and throughfall were not yet understood. In this regard, the possible role of elemental sulphur production in the soil was mentioned. In reply, Dr. Odén expressed the view that soil storage of sulphur was a transient phenomenon, and was not likely to be significant on the time scale of one decade.

A question was raised concerning the magnitudes of the various terms in the box model. The atmospheric input of sulphur, for example, could not be determined to better than $\pm$ 25%. Because the final figures in Dr. Odén's budget were small differences between rather large numbers, uncertainties remained. A task for the next few years would be to refine the estimates of the magnitude of each term, using more than one method wherever possible.

The last two papers in Session A dealt with melting phenomena and with snowpack chemistry. The contribution by H. M. Seip was based mostly on field experiments in open areas in Norway. The data revealed that dry deposition in that geographic area is small as compared with wet deposition. Snowflakes are formed to a large extent by freezing of supercooled water and may include impurities

in the crystal to a greater extent than corresponding equilibrium
conditions. It has been found that the first surge of meltwater
contains higher amounts of ions than the bulk snow. More than one-
half of the pollutants are released when the first third of the
snow melts. The lack of knowledge seems most pronounced with respect
to details of run-off of meltwater both within the snowpack and on
or through the ground. Improved information on these topics is
necessary in order to understand the effects of melting and snow on
ecosystems.

The paper by A. S. Kallend revealed that in remote areas with
low concentrations of SO_2, the washout of the gas is unlikely to
produce marked changes in the pH of rain. Observations in such
areas show a significant correlation between the ions SO_4^{2-}, NO_3^-
and NH_4^+, but no single molecular species determines the acidity
of precipitation.

Observations in Southern Norway show that the surface waters
gain significant concentrations of Ca^{++}, Mg^{++} and K^+ via processes
occurring within the catchment, while NH_4^+, NO_3^- and H^+ are lost.
Another significant constituent of surface waters are the weak
acids present at concentrations five to twenty times higher than the
strong acid content, although their contribution to the H^+ concentrations is obviously less (perhaps 50%). The concentrations of
weak acids are least during the spring snow-melt.

The subsequent discussion brought out the view that although at
some points in Southern Norway, there is a strong correlation
between H^+ and SO_4^{2-}, stations are to be found within the area,
where the correlation is weak. The lack of correlation is the rule
over most of Europe. This fact suggests that sulphur transport
cannot fully explain the phenomenon of soil acidity. Other sources,
sinks and reaction mechanisms may be active, even several of them at
the same time, without being mutually exclusive. More research in
this area should be given high priority. Even the concept of soil
acidity is far from simple and is not understood in the same way
by the chemist, the ecologist and the soil scientist. Consequently,
specialists in various disciplines do look for different explanations
and sometimes reach contradictory solutions. In this connection,
Dr. Odén made a useful intervention on the concept of acidity.

Rapporteurs Summary

B. DIRECT EFFECT OF ATMOSPHERIC DEPOSITION ON PLANT GROWTH

D. Fowler and J. E. Rippon

In the first presentation I. Johnsen described the relation
between lichen distribution and atmospheric pollution in Denmark.
Though the trends on a regional scale were difficult to interpret,
there was a clear reduction in species diversity as one approached
urban areas. This was mainly due to lichen sensitivity to gaseous
pollutants. The effect of acid rain was not thought to be large
in this. Johnsen emphasised the need for long term studies (10 to
20 years) since changes in plant and animal populations often
require such time scales for detection.

In discussion, Dr. Gorham asked whether the direct effects on
lichen number and species diversity in cities could be separated
from effects of the city microclimate on water relations. In reply,
Johnsen said this particular question was the subject of continuing
debate in the literature, though unequivocal demonstrations exist
of harmful effects of sulphur dioxide and fluorides on lichen
physiology. In response to a question from R. Goldstein, Dr. Johnsen
mentioned that documented effects of loss of lichens on some mite
populations exist.

P. Havas and S. Huttenen's paper dealt with winter damage to
Norway spruce and Scots pine on Finland. Visible damage has been
observed in the vicinity of sources of S and F compounds, though no
evidence for harmful effects of acid precipitation was noted.
Dr. Havas emphasised that in a harsh and cold northern climate such
as Finland experiences in the winter, the essentially climatic
factors may increase tree susceptibility to air pollutants. He
presented data showing that foliage buried in the snow in winter did
not show visible injury in the following spring, whereas those parts
of the trees exposed to the atmosphere were damaged. A significant
correlation was shown between the water potential and sulphur content
of needles, with those with a high sulphur content suffering greater
damage.

The discussion centered on reasons for this needle damage.
S.N. Linzon questioned the interpretation of P. Havas, pointing out
that some Canadian forests showed similar effects of above and below
snow foliage comparisons, but in the absence of significant gaseous
pollutants. G. Abrahamsen noted unusually large foliar nitrogen
concentrations in needles of affected plants in the P. Havas study.
Dr. Havas suggested this was due to a large local source of nitrogen
oxides from a fertiliser factory.

Results of four field experiments using artificial acid rain as
the treatment on Norway spruce, Scots pine and lodgepole pine applied
at various growth stages were described from Norwegian experiments
by B. Tveite and G. Abrahamsen. Only the most acidic treatments
have shown effects (pH 2.0 and 2.5). The effects were of an increase
in both height and diameter of young Scots pine. Liming treatments
were also included in the experimental design, though these did not
produce effects nor influence the growth responses of Scots pine.
The authors suggested that the enhanced growth seen at low rainfall
pH was caused by increased nitrogen uptake from the soil, though a
beneficial effect of sulphur application alone or in combination
with increased nitrogen uptake could not be excluded.

In the subsequent discussion S.N. Linzon noted that similar
effects had been noted by F.H. Bormann with Eastern white pine
(Pinus strobus). M. Benarie enquired whether natural rain had been
excluded from the experiment. This had not been done but the
resulting dilutions have been calculated and these calculations are
described in the SNSF publication dealing with the work. T. Roberts
suggested that increased ammonification and decreased nitrification
in acid treated soils may account for the Bormann, Tveite and
Abrahamsen data.

C. INPUTS TO SOIL: THE INFLUENCE OF VEGETATION

L. K. Hendrie and I. A. Nicholson

The task of this session on inputs to the soil was to describe
the various pathways, exchanges and effects involved in the passage
of these deposited elements from the intercepting surfaces to the
ground with precipitation acting as the main vehicle.

The canopy is an effective filtering and buffering system
providing a complexity of exchanges, alternative routes and sub-
cycles that influence the composition of the original incident
precipitation. These processes are illustrated schematically in
the flow chart provided by R. Mayer and B. Ulrich. Because of the
importance of water in affecting a variety of structures and
functions within vegetation, plant-water relationships were a natural
focus for attention. H. Tukey's paper provided a forum for
discussion of fundamental processes involved in these relationships.
The papers presented by G. Likens, R. Mayer and B. Ulrich, and
K. Lakhani and H. Miller considered methods by which the movement of
elements along component pathways could be assessed and evaluated.

H. Tukey discussed the importance of water, as rain and mist,
on the aerial surfaces of plants in relation to leaching/absorption
processes and to plant metabolism. He also drew attention to
relationships at the community level. The most important external
factors with respect to the occurrence of effects due to acid
precipitation are wettability and the nature of the cuticular surfaces. The impact of tissue injury was considered, and it was
pointed out that relatively slight injury can have marked effects.
Attention was also drawn to the importance of community structure
on the pattern of cycling within the vegetation. The effects of
surface moisture on metabolic processes, such as the development
of autumn colours, dormancy, flowering and propagation were
discussed.

An important point emerging from the subsequent discussion was that large quantities of water on plant surfaces are not necessary to produce many of the effects described. In fact, a micro-layer can be enough to ensure that substances from the tissues are available for washing-off. Because of the importance of wettability to processes of leaching and absorption, leaf characteristics affecting this property must be understood. Surface tension is related to pH, but other factors are also involved.

Leaching can lead to the appearance of deficiency symptoms in leaves. This may occassionally happen after spring rains if roots have been damaged by winter frosts. Although the loss of cations is one of the more readily demonstrable effects, the exposure of surfaces to water can produce a variety of effects on the surfaces themselves, and these may be of greater long-term importance. The effect of pH was not extensively examined in the work reviewed by H. Tukey Jr., but studies using solutions of pH 4-5 have shown an accelerated loss of Ca^{2+}. Lesions are sometimes produced, but it was pointed out that the effects of pH must be considered in relation to the chemical content of the bathing solutions as this can affect the critical pH values. When Ca^{2+} ions are lost, exchange may take place with H^+ or K^+.

No detailed understanding is yet available of the mechanisms involving surface moisture that trigger metabolic effects. However, growth regulators or precursors of other active substances controlling metabolism may form as a result of stimuli involving moisture stress. A very brief exposure to a stimulus may be effective.

There was wide agreement that both the effects of surface water on metabolism and on the leaching of elements and metabolites, requires further investigation. It was also recognized that it is necessary in developing new research to bear in mind that effects may depend on the coincidence of several factors, both internal and external to the plant, predisposing to the expression of injury or other effect.

An important point emphasized in the discussion was that precipitation with acid properties is only one of a variety of factors interfering with the relationship described. Acid precipitation should therefore be considered in the context of the intricate relationships that occur naturally in vegetation subjected to changes in the aerial supply of moisture.

G. E. Likens reviewed the ecosystem mass balance approach adopted in the Hubbard Brook research programme to develop element budgets. It was shown that an estimated input by dry deposition of about 6 kgS $ha^{-1}yr^{-1}$ could be obtained as the residual by evaluating the sulphur balance using this small calibrated watershed technique.

Emphasis was given to the necessity of establishing clearly defined
boundaries in order to isolate and quantify inputs and outputs.
The implications of perturbations to the system were made apparent,
using as an illustration the impact of clear-felling upon the
sulphur balance.

In the discussion it was pointed out that the presence of
vegetation greatly affects the input of substances by dry
deposition. Unfortunately, there were no satisfactory data
available on the leaching of leaves, such that the information
could be used in separating leaf leaching and wash-off of material
deposited on surfaces. Potassium was found to be heavily leached.
At Hubbard Brook, sulphur appeared in the throughfall in comparatively large amounts (21 kgS ha^{-1}yr^{-1}) compared with an input of
about 4 kgS ha^{-1}yr^{-1}. It was suggested that much of this may have
been material that had been cycled within the system.

A query was raised about the loss of sulphur from the soil
directly to the atmosphere since this process could affect the
balance. The author suggested that if it occurred at all, it is
probably small in magnitude, and would result in only a minor
adjustment to the dry deposition value. Throughfall and stemflow
values would remain unaffected as they were derived independently.

The importance of ground-water conditions in small watershed
studies was emphasized. This was not a problem at Hubbard Brook,
where an equilibrium state regarding the storage term could be
confidently assumed based upon evidence derived from several sources,
and the gauging weir itself was built on the bedrock. It was
indicated that the successful operation of the small watershed
technique involved the setting of boundaries on the system, such as
in relation to bulk precipitation, output and weathering. Inputs
and outputs could then be identified when the boundaries were
crossed.

It was recognized that it is invaluable to have a detailed
knowledge of the internal functioning of an ecosystem, integrating
terrestrial and hydrological components, in understanding and
interpreting the fluxes into and out of the system. There was
general agreement that a well-understood watershed system provides
a unique and effective means of monitoring the inputs to, and some
important effects of substances deposited on, terrestrial ecosystems. Although costs preclude the extensive use of this technique, further watersheds are needed in carefully selected
situations for monitoring purposes.

R. Mayer and B. Ulrich pointed out that meteorological methods,
up to now, do not permit the assessment of dry deposition to forest
canopies on a long time scale. They described a mass balance
approach applied to stands of beech and spruce in Germany to provide
an estimate of the contribution of elements from the atmosphere to

the forest soils. The large magnitude of the storage terms
necessitated an accurate evaluation of the canopy filtering
efficiencies.

The authors' two main conclusions were:

(1) The importance of canopy leaching can be estimated from the
seasonal pattern of below-canopy precipitation when supported by
suitable leaching experiments.

(2) For some elements, such as S, Cl, Na, Ca, Mg (but probably
not for P, N, K and most heavy metals) dry deposition to a forest
canopy can be estimated from the difference between precipitation
fluxes beneath and above the canopy, provided the requirements
for such an estimate, as described in the paper, are taken into
account and checked individually for each ecosystem.

The intrinsic difficulties in determining the magnitudes
involved in the component pathways to the soil surface are overcome
by the adoption of three simplifying assumptions:

a) there is no significant dry deposition directly to the
   soil,

b) absorption/assimilation by leaves and subsequent
   elemental deposition with litterfall is negligible and

c) internal cycling through leaching is small enough to be
   ignored.

The validity of the underlying assumptions and the relevance of
glasshouse experiments to provide a measure of leaching under natural
conditions were questioned. Under natural conditions greater leaf
leaching might be expected, resulting from such sources as leachates
from dead needles on trees prior to leaf fall. Doubt was expressed
about the use of the same filtering characteristics for summer and
winter, a problem which may be in part resolved if precipitation
in summer and winter is comparable.

In support of their approach Mayer and Ulrich pointed out that
while it cannot be asserted that the observed effect is due solely to
dry deposition, they attempted to discount other possibilities by:

a) trying to exclude leaching,

b) using trees in the same soil, and

c) watering test trees for 14 days with rain water.

Other variables, and particularly climatic parameters, could not be
controlled.

The question of representativeness was raised, using by way of analogy the concept of an extraction column, with the suggestion that it is necessary to define both the surface and solution characteristics to obtain reliable estimates of deposition. However, this is impracticable in the field since the variability in canopy structure would make both the definition of a representative profile and standardisation impossible. Further, even if surface character-istics were held constant, variations in other factors, such as wind speed and concentration gradients, would mean that the canopy is not a constant sink in relation to dry deposition.

The authors considered that at present the approach provides the best available estimate. Since the technique cannot enable dry deposition to be assessed, but only help to eliminate other causes such as leaching, collaboration with meteorologists is felt to be desirable to properly evaluate the data interpretation.

K. Lakhani and H. Miller described a mathematical technique which enables the separation of the crown leaching component from total deposition input to the soil beneath a forest canopy to be achieved. The method has the merit of simplicity and appears likely to be unbiased provided the stated assumptions hold. Only three simple measurements are required; catch in the open using both a standard gauge, one modified by the addition of an inert wind filter, and that of throughfall. The method has scope for further development and refinement, both mathematically and in instrumental/ experimental design.

The nature of the inert wind filter and its relationship to actual forest catch characteristics was questioned. However, the method does not require that this device should simulate absolutely the catch characteristics of the forest, but instead provide a measure that is proportional to it. Departure from this propor-tionality would introduce a bias, a problem which may arise at sites experiencing widely different types of precipitation events. The occurrence of frequent mists, for example, could lead to relatively high catch values for the modified gauge, thus weakening the assumption of proportionality. Further complications could result from the effects of droplet size distribution and the possibility of differences in the chemical composition of rain compared with mist.

It was suggested that the magnitude of the estimated leaching of Na^+ (7.8 kgNa $ha^{-1}yr^{-1}$) was unusually high, but Mr. Lakhani pointed out that with the inclusion of the error term (s_e = 5.6 kgNa $ha^{-1}yr^{-1}$), the null hypothesis of no leaching can not be rejected. This provided a pertinent example of the need to include error terms in such estimates.

Indications of differential catch beneath and outside canopies suggest that the foliar loss (l_i) and gain (g_i) of elements by the precipitation passing through the canopy may be substantial. This problem could be reduced by extending the time period considered, a matter which Mr. Lakhani is currently investigating. In order to reduce the limits of uncertainty the X_i error term ($e_{3i} - e_{2i}$) must be small compared to the Y_i error term ($e_{1i} - e_{2i}$). Since e_{2i} provides a correlate between the two error terms it should be negligible. Also, to keep the X_i error term small, e_{3i} should be kept as small as possible.

The method can be used to provide an approximation of dry deposition, D_i, through the product slope xX_i. However, at the moment this is really only useful as a relative value since

$$\text{slope} \times X_i \simeq (1-a)D_i$$

and a is small and positive. Research presently in progress to improve gauge catches could lead to a reduction in a and thus make the product a useful estimator.

Mr. Lakhani suggested two ways in which the method could be improved. Firstly, increasing the horizontal catch area of the gauges in the open while maintaining the same wind filter device would reduce the catch ratio of the gauges and make $\hat{k}$ tend to unity. Secondly, the method is open to further mathematical development because the parameters of the linear relationship arising between two variables are estimated using simple regression techniques even though both variables are stochastic in nature.

In summary, the session provided a forum for discussion which served to emphasise the importance of understanding the complexity of relationships between plants and their aerial environment, especially with regard to the need to consider the interactions between individual plant responses and those of whole ecosystems. In addition, material presented on problems of the measurement of dry deposition and the separation of crown leaching stimulated an awareness of the main areas requiring further work in developing methodology. Cooperation amongst those using different approaches is clearly desirable. An urgent need was apparent for the establishment of other watershed studies and small-scale detailed measurement sites along the lines of Hubbard Brook, to assist in the clarification of the complex processes involved. Finally, one of the most important areas of agreement was the need to understand interface phenomena between plants and the atmosphere, recognizing the importance of physical, chemical and biological processes involved at leaf surfaces, and the mechanisms of interaction between the external environment, plant tissues and metabolic functions. There was broad, agreement, however, that acid precipitation cannot be considered as a factor in isolation.

Rapporteurs Summary

D. EFFECTS OF ACID PRECIPITATION ON SOILS

H. M. Seip and B. Freedman

B. Bache reviewed the acidification of soils. The various processes which may lead to soil acidification were discussed. These processes are:

1) Carbonic acid formation from CO_2 derived from the respiration of soil fauna and flora.

2) Mineral acid generation by nitrification.

3) Organic acid production during the process of plant residue decomposition.

4) Oxidation of pyrite by aeration of previously anerobic soils and,

5) Inputs of acidifying substances from the atmosphere.

Irrespective of the cause of the acidification, a number of effects can be generally observed, such as 1) loss of basic cations, 2) reduction of cation exchange capacity, 3) mobilization of aluminium ions, and 4) changes in biological activity.

The exchange of Al^{3+} for Ca^{2+} or Mg^{2+}, as well as exchange of H^+ for Ca^{2+} were considered in some detail. Bache stressed the importance of identifying the solution component of the soil acidity as the activity ratio a $Ca^{++}/(H^+)^2$, rather than the hydrogen ion activity alone. He noted that the sum $Ca^{2+} + Mg^{2+}$ would best be used in the activity ratio calculation. This implies that the lime potential (or acidity potential) pH - ½pCa, is an important parameter. A simple comparison between the lime potential of the soil and that of rainfall will show the direction of a change of exchange equilibria by acid solutions. The extent of the change depends on a number of factors, i.e. 1) the base saturation of the soil, 2) the cation exchange capacity, and 3) the relative binding strength of the soil materials for hydrogen and basic cations.

In addition, the mechanism of movement of water in the soil must be considered. The situations of exchange are very different when 1) the initial soil moisture content and structure are uniform with depth, and the wetting front develops a constant shape, or 2) surface runoff or channelled flow within the soil occur. In the latter case, the water is in contact with only a small fraction of the soil. This seems to imply (rap. note) that much acidity from precipitation may pass relatively unchanged through the soil and thus, susceptible water bodies can acidify before the buffering capacity of their watersheds was depleted.

B. Bache concluded 1) that for soils of high pH, a small decrease of pH is not of great biological significance, and 2) for strongly acid soils a further soil acidification is unlikely to occur because ion exchange is limited and the H^+ ions thus remain in solution and drain away. However, there is an intermediate range of susceptible, slightly acid, poorly-buffered soils where relatively little acidification may lead to a striking loss of productivity.

A major point of discussion concerned the usefulness of a lime potential determination including only the Ca^{2+} and Mg^{2+} concentrations. In some soils, K^+, Na^+, or Al^{3+} could be more important. In reply, Bache pointed out that the Na^+ should leach quickly, but that in some cases K^+ and Al^{3+} could be significant. He felt, however, that Ca^{2+} and Mg^{2+} were the important cations in the majority of cases. It was also mentioned that the statement that further acidification of very acid soils does not occur might only be valid for aerobic soils. Under reducing conditions, sulfate inputs would be reduced to sulfide, which would produce a flush of acidity if the soils later became oxidized. The topic of soil water movement and the importance of the time of contact between water and soil were discussed further. The need for further consideration of the infiltration of water in soils and of rates of ion exchange became evident.

M. Schnitzer, in his presentation, considered the effects of low pH on the chemical structure and reactions of humic substances in the soil. Humic substances are the most widely-distributed and abundant carbon-containing compounds on the earth's surface, and affect, to some degree, all soil reactions. Hence, they must be considered in any studies concerning the effects of anthropogenic acidification of soils. Humic substances are partitioned by acidification, and are composed of humic acid (soluble in dilute alkali but precipitated by acid), fulvic acid (soluble in both alkali and acid), and the relatively unreactive humins (insoluble in both acid and base). Several characteristics of acid-soluble fulvic

acids which are affected by pH changes include 1) changes in
structure from a stringy, low density material at low pH to a fine-
grained, high density, plate-like structure at high pH, 2) a decrease
in the electron-spin resonance as pH is raised, which is an estimate
of the relative number of free radicals and of molecular size, and
3) changes in the stability constants of fulvic acid-metal complexes
which make metals relatively available at low pH. In addition, under
the influence of high sulfur loading near a smelter, an increase in
the relative amount of sulfonic acid groups relative to carboxyl
groups in the fulvic acid was seen. This is relevant in that cations
bound by these sulfur-modified fulvic acids have relatively high
stability constants, and are somewhat less available for plant
uptake.

Schnitzer noted that he could forsee severe soil degradation
under very high inputs of acid precipitation if large amounts of
fulvic acid were to go into solution, with the effect that large
quantities of cations could be bound as complexes and lost from the
soil system in conjunction with the leaching fulvic acid. It was
stressed that excessive production of soluble fulvic acid in soils
should be avoided due to this consequence of loss of soil minerals.
A soil pH of 2-3 would be necessary to achieve such a fulvic acid
effect on a large scale. Humic acids would not be so affected
because they are insoluble at low pH.

In the discussion, it was noted that the high degree of fulvic
acid sulfonation had only to date been demonstrated in sulfur-polluted
Sudbury soils, and that locality had received very high SO_2
fumigations, so that it was an exceptional case, of limited significance to the phenomenon of regional acid precipitation with which
the workshop was concerned. M. Schnitzer agreed that this did
indeed probably represent a localized, exceptional case.

The next speaker was T.C. Hutchinson, (paper in Part V),
who considered the effects of acidification on heavy metal mobilization in soils. Examples were presented of extreme situations where
field studies were conducted which were relevant to the mobilization of metals in soil by acid percolating waters. Copper, nickel,
and aluminium in mineral soils close to the Coniston smelter near
Sudbury were shown to be toxic to an array of bioassay species, and
this was felt to be due to the apparently high availability of these
metals in the acid (pH $\sim$ 3.5) soils. In another situation, an
Arctic tundra area known as the Smoking Hills was discussed. This
is frequently fumigated by intense plumes of SO_2, emanating from a
naturally-combusting deposit of high-sulfur bituminous shales exposed
by coastal erosion and slumping. Soils in the most intensively-
fumigated zone were very acid (pH's ranging from 2.2-3.0 in an
area where control pH's were $\sim$ 7.5), and had very high soil sulfur
contents. Depth profiles of the laterally-draining active layer of
the soil show low concentrations of calcium and magnesium at the

surface, compared to deeper soils just above the permafrost, and compared to control profiles, which are not base-depleted at the surface. Certain metals (i.e. Mn, Fe, Al, and Zn as well as Ca and Mg) were present in high concentrations in runoff following rains and in tundra ponds in the acidified area. These have apparently been mobilized at the low soil pH's. Recently published data of G. Tyler dealing with Cu-Zn contaminated soils collected at a smelter near Gusum, Sweden were also discussed. Laboratory soil columns leached with acidified solutions had high cation concentrations in the relatively acid (i.e. pH 2.8) eluents. In summary, the mobilization and speciation of metals in soil or water appears to be strongly pH-dependent. At lower soil or percolate pH's, the residence time of metals in soil is much shorter relative to higher pH's.

In the discussion, it was noted that the examples presented represent worst-case situations, which might be unlikely to occur under the influence of normally-occurring acid precipitation, except perhaps after long time periods. However, elevated aluminium concentrations are now reported from many lakes in areas of acid rains. The question of mercury mobilization was also raised. Mercury appears to be mobilized into acidified lakes at low pH, and can become methylated by bacteria and incorporated into fish when pH is raised by liming.

The next presentation was by L. Petersen, who considered the mechanism and possible effects of acid precipitation on the podsolization of soils. Podsolization is the most important soil-forming process in widespread areas of cool, north temperate regions having 1) an excess of precipitation over evapotranspiration, 2) sandy soils of low buffering capacity, and 3) usually overlain by an ericaceous or conifer-dominated plant cover. Reaction of these soils is acid, they are relatively low in available plant nutrients, they accumulate surface organic matter characterized by a relatively low degree of decomposition, and they have clear, well-differentiated elluvial and illuvial horizons. The translocations are caused by soluble organic compounds produced in the A horizon. These compounds move downwards with leaching water and dissolve iron and aluminium from the inorganic soil constituents by complexation. Once a sufficient amount of these metals has been taken up, a mutual precipitation of the metals and the organic matter takes place.

Acid precipitation is perceived as affecting the process of podsolization in two main ways: 1) since an acid surface soil reaction is a pre-requisite for podsolization, additional increments of acid could start the process at an earlier date on susceptible soils where the vegetation cover is appropriate, and 2) additional acid inputs could amplify podsolization by increasing the thickness of the elluvial A horizons and further depleting surface soils of

nutrients. However, it is not known at present whether the
additional H^+ inputs due to acidic precipitation are actually
sufficient to measurably produce these changes.

In the discussion, Wiklander noted that podsolization implied
a weathering and mineral transformation, and that the Al and Fe
bound by organic matter could be released when this organic matter
decomposes, and be available to form non-swelling montmorillonite,
which should provide some buffering capacity and retard podsolization. Petersen noted that he had not studied this reaction, but
that it seemed feasible.

B. Ulrich discussed soil cation exchange and sulfate absorption.
The buffer ranges of mineral soils were described at the beginning
of this paper. In the pH range of 6.5-8.3 (i.e. circumneutral soils)
there is a carbonate buffering system. Between pH 5 and 6.5 there
is a beginning of Al buffering, which is the important system for
strongly acid soils (i.e. pH 3-5), along with organic matter
(pH 3-6). For extremely acid soils (pH < 3) there is an iron
buffering system.

The input-output balance of the Solling forest was discussed.
The soils act as a sink for H^+, Ca^{++}, and SO_4^{-2}, and as a source for
Mg^{++}, Mn^{++} and Al^{3+}. The loss of Mg^{++} and Mn^{++} arises from the
weathering of minerals. Mg^{++} also appears to be leached from the
exchangeable pool. It's loss due to soil acidification is already
believed to be of relevance to forest growth in parts of central
Europe. However, the exchange and dissolution of Mg^{++} and Mn^{++} by
H^+ with subsequent leaching from the soil is much smaller than the
dissolution and leaching of Al^{3+}. The disintegration of clay
minerals is thus the main process consuming hydrogen ions in the
Solling soils. The sulfate retention is at least partially a result
of the buildup of $AlOHSO_4$. While the acidification of the soil
probably started with human use of the forest 1000-2000 years ago,
the buildup of aluminium sulfate is believed to be a recent process
which started with the large scale emission of SO_2 from coal burning
in the last century.

The question of soil acidification due to forest use was again
raised in the discussion. In particular, the removal of bases due
to tree cropping has a significant soil acidifying effect which must
be considered for managed forests.

The last paper in this session was given by L. Wiklander, who
also dealt with cation-anion adsorption and mobility. The soil has
amphoteric properties and therefore has the capacity of adsorption
and exchange for both cations and anions. Experiments with both
equilibrium and dynamic systems have shown that certain polyvalent
anions of soluble salts added to soils increase the adsorption and
decrease the leaching of cations. Thus, the solubility and retention

of nutrient cations in soils are determined partly by the associated anions. In experimental work, the effectiveness of the anions studied proved to be:

$$Cl^- \sim NO_3^- < SO_4^{-2} < H_2PO_4^- < HPO_4^{-2}$$

The anions are bound by hydrated oxides of Al and Fe or by other minerals with free negative charges. When a solution of Ca^{++}, Mg^{++}, K^+, Na^+ and $H_2PO_4^-$ was added to a podsol B2 soil, only Na^+ was leached. The total leaching in the $H_2PO_4^-$ system was 23% of the added cations, compared to 62% for a Cl^- system. Wiklander concluded that we may improve the capacity of the soils to adsorb cations and reduce the leaching of cations and anions by using more phosphates as fertilizers. To a lesser extent, sulfate inputs may also be useful in this respect.

     The importance of this aspect with respect to tropical soils was raised in the discussion. It was agreed that the possibility of improvements does exist, but that it is largely a question of economics whether the method is used. When asked whether his experiments showed that anion exchange is occurring in soils, Wiklander confirmed that this was the case.

     In summary, the processes which normally lead to soil acidification were described. The possible effects on soils of incremental acid loadings through precipitation were identified and discussed. Physical-chemical soil properties which appear likely to be affected by acidification include base saturation, cation exchange capacity, humic acid chemistry and structure, and binding by soil exchange sites of various cations and anions. However, no hard evidence currently exists from field studies that indicates that acidification of sensitive soils has occurred, or that the above physical-chemical processes have been affected under the influence of currently-measured regional inputs of acidifying substances in precipitation. This may be because such soil degradation is a long-term process that is not easily measureable, due to normally-occurring temporal and spatial variation in soil properties, and that only short-term studies have been completed. The need for long-term studies under realistic acid loadings, and for further shorter-term studies under more severe acid loadings (which may to some extent serve to compress the time scale needed to achieve effects on soils) is apparent.

E.   EFFECTS OF SOIL ACIDITY ON NUTRIENT AVAILABILITY

AND PLANT RESPONSE

T. M. Roberts

I. H. Rorison pointed out that a considerable amount of
literature exists on the relationship between soil acidity and plant
growth, both in an agronomic and ecological context. He emphasised
that these interactions were dynamic and that many species can
alter the acidity of the soil. Many species also have the capacity
to adapt to changes in soil conditions.

Changes in the form and rate of supply of macronutrients and the
increasing solubility of potentially toxic polyvalent ions as a soil
is acidified and leached, are well documented. Concomitant changes
in the species composition of the plant communities, at least in
temperate regions, also occur in a predictable manner. What is
perhaps less well understood are the physiological mechanisms which
result in plants having different tolerances of soil nutritional
factors such as low cation availability and high availability of
polyvalent cations.

Dr. Rorison felt that acid precipitation would have little
direct effect on plant roots – sulphate has low toxicity and H^+ ions
per se only directly affect root growth below pH 3. Ammonium
sulphate would act as a source of nitrogen and more H_2SO_4 would be
generated. Nitric acid would have little acidifying effect as the
nitrate would be utilised by plant roots and bicarbonate or hydroxyl
groups released. This may have beneficial effects in terms of
nitrogen supplies to plants. The author felt that acid precipitation would have little effect on the outcome of plant competition
in areas remote from sources although there is evidence of changes
in species compositions of grass swards due to soil acidification
by SO_2 deposition in some urban parks in the UK.

The paper was followed by a discussion of the rate of evolution of tolerance to stress conditions with particular reference to heavy metal toxicity. It was clear that this adaptation may be rapid and that multiple tolerances to a number of elements may arise in the same species. When asked about the importance of foliar uptake of constituents of acid precipitation Dr. Rorison expressed the opinion that the primary stresses which occur during seedling establishment are edaphic, but that foliar uptake of nutrients could be relevant in the case of mature plants or crop plants.

Rapporteurs Summary

F.   EFFECTS OF ACIDITY ON NITROGEN CYCLING

        T. M. Roberts

    M. Alexander reviewed the microbial processes involved in the
nitrogen cycle and highlighted those processes which may be
sensitive to acidification. He felt that mineralisation is unlikely
to be affected whereas nitrification and, to a lesser degree,
denitrification, may be more sensitive. Autotrophic nitrifying
bacteria in particular, do not function below pH 6 in the laboratory,
although some nitrification has been reported in acid soils in field
studies. This may be due to activity at more alkaline microsites
or nitrification by heterotrophic organisms. In the case of nitrogen
fixation the symbiotic associations with legumes are pH sensitive -
for example, there is little nitrogen fixation by the Rhizobium
strains in clovers below pH 4.5. It is not clear whether nitrogen
fixation in the roots of non-leguminous plants is as sensitive to
high acidity. In the case of free-living nitrogen fixers, it appears
that most are sensitive to low pH whereas a few are quite resistant.

    The ensuing discussion considered the high rates of minerali-
sation in the Tropics. Dr. Alexander felt that the high rates of
nitrogen fixation in places such as rice growing areas may be
sensitive to acidity. In Boreal forests however, nitrogen fixation
is much slower and is unlikely to be reduced by the hydrogen or
nitrate ions in precipitation.

    It was also pointed out that nitrification can occur in acid
soils but Dr. Alexander felt that the rates are generally low and
may be due to heterotrophic nitrification. Acidification can result
from autotrophic nitrification in the laboratory but, in the field,
other metabolic products, such as ammonia may neutralise this
acidity.

    D. Johnston then presented the paper by Turner and Gessel on
nitrogen and sulphur deficiency in Douglas fir stands in the Pacific
North-west Region of the USA. Early observation on the uptake of

599

nitrogen and sulphur by trees in Australia had shown that foliar
sulphur levels normally occurred within a narrow range and could be
used to indicate sulphur deficiency. In many sites in the Pacific
North-west Douglas Fir stands respond to nitrogen fertilisation.
In those areas where there was no nitrogen response, a marked
response to sulphur fertilisation was found. In this region, acid
precipitation has increased the nitrogen input by 2-3 fold and may
also help to alleviate sulphur deficiency.

In the discussion it became clear that there were many areas
of sulphur deficiency in the United States. In Germany however,
there is no response in forest growth to nitrogen fertilisation
due to the high input from the atmosphere. In fact, acid
precipitation may have leached major cations to the point where
magnesium would be limiting forest growth. Dr. Johnson said that
acid precipitation may have caused potassium depletion in some
soils in the Pacific North-west.

G.   EFFECTS OF ACIDIFICATION ON SOIL MICROBES, INCLUDING

DECOMPOSITION PROCESSES

D. W. Johnson and D. S. Shriner

M. Alexander's paper was an overview of the effects of
acidity on micro-organisms and microbial processes. He reported
that microbiologists have long been interested in effects of
acidity, and that there is an enormous literature on the subject.
He began the review by listing the six major microbial groups in
soil (bacteria, fungi, actinomycetes, protozoa, algae, and viruses)
and their essential functions.

He noted that studies in microbiological processes are
typically done in pure culture in the laboratory and that there are
serious problems in trying to directly extrapolate these studies to
the field because of difficulties in identifying the active
population of microorganisms in the field, possibilities of shifting
microbiological populations as soil chemical conditions change, and
possibilities of a variety of microsites which differ in pH and
other conditions within the soil matrix.

In general, nitrifying bacteria are very sensitive to acidity.
Their activity falls rapidly with decreasing pH and is undetectable
below pH 4.5. Soil acidification can also be expected to cause a
general reduction in populations of bacteria and actinomycetes and
an increase in the abundance of fungi. The increase in fungi is
attributed to decreased competition from other heterotrophic
organisms at low pH. Increasing acidity can also cause decreased
N fixation rates. Changes in acidity can affect plant diseases in
various ways, either by inhibiting the pathogen directly (reducing
disease) or inhibiting the antagonist to the pathogen (increasing
disease).

One effect of acidification of special concern is that of the
products of microbial activity. Dr. Alexander gave two examples.

601

N_2O emissions from denitrification increase with decreasing pH, and N_2O is destructive to the ozone layer. Production of nitrosamines (proven carcinogenic compounds) also increases with increasing acidity in laboratory studies. Field verification of the latter is needed.

In reply to questions M. Alexander suggested that acid snowmelt could inhibit nitrogen fixation by aquatic blue-green algae. To E. Gorham's enquiry as to whether sulfuric acid formation was an important process in podzols, Alexander replied that, although soil S oxidation is said to be autotrophic, he suspects fungi play a role in S oxidation and this would be little affected by pH. In reply to a further question of whether there were any permanent changes in microbiological activity induced by pH he replied that most studies are short-term. There are indirect evidences of pH adaption from comparisons of soils of various pH, as well as from Norwegian studies showing pH adaption in one soil.

The paper by T.M. Roberts described the experimental design and preliminary results of a H_2SO_4 irrigation study in a mixed pine stand in England using lysimeters. Soils sampled in a transect showed progressive declines in pH and base saturation with increasing proximity to an urban center. Solution data from the actual study site showed some enrichment of H^+, SO_4^{2-}, and NH_4^+ as precipitation passed through canopy and litter layers.

An experiment was described in which sulfur equivalents of 25 and 50 kg/ha were added to lysimeter plots as H_2SO_4 and as elemental S, in order to determine the feasibility of working with the more convenient elemental form. After six weeks, however, there were no apparent effects from the elemental S applications, while additions of H_2SO_4 resulted in increased leaching of H^+ and $SO_4^=$ beneath the litter layer, but had no effect beneath the mineral soil. Decomposition studies showed little effect of acidification.

Roberts also reported on changes in CO_2 evolution following fumigations of Scots pine litter with sulfur dioxide and nitrogen dioxide. 0.1 ppm SO_2 resulted in a 10% decrease in CO_2 efflux, while 0.1 ppm of NO_2 resulted in a 30% increase in CO_2 efflux. When the two gaseous pollutants were added at the same doses in combination, an intermediate effect on CO_2 efflux was observed. This observation serves to point out the possible importance of compensating effects of mixtures of certain stress factors in complex environments.

In monolith lysimeters transferred from forested to open sites, there was a net loss of SO_4 and NO_3^-. The SO_4^{2-} loss was attributed to release of adsorbed SO_4^{2-} from the soil, and the NO_3^- loss was attributed to nitrification.

Following the paper, B. Freedman asked about acidification effects of the gaseous NO_2 and SO_2 on rates of mineralization of the litter, and Dr. Roberts replied that this data is now being taken. J. Rippon inquired as to when SO_4 might begin appearing in leachate from the mineral soil. Dr. Roberts replied that he did not know, but there was not yet an indication of increased SO_4 leaching from this horizon. G. E. Likens offered a possible explanation to the increased nitrate leaching from the soil monoliths. At Hubbard Brook, nitrification inhibitors have been isolated from tree root exudates, and the elimination of roots from the lysimeters may have allowed nitrification to occur by eliminating the source of these inhibitors.

The presentation by J. Rippon described the experimental design and some preliminary results of acid rain studies on soil and catchments in England. Preliminary results showed that both precipitation and streamwaters in one of the catchments had low pH and high sulfate content. In one instance, stream acidity increased with water flow rate. Marked seasonality in H^+ and SO_4^{2-} fluxes was noted. Assuming that there is no net sulfate retention in the catchment, the calculated dry deposition of sulfate accounts for two-thirds of the total sulfate entering the system. In another catchment, streamwater at its source was more acid and had higher sulfate concentrations than rain. Downstream, however, pH rose and sulfate concentrations dropped, presumably due to interactions with springs and bogs.

Preliminary results of H_2SO_4 irrigation on soil monoliths showed net SO_4 retention in the soil, similar to results obtained by Roberts.

Preliminary results from pot studies showed no significant changes in bacterial populations, but some tendencies toward pH reduction following acid treatment. No significant effects on CO_2 evolution, enzyme activity, or mineralization rate were noted following acid treatments. In contrast to field studies, no soil SO_4 retention was noted in pot studies.

Dr. Alexander suggested that reductions in microbiological activity might be occurring in the very top horizons of the soil. B. Bache suggested that the construction of the monolith lysimeters could in itself create disturbances to soil chemical and microbiological processes. Alexander concurred, suggesting also that a period should be allowed so that stability could be re-established before serious sampling begins. F. Anderson pointed out that cutting off roots eliminates a form of competition for soil nutrients, which can in turn affect decomposition rates in monolith lysimeters or trenched plots. Dr. Rippon agreed and also suggested that vertical water movement could occur in the lysimeters. G. Abrahamsen also agreed on the impact on experiments of lysimeter

construction, and discussed how this varies with soil type. He
also stressed the importance of leaving ground cover vegetation
intact. J.N. Galloway suggested that even the establishment of
equilibrium may not allow direct use of that data, since it may be
an artifact of the lysimeter. B. Ulrich noted that a chemical
equilibration period is also necessary for plate lysimeters.
Dr. Schnitzer argued strongly, however, that no sample can be taken
without disturbances and we must live with this fact. He noted
that the pursuit of the present line of reasoning would not allow
us to ever take a scientific measurement.

G. Abrahamsen presented the results of a Norwegian case study
on the "effects of artificial acid rain and liming on soil organisms
and the decomposition of organic matter". The paper described an
ecosystem approach to the determination of the effects of artificial
acidification on decomposer organisms and decomposition of organic
matter.

Fungi from decomposing litter were isolated and their growth
measured at different H^+ concentrations. Soil animals were
extracted from soil samples taken from laboratory and field
experiments, with applications of simulated acid rain and lime. In
coniferous forests located on acid podsolic soils, the abundance
of many soil animals was found to increase under the more acidic
conditions induced by the simulated acid rain, whereas few animals
were found to become more abundant when soil acidity was reduced
by liming.

The increased abundance of certain species of soil animals
may have been due to an increase in abundance of fungus food
reserves available to the animals as a result of the more acid-
tolerant fungal species out-competing more acid-sensitive bacterial
species for available food reserves. The possibility of acid effects
on predatory species of mites was also suggested as an explanation,
but not established in the studies.

Decomposition studies showed that the initial decomposition
of plant remains was only slightly influenced by acidification.
However, decomposition of raw humus material was pH dependent, and
decreased with increased acidity.

Dr. Abrahamsen concluded his remarks with the caution that the
effects observed were produced with much higher concentrations
of hydrogen ions than found normally in precipitation, and hence
they are not directly applicable to interpretation of the acidic
rain problem. Discussion following the paper concluded that long-
term studies with more realistic inputs of hydrogen ions are
necessary in order to assess the effects of current levels of input
on decomposition processes. Dr. Andersson offered during the
discussion to present results of a study by Bååth et al. on

biological soil properties (experimentally acidified) in a Scots
pine forest. In these studies, acidification resulted from
artificially high acid inputs of 50 and 150 kg/ha yr^{-1} of SO_4 as
H_2SO_4. Decomposition rates and decomposer organisms (bacteria,
fungi, and invertebrates) were all reduced as a result of the acid
treatments.

There was general agreement at the conclusion of the discussion, that decomposition studies must, in the future, include
studies of mineralization, nitrification, etc.

B. Freedman and T.C. Hutchinson presented a case study of forest
litter decomposition near the Sudbury smelter. A series of field
plots at various distances from the smelter were utilized to examine
potential effects of heavy metal and sulfate burdens on litter
decomposition. Decreasing inputs and accumulations of copper,
nickel, iron and sulfur were shown in organic soil horizons and
vegetation along a 60 km SSE transect away from the smelter. There
was, however, no effect on the pH of precipitation, or of soil
litter horizons with distance from the smelter. Rates of litterfall
were determined to be unchanged as a function of distance from the
smelter, but the litter standing crop was increased close to the
source, implying an effect on litter decomposition. Additional
evidence for such an effect included lower rates of acid phosphatase
activity and reduced CO_2 efflux from soils collected at the
contaminated sites. Populations of microarthropods and fungi were
not found to be significantly decreased at the contaminated sites.
A laboratory experiment involving addition of copper and/or nickel
to litter homogenates resulted in depression of litter/mineraliza-
tion and CO_2 efflux at metal concentrations similar to those
observed in the field.

In response to questions, Freedman indicated that while
observed vegetation toxicity near the smelter was likely primarily
an effect of SO_2 and of heavy metals in soils, the effects on litter
decomposition are primarily due to heavy metals, specifically copper
and nickel. In addition, he noted that at sites close to the
Coppercliff smelter, atmospheric SO_2 and soil heavy metal burdens
have been in the past more significant detrimental factors than
is acid rain.

E. Gorham suggested that the reason why it had not been possible
to see a distance relationship with rainfall pH could have been
neutralization by iron particulates near the source. This appears
to be the case at all Sudbury smelters where high particulate
deposition occurs.

In response to an inquiry about the influence of the tall
stack on patterns of precipitation around Sudbury, it was noted that
there may be an increase in the area affected with respect to

acidification of susceptible lakes since the use of the new stack,
although this was a contentious point.

General Comments - Summary

1.   Increases in soil acidity can cause many changes in soil
     microbiological populations and processes. Decreases in
     populations of bacteria are likely, with subsequent increases
     in soil fungi due to decreased competition.

2.   Various case studies have shown that applications of acid
     at 'natural pH's' has had little effect on decomposition of
     litter. This may be due to the relative pH tolerance of
     decomposing fungi. Some effects on humus decomposition have
     been noted, however.

3.   Field studies on podzol soils in Norway have shown that
     invertebrate populations are well adapted to acid conditions
     and are actually increased by further acid application.

4.   We need to know more about effects of soil acidification on
     nutrient mineralization. Acidification could slow minerali-
     zation and thus reduce nutrient availability to plants.

4.   Dry deposition of sulfate seems to be a major pathway of
     S input to many soil systems.

6.   Some soils can adsorb SO_4 and this retards sulfate leaching
     following H_2SO_4 applications.

H.   EFFECTS OF ACID PRECIPITATION ON PLANTS AND
     PLANT PATHOGENS

          B. Bache and J. N. Galloway

     The conclusions of the two speakers, E. Cowling and D. Shriner
can be summarised as follows:-

E. Cowling - An Overview.

     1. Cuticular erosion by acid precipitation has the potential
         to predispose plants both to drought stress and
         susceptibility to biotic pathogens.

     2. Interactions (synergisms) between acid precipitation
         and gaseous pollutants may greatly aggravate well
         established damage by the gaseous pollutants.

     3. Acidity in simulated rain can have both predisposing
         and therapeutic effects on host-parasite systems
         depending on the nature of the pathogen (bacteria vs.
         fungi, for example) and the stage in the disease cycle
         during which the acidity is applied.

D. Shriner - A case study of plant and pathogen effects of
             acid precipitation.

     1. Effects of acidic precipitation on plant pathogens and
         plant disease development appear to be most likely to
         occur currently as a result of exposure of plant surfaces
         to short-term levels of high acidity ($\leq$ pH 3.4). If
         tissue injury resulted, it would predispose the plant
         to increased infection or disease development, without
         an effect on viable pathogenic propagules arriving at
         the leaf surface at a later time.

                                607

2.   Plant pathogenic fungal spores and bacteria are
      frequently exposed to precipitation during the process
      of spore dispersion, deposition, or initial germination
      and growth on leaf surfaces. They are often at the most
      vulnerable phase of their life stages when exposed to
      acidic precipitation, which could therefore inhibit
      disease establishment or further disease development.

3.   Long-term chronic stress of plants by acidic
      precipitation may have the potential to predispose host
      plants to greater disease development by combined
      effects on host nutrition, metabolism, and/or host
      defense mechanisms.

     Clear evidence of documented effects of acid precipitation on
plants and plant pathogens was presented by both speakers.
However, all studies were performed under controlled conditions
where the single variable was the pH of the incident precipitation.
To date there have been no confirmed observations of effects of
acid precipitation on plants or plant pathogens in the field. This
is not to say that such effects do not exist. Rather, it means
that other environmental variables (temperature, radiation, water,
etc.) are large enough to make detection difficult. The point was
made that effects of acid precipitation while real and significant
in terms of productivity, may be subtle and difficult to discern.
For example, a plant community may be stressed by acid
precipitation (e.g. cuticular erosion) to the point that it is more
susceptible to damage by a pathogen or an environmental variation
(i.e. drought). This reason for the original stress (acid
precipitation) will be almost impossible to field document but it
may be real and substantial none-the-less. Therefore, while it
may be impossible to document the effects of acid precipitation on
plants and plant pathogen in the field due to their subtle nature,
it is known that effects can occur and that they have been shown to
be significant in laboratory trials.

I. EFFECT OF ACID PRECIPITATION ON WATER QUALITY AND ON
    EFFECTS VIA SOIL INTO WATER BODIES

J. N. Galloway and B. W. Bache

G. E. Likens reviewed data on the chemical composition of
precipitation and water bodies in the U.S.A. He compared the data
for the composition of precipitation in 1974-77 with the trends
observed for 1955-1974, and stressed the difficulties in making
long-term deductions from these trends. The increases in acidity
of precipitation over the north-east USA correlated best with
increases in nitric acid, although sulphuric acid was still the
main acid component of precipitation at Hubbard Brook. Discussion
centred around the interpretation of trends in relation to the
variability of data. The influence of meterological conditions on
the composition of precipitation was emphasised, and an initial
separation of data into winter and summer episodes was recommended.

S. Odén presented data for a period of more than 50 years on
the chemical composition of lakes and rivers in Scandanavia. He
related increases in acidity, sulphate and cation contents to
reduction in biological activity over this period. The discussion
established that much of the acidification occurred before the
recent (1940+) increase in the use of fossil fuels, and that pH
levels had varied little since 1969. Alternative explanations,
apart from acid precipitation, were the oxidation of sulphur from
reduced environments during a series of dry summers, changes in
land use, and the acidification of surface runoff caused by contact
with acid soils.

B. Ulrich used data from the whole-ecosystem study of a beech
forest at Sölling, W. Germany, to show how natural acid production
in an ecosystem can be calculated from phytomass build up,
decomposition, and export. This can only be done if ion uptake is
considered as well as mineralization. Of significance is whether
nitrogen uptake occurs as NH_4^+ or NO_3^-. Mineralization and uptake
processes in the organic layer of soil are coupled in such a way
that the net change in hydronium ion flux caused by percolating

water is close to zero. The discussion brought out that the degree
of dissociation of the acids in humus was unlikely to be as high as
the calculations of the acidity budget required. Nevertheless,
this was a valuable approach to an otherwise intractable problem,
and it highlighted factors not considered elsewhere.

Professor E. Gorham presented data on chemical composition of
lake waters as a function of distance from a smelter. In the
discussion he emphasised the need to trace the flow of precipitation
through a watershed, and to study soil acidity gradients on
transects from pollution sources, as possible means of understanding
the multiple interactions involved. Effects on the phosphorous
cycle may be an important link in these interactions.

Rapporteur's Synthesis and Recommendations

1.   An urgent need is to summarize the present state of knowledge
by bringing together the data that are available from diverse sources
on chemical composition of precipitation, and of lake and stream
waters. These data should be examined statistically so that the
limits of the validity of hypotheses based upon them can be
estimated.

2.   There is no doubt that the vegetation-soil system has considerable capacity for absorbing pollution. It is difficult to separate
the effects of acidity from that of other anthropogenic compounds.
Estimates of the time scale over which a given system can
effectively absorb a given pollution load with little adverse effect
need to be made. In particular, it is important to know whether a
linear or an exponential time curve operates, or whether sudden
changes can be expected after some critical threshold point has been
reached. Long term studies on transects from pollution sources will
help resolve these questions.

3.   There still appears to be some uncertainty over the role of
soil, which provides the link between terrestrial and aquatic
communities. Papers in Part II described the mechanisms and the
results of interactions between acid inputs and various components
of soil. Detailed quantitative data are needed for specific
situations, especially on the extent to which soil modifies the
composition of the precipitation flowing over it or through it into
streams and lakes.

Rapporteurs Summary

J. IDENTIFICATION OF SENSITIVE SITES AND SOILS

I. Johnsen and B. Freedman

The aim of this session was to provide a means by which sites
with soils sensitive to the effects of inputs of acidifying
chemicals in precipitation could be identified. This is an obvious
necessity in our attempts to predict the terrestrial effects of
acid precipitation, since it is these sensitive sites which might
be expected to exhibit the most immediate and severe responses to
such a pollution stress.

The first speaker, L. Wiklander made the important point that
the effect of acidifying substances in precipitation depended
relatively little on the amount of hydrogen ion inputs, as compared
to the susceptibility of the soils. In general, little or no
effect would be expected on calcareous, well-buffered soils above
$\sim$ pH 6, or on acid well-buffered soils below pH 3.5-4.0, whereas
poorly-buffered soils in the range of about 4-6 are relatively
susceptible. Other factors mentioned that affect susceptibility
include climate (especially the excess of precipitation over
evapotranspiration) and topography and physical soil structure,
both of which would influence the amount of runoff relative to soil
percolation, and thus the time of contact between acid percolating
solutions and cation exchange sites in the soil. Wiklander noted
that atmospheric deposition was not the only source of H^+ inputs,
and indeed, he noted that these inputs were only incremental to
larger amounts of soil acidity due to 1) acid root exudations,
2) carbonic acid produced from the high CO_2 tensions in soil,
3) the oxidation of reduced sulfur and nitrogen compounds, and
4) the presence of acid humic material and plant litter.

In the discussion, S.N. Linzon noted that it was his understanding from the Norwegian SNSF studies that podsolic forest soils
were most susceptible to the effects of acid inputs, whereas
Wiklander had stated that this was not the case. Wiklander replied
that most podsols had been acid for a long time, and had low

quantities of exchangeable cations. Because of this latter point,
H^+ passes through the soil, without exchanging for other cations.
He noted, however, that certain brown forest soils were more
susceptible to acidification because of their higher base saturation,
so that the potential for exchange of H^+ for other cations, and
subsequent acidification, still exists. S. Oden noted that the
differences of opinion between Wiklander and the SNSF data may be
due to fundamental differences in the lysimeter experiments upon
which the relative conclusions are based. The Norwegian SNSF
experiments were made using relatively unmodified rates of infiltration of the acid percolating solutions, and large amounts of
water were not used relative to the background rainfall. Under
these conditions, there was little H^+ output in the percolating
water, and cations were exchanged. In Wiklander's experiments, the
infiltration rates were extremely high relative to those found in
most undisturbed soils, and thus the H^+ did not have time to react
at cation exchange sites, and passed through. L. Wiklander replied,
however, that rapid infiltration of relatively large amounts of
acid percolating solutions was a means of compressing time scales
in these experiments, so that the effects of small changes over
long periods of time could be predicted from relatively short-term
experiments. G. Abrahamsen noted that part of the problem was due
to semantics, and that all of the soils studied in the SNSF project
were podsols, and that what Wiklander is referring to as brown forest
soils are actually incipient podsols. At this point, M. Schnitzer
noted that the lowest pH used by Wiklander in his leaching experiments (i.e. pH 2) would produce a degradation of some mineral soil
components, and not just exchange of cations. On another issue,
Gorham raised the point of the buffering of rainwater by watersheds,
in that in Norwegian studies, streamflow reaching water bodies was
usually much less acid than the incident rainfall. He attributed
this effect to buffering by watershed soils. Dr. Abrahamsen and
Seip agreed that this must occur to some extent, although
Abrahamsen noted that he felt that very little soil acidification
had occurred as a result. It was brought out in the following
discussion that buffering substances in the bedrock, water column,
and lake or stream sediments would also be very important in the
phenomenon under discussion.

The next paper by D. Johnson dealt with various aspects of
site sensitivity to acid sulfate loading. He first commented on the
relative influence of common anion species on promoting cation
leaching from soils. He noted that this effect depended on the
relative anion mobility in soils, i.e. in general phosphate has very
low mobility relative to sulfate, which is low relative to nitrate
or chloride. The relative mobility of particular anion species
also varies with soil type. Sulfate, for example, is strongly bound
in soils with a high sesquioxide content, and thus these soils should
not be susceptible to detrimental effects from acid sulfate loadings
since the sulfate would not be leached from the soils in conjunction

with H^+-displaced cations. In the discussion, Freedman asked whether
it was possible to calculate a time of depletion index from data on
soil sulfate adsorption capacity, and sulfate input in precipitation.
Johnston felt that, at present, the measurements were not reliable
enough to do a meaningful calculation. Ulrich noted that sulfate
adsorption by soil was strongly pH-dependent. Johnston agreed in
general, but he also noted that a new pH-independent sulfate
adsorption mechanism had recently been described for some soils.
Wiklander commented that he supported the method used by Johnston
to determine anion exchange, and thus with the anion exchange series
stated. He noted that in columns of podsolic soil in the laboratory, he had observed approximately 100% phosphate adsorption,
30% sulfate adsorption, and 0% chloride adsorption, and that in
high concentrations phosphate could displace sulfate at anion
exchange sites.

The next presentation was by L. Petersen, who noted that
valuable information about the sensitivity of various soils to acid
precipitation could be derived from interpretation of soil classi-
fications. The important properties to consider are 1) the buffer
capacity, 2) the base saturation status of the soil, and 3) the
relative hydraulic conductivity. Some problems arise because
different soil classification systems are used in different countries. An interpretation based on the soil classification system
used in the United States indicates that the following soil orders
may include soils which are susceptible to the effects of acid
precipitation: Entisols, Inceptisols, Alfisols, Ultisols, Spodosols,
and Oxisols. In the FAO/Unesco system, sensitive soils may occur
within several of the soil units, including Regosols, Lithosols,
Cambisols, Luvisols, Acrisols, Podsols, Ferralsols, and Arenosols.

The next presentation was by G. Abrahamsen, and it concerned
the neutralization of H^+ by watersheds, and the susceptibility of
lakes to acidification. The speaker noted that, although few
studies existed that documented trends of decline in the pH of
water bodies, there was good evidence for decline or loss of fish
populations from remote lakes, and of fish kills following the melt
of acid snow in the spring, both of which (along with other associated evidence) was felt to indirectly document water body
acidification. This is thought to be partly due to the saturation
of cation exchange sites in moderately acid soils, with the result
that not all incident H^+ is adsorbed, and passes through into
susceptible water bodies. This H^+ exchange effect is strongly
dependent on soil properties, and in acid soils there is a low
exchange efficiency because most of the cation exchange sites are
already occupied by H^+. It was noted that very little soil was
required to produce significant changes in rainwater pH in the
field, and that large changes in pH were even observed in rainwater
flowing over bare gneiss rocks with only a sparse lichen cover. A
discussion followed centering on the soil buffering capacity

component due to the relatively large amounts of exchangeable
aluminium and iron in coniferous litter. These are ultimately
taken up from lower soil horizons by roots, translocated to shoots,
and incorporated into leaf tissue, which later contributes to the
forest litter horizons. P. Rennie questioned the significance to
forest growth of accelerated base leaching from soils due to
atmospheric acid inputs, and the probable lifespan of the soil
calcium reservoir under current inputs of acidifying precipitation.
Along these lines, Bache noted that little had so far been
determined regarding the time-dependency of soil buffering capacity
depletion, or the rates of its regeneration by weathering or other
processes. It was noted by Abrahamsen that there was great spatial
variation in water quality in Norway, and that this was due to watershed variation, including factors such as relative H^+ inputs, and
the watershed area: lake surface area ratio of the system being
considered. Freedman noted that, in Ontario, the most important
factor determining whether water bodies on Precambrian bedrock
would acidify appeared to be whether the watershed contained
deposits of calcareous glacial deposits, as this strongly influenced
the buffering capacity of the system. He quoted the work of
J. Kramer, who used a calcite saturation index to identify sensitive
water bodies.

The next paper, by B. W. Bache, noted that the lime potential
of soils (outlined and discussed in a previous paper) was the
simplest available indicator of the acid-base status of soils, and
hence had application as an indicator of sensitivity to acidification. He noted that the extent of change in soil acidity caused by
an incoming solution was due to 1) the change in lime potential
(i.e. the relative lime potential difference between the soil and
the incoming solution), 2) the soil buffering capacity, and
3) the fraction of incoming water volume that actually reacts with
the soil. Of these three, the lime potential is the easiest to
measure.

In the discussion, Schnitzer noted that the lime potential
concept was not useful for very acid soils and thus a different
concept of susceptibility was necessary. He also noted that at
low pH (i.e. pH ∿ 2) soil mineral degradation is monitored in
lysimeter percolates, and not just cation exchange. Bache replied
that the aluminium potential would be more useful for these soils,
but that in the present context the lime potential was relevant
as he was considering slightly acid soils with a possibly high
potential for acidication. The question of contact time between
percolating water and the soil was again raised, and it was noted
that forest soils in general tended to have high hydraulic
conductivities relative to heavy agricultural soils, so that under
most circumstances, the contact time between percolating water and
soil cation exchange sites might be expected to be short. Bache
agreed with this in a general sense.

The final paper, by I. Nilsson, presented a theory of ion adsorption isotherms which had potential use in predicting leaching losses from soils under increased inputs of H^+ ions. He presented data on calcium-based isotherms, derived from chemical potential calculations of Ca relative to Mg and K, for a field site in Sweden which had received experimental perturbations of sulfuric acid and/ or liming over a seven year period. His preliminary data indicated that, although within-treatment variation was large, the concept appeared to have potential as an index of sensitivity, and he felt that further field studies would be useful. In the discussion that followed, the appropriateness of the regression techniques used in the calculations was questioned. To some extent, Nilsson agreed with the criticisms, but he pointed out that the relevant information to be gleaned from the presentation was not the absolute appropriateness of statistical techniques, but the concept of calcium isotherms, which appeared from the preliminary data presented to be useful as a site sensitivity index.

In summary, it was agreed that already acid, or well-buffered circumneutral or basic soils were in little danger of pH-change due to additional inputs of acid at rates currently measured in precipitation. However, soils of intermediate pH which are poorly-buffered, free-draining, and with little anion (especially sulfate) binding capacity appeared likely to be susceptible over the long term. Notably, however, no such soils have yet been identified to have acidified in the field under current rates of H^+ loading through precipitation. The need for geographic identification of sensitive soils is apparent. Several site sensitivity indices have been proposed, but these have not yet been thoroughly tested. The need for such testing and the identification of other viable acidification sensitivity indices is clear, as is the need for further acid perturbation studies in the laboratory and field to better determine rate constants for leaching of both cations and anions and for studying the chemistry of mobilization of soil organic acids and clay mineral components.

CONCLUSIONS AND RECOMMENDATIONS*

T. C. Hutchinson

Although there were a great diversity of views expressed on many topics at the conference, there was nevertheless a good deal of consensus on some fundamental aspects of the symposium topic. It was emphasised that there is little solid evidence to date of the occurrence of visible or even detectable damage to terrestrial ecosystems caused by the acid precipitation events of the past twenty years. This provides a marked contrast with that of lake ecosystems where substantial effects have been recorded. The reasons for this lack of damage to date are believed to be of several kinds. One is that naturally acidic soils are very wide-spread, especially in those areas experiencing lake acidification. Forest growth and the essential soil processes are thus able to function at low pH (3.5-4.5). A second factor is that most soils have a relatively high buffering capacity which is contributed by their base status, by carbonates, organic content and by the clay minerals. Agricultural soils are traditionally limed, as required, to maintain pH in a desirable range for plant growth. The delegates were in agreement that although cation loss may be accelerated by acid rains, fertiliser and lime additions would far outweigh deleterious impacts of H^+ ion additions. The anticipated acidity of precipitation events for the future was not felt to be much greater than those occurring now. In artificial rain simulations direct damage to foliage has only been observed when extreme acidities have been used (i.e. pH < 2.5). Likewise, substantial effects on soil microbial processes seem to be largely confined to similar extremes.

It was generally agreed that because of the very wide annual vagaries of climate, effects of acid precipitation on forest growth would have to be very large before they would be detectable above such climate-induced variation. Also, because of the diversity of species in natural communities and their large turn-over time relative to the recent onset of widespread acid precipitation, any subtle changes which may be occurring will be very difficult to detect. The data base for single species performance in forest

*Based on group discussions.

617

productivity studies is also quite limited compared with agricultural
crops.   The latter generally deal with monocultures harvested
annually.   All of these considerations led to broad agreement on the
desirability of long term studies, and of comparative field and
laboratory studies with proper account being taken of scale-up
factors from the latter.   It was also agreed that the complexities
of the acid precipitation problem necessitate a very broad multi-
disciplinary approach, together with much better integrated planning
at a very early stage than has generally been the case to date.

It was felt that within any isolated component of the ecosystem
impacts of acid rains were often demonstrable but that the importance
of these to the functioning of the whole was difficult to assess.
For example, accelerated foliar leaching of cations appear to occur
but this then enters into a complicated soil system.   Nitrogen oxides
contribute to acidity but transformations in the soil may, in part,
enhance nitrogen availability for plant-growth.   Increased acidity,
on the other hand, appear to reduce the rate of soil nitrification
and also favours the ammonium form of N over nitrate.   This has a
strong influence on root uptake.   Studies of any one system in
isolation need to be put into the broad context and this is, of
necessity, challenging, frustrating but absolutely essential.

While the evidence to date for significant harmful effects of
regional acid precipitation on terrestrial ecosystems was considered
either very limited or non-existent, it was also concluded that the
impacted terrestrial ecosystems may have a substantial influence on
adjacent water bodies.   These effects will include that of modifying
and ameliorating the direct acidity of the incoming precipitation.
This can also be expected to involve changes in solubility and, thus,
of mobility of potentially toxic elements such as aluminium and
manganese, and of toxic heavy metals previously deposited by industrial activity on terrestrial systems.   Examples of copper, nickel,
mercury and lead were cited.

Atmospheric Deposition

In considering our knowledge of atmospheric deposition, the
ambiguity of the expression 'acid-deposition' was repeatedly noted.
A working definition was suggested as follows:

Wet deposition includes all pollutant material reaching the
earth's surface in precipitation (the upper limit of vegetation
defining the level of the surface).   Wet deposition also includes
the process of fog-drip, i.e. the interception of cloud or fog
droplets by vegetation.

Dry deposition includes absorption of gases and capture of
particulate material by land (vegetation, soil, snow etc.) and
water surfaces.

The need was emphasised for the establishment of a comprehensive network of wet deposition stations in North America,
comparable to those of the European air chemistry network, and using
collection and analytical methods to minimise the influence of contamination. Direct measurements of dry deposition have proven more
difficult to achieve and, to date, have only been possible over
rather special, flat and uniform surfaces. Such measurements have
revealed typical rates of dry deposition of SO_2 on short vegetation
and on soil and water surfaces, for which the processes are
reasonably well understood. For more aerodynamically rough surfaces,
such as forests or typical countryside with isolated trees and
uneven terrain, rates of dry deposition of SO_2 are not so well known.

Specific research requirements for dry deposition include
studies of the dry deposition of oxides of nitrogen on a wide range
of natural surfaces and for SO_2 on rough surfaces.

One of the principle applications of deposition rates is the
estimation of inputs of sulphur, nitrogen, and other compounds to
ecosystems on catchment, regional and global rules. For such
applications there is a _general_ requirement for research in the
following aspects of the subject:

a) Physiological, physical or chemical processes at surfaces,
    especially of vegetation, which interact with pollutants
    in air and in rain.

b) More extensive monitoring of concentrations of gaseous or
    particulate pollutants and those in rain with special
    emphasis on those areas distant (> 100 km) from sources.

c) Atmospheric chemical processes involved in the trans-
    formation of pollutants in the atmosphere.

d) Studies of the changes in chemical composition of
    precipitation as it passes through vegetation and its
    relationship with leaching and wash-off phenomena.

The Nature of the Acidity in Precipitation

Although the workshop was called to examine the effects of
acid precipitation on terrestrial ecosystems, the delegates felt
that it was necessary to clarify the current understanding of the
chemical components of acidity in precipitation. The delegates
agreed that sulphate was a major contributor and organic acids were
of little significance. However, considerable discussions were
held on the contribution by nitrate and to a lesser extent, chloride.
This debate was initiated by information in the OECD report on
Long Range Transport of Sulphur Compounds over Europe which showed

that, while there was a good correlation between H ions and $SO_4^=$
ions in precipitation in northern Scotland and southern Norway,
this was not the case over much of Europe. The same situation occurs
for the ratio of H^+ and SO_4^- ions, in that the equivalent ratio is
near 1 in the aforementioned areas but varied greatly over the rest
of Europe. There are three obvious explanations for this
variability: (1) The rates of SO_2 oxidation vary greatly with meteorological and other factors, (2) There is local neutralisation of
the acidity formed – mainly by ammonia in the atmosphere and
(3) There may be a significant contribution from oxidation of NO_x
or emissions of HCl.

It was agreed that deposited sulphur, as well as H_2SO_4, could
lead to acidification of soils and lakes. In addition, although
nitrate deposition will contribute to any effects of acid
precipitation on plant surfaces and to corrosion of materials, it
would not lead to soil or lake acidification as it would be rapidly
metabolised in most ecosystems. In central Europe the equivalent
ratios of $SO_4/NO_3^-/Cl^-$ in precipitation were 73/22/5. 58% of these
anions were neutralised by NH_4^+ and the rest by other cations.
Nevertheless, if NO_x emissions make a major contribution to the
acidity of precipitation, then the successful modelling of sulphur
transport may not offer a full explanation of the amounts of acidity
deposited in remote areas. A full understanding of the relative
contributions of SO_2 and NO_x emissions to rainfall acidity is
clearly essential before control measures can be more effectively
introduced.

An estimate of the relative contributions in the north-eastern
region of the US was obtained by examining the equivalent ratios
of SO_4^{2-} : H^+ and NO_3^- : H^+ in precipitation. Monthly data over a
10 year period at Hubbard Brook showed that the upper limits for
the sulphate contribution were 100% in summer and 75% in winter.
The NO_3^- contribution could be up to 30% in summer and 50% in
winter. It should be borne in mind that the most acid episodes
occur during the summer. In addition, the emissions of NO_x from
combustion processes have increased, as has the nitrate concentration
in precipitation, whilst the corresponding sulphate values have
remained fairly constant or decreased.

Similar computations have not been carried out for Scandinavia
although the data base does exist. Nevertheless, the equivalent
ratio of SO_4^{2-} / NO_3^- is about 5:1 suggesting that nitrate may
contribute about 20% of the H^+ ions in precipitation. In addition,
there may be a small contribution (< 10%) from chloride ions.

In summary, it appears that general values can be given to the
components of acidity in precipitation. More observations on the
processes and rates of oxidation of SO_2 and NO_x are necessary,

however, in order to fully understand the formation and transport of acid precipitation.

Soil Science Considerations

It was agreed that acidification of soil is a natural process and that inputs of acid precipitation must be studied in the context of the large fluxes of acid which are common in soils. The detection of effects of additional hydrogen ion input is likely to be difficult and slow in field situations.

In experimental work, attempts to compress the time scale by increasing acid concentration and/or input rates may produce results that would not occur if the concentration was reduced and spread over time. This should be emphasized in reporting results. It was pointed out that care should be taken to adequately describe and classify soils that are used in experiments in such a fashion that it will have meaning to the international scientific community.

In general terms, the reactions of acid with soils are well-known, but quantification of rates and processes are in need of further study. We need to develop a reliable, and hopefully simple measure of buffer capacity that reflects the rates as well as the extent of acid interaction with soil components. Present techniques do not give adequate consideration to the time dependence of the reactions, especially those involving aluminium release from minerals.

We need measures, or indicators, of soil degradation. There was not complete agreement on what an adequate measure of degradation would be, other than the fact that it should be related to a loss of plant productivity. Aluminium leaching loss may be a useful indicator. Other possible degradation indicators suggested were weathering of secondary minerals and reduction of base saturation.

The point was made that surface horizons, and shallow-rooted plants and organisms that inhabit the surface horizons are likely to be better indicators of acidication than deeper horizons or plants that utilize greater portions of the soil and therefore, these upper layers should be examined early in research.

There was considerable discussion of water flow patterns and the extent of water reaction in soils. Additional work is needed; perhaps simulation modelling would be helpful. There was a consensus that most of the water passing through soils, whether it is in channel flow or otherwise, is reacting with the soil but not always to equilibrium. The occurrence of channel flow plus the time dependence of some of the reactions may help explain the diverse results between lysimeters, field tests, and laboratory experiments. Snow melt water running off frozen ground may enter water bodies without major modification by exchange processes. This can be an especial threat to sensitive lake systems.

Further research is needed to identify criteria that are useful in the classifying of vulnerable soils, and in using these to map and identify the areas that should be given special attention. It is believed that the profile description, analytical data, and hydrologic characteristics that normally accompany modern soil surveys would be sufficient to do this. Circumneutral sandy soils were considered particularly susceptible to rapid changes in pH.

Release of bases from soils by accelerated mineralization and addition of nitrogen and sulphur from the air can have benefits to some soils in terms of plant nutrients, in special circumstances. The loss of cations and anions from soils and especially the mobilisation of potentially toxic heavy metals are likely to be disbenefits of acid precipitation.

There is a need for clarification of the chemical processes following the input, specifically the sink-source relationship of important elements such as phosphorus, sulfur, and nitrogen.

There is also a need for a study of the interaction of heavy metal mobility and acid precipitation. We need to know the extent of nitrogen and sulphur deficiencies in natural ecosystems and effects of acid precipitation in alleviating these.

We also need studies designed to determine the effects of acid precipitation on the dissolution and movement of organic compounds in the soils.

Plant Response to Acid Precipitation

There is a good base of knowledge concerning soil acidity and plant response, and also on nutrient uptake and potential response to metal toxicity. There is no unequivocal evidence either supporting or denying harmful effects of acid rain on forest tree growth or performance. Effects can be documented, from those of enhanced cation loss by leaching from leaves to enhanced availability of cations for root uptake, but the overall trends from the increasingly widespread acid rain and snow episodes of the past 10-20 years are either unclear or sufficiently small to be masked by the large annual variation in climatic influence on growth.

There is a real need to determine whether or not significant effects will occur or are occurring with time. Consideration must include both harmful and beneficial effects. Considerable efforts will be needed to provide evidence from both the field and the laboratory and studies must be designed specifically to provide this

information. Appropriate data are most unlikely to be available
from forest productivity and harvest records. Tree ring analyses
will require a much better knowledge of annual climatic variables
and of the ageing phenomenon of individual species. The need for
inter-disciplinary co-operation from the planning stages in such
studies was emphasised.

In forests, changes in species composition are likely to be
long term and difficult to detect. However, sensitive groups may
occur from which useful data could be derived. Such indicator
species, with particular sensitivity to acid precipitation, may
include bryophytes (mosses and liverworts) and lichens. A major
reason for this is their lack of a cuticle, so that acid leaching
of cellular contents can more readily occur. Some of the widespread
feather mosses of the boreal forest and the epiphytic lichens may be
especially worthy of study. Spanish moss could be a southerly
equivalent in North America. In agriculture, some of the leafy
salad crops such as lettuce and spinach, may serve as sensitive
indicator crops for detecting direct damage - though it is not
certain that this will be the case. Interactions between symptoms
of acid rain damage and soil type need to be studied.

In terms of nutrient uptake by plants from soils including
acid ones, a good deal is known. The problems of deficiences of
essential elements and toxicities of heavy metals are well
understood.

Recommendations were made for detailed screening of species
for plant response. Amongst the suggestions were:

a) Organization of an international field network which
    would use common methodology, share standard species and
    set up some specific field experiments. This network
    could be linked as a subsidiary to existing agricultural
    and forestry experimental stations.

b) Screening selected species for tolerance to relevant
    site factors at individual sites.

c) Laboratory experiments to determine mechanisms of
    tolerance to acid-induced soil conditions and direct
    impacts of acid precipitation, especially utilising
    comparisons between tolerant and susceptible populations
    or varieties of the same species.

d) Determine the importance of life cycle stages,
    especially of seedlings, in natural habitats.

e) In analyses of forest impacts, use mathematical models
    (and of existing programmes) to sort out factors
    operating and those of importance.

f) Examine sites of attack, such as acid precipitation
     impact on foliar leaching, and assess harmful and
     beneficial effects.

g) Determine the contribution of dry deposition to
     canopy leaching and throughfall.

Soil Microbial Activity

It was clear that a considerable amount of information is
available on the differences in dominant microbial groups, metabolic
activities and rates of decomposition and mineralisation rates
between mull and mor humus. However, less information is available
on the changes which may occur as acid soils are further acidified.
A number of field and incubation studies have been carried out to
measure the effects of acid applications. Some of these are
difficult to interpret, however, due to "shock effects" which may
be produced by unrealistically acid solutions. From their
collective experience the group were able to produce a fairly
comprehensive table of the processes susceptible to acidification
and those which are unaffected.

Changes which Occur as a Soil becomes Acidified

Process	pH∿6 (mull) to pH 4 (mor)
(1) Dominant microbial groups	Change from bacteria -> fungi
(2) Decomposition	Marked decrease in activity
(3) Respiration	" " " "
(4) Enzyme activity	" " " "
(5) N-mineralisation	" " " "
(6) Nitrification	" " " "
(7) Denitrification	slow in aerated soils
(8) N-fixation	
a) Legume Symbionts	Inhibited
b) non-legume Symbionts	Some reduced
c) non-symbionts	Some groups sensitive, some resistant
(9) Mycorrhizae	Changes in type but not abundance
(10) Earthworm activity	Marked decrease
(11) Soil Microfauna	Diversity reduced & changes in dominant groups.

There are clearly some areas which need further research. In
particular, these are the effects of acid rain on (a) the rate of
nitrogen cycling, (b) the rate of sulphur transformations and
(c) mycorrhizal development.

Due to the unfortunate inability of the two invited experts on mycorrhizae to attend the meeting, very limited discussion on sensitivity of mycorrhizal association to soil acidification took place. It was nevertheless agreed that in view of the importance of such associations in forest nutrition that this is a vital area for future study.

Plant Pathology Affects

It was agreed that damage to the cuticle of leaves caused by acid rain is likely to pre-dispose plants to attacks by pathogens, especially by fungal pathogens. Interactions between gaseous pollutants and acid rains may also increase the phytotoxic damage by gaseous pollutants alone (synergism).

It was recommended that:

1. Much more research is needed on the effects of acid precipitation on susceptibility of plants to biotic plant pathogens.

2. Mechanistic experiments aimed at the identification of specific biological effects under laboratory or greenhouse conditions should be coupled with ecosystems studies to determine their general significance.

3. Identification of life forms, life stages, and life processes that are particularly vulnerable to acid precipitation under episodic conditions should be a major research effect.

Terrestrial - Aquatic Ecosystem Linkages

A. Areas of Consensus

1. Atmospheric deposition (wet and dry fallout) has been anthropogenically acidified.

2. Runoff waters, streams and lakes, have been anthropogenically acidified.

3. The effects of this acceleration of the rate of supply of acidifying substances can be observed in the chemical and biological parts of the ecosystem. Changes in both cationic and anionic loads have been noted. Potentially toxic substances have also been identified as occurring in acidifying lakes, with a probable soil origin e.g. Al.

4. These effects of acid deposition appear to be proportional
    to a) the rate of acid supply, as well as accompanying
    cations and anions, and b) the acid buffering capacity
    of the drainage system (vegetation, soils, bedrock,
    runoff waters). Runoff waters of low electrolyte
    concentration (i.e., Precambrian Shield drainages, for
    example) will be most susceptible to pH alteration by
    acid precipitation.

5. For regions that have a low electrolyte concentration
    runoff waters, the effect of acid precipitation on lakes
    will be related to the ratio land drainage area: lake
    surface area. Direct precipitation of acidified rain on
    the lake surface will have a greater effect on lakes with
    a low value of this ratio, whereas the soils of larger
    drainage basins will tend to buffer acid rain before it
    reaches the lake.

B. Recommendations for Research

    The problem of acid deposition and its effects must be
viewed quantitatively by designing the observations and/or
experiments to allow chemical and hydrological budgets to be
made. The annual flux terms (rate of supply of H^+, rate of
equilibration and/or buffering of acid water with the vegetation
and soil, rate of infiltration of water, rate of runoff,
evaporation, evapotranspiration) are more likely to lead to
useful hypotheses, compared to tables of concentration data
with no hydrological multiplier. We felt that this recom-
mendation was of high priority, and it embraces the following
details:

1. A variety of methods should be used to characterize the
    nature of the acids in solution. In addition to the
    H^+ electrode, P_{CO_2}, base titration, and other methods
    should be used. Anion balance and mobility should be
    emphasized.

2. Chemical data should include the ionic charge balance.

3. Effort should be made toward the quantitative measurement
    of dry deposition. The reactivity and acidity of dry
    deposition and particulate matter in stream flow should
    be studied.

4. Methods for the characterization and reactivity of strong
    and weak acids in atmospheric deposition and stream
    runoff need emphasis.

5. In the budget for the flux of H^+, the sources for the exchangeable cations should be identified. Possible sources are the parent rock, soils, atmospheric fallout, lake and stream sediment, and vegetation. This is of especial concern in view of reports of increased concentrations of such toxic elements as Al, Mn, Hg, Pb and Cd in acidifying lakes.

Detailed watershed studies of the Hubbard Brook type have proven invaluable to date and such long term investigations are deserving of continued support. In major areas where no such integrated studies exist, there is a case for careful selection of a few further watershed studies to determine similarities and differences of response, and to enable better predictive modelling to be carried out.

List of participants at N.A.T.O. - A.R.I. Symposium, May 22-26, 1979

Dr. G. Abrahamsen

Norwegian Forest Research Institute
Postbox 61
1432 As-NLH
Norway

Dr. M. Alexander

Dept. of Agronomy
708 Bradfield
Cornell University
Ithaca, New York 14853
U.S.A.

Dr. F. Andersson

Barrskoglandskarets Ekologi
Lanllrukshogskolan
Box 7008, 5-750 07, Uppsala 7
Sweden

Dr. B. Bache

Macauley Research Institute
Agricultural Research Council
Aberdeen
Scotland

Dr. M. Benarie

Institute National de Recherche
    Chimique Appliquee
B.P.I. 91710
Vert-le-Petit
France

Dr. G. Brunskill

Freshwater Institute
Department of Environment
University of Manitoba
Winnipeg, Manitoba
Canada

Mr. C. Cogbill

c/o Dr. Maycock
Erindale College
Dept. of Biology
University of Toronto
Toronto, Ontario, Canada

Dr. E.B. Cowling

Department of Plant Pathology
    and Forest Resources
North Carolina State University
Raleigh, North Carolina 27607
U.S.A.

Dr. R.M. Cox

Institute of Environmental Studies
Haultain Building
University of Toronto
Toronto, Ontario, Canada M5S 1A1

Dr. P. Dillon

Ministry of the Environment
Resources Road
Islington, Ontario, Canada

Dr. D. Fowler

Institute of Terrestrial Ecology
Bush Estate
Peniciuk
Midlothian EH26 ORA
Scotland

Dr. J.N. Galloway

Department of Environmental Sciences
University of Virginia
Charlottesville, Virginia 22903
U.S.A.

Mr. W. Gizyn

Mining Building
Department of Botany
University of Toronto
Toronto, Ontario, Canada M5S 1A1

Dr. R.A. Goldstein

Environmental Assessment Department
3412 Hillview Avenue
Post Office Box 10412
Palo Alto, CA 94303
U.S.A.

Dr. E. Gorham

University of Minnesota
Department of Ecology and
    Behavioral Biology
108 Zoology Building,
318 Church Street S.E.
Minneapolis, Minnesota 55455
U.S.A.

Dr. F. Gormley

Manager, Environmental Projects
INCO
1 First Canadian Place
P.O. Box 44
Toronto, Ontario, Canada M5X 1C4

Dr. F. Gravenhorst

Institute of Atmospheric Chemistry
Kernforschungsanlage Julich
D-5170 Julich
Federal Republic of Germany

Dr. P. Grennfeldt

IVL Box 5207
Sten Sturegatan 42
S-40924 Gothenbuorg 5
Sweden

Dr. B. Haines

The University of Georgia
Botany Department
Athens, Georgia 30602
U.S.A.

Miss Magda Havas

Institute for Environmental Studies
Haultain Building
University of Toronto
Toronto, Ontario, Canada M5S 1A4

Dr. P. Havas

Department of Botany
University of Oulu
SF - 90100 Oulu
Finland

Mr. Keith Hendrie

c/o Professor F.K. Hare
Institute for Environmental Studies
University of Toronto
Toronto, Ontario, Canada M5S 1A4

Dr. N.C.B. Hotz

66 Orrin Avenue
Ottawa, Ontario, Canada K1Y 3X7

Mr. L.D. Hudon

Secretary,
Ministry of State
Science and Technology Canada
270 Albert Street
Ottawa, Ontario, Canada K1A 1A1

Dr. T.C. Hutchinson   Chairman,
                                       Department of Botany
                                       University of Toronto
                                       Toronto, Ontario,
                                       Canada M5S 1A1

Dr. J.S. Jacobson   Center for Environmental Res.
                                       Cornell University
                                       Hollister Hall
                                       Ithaca 14853, N.Y.
                                       U.S.A.

Dr. I. Johnsen   University of Copenhagen
                                       Institute of Plant Ecology
                                       Oster Farimagsgade 2D
                                       DK-1353
                                       Copenhagen, K
                                       Denmark

Dr. D.W. Johnson   Oak Ridge National Laboratory
                                       P.O. Box X
                                       Oak Ridge, Tennessee 37830
                                       U.S.A.

Dr. A. Kallend   C.E.G.B.
                                       Central Electricity Research
                                           Laboratories
                                       Kelvin Avenue
                                       Leatherhead, Surrey KT22 7SE
                                       England

Dr. K.H. Lakhani   The Inst. of Terrestrial Ecology
                                       Monks Wood Experimental Station
                                       Abbots Ripton Huntingdon
                                       England PE17 2LS

Dr.   G.E. Likens   Section of Ecology and Systematics
                                       Division of Biological Sciences
                                       Cornell University
                                       Bldg., #6, Langmuir Laboratory
                                       Ithaca, New York 14850
                                       U.S.A.

Dr. S. Linzon   Chief,
                                       Air Resources Branch
                                       Ministry of the Environment
                                       Suite 347
                                       880 Bay Street
                                       Toronto, Ontario, Canada M5S 1Z8

Dr. F. Maine   Ministry of State
                                       Science and Technology Canada
                                       270 Albert Street
                                       Ottawa, Ontario K1A 1A1

Dr. R. Mayer   Institut für Bodenkunde und
                                           Waldernägrung
                                       Universität Göttingen
                                       3400 Göttingen
                                       Bünsgenweg 2
                                       West Germany

Dr. W.W. McFee   Director,
                                       Natural Resources and Environmental
                                          Science Programme
                                       Purdue University
                                       Life Science Building
                                       West Lafayette, Indiana 47907
                                       U.S.A.

Dr. R.E. Munn   Institute for Environmental Studies
                                       University of Toronto,
                                       Haultain Building
                                       Toronto, Ontario, Canada M5S 1A4

Dr. I.A. Nicholson   Institute of Terrestrial Ecology
                                       Banchory Research Station
                                       Hill of Brathens
                                       Glassel, Banchory
                                       Kincardineshire AB3 4 BY
                                       Scotland

Dr. P.I. Nilsson   Dept. of Plant Ecology
                                       University of Lund
                                       Lund
                                       Sweden

Dr. S. Odén   Swedish University of Agricultural
                                        Sciences
                                     Department of Soil Sciences
                                     S-750 07
                                     Uppsala
                                     Sweden

Dr. L. Petersen   Department of Soils
                                     The Royal Veterinary and
                                        Agricultutral College
                                     Copenhagen V
                                     Denmark

Dr. K. Puckett   Environment Canada
                                     Atmospheric Environment Service
                                     905 Dufferin Street
                                     Downsview, Ontario, Canada M3H 5T4

Dr. P.J. Rennie   Environmental Analyst
                                     Canadian Forestry Service
                                     Ottawa, Ontario, Canada K1A 0E7

Dr. J.E. Rippon   C.E.G.B.
                                     Research Division
                                     Kelvin Avenue
                                     Leatherhead, Surrey KT22 7SE
                                     England

Dr. T. M. Roberts   C.E.R.L.
                                     Kelvin Avenue
                                     Leatherhead, Surrey KT22 7SE
                                     England

Dr. I. Rorison   Nature Conservancy Unit
                                     Unit of Comparative Ecology
                                     Department of Botany
                                     University of Sheffield
                                     Sheffield
                                     England

Dr. M. Schnitzer   Soil Science Research Institute
                                     Central Experimental Farm
                                     Agriculture Canada
                                     Ottawa, Ontario, Canada

Dr. H. M. Seip   SI, Forskningsv 1 PB 350
                                     Blindern
                                     Oslo 3
                                     Norway

Dr. David Shriner

Oak Ridge National Laboratory
Post Office Box X
Oak Ridge, Tennessee 37830
U.S.A.

Dr. S. Stevens

Air Resources Branch
Ministry of the Environment
Suite 347
880 Bay Street
Toronto, Ontario, Canada M5S 1Z8

Dr. P.M. Stokes

Institute for Environmental Studies
Haultain Building
University of Toronto
Toronto, Ontario, Canada M5S 1A4

Dr. R.S. Strayer

Dept. of Agronomy
708 Bradfield
Cornell University
Ithaca, New York 14853
U.S.A.

Dr. B. Tveite

The SNSF- Project
Box 61
1432 As-NLH
Norway

Dr.   H.B. Tukey, Jr.

New York State College of
    Agriculture and Life Sciences
Cornell University
Department of Floriculture and
    Ornamental Horticulture
20 Plant Sciences Building
Ithaca, N.Y. 14853
U.S.A.

Dr. B. Ulrich

Institut für Bodenkunde und
    Waldernährung
Universität Göttingen
3400 Göttingen Büsgenweg 2
West Germany

Dr. L. Wiklander

Department of Soil Science
Agricultural College Sweden
Uppsala
Sweden

Species Index

Acer, 146
Acer rubra, 336, 409
Agrostis stolonifera, 295, 409
Agrostis tenuis, 286
Alnus glutinosa, 506
Alternaria solani, 437
Anaptychia ciliaris, 133, 134
Antitrichia curtipendula, 134
Anurida pygmaea, 346, 350

Beta vulgaris, 142
Betula papyrifera, 399, 412
Betula pendula, 506, 510
Brassica napus, 288
Brassica oleracea, 143

Calluna vulgaris, 87, 320
Carya glabra, 336
Carya illinoensis, 336
Cladosporium macrocarpum, 345
Coffea arabica, 143
Cognettia sphagnetorum, 346, 350, 357, 377
Collybia acena, 345
Coniothyrium, 345
Cornus florida, 336

Deschampsia cespitosa, 494
Deschampsia flexuosa, 289, 510
Diervilla lonicera, 409

Elatobium abietinum, 170
Enchytronia parva, 346
Euonymus alatus, 146, 147

Fagus sylvatica, 256, 320, 506

Fusarium oxysporum, 422

Holcus lanatus, 290, 291
Hypnum cupressiforme, 137
Hypogymnia physodes, 137, 139

Ilex aquifolium, 506
Isotoma notabilis, 346, 350
Isotomiella minor, 346, 350

Juniperus chinensis, 143

Larix decidua, 506
Lecanora conizaeoides, 137, 139
Liriodendron tulipifera, 336
Lobaria pulmonaria, 134

Malus domestica, 142
Medicago lupulina, 368
Mesenchytraeus pelicencis, 345
Micromphalia perforans, 345
Mortierella, 421
Mucor, 421
Musa, 143

Nanhermannia, 346
Nothrus silvestris, 346

Oppia obsoleta, 377

Penicillium, 421
Pharbitis nil, 148
Phaseolus vulgaris, 142, 144, 439
Physcietum ascendentis, 133
Physconia pulverulenta, 133, 134

Subject Index